Basic Population Genetics, like its aged
predecessor, is dedicated to my family—
Miriam
Bruce
Roberta

Basic Population Genetics

Bruce Wallace

Columbia University Press
New York, 1981

Library of Congress Cataloging in Publication Data

Wallace, Bruce, 1920–
  Basic population genetics.

  Bibliography: p.
  Includes index.
  1. Population genetics. I. Title.
QH455.W34   575.1'5   80-39504
ISBN 0-231-05042-9

Columbia University Press
New York Guildford, Surrey

Basic Population Genetics

Contents

Preface

A dozen or more years have elapsed since the appearance of *Topics in Population Genetics*. In the meantime much has happened. Electrophoretic techniques transform even single electrical charge differences between protein molecules into millimeter or centimeter distances between stained blobs on slabs of gels. The determination of the amino acid sequences of individual proteins is performed by automatic machines. Far from requiring a super microscope, the determination of the sequences of nitrogenous bases in nucleic acids is made by a combination of enzymatic and electrophoretic procedures. Pages of current journals are now filled—top to bottom, margin to margin—with base sequences of newly analyzed DNA fragments, messages in code still awaiting full translation.

Libraries of DNA, especially of yeast (*Saccharomyces cerevisiae*) and *Drosophila* (*D. melanogaster*), have been assembled in thousands of cultures of the colon bacillus (*Escherichia coli*); each culture contains replicating plasmids into which have been inserted small segments of foreign DNA. Scores of workers are busily identifying these segments according to their site or origin within the chromosomes of their species, and arranging them in proper sequence by matching identical terminal base sequences.

Although they have not been as spectacular as the recent advances in molecular technologies, advances in the mathematical aspects of population genetics, and of population biology generally, have also occurred. Many of these have come from the interest of professional mathematicians in the logical extensions that can be erected on Mendelian inheritance, on birth rates and death rates, and on resource utilization. At least one Nobel Laureate in Economics routinely publishes on the genetics of populations.

Population genetics is the conduit through which advances in genetic knowledge are made known to more ecologically and mathematically oriented population biologists. Population genetics, in brief, are *geneticists* who happen to be interested in the genetics of populations. Without the contact these persons provide with modern genetics, many population models would eventually be discarded because they were based on the genetics of another universe or (worse) of another century.

Basic Population Genetics is designed to serve as a second-level text, to be used by students who have had general biology including some elementary ecology, and an introductory course in genetics. Although quantitative, this book is not mathematical; only twice does it trespass from algebra into elementary calculus. Nevertheless, it does not avoid the realities of the living world. Unfortunately, one person's reality is another one's complexity; consequently, fastidious model builders may claim that this text deals with material too advanced for young minds. I hope these persons are wrong because there *is* more to life than partial differentials, diffusion equations, and Markov chains.

In three aspects, *Basic Populations Genetics* differs from its predecessor, *Topics in Population Genetics*; all three aspects emphasize that it is a classroom text. First, each chapter is introduced by a brief Preview that serves both to reveal what is coming and to relate coming material to that which has gone before.

Second, End Material has been added to each chapter in an effort to stimulate additional thought. At times, these end items are simple enough to pass for homework or practice problems. Generally, they are more complex, and serve to connect the classroom with current research problems (and with the journals within which such research is published). A number of these items, by leading into mares' nests, could be best answered by means of term papers.

Third, Boxed Essays have been included in a number of chapters. These essays (like their counterparts, the "asides" of stage plays) expand upon items that otherwise would only be mentioned in passing, but do so under circumstances where a disruption of the narrative would have been inappropriate.

I wish to emphasize that no book, including this one, stands alone. In an era when the intellectual half-life of a published paper is measured in weeks and when libraries are hard-pressed to provide shelf space for current journals, it is ridiculous to believe that any one textbook can do justice to even its own field. Because I have largely slighted the mathematical aspects of population genetics, I would recommend a series of auxillary texts (in decreasing complexity): Crow and Kimura's *An Introduction to Population Genetics Theory*, Li's *Population Genetics*, Kimura and Ohta's *Theoretical Aspects of Population Genetics*, and Spiess's *Genes in Populations* or Hartl's *Principles of Population Genetics*. Above all, for advanced students especially, would be Sewall Wright's four-volume treatise, *Evolution and the Genetics of Populations*.

On the biological side, *Basic Population Genetics* has a series of now-classical predecessors (many first issued by Columbia University Press): Dobzhansky's *Genetics and the Origin of Species* (now *Genetics of the Evolutionary Process*), Mayr's *Systematics and the Origin of Species* (now *Animal Species and Evolution*), and Lewontin's *The Genetic Basis of Evolutionary Change*. Shortly, Dobzhansky's *Genetics of Natural Populations*, I-XLIII will be added to this list (Columbia University Press).

Two further books deserve special comment here: Harper's *Population Biology of Plants*, published in 1977, has proven to be a mine of information of special value, especially in respect to ideas developed in Part II of *Basic Population Genetics*. Nei's *Molecular Population Genetics and Evolution* has also been of tremendous help. My opinions frequently differ from those expressed by Nei, but that does not make his text less useful as an adjunct to my own; on the contrary, our differences should make it all the more useful.

On a personal note, I want to express my thanks to three persons in particular: Professor Ross J. MacIntyre, who found time despite a busy schedule to read and comment extensively on each chapter; the book is better for his having done so. Barbara Bernstein, who in preparing the illustrations made valuable suggestions concerning nearly every one; to the extent that the figures clarify the text, she is largely responsible. Cathy Tompkins, who managed to convert a nearly illegible manuscript into a professional-appearing typescript. Finally, a statement of general appreciation to the graduate biology students and their professors at the University of Texas, Austin, and the University of Michigan, Ann Arbor, where, as a visiting professor, I experimented with and finally settled upon the content and format of *Basic Population Genetics*.

Bruce Wallace

Ithaca, New York
February 29, 1980

Basic Population Genetics

This section of *Basic Population Genetics* deals with the elementary, but necessary, aspects of population genetics. At the core is the Hardy-Weinberg equilibrium, an algebraic statement of expectations. Nearer the periphery, however, lies the genetic variation to which the Hardy-Weinberg calculations are applied. Also at the periphery—in chapter 1—are the mental images, the models, that each of us must build if we are to retain and understand masses of data. Then, come all of the conditions which the Hardy-Weinberg equilibrium assumes are nonexistent: they do exist, however, and their consequences cannot be ignored.

The bulk of experimental population genetics still rests on the small shoulders of several species of that tiny fly, *Drosophila*. Times do change, however; newly developed techniques of molecular biologists have made it possible to carry out genetic analyses of wild organisms, using only a drop of blood or a small bit of tissue from each tested individual. Thus, although members of the *Sophophora* subgenus of the genus *Drosophila* are still "bearers of wisdom," they no longer enjoy the near-monopoly of earlier times.

About Models

PREVIEW: Population genetics, like any other science, is more than a collection of observations and facts; it is a set of *organized* observations. Organization is achieved by relating observations to mental images (models) that each person builds in an effort to make sense of the world around him, in an effort to *generalize* his or her experiences. In this chapter a number of models, both algebraic and geometric, that are frequently used in population biology will be described. An understanding of these models will be useful. Still more important, however, is the understanding of the assumptions on which models are based. Among these assumptions are implicit ones whose existence may elude even the original model builder, and which can be discovered only with difficulty.

> The existing theories of population genetics will no doubt be simplified and systematized. Many of them will have no more final importance than a good deal of nineteenth-century dynamical theory. This does not mean that they have been a useless exercise of algebraic ingenuity. One must try many possibilities before one reaches even partial truth.
>
> *A Defense of Beanbag Genetics* (Haldane 1964)

POPULATION GENETICS IS a branch of genetics that deals with the breeding structure and genetic composition of populations.

Individuals, the transitory packages that encase the genetic endowment of a population at a given moment, are interesting to the population geneticist only to the extent that their differing reproductive successes mold this endowment. The population geneticist studies the frequencies of different alleles in a population, the interactions of alleles at the same or different loci, and the degree to which such interactions govern gene frequencies themselves. He notes the number of individuals in a population, first, to assure himself that the population exists, but, second, because this number determines the actual range of frequencies with which a given allele can occur and the role chance will play in the transmission of this allele from one generation to the next. He is especially interested in the divergent paths followed by isolated populations, for such divergences are the simplest of all evolutionary changes. Equally interesting are the events which occur when two formerly isolated populations meet once more; from such meeting sometimes come the finishing touches on behavioral differences or habitat preferences that lead to the reproductive isolation of the two groups. Speciation, the splitting of one formerly interbreeding population into two reproductively isolated groups of individuals, involves the perfection of one or another (or an efficient combination of several) isolating mechanisms.

The beauty of population genetics lies not in enormous catalogues of alterations known to have occurred in the genetic composition of various plant and animal species, but in generalizations that allow us to say with some assurance what the genetic composition of a given population is like at a given moment. These generalizations can be viewed as probability statements that refer to the existing situation at a given locus, as statements concerning the distribution pattern of expectations at many loci, or as descriptions of events that are likely to befall a single locus over a long period of time. The beauty of population genetics, in short, lies with its abstractions rather than with descriptions of actual events. Geneticists, as a rule, regard abstract generalization, if not with affection, at least without fear.

Any abstraction involves the use of a model. This chapter is about models. The need for these comments is twofold: many biologists distrust models; to the extent that their distrust reflects misunderstanding, a short discussion at this point will serve a useful purpose. Furthermore, much of what follows in subsequent chapters is expressed in general—frequently algebraic—terms; these chapters may get a better reception if the usefulness of models is made clear now.

The formulation of generalizations is the goal of all research. Ideal descriptions of events are those which include only pertinent factors and from which everything superfluous has been removed. Thus the shape and size of flasks are not generally included in formulas of chemical reactions. Mathematical equations omit all reference to actual "things." The same search for generalization exists in biology, too; unfortunately, the scope of biology is enormous and its subject matter is exceedingly complex. In building models, population biologists (that is, population geneticists and ecologists) may oversimplify in their search for mathematical tractability and, in doing so, risk discussing a world other than the one we know on Earth. Alternatively, they may remain with empirical results on which, at best, only imprecise and trivial predictions can be based (Slobodkin 1968a).

Living things come in a variety of shapes and sizes—plants and animals, vertebrates and invertebrates, warm-blooded and cold-blooded, flying and crawling, aquatic and terrestrial. The environment by virtue of its physical irregularities, compounded by the added irregularities in the distribution of various forms of life, is extremely complex. Small wonder then that a biologist who specializes in the study of one zoological or botanical family or one genus is skeptical of "broad" statements derived from a study of an entirely different organism; differences in the form and habits of organisms are much more obvious than are similarities in the challenges of life which confront them—challenges to which they respond in fundamentally similar ways.

Haldane (1964) has given an eloquent defense of models in a reply to criticisms (Mayr 1963) leveled at the work of early population geneticists. Mayr, noting that genes do not act independently of one another during development, claimed that the treatment of genes as independent units in devising mathematical theories of population genetics was in many respects misleading. Glancing back over the previous two or three decades, Mayr (1959) asked, "But what, precisely, has been the contribution of this mathematical school to the evolutionary theory?" He was referring to the contributions of Fisher, Haldane, and Wright, which began to appear in the early 1920s and ten years later were summarized in three classic works: *The Genetical Theory of Natural Selection* (Fisher 1930), *The Causes of Evolution* (Haldane 1932), and *Evolution in Mendelian Populations* (Wright 1931).

Haldane's defense of population genetics is based on the following points: First, biologists tend to be overly impressed by mathematics; theoretical models that are distrusted are most often those that are poorly understood. Second, he argues that models based on the work of Mendel, Bateson, and Punnett are a great improvement over the earlier statements of Lucretius. Third, he emphasizes the ambiguity of verbal arguments and the improvement that comes from the use of algebraic expressions. Not only is algebraic reasoning exact, but it imposes an exactness of verbal postulates. This, I believe, is the crucial point in the Mayr-Haldane dialogue: Life is complicated, the environment is bewilderingly heterogeneous even for members of a single species, and development itself represents a poorly understood skein of internally regulated, gene-initiated reactions; nevertheless, it does not follow that verbal descriptions based on ill-defined or ambiguous terms are more useful than descriptions that are mathematical in design (if not in symbolism). Levin (1975) has argued as did Haldane: What mathematical models give up in reality, they gain in tightness of logic and precision of mathematics. Folk theorems, in contrast, become more and more garbled with time.

Andrewartha (1963:181) has expressed reservations concerning theoretical (conceptual) models. Without such models, he admits, we may not know what experiment to do. He goes on to say, however, that "it is the experiment which provides the raw material for scientific theory. Scientific theory cannot be built directly from the conclusions of conceptual models." As an example of a faulty model, he would probably cite one that begins by assuming a uniform environment; the environment, as he points out, is patchy and irregular in virtually all respects. In reply, I would say that a *useful* model is one which helps us understand a number of similar observations, while a truly *powerful* one serves as a basis for explaining discrepancies between different sets of observations. An accurate mathematical description of the patchiness of the environment would not only vary from species to species (the environment of the American bison is surely different from that of dung beetles that may live underfoot) but, if formulated, would be complicated beyond comprehension. On the contrary, a model based on a uniform environment may be quite accurate for many purposes and, whenever empirical observations deviate markedly from the model, may still provide a basis for discussing the role of patchiness of the environment in population studies. That this is the role of the model in research can be gathered from the following statement by Sturtevant (1965: 51): "One of the striking things about the early *Drosophila* results is that the ratios obtained were, by the standards of the times, very poor . . . with *Drosophila* such ratios as 3 : 1 were rarely closely approximated." The scientific discoveries of the Morgan school of Drosophilists at Columbia University were based on conceptual models, not on raw data as such. More recently, the model of DNA structure (Watson and Crick 1953) rather than the empirical data concerning the chemical composition of DNA (Chargaff 1950) has given rise to what is now known as molecular genetics.

Every model, even the most successful, contains within it the seeds of its own destruction. It must make predictions with sufficient precision so as to be falsifiable. The ease with which the ramifications of a mathematical model can be extended by subsequent calculations should not detract from the vulnerability of its basic assumptions. Extended calculations that contain predictions which are subsequently confirmed strengthen the original model, of course. Those that are carried out as intellectual exercises without decreasing the model's vulnerability do not affect the validity of the original model. Felsenstein (1975) showed that the assumptions underlying one mathematical model (Malécot 1969) were incompatible; no extension of Malecot's original calculations would have altered this critical fact.

The remainder of this chapter will be devoted to a consideration of models that deal with population growth and regulation, the survival of individuals, and the historicity of populations. Some of these models can be transformed one into the other; others are contradictory. Each, however, illustrates the use of simplified generalizations and mathematical abstractions in understanding the biology of natural populations.

The Growth and Regulation of Populations

The following models relate to the number of individuals of a given species, or better, to the numbers of individuals of that species inhabiting a given locality. The number of starlings near Ithaca, New York can serve as an example. There are several things we can say about a local population with complete assurance that we shall be correct. First, the number of individuals in the local population is not constant from year to year or from generation to generation. In the case of starlings, there were none anywhere in the United States before 1890, the year they were introduced from Europe. Since then their numbers have increased steadily but with fluctuations large enough so that they have been greater nuisances in certain years than in others.

The second thing we can say about populations is that, despite temporal fluctuations, in the long run they tend to remain constant in size. Following their introduction into the United States (with perhaps a lag of several years), starlings increased in numbers very rapidly. Within occupied areas, however, they have ceased increasing at the earlier rate; there has been no spectacular rise in Ithaca, for example, during the past decade. No elaborate argument is required to justify the claim that a species can never increase in numbers geometrically for long; if no other limiting factor intervenes, standing room will be exhausted in a rather small number of generations.

The final statement we can make with certainty is that despite the numbers of individuals actually existing at any one moment, extinction of a population—indeed, of an entire species— is not only possible but, in the long run, is even probable. Most species of the past (some say 99.9 percent), lest we have forgotten, have disappeared. The wholesale extinction of numerous dinosaur species followed immediately the period during which they were the dominant animals on Earth. The extinction of the passenger pigeon in the United States, and the near extinction of the American bison, remind us that mere numbers of individuals are no guarantee of a species' continued existence.

The exponential increase in population numbers is perhaps the simplest of all models of population growth. The multiplication of bacteria within a culture tube (figure 1.1) provides a simple empirical demonstration of this model. Let each individual in a population produce R offspring, and let the number of individuals at time, t, be N_t. Starting with the initial population, N_0, the number of individuals increases as follows:

$$N_1 = RN_0$$
$$N_2 = RN_1 = N_0 R^2$$
$$N_t = N_0 R^t$$
$$N_{t+1} = N_0 R^{t+1} = RN_t.$$

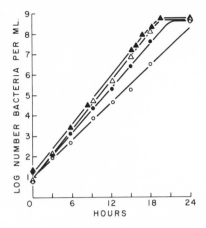

Figure 1.1 The exponential growth of four different strains of *Escherichia coli* in laboratory cultures (each symbol represents one strain). Beyond 5×10^8 bacteria per ml, the exhaustion of certain limiting nutrients terminates the exponential or log phase of growth; death and cell division are approximately equal after this point is reached, thus stabilizing the number of viable organisms. Reproduced with permission: from Atwood et al., *C. S. H. Symp. Quant. Biol.* (1951), 16:345–55; © 1951 Cold Spring Harbor Laboratory.

The change in the number of individuals equals

$$\Delta N_t = N_{t+1} - N_t$$
$$= RN_t - N_t = (R - 1)N_t$$
$$= rN_t, \text{ where } r = R - 1.$$

The data illustrated in figure 1.1 show that under certain circumstances, exponential growth can be sustained for a considerable number of generations; 18 hours, for example, represents some 50–55 generations for *Escherichia coli*. Small differences in the rate of increase, as between the strains illustrated, may lead eventually to considerable differences in the relative numbers of individuals; at 18 hours, a hundredfold difference exists between the densities of individuals in the most and least crowded cultures shown in figure 1.1. Finally, the substitution of a stationary phase for the initial exponentially increasing (also known as the "log") phase illustrates for bacterial cultures the exhaustion of one or more essential nutrients from the culture medium, or the accumulation of inhibitory toxins. The figure suggests that the size of the stable population need not depend upon its initial rate of growth, three of the illustrated cultures have stabilized at the same level by 21 hours, and the fourth culture may do so by 24 hours.

The exhaustion of environmental resources and the role such exhaustion plays in population stabilization is also illustrated in figure 1.2. In this figure the environment, unlike the bacterial culture illustrated previously, is self-renewing although limited in its capacity to support the individual members of the population. This limit (K) is called the *carrying capacity* of the environment.

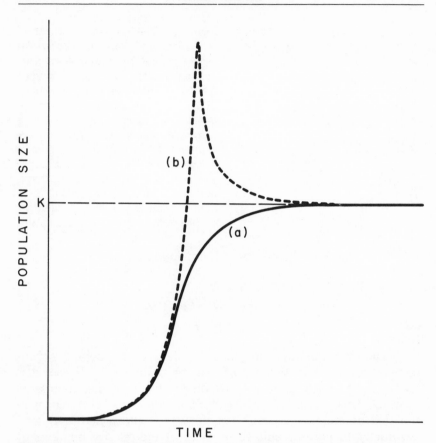

Figure 1.2 Logistic curves showing (a) the smooth growth of a population as it approaches the maximum sustained size (K) permitted by environmental resources, and (b) the decline in size of a population that has overshot the sustained size permitted by a late-appearing resource shortage.

An environment with a finite carrying capacity imposes a limit such that, when population size, K, is reached, population growth ceases. Any mathematical term that reduces ΔN to zero as N approaches K can illustrate this effect. One such term is $(K - N)/K$; another is log (K/N). The first is more widely used; the resulting model (the Lotka-Volterra model) is written:

$$\Delta N = rN \left(\frac{K - N}{K} \right).$$

Because of the parenthetic term, ΔN approaches zero (no further growth in population size) as N approaches K. The resulting growth curve (the *logistic* curve) is illustrated in figure 1.2a.

Many persons (for example Nei 1975: 39) assume that populations normally approach K from below; that is, that population size is initially small and subsequently increases until N equals K. This assumption is also based on the notion that K is a constant. If K is variable, then N may often exceed the momentary carrying capacity of the environment; in that case the population must decrease in number. The potato famine in Ireland (1846–1851) illustrates this point: a population that had previously stabilized at about 3 million persons grew suddenly to nearly 9 million. The failure of the potato crop in 1846 and subsequent years caused the death of 1 million persons, and forced the emigration of an additional 1.5 million Irish—largely to the United States. One might propose (Wallace 1975a) that the most limiting value of K is the last one evoked during a population's growth. Under a dismal prophecy such as that, populations would always overshoot the environment's carrying capacity (figure 1.2b) only to collapse in size again under dire circumstances.

The equation for the logistic curve can be expanded readily to encompass two species that compete for the same environmental resources:

$$\Delta N_1 = r_1 N_1 \frac{(K_1 - N_1 - \alpha N_2)}{K_1}$$

$$\Delta N_2 = r_2 N_2 \frac{(K_2 - N_2 - \beta N_1)}{K_2}.$$

In these equations, α and β are equivalency terms: α measures the extent to which each individual of species 2 utilizes resources that otherwise would support individuals of species 1, and β measures the utilization by each member of species 1 of resources that otherwise would have supported individuals of species 2.

We see in the above equations that ΔN_1 and ΔN_2 become zero (other than in the trivial cases when the r's or N's equal zero) when $K_1 - N_1 - \alpha N_2$ and $K_2 - N_2 - \beta N_1$ equal zero. Setting N_1 equal to zero, we see that $N_2 = K_2$ or, $N_2 = K_1/\alpha$; setting N_2 equal to zero, $N_1 = K_1$ or $N_1 = K_2/\beta$. N_1 does not change (that is, $\Delta N_1 = 0$) at any point along the line which joins $N_1 = K_1$ ($N_2 = 0$) and $N_2 = K_1/\alpha$ ($N_1 = 0$). Similarly for N_2, ΔN_2 equals zero at any point on the line connecting $N_2 = K_2$ ($N_1 = 0$) and $N_1 = K_2/\beta$ ($N_2 = 0$). Any population in which the number of species 1 or 2 lies to the right and above these lines, dwindles in size ($\Delta N < 0$); any population in which the number of either species lies below and to the left of these lines, grows in size ($\Delta N > 0$). Four pos-

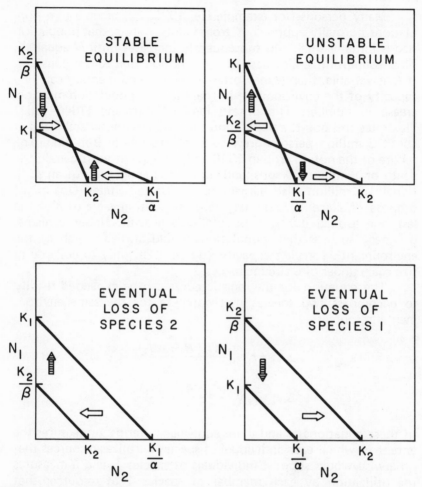

Figure 1.3 Diagrams illustrating the four possible outcomes for two species (#1 and #2) that utilize the same resource, and whose growth fits the logistic curve. The solid lines connect the points at which no change in numbers of either #1 (line connecting K_1 and K_1/α) or #2 (line connecting K_2 and K_2/β) will occur. Either species will decline in number if it lies above and to the right of its line; either will increase if it lies below and to the line's left. Shaded arrows represent the change in number of species #1 (N_1); open arrows represent changes in species #2 (N_2). Thus, as species #1 approaches K_1, N_2 approaches 0; alternatively, as N_2 approaches K_2, N_1 approaches 0. In the case of the stable equilibrium, the number of either species approaches the line at which ΔN equals 0; consequently, the number of each species stabilizes ($\Delta N = 0$) where the solid lines intersect.

sible relationships exist; their consequences are illustrated in the four diagrams of figure 1.3.

For the moment, the important aspect of figures 1.2 and 1.3 (and of the calculations on which they are based) is that the predicted outcomes depend upon assumptions that are not always obvious. The term, $(K - N)/N$, which eventually reduces population growth to zero could, as was mentioned earlier, have taken on other forms such as log (K/N). Furthermore, the equivalency terms, α and β, need not be constant nor need they be restricted to first-order expressions of N. One can imagine expressions such as $K - N - \alpha_1 N_2 - \alpha_2 N^2$, for example (Ayala et al. 1973). Even the term, N, which refers to the number of individuals in a population, suggests that these individuals are similar to one another; in truth, they need not be: some are large, for example, while others are small. In short, the equations and diagrams are models—useful ones, to be sure—based on numerous simplifying assumptions.

The regulation and stabilization of population size can also be represented as shown in figure 1.4, a diagram borrowed from Haldane (1953). The vertical axis in this case represents the ratio of adult, fertile daughters (that is, *new* mothers) to the mothers (*old* mothers) that produced them; the horizontal axis represents population size. As long as the number of new mothers exceeds the number of old ones, the population grows in size (males are simply ignored in this account). There is a population size, however, at which each mother (because of lack of food, of living space, or of any other resource) can successfully rear only one daughter as a replacement for herself in the next generation; the population stabilizes at that point. Should the population accidentally exceed this number, it subsequently decreases in size because each harassed mother leaves fewer than one daughter; if by some chance it falls below the equilibrium number, it then increases once more because each old mother leaves more than one daughter. Curves *a* and *b* in figure 1.4 differ in that *b* suggests that a population may have a critical minimum size: below the point *B*, population *b* would dwindle and disappear; above *B*, it grows to *C*, its stable equilibrium size.

Figure 1.4, like its predecessors, is a model. The axis depicting population size offers no suggestion that individuals may differ from one another. And, aside from raising or lowering the line representing a daughter/mother ratio of 1.00, no method is provided for allowing the equilibrium size to vary. The model says nothing specific about the nature of the limiting resources; in

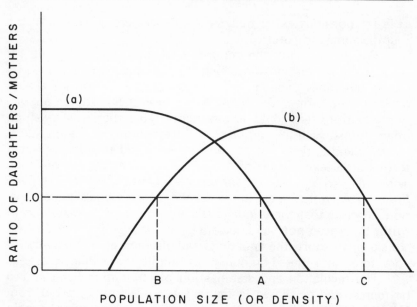

Figure 1.4 Curves illustrating the ratio of daughters to mothers as a function of population size (or density). Curve (*a*) represents a species for which a single (gravid) female is sufficient to found a population; the population would eventually stabilize at size *A*. Curve (*b*) represents a species for which a critical population size (*B*) is needed in order that the population be maintained. Below size *B*, daughters are insufficient in number to replace their mothers. Above size *B* the population grows until it attains size *C* where it stabilizes. (Notice in this diagram that a fourfold increase of daughters to mothers represents a fourfold increase in population size as well. When the number of daughters no longer exceeds the number of mothers, the ratio of daughters to mothers equals 1.00, and the population no longer changes in size.)

effect, it merely emphasizes that the eventual size of a population may represent a *stable* equilibrium.

Individual Survival

Taking the attainment of adulthood as a measure of successful survival, each individual either survives or does not. How can the matter of survival be treated so as to be amenable to mathematical calculations?

One possibility (see, for example, Morton et al. 1956) is to regard survival as the successful outcome of many successive, independent challenges each of which may lead to death. These challenges may be external to the individual as are environmental accidents, or they may be internal as, for example, genetic shortcomings or happenstance errors occurring during development.

Let each challenge (1, 2, 3, . . . , etc.) be characterized by a slight probability of death $(X_1, X_2, X_3, . . . ,$ etc.). The probability of survival, S, then becomes $(1 - X_1)(1 - X_2)(1 - X_3) . . . (1 - X_i) . . . (1 - X_n)$. This may be rewritten as

$$S = \prod_{i=1}^{n} (1 - X_i).$$

If the chance of death is small,

$$S = 1 - \Sigma X_i \text{ or } 1 - n\overline{X}.$$

This can be rewritten as

$$S = e^{-n\overline{X}}$$

This model could describe the probability of one's surviving a motor trip from Chicago to St. Louis. If X_i were the probability of encountering a fatal accident in each successive mile (i), the probability of successfully completing the 300-mile trip would be approximately $(1 - \overline{X})^{300}$ or $e^{-300\overline{X}}$. An essential feature of the model is the independence of the individual probabilities; a faulty tire that progressively worsened under high-speed travel would drastically alter these calculations.

An alternative model relating to individual survival can be constructed according to the diagram shown in figure 1.5. Each hexagonal space in the diagram can, *by definition*, support one individual: these are unit biological spaces. One of two individuals occupying the same space will survive only if the two individuals differ in their ability to survive (in competitive ability or aggressiveness, for example); otherwise both individuals will die. Indeed, if the most competitive phenotype is represented by a single individual, that individual will survive despite the presence of one, two, or even more other, weaker ones.

The proportion of unit spaces that will support surviving individuals depends heavily upon the initial density of these individuals, and on the number of phenotypes possessing differing

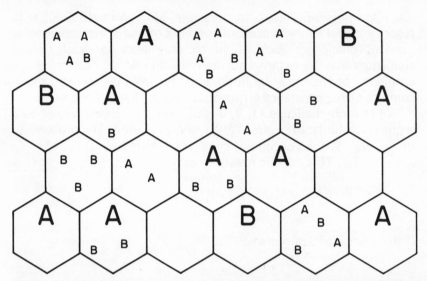

Figure 1.5 The subdivision of an area into unit biological spaces; that is, spaces which by definition can support only one member of a population. The figure shows (1) the survival (large letters) of individuals of phenotypes *A* and *B* when each occurs alone in a unit space, (2) the failure of each phenotype to survive (small letters) when two or more individuals of a single sort attempt to grow in the same unit space, but (3) the survival of a single individual of the more aggressive *A* phenotype when it and one or more *B* individuals compete for the same space. Under this model, phenotypic variation provides a basis for self-thinning among overcrowded individuals.

competitive abilities (table 1.1). Given a single phenotype and a Poisson distribution of competing individuals (such as seeds per unit area), the maximum proportion of successfully occupied areas is about 37 percent; this would be achieved if on the average one seed fell in each space. Empty spaces (also 37 percent; see the

Table 1.1 Proportion of single unit biological spaces containing surviving plants under various conditions in respect to (1) average number of seeds per unit space and (2) numbers of equally frequent, hierarchal competing types (*A* can displace *B*, and *B* can displace *C*, for example).

Number of Competing Types		Average Number of Seeds Per Unit Space			
		1	*2*	*3*	*4*
1	*A*	0.368	0.271	0.149	0.073
2	*A > B*	0.487	0.505	0.411	0.307
3	*A > B > C*	0.533	0.609	0.552	0.468
4	*A > B > C > D*	0.555	0.666	0.637	0.570

Table 1.2 A tabulation of the probabilities that an event will occur k times, given that its average frequency of occurrence is m and that occurrences are distributed according to the Poisson distribution:

$$\frac{m^k}{k!}\, e^{-m}.$$

k \ m	0.1	0.2	0.3	0.4	0.5	1.0	1.5	2.0	2.5	3.0
0	.905	.819	.741	.670	.607	.368	.223	.135	.082	.050
1	.090	.164	.222	.268	.303	.368	.335	.271	.205	.149
2	.005	.016	.033	.054	.076	.184	.251	.271	.257	.224
3	.000	.001	.003	.007	.013	.061	.126	.180	.214	.224
4	–	.000	.000	.001	.002	.015	.047	.090	.134	.168
5	–	–	–	.000	.000	.003	.014	.036	.067	.101
6	–	–	–	–	–	.001	.004	.012	.028	.050
7	–	–	–	–	–	.000	.001	.003	.010	.022
8	–	–	–	–	–	–	.000	.001	.003	.008
9	–	–	–	–	–	–	–	.000	.001	.003
10	–	–	–	–	–	–	–	–	.000	.001
11	–	–	–	–	–	–	–	–	–	.000

Boxed Essay and tables 1.2 and 1.3) and spaces with two or more seeds (the remaining 26 percent of all spaces) would be devoid of mature plants.

Under the unit-space model, increases in the number of competing phenotypes always result in increases in the proportion of successfully occupied spaces: the greater the number of pheno-

Table 1.3 Two examples of events which appear to have occurred according to expectations based on the Poisson distribution; that is, in each example each event seems to have occurred independently of (or, unrelated to) the occurrence of other, similar events. (A) the number of times various city blocks were hit by bombs during the German air raids on London during World War II. (B) The number of soldiers killed by mule kicks per corps per year in ten Prussian cavalry corps over a period of 20 years.

	(A)			(B)	
# Hits	# Blocks	# Expected	# Deaths	# Corps-Years	# Expected
0	229	227	0	109	109
1	211	211	1	65	66
2	93	99	2	22	20
3	35	31	3	3	4
4	7	7	4	1	1
5	0	1	5	–	–
6	0	0	6	–	–
7	1	0	7	–	–

NOTE:

Total number of blocks = 576
Total number of hits   = 537
Average hits/block = m = .93

Total number of corps-years   = 200
Total number of deaths   = 122
Average deaths/corps-year = m = .61

types, the greater the chance that a space occupied by two or more plants will contain only one of the most competitive type. There is always a density of seeds beyond which the proportion of spaces supporting survivors decreases; nevertheless, the density yielding the maximum proportion of occupied spaces increases as

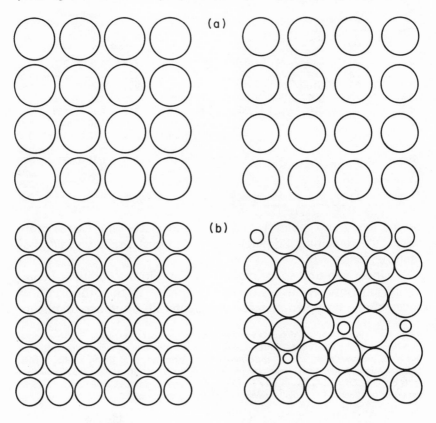

HYBRID   OUTCROSSED

Figure 1.6 A diagram illustrating the growth of F_1 hybrid and outcrossed cabbages under (a) adequate and (b) inadequate spacing. With ample space provided for each plant, both the genetically uniform F_1 hybrids and genetically heterogeneous outcrossed individuals yield rather uniform heads. With inadequate space, the genetically uniform F_1 hybrids remain uniform (although the individual heads are smaller than before) but the more aggressive individuals among the outcrossed plants tend to displace weaker ones, thus causing the cabbage heads to differ considerably in size. In the latter case, the genetically variable individuals compete for what approximates the single unit biological spaces of figure 1.5. (Suggested by H. M. Munger.)

the number of phenotypes increases. Thus, in table 1.1 we see that about two-thirds of all spaces would contain surviving plants if there were four phenotypes and the density of seeds were two or three per space; whereas with a single phenotype this density would reduce survival to at least one space in four, perhaps one in seven.

The concept of unit biological spaces and competition among plants is illustrated in figure 1.6. Here we see a diagram representing the variation in the size of cabbage heads which results when genetically variable individuals are planted too close together; many small plants have been overwhelmed by larger, more aggressive neighbors. Under wider spacing (figure 1.6a), all plants produce heads of about the same size, even those that would do poorly when crowded. In contrast, genetically uniform (F_1 hybrid) plants produce heads of uniform size both when crowded and when widely spaced. With only minor changes, the schematic diagrams of figure 1.6 would have accurately represented the crowns of trees in a forest where certain trees dominate and crowd out less competitive ones. A forest in which the individual trees are phenotypically uniform and in which the trees are planted too densely "stagnates"; no tree grows properly (see Hawley 1946: 222–23). A stagnated forest reflects the effects of overcrowding plants of which there is but one phenotype (table 1.1).

The difference between the multiplicative and the unit-space models of individual survival stems from the assumed independent fatal accidents of the first. In the unit-space model, as illustrated in figure 1.5, the elimination of B by A, when the two seeds fall within the same space, is not the outcome of a series of independent events. On the contrary, hour-by-hour and day-by-day, A grows larger and its leaves and roots overwhelm those of B so that, as A thrives, B wanes (Wallace 1977a).

The Historicity of Populations

The final model to be discussed here is a modification of one developed by Lewontin (1966) in asking whether nature is probable or capricious. Nature is capricious, according to Lewontin's definition, if repeated samples fail to increase one's information regarding a given system. Suppose we have an enormous vat of beans some of which are red and others white, and the two kinds are thoroughly mixed. The analysis of cumulative samples of, say, fifty beans each will lead us to a better and better estimate of the true proportions of red and white beans in this vat. There is no guarantee that this is so, of course; rather, it is highly probable that large numbers give more accurate estimates of true frequencies than do small ones.

The key word in the preceding paragraph is "cumulative." Suppose that in recording the results of successive samples we are limited to recording the last ten samples only. "Repeated" sampling no longer leads to more accurate estimates of the proportions of red and white beans. The record is imperfect because everything that happened eleven or more samples ago is forgotten.. Our recording system has a short memory.

The vat of beans can now be regarded as the source of environmental challenges to a population. Each successive challenge to the population is represented by the proportions of red and white beans in a draw. Let's assume that the true proportions in the vat are 50:50. The population we are studying also contains two alleles, A and a, whose initial frequencies p_0 and q_0 are 50:50. With each draw of beans from the vat, we alter p in the following way (a numerical example is given at the end of this chapter):

    1. Calculate the difference between the proportion of red beans and p (negative if p is the larger of the two).

    2. Multiply the difference by either p or q, whichever is smaller.

    3. Add or subtract (according to its sign) the product to or from p.

    4. Repeat.

Several features of the above procedure are worth noting: First, the population responds poorly to the environmental challenge if p is close to 0 percent or 100 percent because step 2 involves the product of a small fraction. Second, the population responds readily as long as p remains between 30 percent and 70 percent. Third, the fate of p depends upon the *order* with which chance events occur. The same draws taken in the reverse order do not necessarily bring a second population over the same course or to the same final point as do the first. The uniqueness of the historical paths taken by different populations even to the same *overall* influences is shown in figure 1.7.

For our present purposes, the importance of this exercise lies in the use of a vat of beans (or of a table of random digits) in

Figure 1.7 Gene frequencies (q) of successive generations in a fluctuating environment. The solid line is a series of frequencies resulting from a random series of environmental conditions; the dashed line is a series of frequencies generated by the identical environments operating in the reverse order. The differing end points of the two lines illustrate that evolutionary changes are historical changes; the outcome depends not only on specific events but also on the historical sequence in which these events occur. (After Lewontin.) Reprinted with permission: from *BioScience* (January 1966); © 1966 American Institute of Biological Sciences.

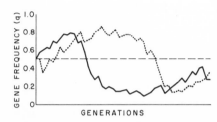

GENERATIONS

illustrating the unique history of any population. A population in which p is nearly 0 percent or 100 percent is a poor recorder of current events. One point has by now been very nearly forgotten: a population is not a vat of beans nor is it a table of random digits. Rather, a clever model has served to illustrate some very complex—but general—interactions between organisms and their environment, and the recording of these interactions in the genetic apparatus of the population.

BOXED ESSAY

THE POISSON DISTRIBUTION

Events that occur sporadically in time or space, each independent of the others, are distributed according to a pattern described by Poisson (1837) as well as by others (for example, "Student" 1907). Here we shall derive the Poisson distribution as a limit to the more familiar binomial distribution.

Suppose two mutually exclusive events (son vs. daughter, success vs. failure) have probabilities of p and q, where $p + q = 1$. The probability of obtaining k successes (or sons) in n trials is given by the general term.

$$\frac{n(n-1)(n-2) \ldots (n-k+1)}{k!} p^k q^{n-k}.$$

Let n grow large and p grow small, but in such a manner that $np = m$ where m equals the mean number of successes (mutant bacteria, weed seeds among samples of grain, insects found within a prescribed area, etc.). Because $np = m$, $p = m/n$ and $q = 1 - (m/n)$.

If n is very large, the alteration of the general term of the binomial distribution given below the brackets is approximately true (be sure that you understand the substitution for each bracketed term):

$$\underbrace{\frac{n(n-1)(n-2)\ldots(n-k+1)}{k!}\left(\frac{m}{n}\right)^k}_{\dfrac{m^k}{k!}} \underbrace{\left(1-\frac{m}{n}\right)^n}_{e^{-m}} \underbrace{\left(1-\frac{m}{n}\right)^{-k}}_{1.}$$

[Notice: $n(n-1)(n-2)\ldots(n-k+1) \approx n^k$]

Thus, the general term of the binomial distribution becomes, in the case of the Poisson distribution, $m^k/k! \, e^{-m}$.

Those who own pocket calculators can develop the Poisson distribution as follows: (1) Register m. (2) Alter its sign. (3) Obtain e^{-m} ($k = 0$). (4) Multiply by m ($k = 1$). (5) Multiply by m and divide by 2 ($k = 2$). (6) Multiply by m and divide by 3 ($k = 3$). Continue until result is negligible. If m is an integer, one pair of adjacent terms must be identical. (Why?)

A brief table of expectations under the Poisson distribution is given in table 1.2. Two examples in which observed distributions are compared with those expected under the Poisson distribution are given in table 1.3. In both instances, the observed and expected distributions are extremely similar.

Although proof lies beyond this essay, the variance (σ^2) of a Poisson distribution equals its mean (m). This fact may be tested empirically by calculating the variances (see page 75) of the distributions shown in table 1.2 or 1.3 (see page 17).

FOR YOUR EXTRA ATTENTION

1. Many populations (as you will see in later chapters) are not uniform for the alleles at individual gene loci; instead (using the A locus as an example), individuals can be of several types: AA, Aa, and aa. To account for this observed genetic heterogeneity by stabilizing forces, three models have been proposed: Some persons claim that heterozygous individuals survive in greater numbers and produce more offspring than homozygotes. Others suggest that different portions of the environment favor one or the other of the two homozygotes, but that heterozygotes do nearly as well under all conditions. Still others believe that organisms of each genotype seek out environments most favorable to themselves.

Try to represent these three possibilities by means of diagrams. Check your representation with one used by Powell and Taylor (1979).

2. Assume that a population consists of asexual females (n), sexual females (N) and sexual males (N); the total population equals $2N + n$ in number. Assume, too, that both asexual and sexual females produce c surviving offspring, those of asexual females being in turn asexual females.

Now, the original proportion of asexual females in the population equals $n/(2N + n)$. The proportion of asexual females among the progeny equals $cn/(cN + cn)$ or $n/(N + n)$. When n is small, the latter proportion is nearly twice the original one.

Thus, it appears that asexuality will rapidly become the predominant means of reproduction in any population. Or will it? This problem (and model) will arise again in chapter 24; it has been discussed thoroughly by Williams (1975), Maynard-Smith (1978) and Lloyd (1980).

3. Mechanical models, like mathematical ones, are based on implicit assumptions. By using a small circulating pump, a major TV network intended to illustrate the effect of doubling the mutation rate on the frequency of mutant genes in a population. Water from a reservoir was pumped through a tube and into a perforated cylinder where it accumulated until the rate of leakage from the cylinder into the reservoir equaled the rate at which water flowed from the tube into the cylinder. Having marked this level, a stopcock valve was turned so that the water now entered the cylinder through two tubes. To the consternation of the program officials, the level of water in the cylinder did not change. Why?

4. Imagine that a particular fly lays its eggs on a flat surface, sparsely and randomly distributed. Imagine, too, that a parasitic wasp wanders aimlessly over this surface seeking fly eggs to parasitize. (Each parasitized fly egg produces a wasp, not a fly.)

The area searched by a female wasp equals the distance to either side within which an egg can be detected times the length of the path which she travels. The number of eggs within that area, the number of eggs she encounters, equals the total eggs on the surface times the fraction of the total area the wasp has searched.

Using this model, Nicholson and Bailey (1935) calculated that an increase in the density of fly eggs (as through selection for higher fertility among flies) would lead to a smaller number of flies and a larger one of parasitic wasps. J. B. S. Haldane (1953) referred to this finding as "a blinding glimpse of the obvious."

5. To illustrate the inefficiency of a random search for food (as that used by the wasp in the previous example), imagine a long tube divided midway along its length by a 100-holed net. This net has been covered by a film of nutrient medium. A small winged organism flies openmouthed from one end of the tube to the other through the holes in the net. If, in passing, it encounters food, it survives to make a return flight; if, however, it passes through an empty hole, it perishes.

Notice that the probability of surviving on successive passages through the net can be written 1.00, 0.99, 0.98, 0.97, . . .

Confirm that the probability of survival for 30 successive passages is less than 1 percent even though 70 percent of the original food remains. Confirm as well that the probability that *two* organisms would survive 20 simultaneous passages is only about 1 percent even though 60 percent of the original food would remain uneaten at that time.

To the extent that food is encountered by chance, individuals speed each other's demise even though seemingly ample food remains to be discovered; this effect of one individual on the survival of another is one aspect of competition—a topic that will appear periodically throughout this book.

6. Geneticists are accustomed to speaking of "Mendel's Laws"—laws, incidentally, that form the basis for population genetics. Mathematicians occasionally refer to the "Mendelian *model*."

Consider the data which Mendel attempted to explain, and the hypothetical assumptions or postulates by which he proceeded. Sutton (1903) and Boveri (1903) were among the first to associate physical bodies to what previously had been intellectual concepts; the process they began continues even today under the name "genetics."

7. Carry out the calculations described on page 20 for the following "draws," proceeding first from left to right, and then from right to left. What evidence emerges suggesting that selection is a historical process?

Draw

1	2	3	4	5	6	7	8	9	10	11	12	13	14	15
.7	.5	.8	.3	.7	.7	.2	.9	.2	.1	.7	.9	.9	.7	.6

Hint: The first two cycles of calculations proceed as follows:

1. $(.5 - .7) = -.2$ 2. $(-.2 \times .5) = -.10$ 3. $(.5 - .10) = .40 = p_1$
2. $(.40 - .5) = -.10$ 2. $(-.10 \times .4) = -.04$ 3. $(.40 - .04) = .36 = p_2$

Patterns of Distribution

PREVIEW: This chapter deals with the actual populations of population genetics. It deals with the distribution of individuals in both the static sense (Where are individuals found?) and the dynamic sense (How did they get there?). Consequently, dispersion patterns as well as those of distribution are discussed. Of necessity, the material covered in this chapter is closely related to that covered later in chapter 4 (Migration); here, however, the movement and position of organisms are important; in chapter 4 their reproductive behavior and the fate of the genes they carry are the centers of attraction.

POPULATION GENETICISTS STUDY populations. Except for those populations that are confined to small cages in laboratories, most populations do not have sharp boundaries. Local populations (the "aggregates" of figure 2.1) shift their positions with time. They merge into one another. Individuals constantly enter or leave as migrants. Even in the case of plant species where mature individuals do not move, pollen and seeds can be transported over considerable distances. Because of these many uncertainties and inconstancies, one might think that the idea of a local population is useless; we shall see, however, that this is not so.

Where are the individuals found? We are all familiar with global or continental maps on which the geographical distributions of various organisms are enclosed within smoothly curving lines

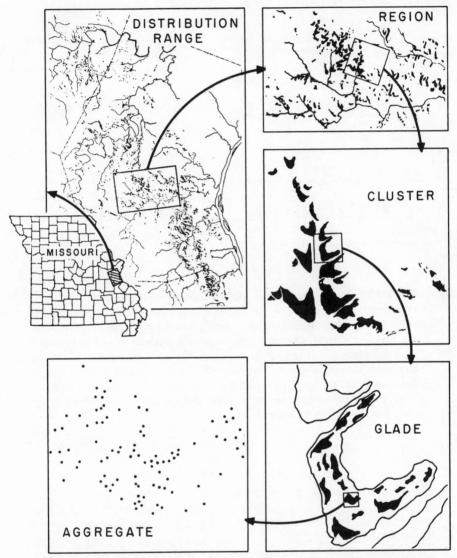

Figure 2.1 The geographical distribution of *Clematis fremonti* var. *riehlii* and its smaller, local distributions of decreasing sizes, including the distribution of individual plants in the *aggregate*, the smallest local population of this species. (After Erickson 1945.)

(see figure 2.2a). We are all also aware that no species is spread uniformly throughout its entire distribution area; rather, its members are associated in groups of various sizes, which are found in an entire heirarchy of patches. Erickson (1945) has illustrated the distribution of *Clematis* exceptionally well (figure 2.1). The successive diagrams of his figure can be regarded as successively larger magnifications; in that case, the area occupied by the aggregate whose individual members are clearly visible represents a 3,000,000-fold enlargement of one small portion of the total distribution range of this species.

The geographical distributions of *D. pseudoobscura* and *D. persimilis* are shown in figure 2.2a. (Omitted from the figure is an area near Bogotá, Colombia, in which an isolated population of *D. pseudoobscura* was found during 1960.) Although the area occupied by *D. persimilis* falls almost entirely within that occupied by *D. pseudoobscura*, the two species are not commingled at random.

A transect some 70 miles in length is designated by an asterisk in figure 2.2a. This transect extends from low-lying valley areas of California into the Sierra Nevadas. The proportions of *D. pseudoobscura* and *D. persimilis* flies that occur at different elevations along this transect, are shown in figure 2.2b. At elevations below 3,000 feet, collections of these flies consist almost entirely of *D. pseudoobscura*; at 10,000 feet or above they consist almost entirely of *D. persimilis*. At intermediate altitudes, mixtures of the two species are generally observed although the collection taken at 4,000 feet which consisted almost entirely of *D. persimilis* warns us that elevation is not the sole variable in determining the proportions of these two species within different localities.

Restricting our attention to *D. pseudoobscura*, figures 2.2c and 2.2d show that these flies are distributed uniformly on a microscale in neither space nor time. Figure 2.2c shows the variation in numbers of flies caught in small traps set at 10 meter intervals along a line some 300 meters in length. The numbers range from well over 100 flies per trap to about 15; this variation might easily reflect the spotty distribution of trees or bushes in the trapping area. Figure 2.2d shows the numbers of flies arriving at traps throughout the day, from 5:00 A.M. until 7:00 P.M. Two peaks of activity can be seen: a small one in the morning and a much larger one in the late afternoon just before sunset. (Until this diurnal activity had been discovered, workers who normally

Figure 2.2a The North American distributions of *D. pseudoobscura* and *D. persimilis*.

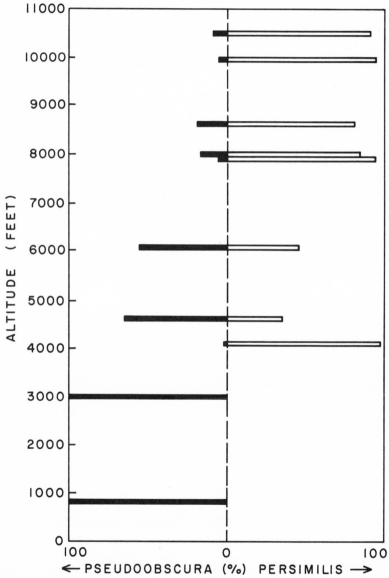

Figure 2.2b The relative proportions of *D. pseudoobscura* and *D. persimilis* at different altitudes along a 70-mile transect in Yosemite, California region (shown as * in *a*).

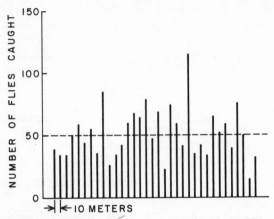

Figure 2.2c The numbers of *D. pseudoobscura* flies caught during two evenings' trappings at 10-meter intervals on Mount San Jacinto, California (the mean number, 51, is shown by the dashed line).

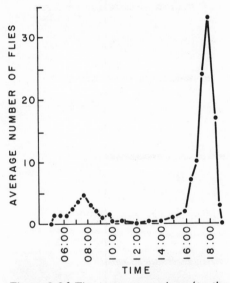

Figure 2.2d The average numbers (to the nearest $\frac{1}{2}$ fly) of *D. pseudoobscura* visiting traps on Mount San Jacinto between 5:00 A.M. and 7:00 P.M. (*a* and *d* after Dobzhansky and Epling 1944; *b* after Dobzhansky 1948a; *c* after Dobzhansky and Wright 1943.)

exposed traps only before and after lunch believed that *D. pseudo-obscura* was a rare species!)

Timofeeff-Ressovsky, a pioneer population and radiation geneticist, studied the local distribution of a number of European Drosophila species. One of his maps (Timofeeff-Ressovsky 1939) is reproduced in figure 2.3. Within this small German farmyard, the three types of *Drosophila* (*D. melanogaster*, *D. funebris*, and *D. "obscura"*—an unidentified assortment of *D. obscura* and *D. subobscura*) have quite different distributions: *D. melanogaster* has most likely aggregated near fermenting fruit or vegetable products; *D. funebris*, on the other hand, seemingly remained

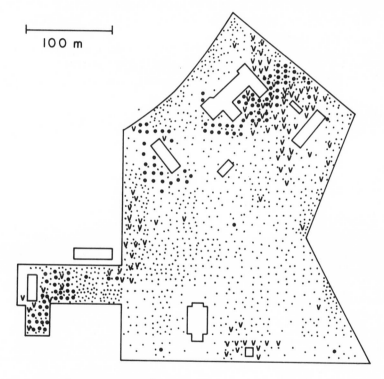

⊢————————⊣
100 m

···" O B S C U R A "

vᵛv F U N E B R I S

•ᵉ• M E L A N O G A S T E R

Figure 2.3 Capture sites for three species of *Drosophila* on a tenement farm near Berlin, summer 1938. (After N. W. Timofeeff-Ressovsky 1939.)

near manure and other animal waste (note the presence of *D. fune-bris* near a small outhouse while the nearby building [a house?] has few flies of any sort).

An even more remarkable pattern of intertwining patches of different Drosophila species has been studied by Richardson and Johnston (1975). An area only 60 meters by 75 meters located on the island of Hawaii is shown in figure 2.4. Within this small area there are multiple, clearly identifiable spots at which one or the other of three Drosophila species may be found (*D. kambysellisi, D. mimica*, and *D. imparisetae*). The turnover from a majority of one species to that of another can occur within a distance of one or two meters. Data of such precision could not be gathered by trapping the flies, because traps themselves have certain attractive radii; the data on the Hawaiian species was obtained by carefully sweeping with a collecting net at designated spots within the test area.

The account of where members of a species are found can be concluded with two examples extending points made earlier.

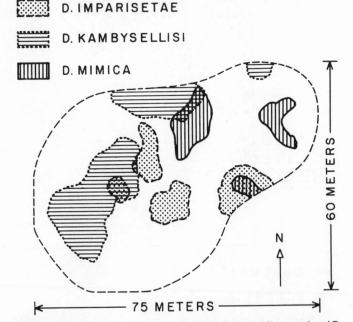

Figure 2.4 The distribution of three Drosophila species (*D. kambysellisi, D. mimica*, and *D. imparisetae*) over a 60m × 75m area on the island of Hawaii. (Adapted from Richardson and Johnston 1975.)

Figure 2.5 shows the diurnal activity periods for *D. simulans* and three other coexisting species (the "medio" group). Although these species occur within the same locality, they exhibit different patterns of temporal activity: members of the "medio" group are much more active in the morning than they are in the evening;

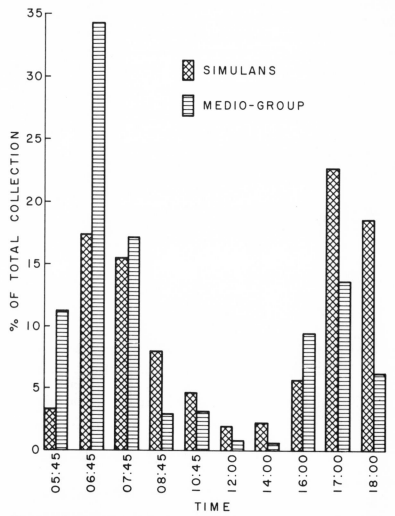

Figure 2.5 Day-long collection records for *D. simulans* and for three *Drosophila* species of the "medio-group" revealing the diurnal activities of these flies, and also the differential activities of the two sorts of flies in the morning and evening. The total flies captured were 1,436 *D. simulans* and 1,402 "medio-group" flies. (After Pavan et al. 1950.)

D. simulans, on the contrary, tends to be somewhat more active in the evening. Different species can subdivide the day in different ways.

Two common sources of food for *Drosophila* inhabiting the tropical rain forests of Brazil are palm fruits and fallen flowers decaying on the forest floor. Figure 2.6 shows the distribution of various Drosophila species (or species groups, since classification is not always easy under field conditions) between these two food sources. As Timofeeff-Ressovsky found (figure 2.3), tropical *Drosophilae* also exhibit preferences: *D. willistoni*, for example, is

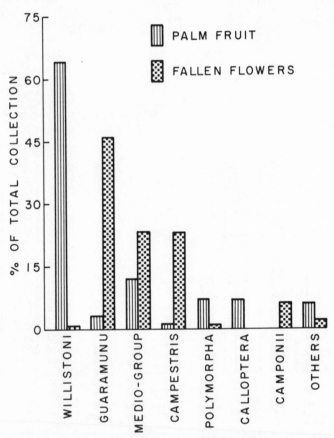

Figure 2.6 Relative proportions of different species or species-groups (where classification in the field is difficult) of *Drosophila* caught on the fruits of the Macauba palm, or on the fallen flowers of the Bombax tree. (After Dobzhansky and Pavan 1950.)

found almost exclusively on fallen fruits; *D. guaramunu*, members of the "medio"-group, and *D. campestris* are found for the most part on fallen flowers. Different species can subdivide food resources in different ways.

The picture that emerges from studies of the sort reported here is one in which a species—within its overall distribution range—exhibits a patchy distribution. Each exhibits, in fact, a hierarchy of patchiness which terminates in collections of individuals that constitute the smallest groups, the local populations. Furthermore, at the level of coexisting local populations, we find that the food sources (palm fruit vs. fallen flowers, fermenting fruit vs. manure and other organic waste) are differentially apportioned among members of different species. Time may also be apportioned, with members of one species moving about at one time (either within individual days or on a coarser scale) and those of another moving about at a different time.

So much for where the members of a species are found. How do they get there? How do they go from one spot to another? An obvious answer to this question might be, "Any way that you can imagine." A fly—one fly—can go anywhere, by almost any means. So can one seed. Either can be carried by plane from one continent to another. Typhoons, hurricanes, or other violent storms can blow individual seeds or small animals such as insects tremendous distances. Even dust devils can pick up small pebbles, pieces of bark and twigs, and organisms of comparable size or weight and carry them surprising distances across a desert area. Migratory birds, as Darwin showed, carry a multitude of small organisms in the mud on their feet. The accidental transport of individual gravid females may account—and most likely has accounted in the past—for the introduction of species into new geographical areas. It may also account for the occasional introduction of a new allele into an existing population where it was not found before. A study of rare, once-in-a-lifetime accidents of transport, however, is unrewarding. Such accidents are capricious in precisely the sense Lewontin defined the term, and in which we used it in chapter 1. We readily admit that accidents of dispersal occur. We readily admit that some of the distribution patterns we see today have had accidental beginnings. But we cannot foretell when or where capricious events will recur; such events, by definition, do not lend themselves to investigation.

Rare accidents cannot be studied, but common ones can be. Although one cannot predict in advance the eventual movements or final resting place of an individual fly (or individual human being), one can describe the distribution of movements and the ultimate resting places of many flies. By marking flies, releasing them, and then recapturing them and noting their position when recaptured, one can say that a fly released at point X will with a certain probability be found, after a lapse of some time, at Y or beyond. Similar statements can, of course, be made in respect to human beings where birthplaces represent points of release, places of marriage (or of death) represent points of recapture, and surnames and given names serve as permanent markings. The data one collects are empirical in the case of both flies or persons, but they can be made quite precise. In the case of flies, the experimental procedures used are now known to influence the results obtained; some care is needed, then, in comparing different experimental studies (see Endler 1979).

Exceptionally thorough and systematic analyses of the dispersion of *D. pseudoobscura* were made by Dobzhansky and Wright (1943, 1947). Their results are summarized in table 2.1 and figure 2.7, according to a method described by Wallace (1966a). Because the summarizing procedure is much simpler than the

Table 2.1 Total collection data for the first three release-recapture experiments of Dobzhansky and Wright (1943). These are "lifetime" dispersal data because, as the summation of many days' collection, they are equivalent to the expected spatial distribution about a given source of flies 1, 2, 3, 4, . . . days old.

Distance	Flies	Trap Days	Average	Log (Avg. × 10)
0	728	23	31.65	2.500
20	993	82	12.11	2.083
40	956	82	11.66	2.067
60	412	83	4.96	1.695
80	464	84	5.52	1.742
100	230	84	2.74	1.438
120	163	84	1.94	1.288
140	148	84	1.76	1.246
160	118	84	1.40	1.146
180	173	84	2.06	1.314
200	125	84	1.49	1.173
220	109	84	1.30	1.114
240	65	58	1.12	1.049
260	40	58	0.69	0.839
280	42	58	0.72	0.857
300	28	44	0.64	0.806
320	15	25	0.60	0.778
340	18	25	0.72	0.857
360	11	25	0.44	0.643
380	1	6	0.17	0.230

$$b = -0.1020 \quad s_b = 0.0052$$

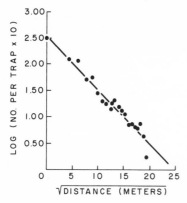

Figure 2.7 "Lifetime" dispersal of *D. pseudoobscura* flies. Circles represent the data listed in table 2.1; the straight line has been fitted to the transformed data.

original mathematical analyses (see Wright 1968), the original papers should be consulted by persons seriously interested in studying the dispersal of organisms.

Drosophila flies of both sexes maintain a rather vigorous reproductive life for at least two weeks; senile and physically debilitated individuals are rarely encountered except in experiments on longevity. Consequently, the recapture data of Dobzhansky and Wright (1943) have been combined over the entire duration of their experiments in order to obtain an estimate of "lifetime" dispersion—the combined distributions of flies aged 1, 2, 3, 4, and more days. In these experiments marked flies were released at the center of a cross (or row) of trapping sites spaced at 10-meter intervals. Beginning the first evening following the flies' release and continuing for a week or more, traps were set out at predetermined sites, the number of marked flies captured in each trap was recorded, and then (as the period of evening activity drew to a close as shown in figure 2.2d), the marked flies were released once more at their point of capture so that they might continue dispersing. Table 2.1 lists the cumulative data for three separate experiments carried out in Southern California. The results are shown graphically in figure 2.7: the logarithm of the number of recaptured flies decreases linearly with the square-root of distance over a considerable range (from 0 to nearly 400 meters).

A linear relationship is a valuable one for visualizing relationships graphically. However, neither the logarithm of numbers nor the square roots of distances correspond to our intuitive notions of numbers and distances; therefore, the data of table 2.1 are replotted on more familiar arithmetic scales in figure 2.8. From this figure a simple conclusion can be drawn: most of the released

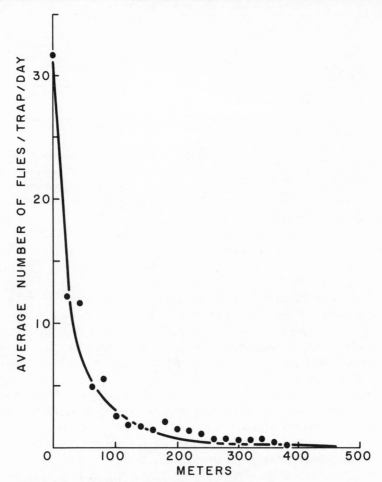

Figure 2.8 A representation of the data of table 2.1 on an arithmetic scale in order to emphasize the distortion of scales needed in order to obtain the straight-line relationship shown in figure 2.7. The line shown here has been calculated from the straight one shown in the previous figure.

flies were recaptured close to the point of release; a few traveled considerable distances. Once the latter flies were in motion, they were nearly as likely to go great distances (400 meters or more) as lesser ones, that is, the tail of the curve becomes nearly horizontal as distance increases.

Data similar to those obtained by Dobzhansky and Wright for *D. pseudoobscura* have been collected for scale insects as well (Willard 1974). Figure 2.9 shows the numbers of migrant scale

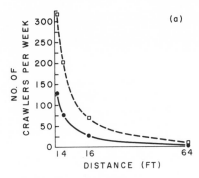

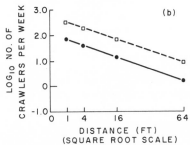

Figure 2.9 The horizontal distribution (numbers of individuals per trap per week) of California red-scale crawlers that are migrating from a source. a. Data plotted on arithmetic scales; b. vertical axis, log number, and horizontal axis, square root of distance. Total data, solid line; week with highest numbers, dashed line. (After Willard 1974.)

insects (crawlers) that have traveled various distances. In this case, distance is measured in feet. Figure 2.9a corresponds to figure 2.8 in showing the dispersion of crawlers on an arithmetic scale; figure 2.9b shows the same data transformed to a plot of log-number versus square-root-distance. The linearity of the data in the latter diagram is remarkably good.

Curves such as those shown in figures 2.8 and 2.9a can serve as a means for estimating the places of origin of individuals that are captured at a given point. We assume that young flies have not had sufficient time to disperse as far as have older ones. We also assume that patterns of arrival correspond to those of dispersal (otherwise all flies would accumulate in certain "hot" spots). And, finally, we assume that the lifetime of flies in nature approximates those of the released ones. With these assumptions, a curve such as the one in figure 2.8 can be rotated about the vertical axis, thus creating a solid of revolution (a favorite subject of study in elementary integral calculus). The proportion of the total volume falling within certain distances of the axis of rotation (the Y axis) can then be calculated. The result for three different Drosophila species is shown in figure 2.10. *D. pseudoobscura* is the most motile of the three species shown in the figure; *D. willistoni* and *D. funebris* move about much less.

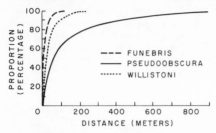

<figure>

FUNEBRIS
PSEUDOOBSCURA
WILLISTONI

PROPORTION (PERCENTAGE)

DISTANCE (METERS)

</figure>

Figure 2.10 Reconstruction of the composition of local populations of flies based on dispersal data. The curves represent the proportion of flies at a given locality that have been born within various distances of that locality. (After Wallace 1966a.)

Flies may reasonably disperse not randomly but in accordance with their needs or with the opportunities presented to them. Endler (1979), for example, has shown that the observed dispersion of flies parallels the distance between the traps set for them by the experimenter. If traps are placed considerable distances apart, flies move farther in a shorter time than if traps are more tightly spaced. When released in arid regions, *Drosophila* will travel amazing distances to find a more suitable habitat: marked *D. melanogaster* flies have been known to travel four or five miles in 24 hours (Yerington and Warner 1961), and *D. pseudoobscura* flies released in Death Valley in the evening have been found the next morning at oases nine miles distant (John A. Moore, personal communication).

Experiments on "microdispersion" have been carried out with both *D. melanogaster* (Wallace 1970a) and *D. aldrichii* (Richardson 1969). In the former study, unplugged culture bottles (sheltered from the sun and rain) served as the source of *sepia*-eyed flies. The capture data for these flies over a distance of seven meters are given in table 2.2. Note that although the trap seven meters from the source of *sepia* flies was well visited by native wildtype *D. melanogaster*, *sepia*-eyed flies neither arrived nor accumulated at that trap.

Table 2.2 The numbers of *sepia* and wildtype *D. melanogaster* captured over a 15-day period at various short distances from unplugged culture bottles that served as a continuous source of *sepia* flies. (After Wallace 1970a.)

Distance	Total Sepia	Total Wildtype
$\frac{1}{2}$ meter	13	128
1 "	14	106
$1\frac{1}{2}$ "	5	56
2 "	1	43
$2\frac{1}{2}$ "	0	37
3 "	1	31
4 "	0	37
5 "	0	54
6 "	2	48
7 "	0	98

Table 2.3 Labeled and nonlabeled flies (*D. aldrichii*) captured at various distances from a continuous point source. (The source consisted of a dysprosium-labeled rotting canteloupe. Dysprosium is a rare-earth element that can be detected in labeled flies by its radioactive isotopes that are induced by the exposure of flies to thermal neutrons.) (After Richardson 1969.)

Distance	Labeled	Total	% Labeled
4.5 meters	82	600	13.7
9.0 "	24	336	7.1
13.5 "	2	307	0.7
18.0 "	3	242	1.2
36.0 "	32[a]	6536	0.5

[a]All flies captured in the many traps at 36 meters had ingested the label at least 3 days before being captured.

The experiment involving *D. aldrichii* was carried out in what I consider to be the proper manner for studies of dispersal: one bait (a rotten canteloupe) was labeled with a rare-earth element and placed in a field of rotting canteloupes. Among the flies that were subsequently captured at various distances, those that had fed on the labeled canteloupe could be identified. The data presented in table 2.3 show that the proportion of labeled flies falls

Table 2.4 The distribution of marked and recaptured *D. nigrospiracula*, a cactus-feeding fly of the American Southwest. "Nearby" refers to experiments in which traps were placed close to rotting cacti on which flies had been marked by dusting with a fluorescent powder. "Distant" refers to later experiments in which flies on different cacti were marked with dusts of different colors; these marked flies were later identified as migrants upon recapture at cacti other than the ones where they had been labeled. (After Johnston and Heed 1975 and 1976.)

Distance (meters)	Nearby (% and no. recaptured)		Distant (% and no. recaptured)	
0	51.2	42	83.6	2714
1.5	18.3	15	—	—
3.0	6.1	5	—	—
6.1	8.5	7	—	—
12.2	6.1	5	—	—
18.3	2.4	2	—	—
24.4	1.2	1	0.7	22
30.5	1.2	1	—	—
36.6	1.2	1	—	—
42.7	2.4	2	—	—
48.8	1.2	1	—	—
55.0	—	—	2.9	96
95.0	—	—	0.9	31
145.0	—	—	3.2	106
255.0	—	—	1.3	44
395.0	—	—	1.1	36
475.0	—	—	1.1	37
655.0	—	—	5.2	172
900.0	—	0	0.9	28

off rapidly as the distance from the labeled bait increases; the proportion of labeled flies approaches zero 10 meters or more from the marked bait.

Summarized results of two important studies (Johnston and Heed 1975 and 1976) on the dispersal of *D. nigrospiracula*, a cactus-feeding species of the American Southwest, are shown in table 2.4 and figure 2.11. In an early experiment, traps were placed at various (nearby) distances from a rotting, fly-infested cactus. The number of marked flies recaptured decreased rapidly with distance from the cactus where they had been marked for identification. Thus, it seemed that these flies were incapable of traveling from one rotting cactus plant to another because such plants are widely scattered in the desert. The later experiments resolved this apparent difficulty. By marking the flies on one cactus, Johnston and Heed were able to show that these flies did

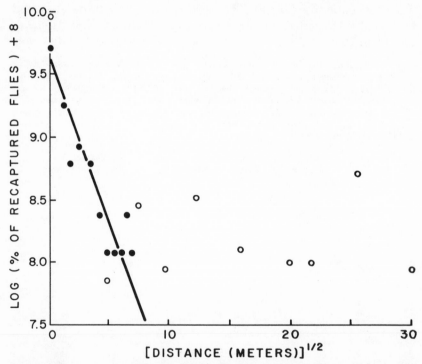

Figure 2.11 The dispersal of labeled *D. nigrospiracula* near (●) and between (○) cactus plants, whose rotting tissue is the fly's only source of food. The solid line is based on all data collected within 50 meters of the source of labeled flies. (After Johnston and Heed 1975, 1976.)

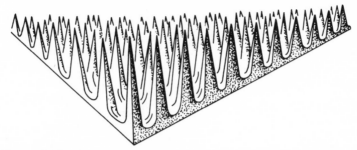

Figure 2.12 A "bed-of-nails" view of local population structure. This representation emphasizes both the very limited dispersal of most flies under normal circumstances, and the unpredictably great distances traveled by those individuals that do disperse (compare with figure 2.11). The genetic consequences of this type of population structure is complicated, as we shall see later, by the mating advantage commonly possessed by rare, immigrant male flies.

indeed travel to other rotting cacti—even to one 900 meters from the center of dispersion.

The results of studies on dispersal can be represented (as in figure 2.12) by means of a figure that resembles a bed of nails. Many individuals remain close to their birthplaces, thus creating the spikes in the figure; the patchiness of the distribution of local populations of individuals accentuates the spikes while causing them to be distributed much less uniformly than either the figure or the bed-of-nails analogy suggests. Beneath the spikes, however, is the board, the combined brims of all the solids of revolution created by rotating curves such as that shown in figure 2.8. This board represents the distribution patterns of the far-ranging migrant individuals—individuals who, though low in frequency, travel surprisingly great distances to find water or fresh sources of food. These individuals, of course, are the ones that provide the genetic connections between neighboring local populations, and between the more widely scattered populations of the species, as well.

Earlier, the birthplace and place of marriage for human beings were compared to the points of release and recapture in the Drosophila experiments. That this was not an idle comparison is shown in figure 2.13. In this figure are plotted the probabilities of marriage as a function of the distance between the birthplaces of the marriage partners. Curve (a) shows the results for a study carried out in Italy; curve (b) for a corresponding study made in

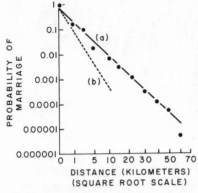

Figure 2.13 Matrimonial migration in (a) Italian and (b) Japanese populations. These curves are analogous to those in figures 2.7 and 2.9; the logarithm of the probability of marriage decreases linearly with the square root of the distance between the birthplaces of the marriage partners. (Points and curve (a) after Cavalli-Sforza 1959; theoretical curve (b) after Yasuda 1967.)

Japan. In both cases, the probability of marriage falls off linearly with the square root of the distance between the bride and groom's birthplaces.

In our consideration of the distribution of a species, we have discussed where individuals are found (distribution in the strict sense), and how they move out or disperse from a source. The chapter may conclude, then, with a brief glance at an intermediate situation: the relatively unimpeded spread of an organism as it invades new but favorable territory. Dobzhansky (1973a) raised the following question: Is the geographical distribution of an organism such as *D. pseudoobscura* the consequence of active dispersal or passive transport? Are the active dispersal movements of the sort he observed in his studies with Wright (see table 2.1), for example, sufficient to explain a geographical distribution that covers millions of square miles? Or, must one assume that the wind or other accidental means of transport was responsible for establishing new foci as gravid females were dropped here and there in the otherwise empty (of *D. pseudoobscura*) but hospitable landscape?

Dobzhansky concluded that active dispersal is too slow, and therefore that passive transport is the more likely cause for geographic distribution patterns. I think that in part his argument is wrong. Dobzhansky's interpretation of active dispersal was based on data obtained by releasing and recapturing marked flies. In these experiments, the variance of the distributions of released flies increased linearly with time: each day, in one set of experiments, added 8,000 m.^{2} to the mean-squared distance the flies had traveled. The released flies, in a sense, occupied territories on successive days whose areas were proportional to the number of days following their release. This pattern of spread, however, does

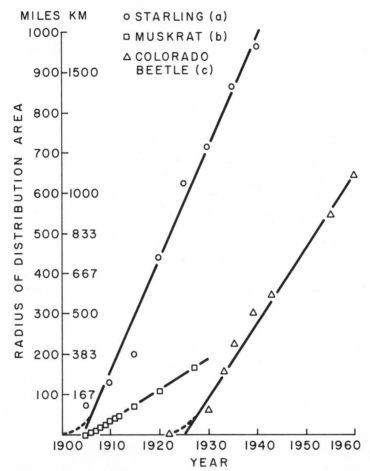

Figure 2.14 (a) The spread of the European starling (*Sturnus vulgaris*) in the United States from its introduction (1890) until 1950. (After Wing 1943.) (b) The spread of the muskrat (*Ondatra zibethica*) in Central Europe from its introduction (1905) until 1927. (c) The spread of the Colorado beetle (*Leptinotarsa decemlineata*) in Europe from its discovery in 1922 until 1960. (Curves b and c are after Nowak 1971.) The curves show the increase in radius of the distribution, calculated as if the total area occupied at successive intervals were circular.

not imply that *D. pseudoobscura*, having occupied an area 1,000 kilometers in radius, disperse in such a manner that only 8,000 m.^{2} are added to the total distribution area in a single day; a pattern such as that would restrict the further spread of these flies to but two or three millimeters per day (see Richardson 1970, for a fuller discussion of dispersal patterns).

The front of the distribution area of a species invading a new region advances nearly linearly with time. Three instances are illustrated in figure 2.14: the spread of the European starling (*Sturnus vulgaris*) in the United States, and that of the muskrat (*Ondatra zibethica*) and Colorado beetle (*Leptinotarsa decemlineata*) in Central Europe. In each case, the total distribution area has been treated as if it were circular; the graphs show the linear increase of the radius of distribution with time. These curves suggest that *D. pseudoobscura*, upon encountering a previously vacant but ecologically suitable geographic area, might have advanced nearly a kilometer each generation or several kilometers annually by active dispersal alone.

Drawing upon our earlier discussions of K, the carrying capacity of the environment; the excess number of zygotes, r, which are destined to be eliminated in a stationary population; and the dispersal pattern of individuals, an attempt has been made in figure 2.15 to diagram the events which occur at, immediately

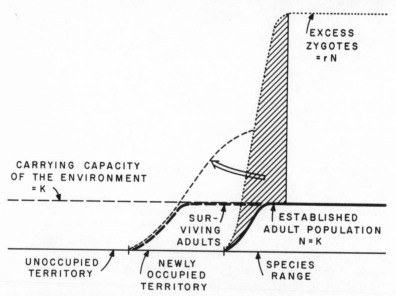

Figure 2.15 Hypothetical scheme illustrating why (unlike the dispersal of individuals in a release-recapture experiment) the border of a species that is spreading into unoccupied (but hospitable) virgin territory should advance at a relatively constant rate.

behind, and beyond the advancing front of an invading species. At some distance behind the front, the species has attained the number which corresponds to the carrying capacity of the environment. At the front itself, excess progeny exist which, if they were to hatch within the interior of the distribution area, would be doomed to extinction; however, because they are at the advancing front itself, those that disperse forward encounter favorable sites for reproducing. The pattern of dispersal should resemble figures 2.7 and 2.8; each point on the species's boundary should generate such a distribution in front of it. Progeny of the dispersing individuals can survive within the limits imposed by K. The distribution range, then, advances linearly within a single generation according to the linear dispersal pattern of the dispersing (migrant) individuals.

BOXED ESSAY

The description of population structure that has been presented informally throughout this chapter has been formalized in contrasting models by Levins (1970) and Boorman and Levitt (1972, 1973); "metapopulation" is the term used by these persons in referring to the large aggregation of small, local populations.

In addition to the local populations that actually exist at a given moment, Levins visualizes other localities that might have been occupied but for some reason—including accidents—are not. At a second moment, some of these "empty" localities might be filled but (if the area is at equilibrium) an equal number of originally occupied localities would now be empty. Thus, the colonization of empty localities and the extinction of local populations can be compared to the twinkling of small bulbs on a large decorative panel of lights.

Boorman and Levitt visualize the colonization of unoccupied localities and the extinction of local populations as events that are restricted primarily to the region surrounding a more-or-less permanently occupied species area (this notion will reappear in figures 23.13 and 23.14).

Wilson (1975: 108) shows that the "permanently" occupied area of Boorman and Levitt's model may appear to be so only because the observer does not possess the visual acuity required to see it otherwise; on a fine scale this "permanently" occupied area would also exhibit the characteristics of Levin's metapopulation.

FOR YOUR EXTRA ATTENTION

1. Dispersion experiments have been notoriously difficult to design: because an organism often moves so as to improve its lot, the baits set by the investigator may themselves influence the results which are obtained.

Butterflies are large enough to be seen, collected, marked, and released without the use of artificial baits. A number of excellent studies have been made by such capture-release-recapture methods (see Ehrlich 1965, and Ford 1971: 6-8). Snails can also be marked for recapture (see LaMotte 1951: 56, Ford 1971: 185, and Lomnički 1969). The movement of a number of animal species has been discussed by Emlen (1973). A popular account of the fate of alien animals following their transport from one continent to another has been prepared by Laycock (1970).

2. Endler and Johnston (1980) have shown that the dispersal rate for *Drosophila* is dependent upon the number of food sites visited rather than on the actual distance traveled. In what sense can they claim that the neighborhood size is greatly inflated by using baits to estimate dispersal and numbers of these flies?

3. The numbers of flies per trap illustrated in figure 2.2c clearly differ; this can be shown by a simple Chi-square test. Suppose for example that the numbers of flies per trap are as follows:

Trap	1	2	3	4	Total
Flies	10	100	20	30	160
Expected	40	40	40	40	160
difference $(O-E)$	-30	60	-20	-10	0
d^2	900	3,600	400	100	

Clearly the sum of d^2/E = chi-square will be enormous, with only three degrees of freedom.

The Chi-square test reveals whether differences in the distribution of flies among traps is greater than could be reasonably expected by chance events alone. Other events that affect the distribution of flies may themselves affect first this trap and then that one in a random pattern. In that case, a series of daily collections as follows might provide a more appropriate analysis of the collection data:

Trap	1	2	3	4
Flies: Day 1	10	100	20	30
Day 2	40	10	50	0
Day 3	60	20	50	70
Total	110	130	120	100

Under these circumstances, repeated sampling may reveal the uniformity of flies per trap even though the distribution in any one day is clearly heterogeneous. Certainly, the data on which figure 2.2c is based (and which include collections made during two evenings) would not be repeated if one were to expose traps in precisely the same sites today, 35 years later.

Detecting and Measuring Genetic Variation: Classical Techniques

PREVIEW: This chapter deals with the third and fourth elements that are essential for a science of population genetics: naturally existing variation, and the tools or experimental procedures required for revealing and measuring this variation. Genetics (in the classical, although not necessary in the molecular, sense) is a science whose existence depends upon heritable variation, upon dissimilar segregating alleles. Observational data concerning genetic variation together with a concept of the local population and theoretical models that assist in organizing and understanding these data are the stuff of population genetics. The procedures discussed in this chapter are those that have grown up with the science of genetics, especially since the late 1920s when H. J. Muller developed his now classical *C1B* technique.

THE EMPIRICAL STUDY of population genetics, as Lewontin (1974a) has said, has always begun with and centered around the characterization of the genetic variation in populations. The population geneticist is concerned with gene frequencies, and the comparison of these with some preconceived hypothesis. The hypothesis may state that zygotic ratios are related to gene frequencies as the Hardy-Weinberg equilibrium predicts; the comparison of observed and expected frequencies would then be made by means of a Chi-square or other appropriate statistical test. Depending upon the results of this comparison, a second hypothesis may say that two populations have the same gene frequencies, the same zygotic frequencies, or both; appropriate procedures exist for testing hypotheses of these sorts. Or, again, the hypothesis may say that gene frequencies have remained constant from one generation to another as the Hardy-Weinberg equilibrium also predicts; this hypothesis, too, is testable once gene frequencies have been determined.

In this chapter, we shall describe some of the classical procedures used in the study of populations; "classical," in this sense, means techniques other than those recently developed by molecular biologists which probe the nature of DNA and of protein molecules themselves. A description of procedures without a corresponding description of experimental data would be highly artificial; therefore, both experimental data and some procedures used in their analysis will be described. As we shall see, populations—when examined by either classical or molecular procedures—reveal a storehouse of concealed variability.

Direct Observation

Many mutations that are useful in population studies are recessive. Unfortunately, many of these are also rare in the sense that few individuals carry them; consequently, the frequency of homozygous individuals which exhibit the mutant phenotype is exceedingly low. Occasionally, however, contrasting heritable phenotypes do occur within populations of a species, and these can be used in searching for either spatial or temporal changes in the constitution of local populations. A simple example is given in table 3.1. Here are listed the frequencies of dorsal-striped red-backed salamanders observed in monthly samples on the Del-Mar-Va peninsula over a 21-year period. These data have been chosen because they illustrate a commonly neglected aspect of population genetics: no evidence exists here that the frequency of dorsal-striped individuals in this region changed from an overall annual mean of 77.8 percent during any month or season of the year. Published

Table 3.1 The frequency of dorsal-striped red-backed salamanders (*Plethodon cinereus*) in monthly samples taken over a 21-year period on the Del-Mar-Va Peninsula (U.S.). (After Highton 1977.)

Month	Number Striped	Total	% Striped
January	108	135	80.0
February	105	136	77.2
March	107	143	74.8
April	260	345	75.4
May	442	582	75.9
June	126	165	76.4
July	187	240	77.9
August	118	147	80.3
September	492	616	79.9
October	556	704	79.0
November	100	127	78.7
December	102	133	76.7
Total	2,703	3,473	77.8

NOTE: Chi-square (11 degrees of freedom) = 6.46

$$.75 < p < .90$$

data on the genetics of populations are unquestionably biased towards those studies in which significant changes were observed; data from an unchanging population (to many persons) correspond to negative data.

Data shown in table 3.2 illustrate genetic differences that distinguish populations (in this case, of the ladybird beetle) inhabiting different geographic locations. In the 1920s it would have been fashionable to *name* the local races according to the preponderance or lack of preponderance of black individuals. Today, one seeks instead to explain in terms of natural selection why one allele might displace another in a given locality. The information presented in table 3.2 shows that percentages of black forms

Table 3.2 The proportion of the black form of the two-spot ladybird beetle (*Adalia bipunctata*) near various cities in Great Britain, together with the micrograms of soot per cubic meter of air in those same cities. (After Creed 1971.)

Locality	Number of Black	Total	% Black	Soot
Birmingham	49	153	32.0	43
Cambridge	6	137	4.4	28
Edinburgh	306	341	89.7	50
Glasgow	106	123	86.2	78
Leeds	109	128	85.2	52
London	0	80	0	27
Newcastle	105	141	74.5	96
Oxford	6	216	2.8	19
Reading	6	227	2.6	19
Swansea	0	45	0	19
York	15	46	32.6	45

exceed 30 percent only when the amount of soot per cubic meter of air exceeds $40\mu g/m^3$. Soot, obviously, is not a direct cause of blackness of ladybird beetles; the pigmentation in these insects is known to have a genetic basis. The available data do not permit one to decide, however, between the cryptic nature of black beetles on a dark (sooty) background, the absorbtion by black beetles of sunlight that is otherwise partially obscured by polluted air, or still some other selective force. The data listed in table 3.2 are clear, however, on one point: geographically separate populations of one species may differ genetically.

Color morphs (such as dorsal-striped salamanders, black ladybird beetles, red and silver foxes, black and agouti hamsters, and a host of others) are easily identified and scored. Still other levels of genetic variation can be scored, however, without resorting to other than simple culturing techniques. Table 3.3 lists the resistance of the progeny of single-caught wild females of *Drosophila melanogaster* to heavy exposures of X-radiation (Parsons et al. 1969). Obviously, nearly all of the progeny of certain females are killed by the radiation exposure, virtually none of the progeny of other females are killed, however. By performing reciprocal matings between individuals of the resistant and sensitive strains (table 3.4), Parsons and his colleagues were able to show that resistance to X-radiation has a genetic basis. In this case, the genetic variation could not be seen directly, but it could be revealed by subjecting the progeny of wild-caught females to a stressful condition. X rays are but one such stress; Parsons and his colleagues have been especially successful in identifying many sorts of genetic variation by these and similar procedures (Parsons 1980).

Table 3.3 Resistance of female progeny of individual wild-caught female *D. melanogaster* to extremely heavy exposures (110,000 r) of gamma-radiation. Numbers are the percent of exposed females dying within two days. (After Parsons et al. 1969.)

Strain	% Dead
1	54
2	100
3	6
20	46
21	36
22	22
23	60
24	60
26	8
29	0
32	94

Table 3.4   Evidence that resistance to radiation in *D. melanogaster* has a genetic basis. Numbers in the body of the table are percent females dying within two days following an exposure to 110,000 r gamma-radiation. (After Parson et al. 1969.)

		Male Parent Strains[a]			
		3	29	23	2
Female Parent Strains[a]	3	12	16	44	32
	29	28	28	40	88
	23	36	68	92	96
	2	16	84	92	88
Low mortality X low mortality:				Average	21
High mortality X high mortality:				Average	92
Average:					56
Low mortality X high mortality:				Average	51
High mortality X low mortality:				Average	51
Average:					51

[a]See table 3.3 for the resistance of these strains to radiation exposure.

The direct observation of heritable variation need not, of course, be limited to what can be seen by the unaided eye. The cytological examination of the giant chromosomes of many Dipteran species has revealed a wealth of chromosomal rearrangements that are not revealed during the examination of standard mitotic metaphase figures.

The simplest of all chromosomal rearrangements (other than a mere loss or gain of a chromosomal segment, a gain or loss that disrupts the normal content of a genome) is the rotation through 180° of a chromosomal segment with the result that the included gene loci now occur in a reversed sequence. The point-by-point pairing of the "normal" and its "inverted" gene sequence results in chromosomal loops such as those illustrated in figure 3.1 (Wallace 1966d). In this figure are shown the pairing configurations of the heterozygous combinations of three different gene arrangements of the third chromosome in *D. pseudoobscura*. Such inverted sequences are inherited, of course, as units comparable to alleles at a single locus. In thoroughly studied species, the different gene arrangements are identified by name; in species whose giant chromosomes are less well-known, numbers of inversion loops per individual can be scored as a quantitative measure of genetic variation in a population. For example, in different geographic localities, populations of *D. willistoni* exhibit different average numbers of inversion loops per larva, varying from 1 or 2 (or fewer) to 9 depending, apparently, upon the diversity of the physical environment (Cunha et al. 1959). A study of inversions in *D. melanogaster* (Stalker 1976) suggests that more individuals than expected are heterozygous for intermediate numbers of inversion loops (table 3.5).

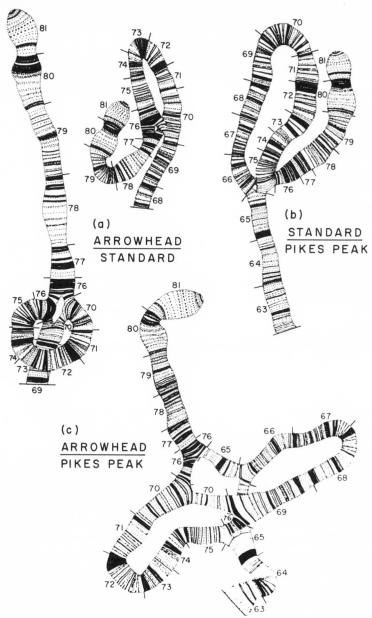

Figure 3.1 Several pairing configurations of giant chromosomes of the salivary gland cells of *D. pseudoobscura* larvae. Loops such as these occur during the pairing of chromosomes whose gene arrangements differ. The numbers serve to identify small segments of the chromosomes; by convention the entire genome is subdivided into 100 such segments. (After Dobzhansky and Epling 1944.) Reprinted from B. Wallace, *Chromosomes, Giant Molecules, and Evolution*; © 1966 W. W. Norton.

Table 3.5 The observed (and expected) numbers of inversions in paired gamete samples of *D. melanogaster* from Texas. The expected numbers listed here are obtained as sums of individually calculated expectations, and cannot be calculated from the observed values listed in the table. (After Stalker 1976.)

Number of Inversions	Observed	Expected	O – E
0	35	50.6	–
1	63	53.9	+
2	132	125.3	+
3	73	63.3	+
4	61	61.8	–
5	9	15.4	–
6	9	10.3	–
7	0	0.9	–
8	0	0.6	–

Pericentric inversions, those whose two break points fall on either side of the centromere, may detectably alter the shape of even small metaphase chromosomes. Figure 3.2 shows the interrelations between the size of the grasshopper, *Moraba scurra*,

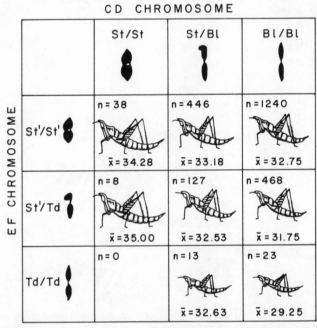

Figure 3.2 Variation in the chromosomal constitution of the grasshopper *Moraba scurra* and concomitant variation in the sizes of individuals of the different genotypes. n = number of individuals studied; $\bar{x}$ = average wet weight (mgs.) per individual. (After White and Andrew 1962.)

and its genetic constitution in respect to normal and (pericentric) inverted sequences for two chromosome pairs. The sizes that are associated with these karyotypic combinations are in many instances significantly different.

Inbreeding

The direct observation of genetic variation is feasible when contrasting morphological types are reasonably frequent or, as in the case of cytological analyses, when both homozygotes and the heterozygotes can be recognized visually. Rare recessive alleles cannot be easily studied by direct observation because the affected homozygotes are exceedingly rare: if the frequency of a mutant allele equals q, the frequency of homozygous individuals carrying two such alleles in a randomly mating population equals $q \times q$, or q^2 (for reasons to be explained in chapter 5).

Procedures exist, however, by which the investigator can avoid the restriction imposed by *random* mating; he can force organisms to inbreed. In this manner, the frequency of *families* exhibiting recessive mutant characteristics may approach q, rather than q^2. One such procedure (Spencer 1947) is shown in figure 3.3. Gravid *Drosophila* females (wildtype in appearance) are captured in the field and placed in laboratory cultures where they

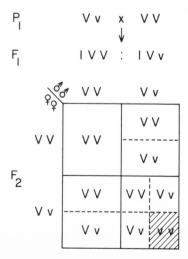

Figure 3.3 Breeding procedure for the extraction of visible mutations (*v*) from wild populations of *Drosophila*. (After Spencer 1947.)

Table 3.6 The distribution of mutations recovered among the F_2 progeny of 736 pair matings of wild-caught *D. mulleri*. (After Spencer 1947.)

	Number of Mutations Observed					Total Cultures
	0	1	2	3	4	
Number of cultures	513	189	28	6	0	736

Average number of mutations per culture: 0.36
Variance: 0.36

produce F_1 progeny. If either the original female or her mate were heterozygous for an autosomal recessive, visible (*v*) mutation, one-half of the F_1 males and females are heterozygous for this same allele. As the checkerboard in figure 3.3 shows, one sixteenth of the F_2 generation obtained by mass culturing F_1 individuals (*vv*) should then exhibit the mutant phenotype (provided that those homozygous mutants are viable). Spencer used this procedure in raising 736 F_2 cultures of *D. mulleri* (Table 3.6), which were examined for the presence of one or the other of numerous mutant phenotypes; the results suggest (in contrast, for example, to the inversions of table 3.5) that these mutations occurred in parental females in a random fashion: the mean number of mutations per female (0.36) equals the variance in the distribution of mutations among females (also 0.36), a characteristic of the Poisson distribution. Not all studies have yielded such random-appearing results as those obtained by Spencer; in table 3.7, for example, are shown the results of a later study (Pentzos-Daponte et al. 1967) on *D. subobscura* in which the variance is significantly smaller than the mean; these results do resemble those of Stalker shown in table 3.5.

Table 3.7 The numbers of mutant genes carried in heterozygous condition by female *D. subobscura* near Thasos, Greece. (After Pentzos-Daponte et al. 1967.)

Number of Mutant Genes/Female	Observed	Expected	O – E
0	0	5.7	–
1	5	19.9	–
2	30	34.9	–
3	59	40.7	+
4	62	35.7	+
5	25	24.9	+
6	4	14.6	–
7	2	7.2	–
8	1	3.1	–
9	1	1.2	–
10+	0	0.4	–

NOTE: Total flies: 189; Total mutations: 670; Mutations/fly: 3.54; Variance: 1.54[a].
[a]Variance is significantly smaller than expected.

Marriages between relatives provide the means for studying genetic variation in human populations. As shown in figure 3.4 (in respect to cousin marriages), the probability that a particular one of the four grandparental alleles at a given locus becoming homozygous equals $\frac{1}{64}$; consequently, the probability of homozygosis for any of the four alleles equals $\frac{1}{16}$. Similar probabilities can be calculated for marriages between relatives of other sorts (see Bodmer and Cavalli-Sforza 1976, ch. 11).

The survival of children whose parents have common ancestors have been used by several workers to measure what is in effect (although not necessarily in actuality) the frequency of lethal genes in populations. A more precise term for what is measured is *lethal-equivalent* because the sole observation is the mortality of inbred children. The number of mutant alleles at various loci that are needed to cause the death of any one infant is, of course, unknown.

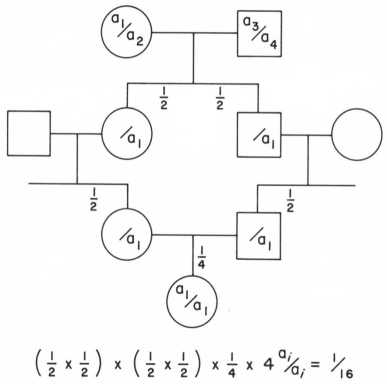

$$\left(\frac{1}{2} \times \frac{1}{2}\right) \times \left(\frac{1}{2} \times \frac{1}{2}\right) \times \frac{1}{4} \times 4 \, {}^{a_i}\!/_{a_i} = {}^{1}\!/_{16}$$

Figure 3.4 The descent of one of four alleles from the great-grandparents to a homozygous child of a cousin marriage. (After Wallace 1970b.)

Table 3.8 The use of cousin marriages to estimate the frequency of viability modifiers in a South Brazilian population. (After Marçallo et al. 1964.)

Total mortality:	
Cousin marriages:	.2877
Unrelated marriages:	.1774
Survival (= S, or e^{-m})	
Cousin marriages:	.7123
Unrelated marriages:	.8226
Estimate of m (= $-\ln S$)	
Cousin marriages:	.3393
Unrelated marriages:	.1953
Difference	.1440
Difference X 16	2.3040[a]

[a]Difference represents the lethal equivalents revealed by homozygosis for $\frac{1}{16}$th of all gene loci; total lethal equivalents per haploid genome is estimated by multiplying by 16. The lethal equivalents per *diploid* chromosome set equals 2.304 X 2 = 4.608.

Data suitable for estimating the frequency of lethal-equivalents in human populations are given in table 3.8; these data concern the infant mortality of offspring born to unrelated parents and to parents who were cousins. The total mortality of the inbred children is higher than that of children of unrelated parents; conversely, the survival of the inbred children is lower. Recalling the earlier account of the Poisson distribution, one can imagine that the proportion of survivors corresponds to the proportion of individuals who received no lethal genes from a randomly segregating pool of such genes. If this were true, the proportion of survivors, S, equals e^{-m} where m equals the mean frequency of such lethals (in the portion of the genome, $\frac{1}{16}$, that is rendered homozygous by cousin marriages). The procedure which is followed in estimating frequency of lethal equivalents in $\frac{1}{16}$th of the genome (0.144) can be adjusted to include the entire genome (2.304). Because human beings are diploid and thus possess two genomes, the frequency of lethal equivalents per diploid combination can also be calculated (4.608). That is, the ordinary person of the studied population carried a store of mutant genes such that, if they were packaged in units just sufficient to cause death (1 lethal = 2 semilethals = 10 tenth-lethals = 1 lethal-equivalent), each diploid combination would be found to contain about 4.6 lethal equivalents.

A second study illustrating the calculation of lethal equivalents is presented in table 3.9 and figure 3.5. In this case, data are available not only for children of first cousin but also for second cousin and $1\frac{1}{2}$ cousin marriages as well. With the additional information, a regression (that is, a line best fitting the four data

Table 3.9 Data on stillbirths and neonatal deaths among babies born in two provinces of France. (After Sutter and Tabah, from Morton et al. 1956.)

Parents	Unrelated	2d Cousins	1½ Cousins	1st Cousins
Homozygosis	0	.0156	.0313	.0625
Stillborn and				
neonatal deaths	108	34	9	69
Total births	2,745	549	183	743
Proportion				
surviving	.9607	.9381	.9508	.9071

points) can be calculated for the natural logarithm of the surviving fraction. That is, $S = e^{-A-BF}$ or $-\ln S = A + BF$ where S = the proportion of survivors, A = deaths not depending on homozygosis (including those caused by accidents and other external influences), and F = the proportion of homozygous loci.

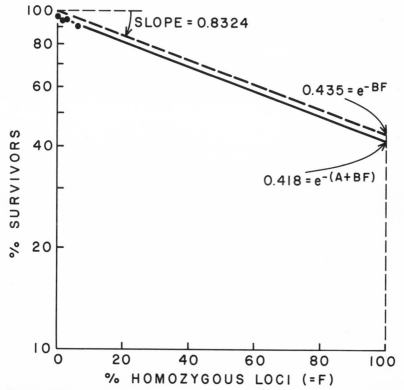

Figure 3.5 Decrease in the proportion of human zygotes successfully avoiding being stillborn or dying at birth among French children born of parents who were unrelated, or were 1st, $1\frac{1}{2}$, or 2nd cousins. Further explanation in text. (After Sutter and Tabah, from Morton et al. 1956.)

Calculations based on the data in table 3.9 yield the following: $A = -0.0402$ and $B = -0.8324$. Consequently, when $F = 0$ (the Y-intercept), the proportion of survivors equals 0.96; about 4 percent of these newborn French babies die from causes other than deleterious recessive mutations. The slope of the regression (-0.8324) represents, with its sign reversed, the number of lethal equivalents per haploid genome. If homozygosis were complete $(F = 1.00)$, the proportion of newborn babies escaping genetic death would be 0.435 $(= e^{-0.8324})$; the proportion escaping both genetic and accidental deaths would be 0.418 $(= e^{-0.8726})$.

If the number of lethal equivalents equals 0.83 per genome in this French population, the number per diploid chromosomal set equals 1.66. The lower proportion of lethal equivalents in the French than in the Brazilian population most likely reflects the level of medical treatment available to newborn babies in France. If all medical practices were perfect, presumably no babies would be lost, and there would be no lethal equivalents.

Modified *ClB* Procedures

In the case of many *Drosophila* species, standard procedures exist for rendering individuals homozygous for virtually all genes carried by an entire chromosome. A variety of terms are used by different workers to describe these procedures; "modified *ClB*" is a term that gives credit to the original technique of this sort, the one devised by H. J. Muller for use in his early studies on mutation in *D. melanogaster*.

Figure 3.6 represents *ClB*-type crosses used in the analysis of chromosome 2 in *D. melanogaster*. The source of the tester females (*CyL/Pm* P$_1$ ♀♀) is a true breeding stock of *CyL/Pm* (read "*Curly-Lobe/Plum*") flies. Because the two dominant genes, *Curly* and *Plum*, are recessive lethals, the only viable offspring of *CyL/Pm* X *CyL/Pm* matings are themselves *CyL/Pm*.

Wildtype males (P$_1$) whose chromosomes are to be analyzed are mated individually with *CyL/Pm* virgin females. *One CyL/+* son is chosen from each of the original crosses and is mated once more with *CyL/Pm* virgin females (F$_1$). Only one son for each cross is used in order to assure that only one or the other of the two parental wildtype chromosomes will be tested, *not both*. Be-

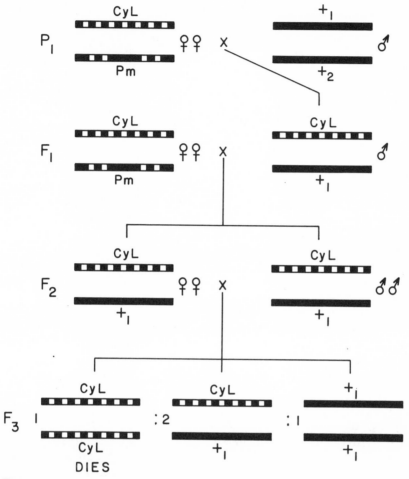

Figure 3.6 A series of crosses used to obtain flies that are homozygous for wildtype chromosomes in *D. melanogaster*. Wildtype chromosomes are identified as $+_1$ and $+_2$; chromosomes carrying crossover suppressors are genetically marked with the dominant mutations *Curly (Cy)*, *Lobe (L)*, and *Plum (Pm)*. Similar crosses are used to manipulate autosomes in many *Drosophila* species; only the nature of the dominant genetic markers differ.

cause the F_1 male is *Curly-Lobe*, he must carry one genetically marked chromosome; the other chromosome then, must be either $+_1$ or $+_2$.

The phenotypically *Curly-Lobe* flies of the F_2 generation must (except for new mutations) carry identical wildtype chromosomes because only one was brought in by the F_1 *CyL*/+ male.

Consequently, when virgin $CyL/+$ F_2 females are mated with their $CyL/+$ brothers they produce the F_3 progeny shown in figure 3.6: $\frac{1}{4}$ of the progeny (CyL/CyL) die because they are homozygous for the lethal gene, Cy; $\frac{1}{2}$ are phenotypically $Curly\text{-}Lobe$ and are heterozygous for the wildtype chromosome being tested; and $\frac{1}{4}$ are homozygous for the tested, wildtype chromosome. The expected ratio of viable genotypes, consequently, is $\frac{1}{2} : \frac{1}{4}$, or $2:1$, or $66\frac{2}{3}$ percent : $33\frac{1}{3}$ percent.

Because the flies that are homozygous for wildtype chromosomes are homozygous at every locus on that chromosome, a lethal allele at any locus eliminates this class of F_3 progeny flies. In this case, the test culture contains only $CyL/+$ flies. Alternatively, if the tested chromosome carries a recessive allele (at any locus) that lowers the probability that its homozygous carrier will survive to adulthood, the proportion of $CyL/+$ flies will exceed $66\frac{2}{3}$ percent but will fall short of 100 percent (the remainder are flies homozygous for the tested chromosome). The presence of sterility alleles on the tested chromosome can be detected by attempting to raise offspring from the homozygous flies. Finally, the non-$Curly\text{-}Lobe$ flies may not appear to be "wildtype" at all; in this case, the tested chromosome carries a recessive visible mutation that may affect any conceivable aspect of the flies' morphology.

Needless to say, the dominant mutant marker genes used in the study of different Drosophila species differ. The scheme shown in figure 3.6 could be generalized by substituting the symbol D_1 for CyL and D_2 for Pm. Either D_1 or D_2 must be associated with chromosomal inversions that prevent crossing over between the marked chromosome and the one being tested in heterozygous females (crossing over does not occur in males of most Drosophila species). For convenience in raising D_1/D_2 virgin females, it is also best that both homozygotes $(D_1/D_1$ and $D_2/D_2)$ be inviable.

The CIB-type procedure is amendable to diverse modifications that are suitable for answering a variety of questions: What sorts of recombinant chromosomes are produced by a female $(+_1/+_2)$ heterozygous for two different wildtype chromosomes? Mate such a female to CyL/Pm males, collect many $CyL/+$ sons, mate each son individually with CyL/Pm virgin females, and proceed into the F_2 and F_3 generations of figure 3.6. What is the zygotic constitution of the P_1 male designated $+_1/+_2$? Collect not one but 7–10 $CyL/+$ F_1 sons and mate each one individually with CyL/Pm virgin females. Because each son has a 50:50 chance of obtaining either of the two wildtype chromosomes, a test involving 7 F_1 sons has only one chance in 64 of *not* involving both chromosomes as well. How can one estimate the viability of $+^1/+^2$ flies, those that are heterozygous for two chromosomes of different origin rather than homozygous for either? Heterozygous combinations of wildtype chromosomes can be obtained by mating $CyL/+_1$ F_2 ♀♀ from one test culture to $CyL/+_2$ F_2 ♂♂ from another. This cross yields 2 $CyL/+$: $1+_1/+_2$ where the $CyL/+$ flies are of two sorts: $CyL/+_1$ and $CyL/+_2$. Finally, how might one measure the effect of the tested chromosome on its heterozygous carriers? The answer to this question requires that the test cultures contain a class of flies *not* carrying the tested wildtype chromosome. Such cultures can be obtained by mating $CyL/+_1$ virgin females of one F_2 culture with $Pm/+_2$ males from another. The resulting progeny—CyL/Pm : $CyL/+_2$: $Pm/+_1$: $+_1/+_2$—offer a basis for evaluating the effect of $+_1$ on $Pm/+_1$ carriers and $+_2$ on $CyL/+_2$ carriers. Round-robin matings insure that $+_2$ will also occur in $Pm/+$ flies and $+_1$ in $CyL/+$ ones.

Data characteristic of the sort obtained by the use of the modified *CIB* technique are shown in tables 3.10 and 3.11. Table 3.10 lists the numbers (and proportions) of 1063 tested wildtype chromosomes that yielded F_3 cultures in which the homozygous wildtype flies were found in various frequencies ranging from 0 percent (lethals) to 40 percent or more. Lethals (0 percent–4 percent wildtype flies) accounted for 15.6 percent of all tested

Table 3.10 Numbers and proportions of chromosomes sampled from a natural population of *D. pseudoobscura* that when homozygous produce various percentages of wildtype flies (the expected percentage is 33.33). (After Dobzhansky and Spassky 1963.)

Frequency of Wildtype Flies in F_3 Culture (%)	Number of Cultures	Proportion of Total Cultures Tested (%)
0	134	12.6
0–4	32	3.0
4–8	41	3.9
8–12	26	2.4
12–16	45	4.2
16–20	52	4.9
20–24	87	8.2
24–28	172	16.2
28–32	282	26.5
32–36	160	15.1
36–40	28	2.6
40–44	4	0.4
Total tests	1,063	100.0

Table 3.11 Numbers and proportions of pairwise combinations of two different chromosomes sampled from a natural population of *D. pseudoobscura* that produce various percentages of wild-type flies (expected percentage is 33.33). (After Dobzhansky and Spassky 1963.)

Frequency of Wildtype Flies in F_3 Cultures (%)	Numbers of Cultures	Proportion of All Cultures Tested (%)
0	1	0.1
0-4	0	0
4-8	0	0
8-12	1	0.1
12-16	0	0
16-20	5	0.5
20-24	14	1.4
24-28	106	10.3
28-32	327	31.6
32-36	406	39.3
36-40	154	14.9
40-44	18	1.7
44+	2	0.2
Total tests	1,034	100.1

chromosomes. Some 10.5 percent of the tested chromosomes markedly reduced the proportion of wildtype homozygotes (4 percent–16 percent wildtype F_3 flies). The remaining 785 chromosomes produced F_3 cultures in which the proportions of homozygous wildtype flies form a bell-shaped distribution (figure 3.7) with a modal class in the interval 28 percent–32 percent, somewhat lower than the theoretically expected $33\frac{1}{3}$ percent.

Table 3.11 presents the "control" data for table 3.10; that is, data obtained by mating genetically marked virgin F_2 females of one culture with similarly marked F_2 males of another. The F_3 progeny of each culture in this case consist of $2 D/+ : 1+_1/+_2$ flies. These data (consisting of 1,034 cultures) differ from those in the previous table in two important ways: Only two cultures contain fewer than 16 percent wildtype (heterozygous) flies; this fact implies that the 26.1 percent of the tested chromosomes which yielded low proportions of homozygous flies were of many sorts each of which was rare, thus making it unlikely that randomly generated heterozygotes would also be homozygous for such factors by chance. Second, the modal class (see figure 3.7) for heterozygous chromosomal combinations falls in the 32 percent–36 percent interval, a whole interval above the modal class of homozygotes.

Data such as that presented in tables 3.10 and 3.11 (or even diagrams prepared from such data as in figure 3.7) are not dis-

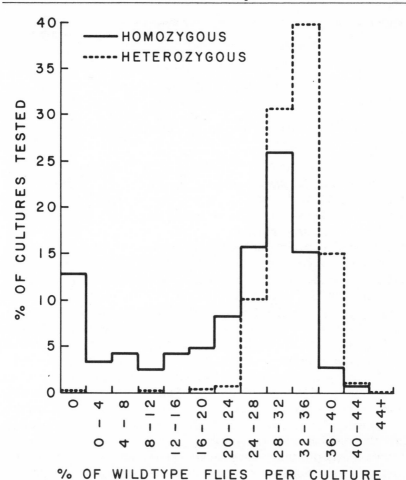

Figure 3.7 Distributions of viabilities of flies homozygous or heterozygous for third chromosomes obtained from a wild population of *D. pseudoobscura*. The data on which this figure is based are presented in tables 3.10 and 3.11 (Dobzhansky and Spassky 1963.)

cussed conveniently; the communication of facts requires that they be summarized. Each summary, of course, sacrifices some of the information originally present in the raw data.

Two types of summaries have been used in discussing data of the sort shown in tables 3.10 and 3.11. The first of these is the pigeonhole summary; it consists of listing the proportions of tested chromosomes that fall into certain arbitrarily defined categories. Summaries of this sort which describe the genetic variation characteristic of the major autosomes of three Drosophila species

Table 3.12 Estimates of the concealed genetic variability of three *Drosophila* species: *D. prosaltans*, *D. pseudoobscura*, and *D. persimilis*. (After Dobzhansky and Spassky 1953; Sankaranarayanan 1965.)

	Prosaltans		Pseudoobscura			Persimilis		
	2d	3d	2d	3d	4th	2d	3d	4th
Lethals and semilethals (%)	33	10	33	25	26	26	23	28
Subvitals (%)	33	15	94	78	70	67	80	98
Supervitals (%)	0	3	0	0	0	0	3	0
Sterility (%)	20	11	19	24	16	32	30	27

(*D. prosaltans*, *D. pseudoobscura*, and *D. persimilis*) are listed in table 3.12. Here the chromosomes have been classified as *lethal* (fewer than 10 percent of the expected proportion of homozygous wildtype flies), *semilethal* (between 10 percent and 50 percent of the expected proportion), *subvital*, *supervital*, or *sterile* (homozygous flies incapable of producing offspring). The method by which chromosomes are classified as *subvital* or *supervital* (or *normal*) will be described immediately. The points to make in passing, however, are (1) that lethals and semilethals are not uncommon among chromosomes obtained from wild populations and (2) of those chromosomes that are not lethal or semilethal, the vast majority are demonstrably deleterious when homozygous in respect either to viability or to fertility. A further analysis (see Dobzhansky et al. 1942) would have also shown that the development of homozygotes is slower than that of control heterozygotes.

Figure 3.8 illustrates the basis on which tested chromosomes are classified as subvital, normal, or supervital. The bell-shaped (normal) distributions have mean values corresponding to the average proportions of wildtype flies (homozygous or heterozygous) in the nonlethal, nonsemilethal cultures (those exceeding 16 percent in figure 3.7, for example). The dispersal of frequencies of wildtype flies in the tested cultures (both homozygotes and heterozygotes) about the mean value has three sources: (1) statistical errors (binomial variance) resulting from the finite numbers of flies hatching in individual cultures, (2) environmental differences between cultures (environmental variance), and (3) genetic differences between chromosomes and chromosomal combinations (genetic variance). Two of these variances—binomial and environmental—can be evaluated directly or by experiment. The first is related to n, the number of flies counted per culture, by the equation $\sigma^2 = pq/n$ (see Boxed Essay). The second can be estimated by the study of replicate cultures in which the flies being tested are

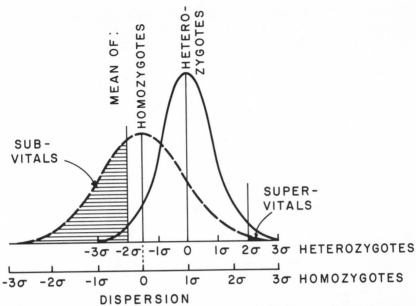

Figure 3.8 Diagram illustrating a technique for calculating the relative proportions of subvital, normal, and supervital chromosomes (as revealed in homozygous tests). Normal viability is defined as that which falls within 2 standard deviations (all calculations based on genetic variation only) of the average frequency of flies heterozygous for two different wildtype chromosomes. (After Wallace and Madden 1953.)

genetically identical. The total variance is calculable, of course, from the raw data themselves. Therefore, genetic variance can be estimated by subtracting the binomial and environmental variances from the total variance.

The diagrams in figure 3.8 are based on the dispersions of homozygous and heterozygous viabilities that are ascribable to *genetic* variance; binomial and environmental variances have been removed. *Normal* viability is now defined as that which produces frequencies of flies carrying heterozygous chromosomal combinations lying between plus and minus two standard deviations of the mean viability of heterozygous test cultures ($\overline{X} \pm 2\sigma_X$). This range includes about 95 percent of all heterozygous combinations; $2\frac{1}{2}$ percent lie below this range (and, consequently, are considered to be *subvital*) and a corresponding $2\frac{1}{2}$ percent above (supervital). The area under the homozygous distribution curve lying to the left of the lower cutoff point for normalcy corresponds to the proportion of subvital homozygotes; this area can easily exceed 80 percent or 90 percent (see table 3.12). Similarly, the area under the homozygous curve lying above the upper cutoff point for normalcy corresponds to the proportion of supervitals; this proportion commonly approaches zero (table 3.12) and seldom exceeds the $2\frac{1}{2}$ percent which is the proportion automatically included among heterozygous combinations.

In contrast to the pigeonhole method of classifying chromosomes according to their effect on the viability of their homozygous carriers, another type of summary can be prepared that is based solely upon mean frequencies. From the data presented in tables 3.10 and 3.11, the following mean proportions of wildtype flies can be calculated (.02, .06, .10, etc. have been used as mid-values for the various class intervals):

Heterozygotes	0.324
Homozygotes (all)	0.223
Homozygotes (quasi-normal)	0.286

If the 0.324 average of heterozygotes is adjusted to 1.000, the relative values of homozygotes become 0.688 (all) and 0.883 (quasi-normal). These values can be regarded as proportions of chromosomes that escaped randomly segregating lethal mutations and, hence, correspond to e^{-m} of the Poisson distribution; m in this case would correspond to the average number of lethal equivalents per chromosome. If e^{-m} = 0.688, m equals 0.374; if e^{-m} = 0.883, m equals 0.124. Consequently, to describe the data contained in tables 3.10 and 3.11 in terms of lethal equivalents one would say that the average chromosome contains 0.374 lethal equivalents, that "normal" chromosomes carry an average of 0.124 lethal equivalents, and hence that the lethal and semilethal chromosomes carry 0.250 (= .374 - .124) lethal equivalents. An examination of table 3.10 shows that 26.1 percent of all chromosomes yielded fewer than 16 percent homozygous carriers per culture; the lethal-equivalent calculation confirms that these chromosomes carry nearly one lethal equivalent each.

Table 3.13 reveals a poorly understood (and seldom noticed) aspect of the viability effects of quasi-normal chromosomes. Using the lethal-equivalent technique, the average viability of flies carrying quasi-normal chromosomes can be converted to lethal equivalents per chromosome. Using the pigeonhole procedure (being careful to eliminate binomial and environmental variances from the viability distributions), one can calculate the proportion of

Table 3.13 The estimation of the average degree of subvitality of subvital chromosomes measured in lethal equivalents. Lethal equivalents for quasi-normal chromosomes have been computed according to the method of Crow and Temin (1964), the frequencies of subvitals (see table 3.12) by the method of Wallace and Madden (1953).

	D. prosaltans		D. pseudoobscura			D. persimilis		
Average viability of quasi-normals	0.917	0.966	0.750	0.771	0.856	0.879	0.834	0.779
Lethal equivalents	0.087	0.035	0.288	0.260	0.155	0.129	0.182	0.250
Frequency of subvitals	0.33	0.15	0.94	0.78	0.70	0.67	0.80	0.98
Lethal equivalents per subvital chromosome	0.26	0.23	0.31	0.33	0.22	0.19	0.23	0.26

subvital chromosomes among those that are neither lethal nor semilethal. One can now ask, "How many lethal equivalents are there on the average subvital chromosome?" The answer is remarkably constant: about 0.25. The meaning (if any) of this relationship between lethal equivalents and subvitals is unknown.

Because the techniques that we have been discussing can reveal (in a quantitative form) the genetic variation within populations, they can also be used to compare different populations in order to determine whether they differ genetically. Such comparisons can be made between populations isolated in space (over any desired distance) or within the same population but at different times. Tables 3.14 and 3.15 present data showing that the genetic composition of Drosophila populations can change with time.

Table 3.14 Changes in the frequencies of lethal second chromosomes during four years' sampling of a South Amherst (Mass.) population of *D. melanogaster*. (After Ives 1970.)

Year	Number of Chromosomes Tested	% Lethal (+ semilethal)
1966	704	17.8
1967	2,133	24.5
1968	755	29.6
1969	525	32.8

Table 3.15 Temporal changes in the frequencies of second chromosomes in *D. melanogaster* that cause lethality, or the sterility of male or female homozygotes. (After Oshima and Watanabe 1973.)

Collection Date	Number of Chromosomes Tested	Lethals	Sterility male	Sterility female
October 1968	475	15.2%	9.3%	5.3%
July 1970	121	35.2	15.7	9.9
October 1970	221	25.8	10.4	11.3

Table 3.14 shows that the frequency of lethal second chromosomes increased nearly twofold from 1966 to 1969 in an Amherst, Mass., population. Table 3.15 shows sizable fluctuations in the frequencies of both lethal and sterility genes in a Japanese population of *D. melanogaster*. The frequencies of chromosomes causing these pathological conditions in their homozygous carriers seemingly rise and fall independently of each other in the latter population.

Revealing Genetic Variation by Artificial Selection

Among the classical techniques for revealing and measuring genetic variation within populations, artificial selection for altered morphological or physiological characteristics must be included. Underlying such selection schemes is the assumption that no long-term, heritable change can occur in a population subjected to selection unless genetic variation in respect to the selected trait preexists in that population. A number of studies (Luria and Delbruck 1943, Lederberg and Lederberg 1952; Bennett 1960) have shown that this assumption is valid in the sense that the experimental conditions themselves do not *cause* the genetic variation needed for the population's (bacteria and *Drosophila*) response; a possible exception has been described by Hartman and Hulbert (1975). The schistosomicidal drug, hycanthone, is known to be mutagenic; it is possible, therefore, that mutations conferring resistance to this substance (mutations that have arisen repeatedly during the use of this medicine) arise in part as hycanthone-induced mutations.

In addition to the entire history of the domestication and selective improvement of agricultural plants and animals, numerous experimental studies on the selective response of field and laboratory organisms have been carried out as a means for probing and understanding existing natural variation. The number and variety of selection experiments performed on *Drosophila* alone, as Lewontin (1974a) has pointed out, are so extensive that a morphological or behavioral trait which has not been studied is difficult to find.

An example of one such experiment is given in figure 3.9. In this study, flies (*D. melanogaster*) were exposed for many generations to an aerosol mist of DDT; the parents for each

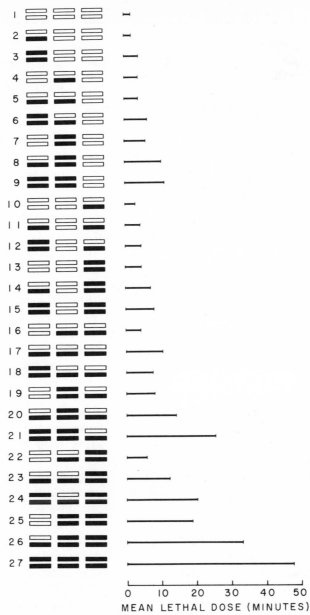

Figure 3.9 The localization accordng to chromosome of genes conferring resistance to DDT in an artificially selected, laboratory strain of *D. melanogaster.* Chromosomes of the original, sensitive strain are represented as open bars; the corresponding chromosomes from the selected strain are represented as solid ones. The resistances of the two extreme classes (all open and all solid bars) closely resemble those of the original sensitive and the resistant strains of flies. (After J. C. King 1971.)

succeeding generation were the flies that survived exposure to DDT in the previous one. From flies that originally suffered a heavy mortality following a two-minute exposure to the aerosol, a resistant strain was obtained which exhibited only 50 percent mortality following nearly an hour's exposure to the same DDT mist.

The diagram presented in figure 3.9 offers proof that the resistance to DDT exhibited by the selected strain of flies has a genetic basis and, in addition, demonstrates that each chromosome from the resistant strain contributes to this resistance. Using *ClB*-like procedures for the X, second, and third chromosomes simultaneously, flies carrying all 27 combinations of chromosomes from the parental (sensitive) strain of flies and the resistant one were constructed and tested for DDT resistance. The number of resistant chromosomes increases from the top of the diagram toward its bottom; with some irregularity, resistance to DDT (measured as LD_{50}, the length of the exposure time required to kill one-half of the exposed flies) increases as well. A careful examination of figure 3.9 reveals that, with one exception, each time a chromosome from the resistant strain is added to any combination of chromosomes, the LD_{50} of the new combination is greater. Consequently, we can conclude (1) that the exposed flies possessed genetic variation in respect to DDT resistance on every chromosome, and (2) that the selection of the survivors following an exposure to DDT, generation-by-generation, increased the proportion of genes conferring resistance to this insecticide.

The type of selective response illustrated here is not unimportant in the fight against insect, animal, and plant pests. Every one of the targets—bacteria, insects, and mammals—of pesticidal agents has developed resistance to these agents as a result of what may be regarded as "natural" selection. A once promising rat and mouse poison, Warfarin, has now proven ineffective in controlling these pests. Within rat populations a mutant allele has proven capable of rendering Warfarin relatively harmless (Bishop et al. 1977). Interestingly, males homozygous for this allele have markedly reduced viability as embryos. Here, for the first time in this book, we encounter a situation where one homozygote is harmed by one agent (Warfarin), a second homozygote by another (lowered embryonic viability), while individuals heterozygous for the two alleles largely avoid both dangers (see page 227).

BOXED ESSAY

SOME ELEMENTARY STATISTICS

A number of statistical terms recur throughout this book, not that population genetics is unique in this respect, for it is not; the same terms are used wherever biological data are treated quantitatively.

The average or mean value $(\overline{X})$: the value obtained by summing all observations (X_i) and dividing by the total number (N) of all observations; for example, $(X_1 + X_2 + X_3 + \ldots)/N$. If there are N_i observations whose value (X_i) is the same $(N_i$, where $N_1 + N_2 + N_3 + \ldots = N)$ they can be summed as follows: $(N_1 X_1 + N_2 X_2 + N_2 X_3 + N_i X_i)/N = \overline{X}$. The last expression can be rewritten

$$f_1 X_1 + f_2 X_2 + f_3 X_3 + \ldots = \overline{X} = \Sigma f_i X_i$$

The most commonly used measure of the dispersal of observations about their mean value is variance (σ^2), the mean-squared deviation of each observation from the mean: $f_1 (X_1 - \overline{X})^2 + f_2 (X_2 - \overline{X})^2 + \ldots = \sigma_X^2$ (read, the variance of X).

The above expression is true only if the observations being analyzed constitute the entire universe of such observations, or if the number of observations is large. If the observations represent a *sample* that is to be compared to other similar samples, the variance is estimated as $[\Sigma (X_i - \overline{X})^2]/(N - 1)$. For ease in computation, $\Sigma (X_i - \overline{X})^2$ is commonly calculated $\Sigma (X_i^2) - (\Sigma X)^2/N$.

The *normal* distribution, one that closely approximates many empirical observations, is constructed using the standard deviation $(\sigma$, the square root of variance) as the unit of dispersal in a two-dimensional diagram. In such a diagram, approximately 95 percent of the area under the curve lies between $\overline{X} \pm 2\sigma$, about 99 percent between $\overline{X} \pm 3\sigma$.

An important feature of variances is that those of different, independent origins are additive (like the squared sides and the hypotenuse of a right triangle). In this chapter, for example, the total variance in the proportions in wildtype flies among a number of test cultures was ascribed to three sources: binomial (sampling) variance, environmental variance, and genetic variance.

The following numerical example (chosen as the most instructive from a number of attempted ones) was constructed with random digits generated by a pocket calculator; it illustrates the calculation of means and variances, and the additivity of the latter. Column #1 consists of digits, randomly generated, ranging from 0 to 9; column #2 consists of similar digits from each of which 5 was subtracted.

	#1	#2	Total
	6	-4	2
	6	-4	2
	3	-5	-2
	3	3	6
	4	-4	0
	5	0	5
	4	-2	2
	8	2	10
	2	0	2
	2	-2	0
Sum	43	-16	27
$\overline{X}$	4.3	-1.6	2.7
Variance	3.79	7.60	12.01

The expected variance of the "total" column equals $3.79 + 7.60 = 11.39$. The variance of column #1 can be calculated as $[\Sigma X^2 - (\Sigma X)^2/N]/N - 1) = (219 - 184.9)/9 = 3.79$; the others can be calculated in a similar manner.

FOR YOUR EXTRA ATTENTION

1. The success of Muller's *CIB* technique lies in its capacity to reveal the presence of rare alleles at many loci (as opposed to electrophoretic techniques which are best at revealing common variants at individual loci). Comparable techniques have been developed in *Oenothera* (see Epp 1974), fish (see Leslie and Vrijenhoek 1978), and yeast (see James 1959 and Wills 1968).

As a practice exercise, think of some genotypes that would be interesting to compare. By proceeding backward from the desired end, devise mating procedures adequate for generating the desired genotypes and, thus, for carrying out your proposed study (see Wallace, 1968a for descriptions of several special uses of *CIB* techniques in *Drosophila*).

2. The color morphisms of various spittlebugs have been intensively studied by Halkka and his colleagues. The following data (from Halkka et al 1976) were obtained from samples of *Philaenus spumarius* captured in localities separated by five meters.

		Phenotypes		
Population	A	B	C	Total
1.	2,503	-	126	2,629
2.	849	4	74	927

Are these samples likely to have been drawn from a single local population?

3. Ahmad et al. (1978) studied the formal genetics of two allozyme loci (octanol dehydrogenase, *Odh*, and acid phosphatase, *aph*) in the mosquito (*Anopheles culicifacies*) by means of test crosses between heterozygotes and homozygotes. Do these results fit Mendelian expectations?

Odh (MS X SS)
Offspring

Cross	*MS*	*SS*	(MS–SS)
1. ♀♀	134	128	+
♂♂	132	120	+
2. ♀♀	36	33	+
♂♂	32	29	+
3. ♀♀	182	144	+
♂♂	150	140	+
4. ♀♀	352	317	+
♂♂	322	299	+

Hint: In addition to performing a Chi-square test, determine the probability of obtaining eight pluses out of eight differences if the numbers of MS and SS individuals were expected to be equal.

aph

Cross	(FS X FF)			Cross	(FS X SS)	
	FF	*FS*			*FS*	*SS*
1. ♀♀	18	18		5. ♀♀	61	49
♂♂	16	20		♂♂	63	47
2. ♀♀	55	66		6. ♀♀	57	62
♂♂	61	53		♂♂	64	71
3. ♀♀	17	25		7. ♀♀	39	24
♂♂	23	25		♂♂	40	38
4. ♀♀	51	49		8. ♀♀	96	113
♂♂	66	60		♂♂	99	102

4

Detecting and Measuring Genetic Variation: Electrophoresis

PREVIEW: This chapter continues the task that was begun in chapter 3. Technological advances in the separation by means of high-voltage electrical fields of protein molecules according to their size, shape, and net electrical charge have allowed geneticists to extend their studies of genetic variation to the primary gene products themselves. Here we examine the uses to which these new techniques have been put, and the data which they have yielded.

THE EMPIRICAL STUDY of population genetics, quoting Lewontin (1974a) once more, begins with and centers around the characterization of genetic variation within populations. A population cannot be described genetically except in terms of genotypic frequencies and, calculating from these, frequencies of the various alleles at different loci and their patterns of association (linkage patterns) among loci.

The techniques described in chapter 3 depended almost entirely upon gross, readily observable genetic differences. Some of these were color patterns affecting the external appearance of organisms; in this category would fall dorsal-striped salamanders, black hamsters, red or silver foxes, mimetic butterflies, melanic moths, black ladybirds, and a nearly inexhaustible list of other examples.

Cytological variation is not gross but is rendered so by modern microscopes. The giant chromosomes of Dipteran larvae (*Drosophila*, mosquitos, midges, gnats, houseflies, black flies, and others) magnified 1,000 times appear to be one or more meters long. Consequently, the detection of inverted segments or transpositions of chromosomal regions, even though these aberrations may involve only 1 percent or less of the entire genome, is relatively simple. Changes at individual gene loci, of course, lie beyond microscopic resolution.

By the use of modified *CIB* techniques, the detection of other than lethal mutations became possible. Timofeeff-Ressovsky (1935), for instance, showed that mutations with slight effects on the viability could be revealed in each of two or more successive generations; Dobzhansky and Queal (1938) confirmed this observation. Nevertheless, the greater the effect of a mutant allele on viability, the easier it is to detect. And, because it seems unlikely that the important genetic changes in populations have frequently, or ever, involved those that were lethal to their homozygous carriers, one is forced to conclude with Lewontin (1974a) that what can be measured by classical techniques is uninteresting; and what is interesting, they cannot measure.

The proper study of the genetics of populations requires that individual alleles at individual gene loci be recognizable as such. It requires technical procedures that magnify variations at the gene level in a manner comparable to the magnification of chromosomal changes by the compound microscope, or of gene action by characteristic pigmentation patterns of individuals. This requirement has been met, in respect to primary gene products, by electrophoretic techniques whose origins can be traced to Abramson and Gorin (1940), Smithies (1955), and Hunter and Markert (1957).

Electrophoresis: Introduction

The central dogma of modern genetics postulates that deoxyribose nucleic acid, DNA, contains the information essential for its own precise replication and for the specification of amino acid sequences in enzymes and other protein molecules (figure 4.1). The essential truth of the dogma has not been weakened by the discovery (Temin 1972) of an enzyme, reverse transcriptase, capable of synthesizing DNA from an RNA molecule, nor that faulty DNA polymerase molecules (proteins) can cause errors in DNA replication (Commoner 1964).

The means by which DNA controls its own synthesis lies in its double-stranded structure: the DNA molecule consists of two sugar-phosphate backbones of opposite polarity from which extend, in any sequence, purine (adenine and guanine) and pyrimidine (thymine and cytosine) bases; the two bases at the corresponding level in the two strands form specific pairs: if adenine occurs on one strand, thymine occurs on the other; alternatively, if cytosine is found on one, guanine occurs on the other. Consequently, the two strands of a DNA molecule are complementary, not identical; however, each strand, when isolated from the other, is capable of controlling the synthesis of a new, complementary strand identical to the missing one. This ability is the basis of "self-replication," a characteristic that was recognized as a fundamental property of genes from the earliest days of genetics.

Information regarding the structure of protein molecules resides in the sequence of purines and pyrimidines along the length of the DNA molecule. Both DNA and protein molecules are fundamentally long, linear molecules. It is now known that the sequence of amino acids in protein molecules is specified by corresponding sequences of three-base units on the DNA molecule. Protein is not made directly from DNA, however. One strand of the DNA in a

Figure 4.1 The central dogma of molecular biology. DNA controls its own replication, thus insuring the synthesis of accurate copies of itself; DNA specifies the sequence of purine and pyrimidine bases in RNA; the sequences of bases in specific mRNA molecules specify the sequences of amino acids in the corresponding protein molecules that are assembled under the control of these RNA molecules. Information is transmitted (with few, and acceptable, exceptions) from left to right, from DNA to proteins. Consequently, the systematic production of an altered protein molecule is evidence of an alteration in the responsible segment of DNA.

region which can be called the *structural* gene is copied (transcribed) in the form of an RNA molecule (messenger-RNA or mRNA) which serves, in turn, as the template for protein synthesis (translation). Starting at a prespecified point, the protein-making machinery reads successive sets of three bases (codons), translating each codon into an amino acid which is inserted into the growing protein molecule (or reading certain codons as termination signals that bring protein synthesis to a stop). A thorough discussion of the genetic code which relates each codon to an amino acid (or to a termination signal) and of the protein-making machinery consisting of ribosomes, transfer-RNAs, and an energy source lies beyond the scope of this book. It is essential, however, for any serious population geneticist to understand the nature of inheritance and hereditary processes; accounts can be found in books such as Watson (1976), Lehninger (1970), Goodenough and Levine (1974), or a host of other modern genetic and biochemical textbooks.

Electrophoresis: Protein Structure

The aspect of protein structure that concerns the detection and measurement of genetic variation is that which permits small differences, single amino acid substitutions, to be detected with ease. Because the means for detecting differences, electrophoresis, relies on alterations in the electrical charge of protein molecules, our concern is centered largely on the electrical charge of amino acids.

Proteins are constructed from twenty common amino acids. Of these twenty, sixteen do not carry an electrical charge within the physiological range of pH's; consequently they do not greatly influence the migration of a protein molecule in an electrical field.

The non-neutral amino acids (see figure 4.2) are arginine (Arg), aspartic acid (Asp), glutamic acid (Glu), and lysine (Lys). The electrical charges reside in the carboxy ($-COOH$) and amino ($-NH_2$) side groups of these amino acids. A $-COOH$ group becomes charged (ionized) only in alkaline solutions of high pH containing few H^+ but many OH^- ions; under these circumstances $-COOH$ loses H^+ ions to the solution, thus becoming charged: $-COO^-$. The $-NH_2$ side group, on the other hand, in the same

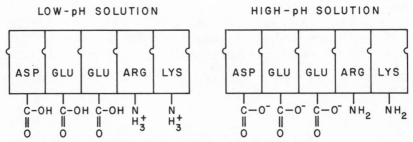

Figure 4.2 Diagrammatic representation of those amino acids which may carry an electrical charge, and an illustration of the net electrical charge characterizing a protein molecule in an acidic (low pH) or basic (high pH) solution.

alkaline solution remains unchanged and uncharged. In an acidic solution of low pH containing many H^+ but few OH^- ions, the $-COOH$ side group remains uncharged. In this case, however, the $-NH_2$ side groups acquire H^+ ions, thus becoming charged, $-NH_3^+$. As the lower portion of figure 4.2 suggests, the net charge of a protein molecule depends both upon the numbers of acidic and basic amino acids it contains, and upon the pH of the surrounding medium. The normal procedure for electrophoretic studies has involved the use of somewhat alkaline buffer solutions, thus causing most protein molecules to acquire net negative charges which, in turn, cause them to migrate toward the positive pole of the electrophoretic apparatus.

The separation of two types of protein molecules, one bearing a net negative charge of 1 and the other of 2, in an electrical field is illustrated in figure 4.3. The matrix is supposed to be of starch or some other gel-like material (hence the term, gel electrophoresis) moistened with the (slightly alkaline) buffer solu-

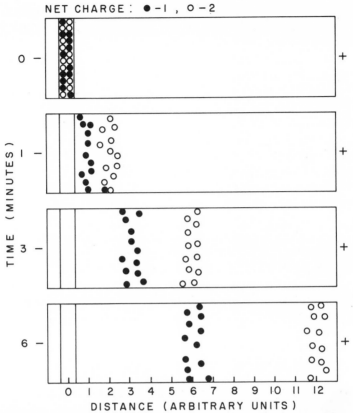

Figure 4.3 The separation of two water-soluble protein molecules having different net negative electrical charges (-2, -1). When in an electrical field, both types of molecules migrate toward the positive pole but, because of the charge difference, one does so about twice as rapidly as the other.

tion. The two protein molecules at time zero are intermixed in a narrow zone, perhaps on a piece of moist filter paper that has been inserted edgewise into the gel.

Because of their negative electrical charge, both types of molecules begin migrating toward the positive pole once the gel has been connected to a high-voltage apparatus. The molecules with the greatest electrical charge (open circles in the figure) migrate twice as fast, however, as the others (solid circles). The longer the current flows through the gel, the farther apart the two bands of migrating proteins become. Here, then, is the beauty of the electrophoretic procedure: single amino acid substitutions in

the (protein) products of single genes are transformed into millimeter or even centimeter distances between these proteins in a
supporting matrix (starch or acrylimide gel).

A living organism is not a mixture of two proteins, or of even
a small number of proteins; each fly, each oat seedling, and each
small piece of mammalian tissue contains thousands of proteins.
If a fly, for example, is squashed onto a small piece of filter paper
(as shown in figure 4.4) and subjected to electrophoresis in a high-
voltage current, its electrically charged, water-soluble protein molecules will normally migrate toward the positive pole if the buffer
solution is slightly alkaline. Many different proteins will have the
same electrical charge and will migrate at the same rate, except for
the minor influences of the size and configuration of each type of
molecules. After several minutes, streaming from the piece of filter
paper towards the positive pole, lies a complex protein soup or
goulash within which each protein has taken up a position appropriate for its electrical charge and physical configuration.

Certain proteins are present in sufficient quantities in cells that
their presence in an electrophoretic gel can be tested directly by a
simple protein stain; locating hemoglobin molecules following the
electrophoresis of red blood cells is not difficult. Most proteins,
however, occur in trace amounts. How are they to be revealed?

Fortunately, many of the proteins that are present in minute
quantities are enzymes. Consequently, the activity of enzyme
molecules can be used in their own detection: enzymes can be
forced to reveal their whereabouts. By placing the gel in a solution
containing (1) a substrate on which the enzyme will act and (2) a
(colorless) dye which will precipitate as a colored substance in the
presence of a product of the enzymatic reaction, the position of
the enzyme molecules in the gel can be determined. Figure 4.4
shows how the use of five substrates reveals the presence of six
enzymes, two of them being enzymes that act upon the same
substrate. Each enzyme is revealed by a colored band which
appears at a characteristic position in the gel. According to the
diagram in figure 4.4, enzyme 5 and one form of enzyme 3 have
identical mobilities; however, each is revealed only by the proper
substrate. Otherwise, each enzyme in the figure has been shown as
possessing its own characteristic mobility.

The only restriction on the enzymes (and, consequently, on
gene loci) that can be resolved by electrophoretic techniques is the
number of "developers" that can be worked out. Table 4.1 lists
the enzymes that are routinely studied by Drosophila geneticists.

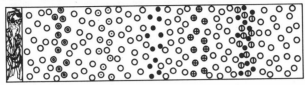

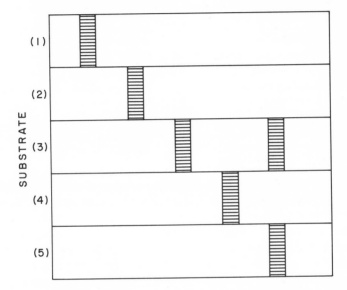

ENZYME NO.		ATTACKS SUBSTRATE NO.
◉	(1)	(1)
◎	(2)	(2)
●	(3)	(3)
⊕	(4)	(4)
⊕	(5)	(5)

Figure 4.4 By the use of specific substrates and appropriate dyes, the location of specific enzyme molecules in the spread-out band of proteins formed by the electrophoresis of a squashed fly is revealed. Note: (1) The enzymes capable of attacking substrate 3 differ in migration rates. (2) The fast-moving enzyme that attacks substrate 3 and the one that attacks substrate 5 have identical migration rates, but are revealed independently by the different substrates.

Table 4.1   An analysis of allozyme variation in plants and animals can now be made for a wide variety of different enzymes, some of which are listed below. The list can be extended almost at will because an appropriate test for the presence of any enzyme consists of presenting the gel with a substrate (preferable, *the* substrate) which the enzyme can modify, together with a "developing" system that reveals the position in the gel of the modified end product. (Adapted from Barker and Mulley 1976; Powell 1975.)

Acetaldehyde oxidase	Hexokinase
Acid phosphatase	Hydroxybutyrate dehydrogenase
Adenate kinase	Isocitrate dehydrogenase
Alcohol dehydrogenase	Lactate dehydrogenase
Aldehyde oxidase	Leucine-amino peptidase
Aldolase	Malate dehydrogenase
Alkaline phosphatase	Malic enzyme
Amylase	Octanol dehydrogenase
Esterase	Phosphoglucoisomerase
Fumerase	Phosphoglucomutase
Galacto-6-phosphate dehydrogenase	6-phosphogluconate dehydrogenase
Glucose-6-phosphate dehydrogenase	Pyranosidase
Glucose phosphate isomerase	Sorbitol dehydrogenase
Glutamate dehydrogenase	Tetrazolium oxidase
Glutamate-oxaloacetate aminotransferase	Triosephosphate isomerase
Glyceraldehyde-3-phosphate dehydrogenase	Xanthine dehydrogenase
α-glycerophosphate dehydrogenase	

This list is not complete; human and mouse geneticists routinely study well over 100 enzymes, and the list increases daily. Because the number of enzymes is enormous, and because a corresponding number of gene loci are involved in specifying these enzymes, electrophoretic techniques promise to reveal the variation—or lack of it—at a multitude of gene loci, and for any organism. The variant forms of an enzyme whose structures are specified by alleles at one gene locus are called *allozymes*. (*Isozyme* is a term that includes allozyme variation but includes as well the array of enzymes that attack the same substrate even though the enzymes are products of different gene loci. The two forms of enzyme 3 in figure 4.4 are isozymes.)

A hypothetical example based on the principles illustrated in figures 4.2, 4.3, and 4.4 is shown in figure 4.5. Here are shown the colored bands that were obtained by squashing 15 flies, subjecting them to electrophoresis in a suitable gel, and testing for the presence of an enzyme (or enzymes) capable of degrading a hypothetical substrate. At first, the patterns seem to be bewilderingly complex, even though some flies give similar ones (for example, 1 = 6 = 12, 2 = 10, and 7 = 15). Closer examination reveals, however, that the bands appear to fall in three groups: in the center, individuals possess either one or both of two bands; at the top of the diagram, individuals possess either a fast-moving band, a much slower one, or both of these plus a third, intermediate band; and

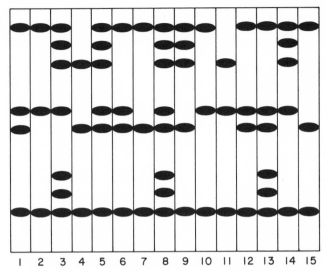

1 2 3 4 5 6 7 8 9 10 11 12 13 14 15

Figure 4.5 A hypothetical example illustrating an electrophoretic analysis of 15 wild-caught flies of a given species for gene-enzyme variation in respect to enzymes capable of degrading an unspecified substrate. Note that the gels of the individual flies differ in having 3, 4, 5, 6, 7, or 8 bands. The observed patterns can be explained on the basis of two alleles at each of three gene loci (see text).

at the bottom of the diagram, all individuals possess a slow-moving band but 3 of the 15 have two additional ones.

Considerable genetic experience or carefully designed test crosses would be needed to interpret figure 4.5 correctly, but the diagram was constructed according to the following rules: Three enzymes capable of degrading the substrate are specified by genes at three separate loci—say, A (top), B (center), and C (bottom). At each locus exist two alleles, one specifying a faster moving molecule (allozyme) than the other (A_F, A_S; B_F, B_S; C_F, C_S). The enzymes specified by the A and C loci are dimeric in structure; the enzyme molecule consists of two subunits, each of which is a product of that gene. Thus, the enzyme, α, specified by the A-locus exists in three possible forms—$\alpha_S\alpha_S$, $\alpha_F\alpha_F$, and $\alpha_S\alpha_F$; the situation in respect to the C-locus is identical. Homozygotes at either of these loci can possess but one band, fast or slow, depending upon the allele carried by the individual; heterozygotes, on the contrary, possess three bands. The enzyme specified by the B-locus is a monomer; heterozygous $B_S B_F$ individuals have slow and fast bands but no additional one. The genotype, then, of

Table 4.2 An analysis of the inheritance of the fast and slow esterase-6 bands in *D. melanogaster*. Reciprocal crosses have been combined in preparing this summary. (After T. R. F. Wright 1963.)

Parents	Offspring		
	F/F	F/S	S/S
F/F × F/F		not tested	
F/F × F/S	20	21	0
F/F × S/S	0	59	0
F/S × F/S	14	25	7
F/S × S/S	0	39	43
S/S × S/S	0	0	12

individual 13 would be $A_F A_F$ $B_F B_S$ $C_F C_S$: those of other individuals can also be reconstructed from the diagram. No $C_F C_F$ individuals have been included in the figure.

The genetics of allozyme variation was first described by T. R. F. Wright (1963) for the esterase-6 locus in *D. melanogaster*. Gels stained for esterases show many bands that occupy different positions on the gel. These can, however, be grouped into clusters (1, 2, 3, etc.); the bands at these more restricted locations may still vary in their position as did those in figure 4.5. The cluster designated esterase-6 can be occupied by a fast-moving band, by a slower-moving one, or by both. Wright designated these F/F, S/S, and F/S in genotype. The correctness of these designations was verified by making a number of single pair matings in which the two parents and their progeny were tested by gel electrophoresis. The results are shown in table 4.2. No patterns are seen that are contrary to expectations based on Wright's classification: each type of mating produced only those types of individuals expected under Mendelian inheritance, and these appear in the correct proportions. Subsequent to Wright's pioneer analysis, his findings have been confirmed by numerous others. Today, most workers *assume* that the enzyme variation they observe is genetic in origin, and that it represents variation at individual gene loci.

Spatial and Temporal Variation

Table 4.3 presents the results obtained by testing *D. melanogaster* flies descended from samples taken at four localities in the eastern United States from New York to South Carolina. Five allozyme loci were included in this study. The populations of flies inhabit-

Table 4.3 Variation in allozyme patterns of a number of gene-enzyme systems among flies captured at (1) Ceres, N.Y., (2) Mt. Sterling, Ohio, (3) Mammoth Cave, Ky., and (4) Columbia, S.C. The five systems listed are (*Pgd*) 6-phosphogluconate dehydrogenase, (*G-6-pd*) glucose-6-phosphate dehydrogenase, (*Est-6*) esterase-6, (*Aph*) larval alkaline phosphatase, and (*Lap-D*) leucine aminopeptidase-D. (After O'Brien and MacIntyre 1969.)

Locus	Gel "phenotype"	Locality 1	2	3	4
Pgd	A	10	6	5	0
	A/B	0	2	2	2
	B	0	5	5	7
G-6-pd	A	3	4	2	0
	A/B	3	7	4	3
	B	4	3	7	6
Est-6	A	8	5	3	6
	A/B	0	9	4	0
	B	0	2	2	0
Aph	A	9	13	3	0
	A/B	0	8	9	0
	B	0	1	6	10
Lap-D	A	4	13	19	5
	A/B	3	4	2	2
	B	1	1	2	0

ing different geographic areas exhibit clearcut genetic differences: Locality I seems to be homozygous (or nearly so) for one of the *Pgd* alleles; all other localities listed possess both alleles *A* and *B*. Although flies from locality 1 are also homozygous (or, again, nearly so) at the *Est*-6 and *Aph* loci, they were not homozygous at the *G-6-pd* and *Lap*-D loci—indeed, none of the sampled populations was homozygous for alleles at these latter loci. Finally, flies from South Carolina were homozygous for the same *Est*-6 allele (A) as were those from New York; in the case of the *Aph* locus, flies from South Carolina were homozygous for *Aph*-B, not for *Aph*-A as were those from New York. Clearly, then, a great deal of variation exists in respect to allozyme variation that can be revealed by gel electrophoresis.

Geographic genetic variation may be seemingly haphazard, or it may occur in an orderly manner paralleling geographic location. The data presented in table 4.4 on the variation at the lactate dehydrogenase locus in the fathead minnow (Merritt 1972) illustrates the second sort: collections of minnows taken in Nebraska, Kansas, and Oklahoma range from one in which nearly all individuals are homozygous, *A'A'*, to one in which all are *AA*. Throughout Kansas, which lies between Nebraska and Oklahoma, samples

Table 4.4 The numbers of fathead minnows, *Pimephales promelas*, from different geographic localities that are homozygous and heterozygous for the lactate dehydrogenase allozymes, *A* and *A'*. (After Merritt 1972.)

	AA	*AA'*	*A'A'*	*Total*
Nebraska	0	3	139	142
Kansas (a)	45	25	5	75
Kansas (b)	13	22	17	52
Kansas (c)	123	113	17	253
Oklahoma	15	0	0	15

Table 4.5 Temporal changes in the allozyme patterns for the alkaline phosphotase-2 locus of *Daphnia magna* captured in a permanent pond near Cambridge, England. (After Hebert 1974.)

	Pattern			
Date of Sample	*S/S*	*S/F*	*F/F*	*Total Number*
5/18/71	0	120	0	120
9/9/71	22	56	18	96
10/18/71	8	41	8	57
4/29/72	12	81	7	100

show sizable numbers of both homozygous (*AA* and *A'A'*) and heterozygous (*AA'*) individuals.

Populations, if they exhibit any genetic variation, might be expected to vary in time as well as space. Individuals occupying two geographic localities represent descendants of an earlier common ancestral population, descendants which have been isolated through time as well as in space. Table 4.5 lists the genetic composition of individual *Daphnia* collected at various times over the period of about one year from a pond near Cambridge, England. The genetic composition of the *Daphnia* in this pond changed considerably in respect to the three genotypes at the *alkaline phosphotase-2* (*Aph-2*) locus. These changes make the desired point. It is necessary to mention, however, that *Daphnia magna* does not breed entirely by means characteristic of bisexual, crossfertilizing organisms; *Daphnia* does, at times, reproduce parthenogenetically.

Summarizing Comments

Two brief sets of comments seem appropriate at this point: one concerning the technique of gel electrophoresis that we have described in some detail in this chapter, and the other summarizing where we stand at the end of chapter 4.

Gel electrophoresis has done for the genetic analysis of populations what the compound microscope and the oil immersion lens did for cytological studies: it has provided investigators with a means for detecting even single amino-acid substitutions in proteins (polypeptide chains) consisting of 150 amino acids or more. Not all substitutions are detectable, but this is not nearly as important as that many of them are. Because 16 amino acids are neutral, 2 are positively charged, and 2 are negatively charged, and because a change in electrical charge occurs only when a noncharged amino acid substitutes for a charged one, or a positive one for a negative one (or vice versa), only some 36 percent $[(\frac{2}{20} \times \frac{18}{19}) + (\frac{2}{20} \times \frac{18}{19}) + (\frac{16}{20} \times \frac{4}{19}) = \frac{136}{380}]$ of all substitutions can be detected. Furthermore, because 2, 3, 4, or 6 codons may specify the same amino acid (only methionine is specified by a unique codon), not all changes in DNA structure are accompanied by a corresponding alteration in the amino-acid composition of the protein molecule (these are known as *synonymous* mutations).

Nevertheless, the technique allows the investigator to test individual members of a population for genetic differences at individual loci in what appears to be a random sample of gene loci. For those who are studying the genetics of populations, the resolution provided by these new techniques simply cannot be compared to that provided, say, by *CIB*-type techniques. The *CIB* technique treats hundreds of gene loci as a single unit, one that is studied for its combined effect (one locus? two loci? interacting loci?) on the viability of individuals; however, the viability differences between carriers of different chromosomes become undetectable in the very range where they are most interesting. The development of gel electrophoresis and its application to the genetic studies of populations has provided geneticists with the means to carry out studies long dreamed of but formerly technically impossible. If controversy still exists (and it does) over the role of genetic variation in populations, gel electrophoresis and the studies it has made possible are not at fault; rather, it appears that the genetics of populations of cross-breeding organisms is more complex than originally believed. Studies at the level of allozyme variation *had* to be made, and *have* been made. The results have consistently revealed a wealth of variation not previously suspected: the proportion of loci possessing 2 or more contrasting alleles commonly falls between 15 percent to 50 percent (or more) of all those tested; individuals of many species are heterozygous for two different alleles at 5 percent to 20 percent of the tested loci (see Selander 1976). Recently, high-resolution techniques that subject seemingly identical electrophoretic morphs to additional screening (such as temperature sensitivity) have revealed even more genetic variation (see Singh et al. 1976): the frequency of individuals heterozygous for different alleles at the xanthine dehydrogenase locus alone may exceed 70 percent.

Genes not only specify the amino acid sequences of proteins, however; they direct the construction of proteins at specified places and times. The means for studying naturally occurring variation in gene-control mechanisms has not yet been perfected. A number of persons have suggested that variation in gene-control systems is of even greater interest to population geneticists, ecologists, and evolutionists than is the now-recognized variation in enzyme structure (Wallace 1963a, 1976, Wallace and Kass 1974, A. C. Wilson 1975).

Returning to a more prosaic subject, and one of more immediate concern, the accomplishments of chapters 1 through 4 can be summarized as follows: We have learned

- that individual members of population differ genetically from one another,
- that the patterns of these differences vary both in time and in (geographic) space,
- that individuals and their offspring disperse, often for considerable distances, but that this dispersal does not lead to the genetic homogeneity of a species,
- that the task of the population geneticist is to understand (1) the origin and maintenance of the existing genetic variation, (2) its distribution in and among populations, and at and among different gene loci, and (3) the nature of evolutionary changes in the genetic composition of populations through time, and, finally,
- that these tasks are to be undertaken within the framework of genetic and ecological models because these models, together, attempt to depict and account for both *kinds* and *numbers* of individuals.

FOR YOUR EXTRA ATTENTION

1. Gooch and Schopf (1970) determined the leucine aminopeptidase (*Lap*-3) genotypes of 43 individual ectoprocts (*Schizoporella unicornis*) at

Green Pond, Maine. The genotypes on the two sides of a floating dock are shown below. Is there reason to believe that the two samples differ?

	Genotypes			
	.94/.94	.94/.98	.98/.98	Total
Open harbor	20	7	3	30
Shore side	5	8	0	13

2. Electrophoretic analyses have been extended from the one-dimensional procedures illustrated in Figures 4.3 and 4.4 to two-dimensional ones in which, after being dispersed along one side of a gel slab, the slab is rotated 90° and the dispersed proteins are redispersed (under altered conditions) at right angles to their original direction (O'Farrell 1975).

Brown and Langley (1979) and McConkey et al (1979) have used two-dimensional techniques on strains of wildtype *D. melanogaster* and human tissue-culture cells. Less than 1 percent heterozygosity was found in the human cells; only 6 of 54 loci examined in the flies were polymorphic.

Recognizing that the two-dimensional electrophoretic technique is used in analyzing "bulk" proteins, consider some of the contrasts between this technique and the histochemical ones that reveal the presence of even trace amounts of enzymatically active proteins.

Hint: It may be useful to draw two parallel lines of equal length: one representing the total protein in the human body, and the other the total DNA. Beginning with the proteins that occur in large quantities—hemoglobin, myoglobin, collagen, histones, etc., connect the corresponding points representing the proportion of all protein to the proportion of DNA needed to specify their structure.

3. A good deal of this and the previous chapter dealt with techniques, procedures suitable for providing answers to specific questions.

Tissue transplants provide a means by which the genetic similarity of individuals can be assessed. Such transplants are accepted by the host animal only if it is genetically identical to the donor; besides autographs, only those between identical twins or members of highly inbred strains (as in mice) are not quickly rejected.

Frame one or more questions concerning asexually reproducing animal species that might be answered by tissue-graft techniques. Compare your questions with those raised by Angus and Schultz (1979); consult the references in that paper for additional uses of tissue grafts.

4. From the following information (see Vrijenhoek et al. 1978), reconstruct the origin of the unisexual fish, *Poeciliopsis monacha-lucida*.

This fish arose (arises?) by hybridization between *P. monacha* and *P. lucida*. Only females are known; they reproduce by mating with *P. lucida* males. Only maternal (*monacha*) chromosomes are transmitted to eggs during oogenesis; paternal (*lucida*) chromosomes are eliminated. Recombination does not occur.

Strains of *P. monacha-lucida* females descended from 37 wild-caught females were mated with *P. lucida* males of known genotypes in respect to 22 allozyme loci. This knowledge permitted the reconstruction of the alleles of the maternal genomes (haplotypes). Eight distinct haplotypes were identified; two of these were obtained from widely separated localities. The unique haplotypes occur in localities where *P. monacha* and *P. lucida* coexist.

Has *P. monacha-lucida* had a mono- or polyphyletic origin? If the latter, might it be fair to use the word "arises" in respect to the origin of this hybrid fish?

The Hardy-Weinberg Equilibrium

PREVIEW: The Hardy-Weinberg equilibrium (an extension of Mendel's laws of inheritance) describes the expected relationship between the frequencies of alleles in local populations and the frequencies of individuals of various genotypes in these same populations. In this chapter, the Hardy-Weinberg equilibrium is described first for two alleles at a single locus, then for multiple alleles and multiple loci. Because the Hardy-Weinberg equilibrium represents the *expected* relationships between gene and zygotic frequencies, it serves as the basis for performing statistical tests of goodness-of-fit between observations and expectations; proper and improper interpretation of such tests are discussed in this chapter.

A HEN IS only an egg's way of making another egg. The modern version of Samuel Butler's claim has been stated by E. O. Wilson (1975): "The individual is DNA's way of making more DNA." To those who think solely in terms of individuals, the view that DNA and genes are *the* important units of life appears shocking— as Wilson might attest from several unpleasant personal experiences suffered during public appearances. Nevertheless, in understanding populations—the *genetics* of populations—each population might best be thought of as a collection of genes. Genes can be described in terms of the frequencies of various alleles (gene frequencies); collections of individuals must be described in terms of the frequencies of *combinations* of alleles (zygotic frequencies). The latter is considerably more difficult than the former. To list the 52 cards of an ordinary deck of playing cards is not difficult; to list the 1,378 possible pairs of these same cards, however, is not an easy task.

Gametic and zygotic frequencies are not unrelated. In this chapter we shall see how each is computed, and learn both how the two should be related and the assumptions upon which this expected relationship rests. In doing this, we shall be reconstructing arguments and arriving at conclusions reached by Pearson (1904), Castle (1903), Hardy (1908), Weinberg (1908), and Chetverikov (1926). That a phenomenon which is really a corollary of Mendel's law of segregation was not formally described until eight or more years after the rediscovery of Mendel's paper (and then by one of the world's foremost mathematicians!) seems strange, indeed. That the significance of the Hardy-Weinberg equilibrium was not generally appreciated until it was "popularized" in the *Genetics and the Origin of Species* (Dobzhanksy 1937) is stranger still.

The Calculation of Gene Frequencies

In chapter 3, frequencies of various kinds of chromosomes were discussed. These frequencies were arrived at by testing a series of chromosomes by modified *ClB* techniques (one chromosome from each of a large number of sampled individuals) and, having classified them according to some scheme (the pigeonhole technique illustrated in table 3.12, for example), by calculating the proportion of chromosomes found in each class.

Gene frequencies are computed in a similar way. In this case, however, gene frequencies are based upon the genotypes of the sampled individuals themselves. An assumption in what follows is that both homozygous and heterozygous individuals can be recognized and scored.

Let the zygotic frequencies of *AA*, *Aa* and *aa* individuals be *X*, *Y*, and *Z* for both males and females. The frequency of *A* (*gene* or *allele* frequency) is defined as $X + \frac{1}{2}Y$, and the frequency of a is defined as $\frac{1}{2}Y + Z$. If A and a are the only alleles, $X + Y + Z$ equals 1.00; consequently, $X + \frac{1}{2}Y + \frac{1}{2}Y + Z$ also equals 1.00. The *defined* frequencies may be rationalized in either (or both) of two manners: only one of the two gametes that gave rise to each Aa individual carried allele A; the other carried a. Or, thinking now of the gametes that an individual will produce, Aa individuals produce gametes only one-half of which carry A; the other half carry a. The justification is not important; that the gene frequencies are as defined, on the contrary, *is* important.

The Derivation of the Hardy-Weinberg Equilibrium

Assume that the AA, Aa, and aa males and females of the preceding section mate at random. The frequencies of the different types of matings as well as the proportions of the various types of offspring produced by each can be represented by a "checkerboard" such as that shown in figure 5.1. The margins, top and side, represent the two sexes; each margin is divided into sections of size X, Y, and Z (total length of a side equals 1.00) representing the genotypes, AA, Aa, and aa. The solid lines marking the sections intersect to form nine compartments; the area enclosed within each represents the probability of one type of mating, while the sum of the nine probabilities equals 1.00, the total area of the square. The nine compartments have been subdivided (dashed lines) according to the proportions of offspring each type of mating produces; some are left intact, others are divided in half, and one ($Aa \times Aa$) is divided into quarters. The subdivisions represent the proportions of offspring of various genotypes, and they, too, add up to 1.00.

The values, X, Y, and Z had no relation to one another except that their sum equaled 1.00. Consequently, the offspring produced have no simple proportions: the proportion of AA individuals (see figure 5.1) equals $X^2 + XY + \frac{1}{4}Y^2$. This equals $(X + \frac{1}{2}Y)^2$ and, in fact, the offspring of the three genotypes *do* exhibit a simple relationship in terms of gene frequencies. Recalling that $p = X + \frac{1}{2}Y$ and $q = \frac{1}{2}Y + X$, the subdivisions of Figure 5.1 formed by dashed lines can be seen to equal

$$p^2 AA : 2pq\ Aa : q^2\ aa.$$

The proportions p^2, $2pq$, and q^2 need not equal the original proportions X, Y, and Z; they can, however, be regarded as representing new values: X_1, Y_1, and Z_1. In this case, because $p_1 = X_1 + \frac{1}{2}Y_1$, p_1 must equal $p^2 + pq$, or $p(p + q)$, or p (because $p + q = 1$). Similarly, $q_1 = \frac{1}{2}Y_1 + Z_1 = pq + q^2 = q(p + q) = q$.

The population that has been described above can be represented in the zygotic and gametic stages as follows:

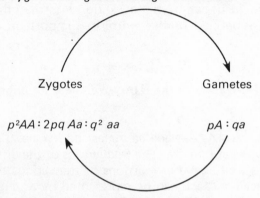

Zygotes   Gametes

$p^2 AA : 2pq\,Aa : q^2\,aa$ $pA : qa$

Undisturbed, each generation of zygotes produces A and a gametes in the proportions p and q; these gametes unite to form AA, Aa, and aa zygotes in the proportions $p^2 : 2pq : q^2$. The cycle can be repeated indefinitely.

The implications of the Hardy-Weinberg equilibrium are less obvious, perhaps, to those who think primarily in terms of individuals than to those who look upon individuals as transient beings, and who concentrate, instead, upon gene (or allele) frequencies. Dominance and penetrance, for example, are terms that describe an aspect of an allele in respect to individual development; these terms are misleading if they are applied to genes in populations: dominant genes do not *dominate* in respect to the relative frequencies of alleles. Persons who think of, or deal with, individuals know that many young plants and animals die before reaching maturity; such knowledge may suggest that populations are in a state of perpetual change. This need not be so; the Hardy-Weinberg equilibrium—as the term itself suggests—emphasizes a static condition rather than a state of change.

By concentration upon *gene* frequencies and by avoiding any reference to differential survival or reproduction, the constancy of gene frequencies from generation to generation in a population becomes obvious. The proportions of AA, Aa, and aa individuals are simple binomial probabilities that describe the likelihood of

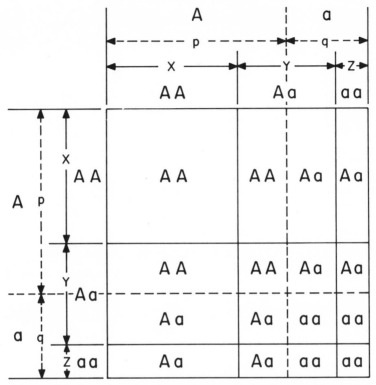

Figure 5.1 A geometric demonstration that whatever the initial proportions of *AA*, *Aa*, and *aa* individuals in a crossbreeding population (provided only that the proportions are the same in the two sexes), these proportions become p^2 *AA*, $2pq$ *Aa*, and q^2 *aa* in the next and all subsequent generations.

picking 2*A*'s if the frequency of *A* equals *p*, of picking 1*A* and 1*a* if the frequency of one is *p* and the other *q* (= 1 − *p*), and of picking 2*a*'s if the frequency of *a* equals *q*. These are merely patterns of association and, when these associations are dismantled once more (provided that none are subjected to disproportionate loss), the simple elements are regained in their original proportions.

Calculating Gene Frequencies: Examples

Table 5.1 illustrates two methods for calculating gene frequencies from known numbers and proportions of zygotes. In the hypo-

Table 5.1 Alternative procedures for calculating the frequencies of alleles *A* and *a*.

Hypothetical Sample

Genotype	Number	Frequency
AA	150	0.25
Aa	420	0.70
aa	30	0.05
Total	600	1.00

Calculations as suggested in text:

$$\text{Frequency of } A = X + \tfrac{1}{2} Y = p$$
$$\text{Frequency of } a = \tfrac{1}{2} Y + Z = q$$
$$p = 0.25 + 0.35 = 0.60$$
$$q = 0.35 + 0.05 = 0.40$$

From the *numbers* of alleles:

Number of *A* alleles = $(2 \times 150) + 420 = 720$
Number of *a* alleles = $420 + (2 \times 30) = 480$
Total number of alleles = $2 \times 600 = 1200 = 720 + 480$
$$p = 720/1200 = 0.60$$
$$q = 480/1200 = 0.40$$

thetical example, both the number and the frequencies of *AA*, *Aa*, and *aa* individuals among a total of 600 are listed. The frequencies of the two alleles, *A* and *a*, are calculated as $X + \tfrac{1}{2}Y$ $(= p)$ and $\tfrac{1}{2}Y + Z$ $(= q)$; the values of p and q equal 0.60 and 0.40. Alternatively, p and q can be calculated as $(2X + Y)/2N$ and $(Y + 2Z)/2N$ with, of course, the same result. The two methods are not only equivalent but also equally simple.

Table 5.2 offers an opportunity to follow gene frequencies in a human population from one generation to the next. The *M*, *MN*, and *N* human blood groups are phenotypes associated with $L^M L^M$, $L^M L^N$, and $L^N L^N$ genotypes. Readers possessing pocket calculators should confirm the following: The frequencies of these phenotypes in the successive generations are 0.31, 0.49, 0.20 and 0.31, 0.50, 0.19, respectively—virtually identical arrays, as the

Table 5.2 The proportions of *M*, *MN*, and *N* individuals in two generations (parents and offspring); the parents have been grouped according to marriage patterns.

Families		Children			
Type	No.	M	MN	N	Total
M × *M*	119	272	1	0	273
M × *N*	142	1	315	0	316
N × *N*	33	0	0	72	72
MN × *M*	341	408	387	1	796
MN × *N*	250	3	334	300	637
MN × *MN*	275	163	317	160	640
Total	1,160	847	1,354	533	2,734

Hardy-Weinberg equilibrium predicts. The frequency of L^M (p) equals (among children) 0.31 + 0.25, or 0.56. The frequency of L^N (q) equals 0.25 + 0.19, or 0.44. The sum of the two frequencies, $p + q$, equals 1.00. The expected frequencies of M, MN, and N children, consequently are p^2 (= 0.31), $2pq$ (= 0.49), and q^2 (= 0.19); these frequencies are also nearly identical to those observed.

Corresponding Gene and Zygotic Frequencies

The Hardy-Weinberg equilibrium packs a great deal of information into a single number. If, for example, two alleles exist at one locus, and the frequency of one (A) in a randomly mating population is known to be 0.80, the following is automatically known:

- the frequency of a equals 0.20,
- the expected frequency of AA individuals equals 0.64,
- the expected frequency of Aa individuals equals 0.32, and
- the expected frequency of aa individuals equals 0.04.

We know even more if we include the frequencies of different parental combinations as well. The numbers of family types listed in table 5.2 are not as they are by accident; each type can be estimated by multiplying the frequencies of the genotypes of the two persons involved: the probability of $M \times N$ marriages, for example, equals 0.31 $\times$ 0.20 $\times$ 2 = 0.124 (Why times 2?); the observed frequency (= 142/1,160) equals 0.122. In this manner, knowing that the frequency of one of two alleles equals 0.80, one can estimate not only zygotic frequencies but also expected frequencies of mating (or marriage) combinations.

Figure 5.2 presents in diagrammatic form the proportion of zygotes that are expected for various values of p; two features of this diagram should be commited to memory: (1) p^2 and q^2 vary exponentially with p and q, when either p or q becomes small, the frequency of homozygous individuals becomes extremely small; and (2) the maximum value for $2pq$ is 0.50 which is attained when $p = q = 0.5$ (notice that the frequency of heterozygous individuals exceeds 0.40, however, for all values of p lying between 0.72 and 0.28).

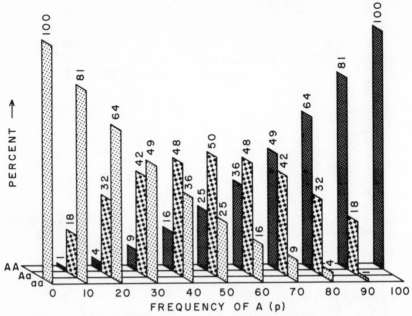

Figure 5.2 The Hardy-Weinberg proportions of homozygous (*AA* and *aa*) and heterozygous (*Aa*) individuals (i.e., *zygotic* frequencies) that correspond to various frequencies of *A* (=*p*).

Evolution and the Hardy-Weinberg Equilibrium

It is unfortunate that Darwin was unaware of Mendel and his work. According to Crew (1966), Mendel was familiar with Darwin's publications; at least three of Darwin's books were included in Mendel's personal library. During much of Darwin's scientific life, he was concerned with the mechanism of inheritance. Between 1862 and 1879, Darwin published three books on hybridization, fertilization, and variation in plants. That Mendel's paper was available but unused by Darwin during this period is one of the ironies of evolutionary biology.

Darwin's interest in the mechanism of heredity arose directly from his theory of evolution by natural selection. For natural selection to operate, a number of conditions must be met. First, there must be variation upon which selection might act. Second, this variation must be heritable. Third, survival of individuals with differing characteristics must be differential rather than random. These three conditions will bring about evolutionary changes in a population but only as long as heritable variation exists. Darwin was well aware of this last point; his interest in heredity was an outcome of his search for the source of and the means for preserving heritable variation.

An illustration of the nature of variation in wing length in *D. melanogaster* and a demonstration that this variation is indeed heritable are shown in figure 5.3. The regression line that summarizes the relation between the wing lengths of offspring and those of their parents runs from lower left to upper right in the chart (positive slope): the experimental data show that short-winged parents tend, on the average, to have short-winged offspring, whereas long-winged parents have long-winged progeny.

The resistance of flies to the insecticide DDT can also be cited as an example in which offspring tend to resemble the mean value of their parents. Table 5.3 lists data from four experiments in which strains of *D. melanogaster* that had been selected for resistance to DDT were crossed to an unselected, sensitive strain. If the resistance to DDT is measured as the exposure (in minutes) required to kill one-half of the exposed flies (LD_{50}), it is seen that the LD_{50} of the hybrid flies is about the same as the average LD_{50}'s of the parental strains. The agreement is somewhat im-

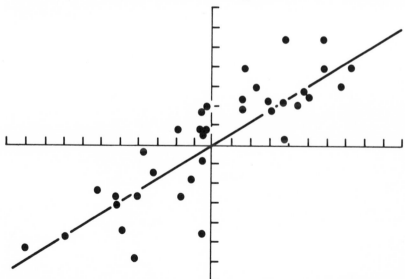

Figure 5.3 Regression of offspring (vertical axis) on parents (horizontal axis) for wing length in *D. melanogaster*. Axes are marked in 0.01 mm intervals; the origin represents the mean of both parents and offspring. Reproduced with permission: from D. S. Falconer, *Introduction to Quantitative Genetics*; © 1960 Longman Group; figure based on data provided by E. C. R. Reeve.

Table 5.3 DDT resistance of hybrid flies obtained by crossing selected by nonselected strains of *D. melanogaster*. The figures are the minutes of exposure to a DDT mist needed to kill one-half of the exposed individuals (lethal dose 50 percent, or LD_{50}). Note the rough similarity between the LD_{50} of the F_1 hybrids and the average LD_{50} of the parental strains. (After King 1955.)

| | Parental Populations | | | |
Experiment	Females	Males	Parental Average	F_1 Hybrids
X1	10.5	2.5	6.5	6.5
X1R	2.5	10.5	6.5	6.8
X5	17.6	2.5	10.1	7.8
X5R	2.5	17.6	10.1	7.4

proved if the logarithm of minutes is used in the calculations rather than minutes themselves.

In the absence of information on the particulate nature of heredity, Darwin accepted the notion of blending inheritance held by his contemporaries. "Blending inheritance" in this sense means a blending of genetic material such that the gametes of children of dissimilar parents uniformly contain average or intermediate hereditary material rather than being divided into discrete classes into which the original material has segregated (as Mendel claimed in his first law) without contamination. If genetic material did blend, as Darwin believed, then the mere process of reproduction—the creation of one generation of individuals by the preceding one—would reduce existing variation by one half. This halving process is illustrated in figure 5.4. A hypothetical population is shown as consisting of individuals whose value in some quantitative sense is said to be 2, 4, and 6; one third of each sex has each value. The mean value of these individuals is 4; the variance (calculated in the figure; see Boxed Essay, chapter 3) is $\frac{8}{3}$. Now, if matings occur at random and if the offspring of each mating have values equal to the averages of their parents, the mean of the second generation is still 4 but the variance is only $\frac{4}{3}$. The variance of the offspring, consequently, is only one-half that of their parents.

A loss of one-half of the heritable variance of a population each generation is a tremendous loss. As Darwin was well aware, it is a loss that could be fatal to his theory of natural selection. If variation were lost in this fashion, that which exists in a population at any moment can be represented as in figure 5.5: One-half of the existing variation must be completely new, one-half of the remainder (one-fourth of the total) is one generation old, one-eighth of the total is 2 generations old, and so forth. Less than one thousandth of the observed variation would be nine or more generations old. In the absence of a mechanism for storing varia-

MALE PARENTS

	2	4	6
2	2	3	4
4	3	4	5
6	4	5	6

FEMALE PARENTS

PARENTS

MEAN : $2(1/3) + 4(1/3) + 6(1/3) = 12/3$ OR 4

VARIANCE : $(-2)^2(1/3) + (0)^2(1/3) + (2)^2(1/3) = 8/3$

OFFSPRING

MEAN : $2(1/9) + 3(2/9) + 4(3/9) + 5(2/9) + 6(1/9)$

$= 36/9$ OR 4

VARIANCE : $(-2)^2(1/9) + (-1)^2(2/9) + (0)^2(3/9) +$

$(1)^2(2/9) + (2)^2(1/9) = 4/3$

Figure 5.4 The halving of a population's genetic variation (in respect to an arbitrary trait) under one generation of blending inheritance.

tion, evolution would necessarily proceed on the basis of that which had arisen in only the past one or two generations! Furthermore, if variation seems to remain constant, then *there must be a source of new variation each generation that equals one-half of the total of existing variation*. These problems were so formidable and their solution so necessary (but so elusive) that Darwin once supposed that fertilization might be a *mixing* rather than a *fusion* of individuals (Fisher 1958:1). It is unfortunate indeed that he did not encounter a paper entitled "Experiments in Plant Hybridization."

The Hardy-Weinberg equilibrium, of course, removes the difficulties described above. Hereditary variation is not lost from a population as a consequence of one generation's giving rise to the next. If (as we have seen) the frequencies of the alleles A and a are p and q, the frequencies of AA, Aa, and aa individuals are p^2, $2pq$, and q^2. In some aspect of their phenotypes let these individuals have values X_{AA}, X_{Aa}, and X_{aa}. Then, the average value, $\overline{X}, = p^2 X_{AA} + 2pq X_{Aa} + q^2 X_{aa}$. As long as the frequencies of

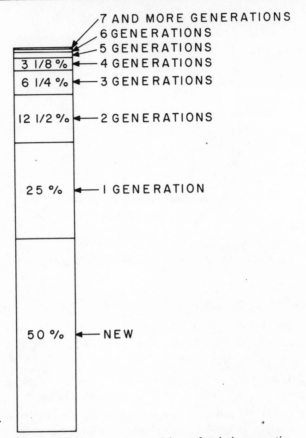

Figure 5.5 The age composition of existing genetic variation in a population under blending inheritance. Less than 2 percent of all existing variation would be more than five generations old under this pattern of inheritance; consequently, evolution would take place primarily by differential survival and reproduction of recently arisen variants.

A and a remain unchanged, both the mean value of X and its variance remain unchanged as well.

Uses of the Hardy-Weinberg Equilibrium

The great theoretical importance of the Hardy-Weinberg equilibrium is that described above: under Mendelian inheritance genetic variation does not necessarily disappear from a population, generation-by-generation. It is also useful in a number of less grandiose respects. First, it provides a base against which to compare empirical data obtained by sampling populations. Do observed zygotic frequencies agree with observed gametic frequencies? Because a number of assumptions are implicit in the calculation of zygotic frequencies by the binomial expansion, agreement with Hardy-Weinberg expectations implies (but by no means proves) that these assumptions have been met. A poor fit of observed and expected zygotic frequencies, on the other hand, suggests that one (or more) of these assumptions is not satisfied with the sampled population. Second, in calculating gene and zygotic frequencies, information is obtained that permits a comparison of samples taken from different populations or from one population at different times; significant differences between such samples reveal microevolutionary changes in space and time.

Testing goodness-of-fit

The Hardy-Weinberg equilibrium was illustrated by means of data on the MN blood group system in human populations (table 5.2). In the earlier discussion of these data, it was enough to point out that observations and expectations were in excellent agreement. Here we shall be more formal in demonstrating this agreement.

Of the 2,734 children, 0.3098 are $L^M L^M$, 0.4952 are $L^M L^N$, and 0.1950 are $L^N L^N$. The frequency of L^M equals 0.5574 ($= p$); that of L^N equals 0.4426 ($= q$). Thus, the expected proportions of type M, MN, and N persons equal 0.3147, 0.4934, and 0.1959, respectively.

These calculations can be extended and tabulated as follows:

Bloodtype	M	MN	N	Total
Observed numbers (O)	847	1354	533	2734
Expected numbers (E)	849.5	1349.0	535.5	2734
Difference (O − E = d)	− 2.5	5.0	− 2.5	
d^2	6.25	25.00	6.25	
d^2/E	.0074	.0185	.0117	
Sum of d^2/E = Chi-square = .0376				

The Chi-square of these calculations has a single degree of freedom (see below); therefore, the probability of encountering deviations as large or larger than those actually observed is 0.90 or more. Thus, in agreement with the earlier conclusion, the observed frequencies are nearly identical to those expected under the Hardy-Weinberg equilibrium.

To those accustomed to testing observations against expectations based on a theoretical model, the single degree of freedom assigned to the Chi-square may seem too small. There is only one degree of freedom in the above test because *the frequencies of alleles in a population have no theoretical expectation*; consequently, *p* (and *q* = 1 - *p*) must be estimated from the observations themselves. Now, one degree of freedom is lost for every parameter (*p* in the present case) that must be calculated from the data. This loss is in addition to the loss of one degree of freedom that reflects the "known" size or frequency of one class of individuals in a sample. In a sample of 300 individuals consisting of classes *A*, *B*, and *C*, if the number of *A* = 100 and that of *B* = 100, then the number of *C must* be 100, also.

Testing Hardy-Weinberg expectations may by now appear to be an exercise in circular reasoning: zygotic frequencies are observed; these are used to calculate gene frequencies; the gene frequencies are used in turn to calculate expected zygotic frequencies. Is it possible for the observed and expected frequencies ever to differ? The following data, taken from Dobzhansky and Pavlovsky (1955), show that the procedure is by no means circular. In this case, *A* and *D* refer to two gene arrangements of an autosome of *D. tropicalis*. Alternative gene arrangements are inherited as alternative alleles: individuals can be homozygous for either arrangement (*A/A* or *D/D*) or heterozygous for the two (*A/D*). The observations and calculations based on them are tabulated below:

Genotype	A/A	A/D	D/D	Total
Observed number (O)	3	134	3	140
Observed frequencies	.02	.96	.02	1.00
Frequency of A = p = 0.50				
Frequency of D = q = 0.50				
Expected frequencies	0.25	0.50	0.25	
Expected numbers (E)	35	70	35	140
d = O - E	-32	+64	-32	
d^2/E	29.26	58.51	29.26	

Chi-square in this example exceeds 116; with one degree of freedom the probability of obtaining such large deviations is virtually zero. Ostensibly, there appears to be a shortage of homozygotes and an excess of inversion heterozygotes (*A/D*), suggesting that the sampled population contains what is essentially a balanced lethal system. Subsequent, independent studies have shown that these homozygotes are, indeed, highly inviable.

Geographic variation in gene frequencies

Table 5.4 lists the numbers of individual fathead minnows homozygous and heterozygous for two alleles (A and A') at the lactate dehydrogenase locus sampled at three separate Western localities. Are these samples random samples from a large population of minnows that occurs throughout Missouri and Kansas? Or, is there evidence that the local populations from which the samples were drawn differ from one another? A comparison of localities #9 and #10 will illustrate both the calculations and the logic on which they are based:

Genotype	AA	AA'	A'A'	Total
Locality #9	2	18	48	68
Locality #10	37	35	8	80
Total	39	53	56	148

If these two localities are inhabited by a single large population of fathead minnows (and that is the hypothesis which is being tested), then the combined samples (Total) provides the best estimate of the relative numbers of the three genotypes in that population. Expected numbers are obtained by apportioning the samples of 68 and 80 according to the proportions $39:53:56$ ($68 \times 39 \div 148 = 17.9$, for example); these expected numbers and (in parentheses) the differences (O – E) are listed below:

	AA	AA'	A'A'
#9	17.9(–15.9)	24.4(–6.4)	25.7(22.3)
#10	21.1(15.9)	28.6(6.4)	30.3(–22.3)

The sum of d^2/E in this case equals 65.0; this equals a Chi-square with 2 degrees of freedom (Why 2?). The probability that these represent chance deviations ascribable to sampling errors alone is

Table 5.4 The numbers of lactate dehydrogenase allozyme patterns among electrophoretic gels of fathead minnows (*Pimephales promelas*) captured at localities (identified by number only) in Missouri and Kansas. (After Merritt 1972.)

Locality (State)	AA	AA'	A'A'	Total
9 (Missouri)	2	18	48	68
10 (Kansas)	37	35	8	80
16 (Kansas)	22	33	9	64

nearly zero. Consequently, it appears that local differentiation of fathead minnow populations has occurred. Note that these calculations have been made using numbers of individuals of the three genotypes; gene frequencies were not calculated. Because populations with identical gene frequencies can differ nevertheless in zygotic proportions, the test performed above is more sensitive—and less work—than one based on numbers (or frequencies) of the alleles, A and A', alone.

That populations of minnows living in Kansas and Missouri might differ genetically may not surprise many; nevertheless, the implication of differences of this sort should not be underestimated. For some reason, the constancy of gene frequencies predicted by the Hardy-Weinberg equilibrium has failed; otherwise, all minnow populations would be identical. During the time that the populations in the two sampled localities have been separated, having descended from a common ancestral minnow population, gene frequencies have changed so that the frequency of A in locality #9 is 0.162, and that in locality #10 is 0.852.

More striking (table 5.5) perhaps are the frequencies of two alleles (1.00 and 1.36, so-called because of the migration rates in electrophoretic gels of the enzyme molecules each allele produces) at the phosphoglucomutase locus in *D. persimilis* in *adjacent* localities as reported by Taylor and Powell (1977). A Chi-square analysis of these data reveals that the five samples are not likely drawn from one uniform population. The Chi-square calculated from these observations equals 13.41; there are four degrees of freedom. The probability that these alleles would be distributed as they are among the five sampled localities (all, incidentally, within the daily flight distance of these flies) is 0.01; thus, it appears that *D. persimilis* flies in this California locality distribute themselves nonrandomly in space according to preferences based on their phosphoglucomutase genotypes.

Table 5.5 The numbers of two alleles (identified according to their migration rates in an electrophoretic gel as 1.00 and 1.36) at the phosphoglucomutase locus in *D. persimilis* flies sampled at each of five ecologically different but adjacent localities at Mather, Calif. Distances between these localities do not exceed the average daily dispersal distance of these flies. (After Taylor and Powell 1977.)

Locality	1.00	1.36	Total
A	117	170	287
B	214	274	488
C	225	289	514
D	79	137	216
E	23	66	89
Total	658	936	1,594

Temporal changes in gene frequencies

Repeated samples of individuals inhabiting a certain locality frequently reveal that the genetic composition of a population does not remain constant. At times, the variation is cyclic; the Piñon Flats (California) population of *D. pseudoobscura* goes through a well-defined cycle of gene-frequency changes each year (table 5.6). Other populations of the same species have shown relatively long-term changes of frequencies which are even yet poorly understood (table 5.6).

In addition to samples taken at different times, individuals of long-lived species when arranged by size may also yield information on temporal changes in gene and zygotic frequencies. Table

Table 5.6 Temporal changes in the frequencies (percent) of different gene arrangements in two populations of *D. pseudoobscura*. (Dobzhansky 1947a.)

Cyclic Seasonal Changes at Piñon Flats, Calif.

Month	ST	AR	CH	TL	No.
March	52.1	18.2	23.1	6.5	1,054
April	40.3	27.7	28.2	3.7	801
May	33.6	29.0	31.3	6.1	642
June	27.9	27.6	39.4	5.1	1,130
July	41.9	21.8	30.6	5.6	124
August	42.4	28.4	25.8	3.4	264
September	47.6	22.5	25.7	4.3	374
October—					
December	49.6	26.3	19.8	4.3	464

Directional Changes of Prolonged Duration at Keen Camp, Calif.

Year	ST	AR	CH	TL	No.
1939	27.8	30.4	38.3	3.5	1,986
1940	31.5	22.6	41.9	4.0	2,382
1941	34.7	24.1	37.1	4.1	764
1942	36.0	16.4	40.3	7.2	414
⋮					
1945	41.0	22.2	29.2	7.6	288
1946	50.0	15.3	28.1	6.7	800

Table 5.7 The numbers of bleak (*Alburnus alburnus*, a fish) of various sizes showing various allozyme patterns generated by three alleles (1, 2, and 3) at an esterase locus. Size is also an indication of age. (After Handford 1971.)

Approximate Length (mm.)	Genotypes						Total
	11	12	13	22	23	33	
39	2	9	25	2	26	8	72
60	9	21	99	6	108	20	263
86	6	10	43	1	41	3	104
101+	1	6	16	3	14	1	41

5.7 lists information on zygotic frequencies for three alleles of an esterase locus in the bleak (*Alburnus alburnus*). In each size class the observed distribution differs considerably from the Hardy-Weinberg expectation. The discrepancy between observation and expectation seemingly arises early in the life of this fish because the observed zygotic distributions do not differ among the four size classes.

Limitations to the use of Hardy-Weinberg calculations

In the past, not satisfied with simply testing the goodness-of-fit between observed and expected zygotic frequencies, population geneticists would use observed discrepancies to decide which genotypes were more, and which less, frequent than expected. This comparison—genotype by genotype—cannot validly be made without additional information: the detailed comparison assumes that the population is in fact at equilibrium. If gene frequencies change between zygote formation and the time when zygotic frequencies are scored, heterozygous individuals very often appear to be in excess; for an apparent excess of heterozygotes in the case of two alleles, it is only necessary that the probability of survival of heterozygotes exceed the geometric mean of the corresponding probabilities of the two homozygotes.

The Hardy-Weinberg equilibrium is insensitive and cannot be used, for instance, to verify that one's classification of genotypes is or is not correct. Ford (1971:133), for example, mentions that the heterozygote (*medionigra*) of *Panaxia dominula* might be misclassified as the typical dominant form, *dominula*; it seemed unlikely to Ford that the error was frequent, however, because the three genotypes closely fit Hardy-Weinberg expectations.

Because misclassification alters gene frequencies as well as zygotic frequencies, the agreement between observed and expected zygotic frequencies offers little evidence concerning the frequency of mistakes. Consider the following example where one-half of all heterozygotes are misclassified:

Genotype	AA	Aa	aa
True zygotic frequencies	0.81	0.18	0.01
Apparent zygotic frequencies	0.90	0.09	0.01
Apparent gene frequencies A: 0.945; a: 0.055			
"Expected" zygotic frequencies	0.893	0.104	0.003

An enormous sample of individuals would need be tested in order to reveal that the (erroneous) expected frequencies deviate significantly from the (erroneous) observed ones. This does not mean that *Panaxia dominula* genotypes were misclassified; the calculation merely shows that a good fit with the Hardy-Weinberg expectation does not rule out the possibility of observational errors.

Finally, although the matter will recur in a later chapter, it should be noted that the pooling of data before calculating Hardy-Weinberg expectations will lead to erroneous results if gene frequencies differ among the pooled subsamples. Suffice it to cite an extreme example: Suppose two neighboring populations of some organism differ; the frequency of *A* in one is 100 percent whereas its frequency in the other is 0 percent (frequency of *a* = 100 percent). Two equal-sized samples taken from these populations and pooled would yield 50 percent *AA* and 50 percent *aa* individuals. The frequencies of *A* and *a* would then be computed as 0.50 each, and the expected frequencies of *AA, Aa,* and *aa* individuals as 0.25 : 0.50 : 0.25. No heterozygotes would be observed; the calculation, however, says that their frequency should be 50 percent. Whenever isolated or partially isolated populations differing in gene frequency are lumped and treated as one large, randomly mating population, the expected frequency of heterozygotes arrived at by Hardy-Weinberg calculations is spuriously high relative to the frequency actually observed.

The Hardy-Weinberg Equilibrium and Multiple Alleles

Hardy-Weinberg calculations are not restricted to loci occupied by two alleles only; they can be extended to cover cases where three or more alleles exist.

The gene frequency of a particular allele in the case of multiple alleles is calculated as the sum of the frequency of the homozygous class plus one-half that of every heterozygous class carrying the allele in question. The sum of the gene frequencies— $p, q, r, s, \ldots$ —equals 1.00. If these are the frequencies of alleles $a_1, a_2, a_3, a_4, \ldots$, the expected zygotic frequencies obtained by the binomial expansion are:

$$a_1 a_1 \quad a_2 a_2 \quad a_3 a_3 \quad a_4 a_4 \ldots a_1 a_2 \quad a_1 a_3 \quad a_1 a_4 \ldots a_2 a_3 \quad a_2 a_4 \ldots$$
$$p^2 \quad q^2 \quad r^2 \quad s^2 \quad \ldots 2pq \quad 2pr \quad 2ps \ldots 2qr \quad 2qs \ldots$$

A comparison of observed and expected numbers can be made using data given in table 5.7. Using the 60 mm. sample because it is the largest of the four (263 individuals tested), the analysis can be tabulated as shown. Sum of d^2/E = 89.49 = Chi-square with 3

Genotype	11	12	13	22	23	33	Total
Observed Number	9	21	99	6	108	20	263

Calculation of allelic frequency

	1	2	3	
	18	21	99	
	21	12	108	
	99	108	40	Total
	138	141	247	526
Frequency	.262	.268	.470	1.000

Genotype	11	12	13	22	23	33	
Expected Number	18.1	36.9	64.8	18.9	66.3	58.1	263.1
(O – E) = d	–9.1	–15.9	34.2	–12.9	41.7	–38.1	
d^2/E	4.58	6.85	18.05	8.80	26.23	24.98	

degree of freedom; the probability that the observed array of genotypes represents a chance deviation from the one expected under the Hardy-Weinberg equilibrium is nearly zero. The tabularized calculations show, as several earlier ones did, that the fit between observations and expectations is poor, indeed. Of more immediate interest, however, is the matter of degrees of freedom. Two gene frequencies (p and q) were estimated from the data (there was no need to estimate r which equals $1 - p - q$); therefore, the traditional 5 degrees of freedom (one less than the number of classes) must be reduced still further to 3.

The Hardy-Weinberg Equilibrium in the Case of Two Loci

If two alleles exist at each of two gene loci, the expected frequencies of the nine possible genotypes are obtained by multiplying the two individual binomial expansions. Thus, if the expected frequencies of AA, Aa, and aa individuals are p^2, $2pq$, and q^2 and those of BB, Bb, and bb are r^2, $2rs$, and s^2, then the expected frequencies of the nine genotypes are:

AA BB	$p^2 r^2$
AA Bb	$2p^2 rs$
AA bb	$p^2 s^2$
Aa BB	$2pqr^2$
Aa Bb	$4pqrs$
Aa bb	$2pqs^2$
aa BB	$q^2 r^2$
aa Bb	$2q^2 rs$
aa bb	$q^2 s^2$

Lewontin and White (1960) have published data on the frequencies of chromosomal rearrangements in the Australian grasshopper, *Moraba scurra* (see figure 3.2); two different chromosomes were analyzed. A portion of their data is given in table 5.8, illustrating both the observed and expected (in parentheses) numbers of the different zygotic types in the case of these two independent "loci." The expected number of ST/TD ST/BL double heterozygotes, for example, is calculated from the data as (126 X 143) ÷ 584 = 31; the observed number is 22.

Linkage and "Independence"

Unless linkage is absolute, even genes lying on the same chromosome are expected to behave in populations as if they were independent. They are independent in the sense that, in the absence of selection, the zygotic frequencies for various combinations of alleles at the two loci are given by the product of the frequencies of the two separate loci.

Suppose that at each of two loci on the same chromosome there exist two alleles: *A* and *a*, *B* and *b*. The frequencies of these alleles are *p* and *q* (*p* + *q* = 1.00) and *r* and *s* (*r* + *s* = 1.00). These alleles will be associated with each other at random if chromosomes bearing *AB*, *Ab*, *aB*, and *ab* have frequencies *pr*, *ps*, *qr*, and *qs*, respectively. We shall now show that a population at

Table 5.8 The numbers of individuals in one local population of *Moraba scurra* (an Australian grasshopper) observed carrying different combinations of standard and inverted gene sequences of two different pairs of chromosomes (expected numbers are given in parentheses). *ST*, standard arrangements for each chromosome; *BL*, inverted sequence of the CD chromosome; *TD*, inverted sequence of the EF chromosome. (After Lewontin and White 1960.)

	Chromosome CD			
Chromosome EF	ST/ST	ST/BL	BL/BL	Total
ST/ST	7(7)	100(93)	324(331)	431
ST/TD	3(3)	22(31)	118(109)	143
TD/TD	0(0)	4(2)	6(8)	10
Total	10	126	448	584

equilibrium in the absence of selection does in fact contain these four possible combinations in precisely the frequencies listed above.

In a randomly mating population, let the frequencies of AB-, Ab-, aB-, and ab- bearing chromosomes equal f_1, f_2, f_3, and f_4 (not necessarily the expected equilibrium frequencies). These chromosomes can exist in ten diploid combinations: AB/AB, AB/Ab, AB/aB, AB/ab, Ab/Ab, Ab/aB, Ab/ab, aB/aB, aB/ab, and ab/ab. The gametes produced by eight of these ten genotypic classes are not affected by recombination; individuals homozygous at one or both loci produce the same gametes (in respect to these two loci) whether recombination does or does not occur. In the case of the double heterozygotes, recombinant chromosomes differ from nonrecombinant parental ones. If the total recombination frequency equals X, the gametes produced by the double heterozygotes will be

Genotype	Noncrossovers		Crossovers	
	AB	ab	Ab	aB
AB/ab	$.5 - .5X$	$.5 - .5X$	$.5X$	$.5X$
	Noncrossovers		Crossovers	
	Ab	aB	AB	ab
Ab/aB	$.5 - .5X$	$.5 - .5X$	$.5X$	$.5X$

The frequency of a given crossover chromosome newly produced within the population is given by the product of its frequency among gametes of the appropriate double hybrid and the frequency of the hybrid within the populations. Thus, as a result of recombination in AA/ab individuals, new Ab chromosomes arise with a frequency equal to $(0.5X)$ $(f_1 f_4)$. However, as the result of recombination within Ab/aB individuals, Ab chromosomes are lost with a frequency equal to $(0.5X)$ $(f_2 f_3)$. At equilibrium, the gain and loss of Ab chromosomes must be equal; therefore, $(0.5X)$ $(f_1 f_4) = (0.5X)$ $(f_2 f_3)$, or $f_1 f_4 = f_2 f_3$. The same conclusion would have been reached whichever of the four chromosomal types had been chosen as an example.

From the definitions of f_1, f_2, f_3, and f_4 the following relations hold:

$$f_1 + f_2 = p \quad \text{or} \quad f_1 = p - f_2$$
$$f_3 + f_4 = q \qquad\qquad f_3 = q - f_4$$
$$f_1 + f_3 = r \qquad\qquad f_3 = r - f_1$$
$$f_2 + f_4 = s \qquad\qquad f_4 = s - f_2.$$

This tabulation contains expressions for f_2 in terms of f_1 and f_4. A further expression relating f_2 to f_3 can be obtained as follows:

$$f_1 + f_2 = p$$
$$f_1 + f_3 = r$$
$$f_2 - f_3 = p - r$$
$$f_3 = r - p + f_2.$$

By substitution into $f_1 f_4 = f_2 f_3$ we obtain

$$(p - f_2)(s - f_2) = f_2(r - p + f_2)$$

or, upon solving:

$$ps = f_2(r + s) = f_2.$$

That is, the equilibrium frequency of chromosome Ab equals ps, the product of the frequency of A times that of b. Despite their linkage, A and b are associated at random in an equilibrium population. Similar calculations would reveal that $f_1 = pr$, $f_3 = qr$, and $f_4 = qs$. Alleles, whether linked or not, are expected to be associated at random within populations of cross-fertilizing individuals; this expectation can be thwarted by absolute linkage or by an event such as differential selection that prevents the attainment of equilibrium frequencies. The study of linkage disequilibria is possible in analyzing allozyme (electrophoretic) variation; Hedrick et al. (1978) have reviewed multilocus systems and their importance in evolutionary studies.

Assumptions Underlying the Hardy-Weinberg Equilibrium

The Hardy-Weinberg equilibrium is based on a number of assumptions. Population samples are compared with theoretical expectations to see if there is reason to believe that these assumptions do not hold. The tabulated data presented in this chapter suggest that observations often do *not* fit Hardy-Weinberg expectations. To some, this suggests that the Hardy-Weinberg equilibrium is a useless theoretical concept. Actually, the reverse is true: deviations from a theoretical expectation identify problems worth pursuing; they tell us that one or more of the assumptions underlying our calculations are not true. These assumptions, which are built into the checkerboard of figure 5.1, are as follows:

- mating occurs at random between individuals of various genotypes,
- migration of individuals into or from the population does not occur,
- selection in the form of differential survival or fertility of the different genotypes does not occur,
- mutation of one allele into another does not take place, and
- sampling errors resulting from the finite size of the population do not exist.

These assumptions are, of course, unrealistic: mating need not occur at random; migration (as we saw in discussing dispersal) does occur; selection, as some of the data have already suggested, does occur; mutation is a well-known genetic phenomenon; and, all populations must be finite—some, in fact, can be surprisingly small. The following chapters consider, one by one, the consequences of these ubiquitous exceptions to the Hardy-Weinberg assumptions.

FOR YOUR EXTRA ATTENTION

1. Data on allozyme variation in a Colorado population of ponderosa pines (Mitton et al. 1979) serve as a means for practicing the calculation of gene frequencies and expected genotypic (zygotic) ones. (The alleles have been identified here only as 1, 2, and 3.)

	Genotypes						
	11	12	22	13	23	33	Total
Peroxidase	0	0	11	0	36	4	51
Esterase-A	1	13	13	4	17	3	51
Esterase-B	24	7	0	9	0	0	40
Phosphoglutamase-A	0	4	47	0	0	0	51
Phosphoglutamase-B	2	6	38	0	5	0	51

2. Comment on the following statement once made in respect to a sample consisting of 480 *AA* individuals, 140 *Aa*, and 5 *aa*: "The error in the calculation of Hardy-Weinberg expectations is great when one homozygous class includes only five specimens." Show that the error is no greater, for

example, than it would have been if calculated from a sample consisting of 550 *AA*, 0 *Aa*, and 75 *aa* individuals.

3. Fujio et al. (1979) analyzed the variants (*A* and *B*) at the catalase locus in the Pacific oyster (*Crassostrea gigas*). What conclusions can (and cannot) be based on the following data?

Locality	No.	Zygotes			Allele Frequency	
		AA	*AB*	*BB*	*A*	*B*
Oudo	120	5	103	12	.47	.53
Hamanako	80	2	73	5	.48	.52
Mangokuura	507	48	346	113	.44	.56

4. The following data (Valenzuela and Harb 1977) list (1) the ABO blood-group (phenotype) frequencies among blood *donors* at a large public hospital in Santiago, Chile, and among blood *receptors* at a private clinic in the same city, and (2) the blood-group frequencies of matched mother-child pairs taken from records in these same two hospitals. Draw as many conclusions as you can concerning Chileans living in Santiago.

Hints: (1) Patients at the public hospital tend to be considerably poorer than those at the private, more expensive clinic. (2) Pool the data on donors and receptors and test for goodness-of-fit to the Hardy-Weinberg equilibrium. Are the results as you expected?

	Blood type				
	O	*A*	*B*	*AB*	*Total*
Donors	1,511	696	215	50	2,472
Receptors	472	347	94	20	933
Hospital					
Mothers	589	279	83	17	968
Children	592	275	88	13	968
Clinic					
Mothers	1,210	856	235	60	2,361
Children	1,187	884	240	50	2,361

5. Explain why Haldane (1965) could claim that, if 2 percent of all babies are homozygous for an unwanted recessive allele, this would only mean that 8 percent of intended marriages would be contraindicated. (A contraindicated marriage is one that might produce one or more homozygous babies.)

6. Johnson et al. (1966) studied the allozyme variation in two strains ("dark" and "light") of *D. ananassae* captured on each of two Samoan islands. Three alleles (fast, medium, and slow) were detected at the esterase-C (*Est*-C) locus. A portion of their data is summarized as follows:

Strain	Island	Frequency*		
		Fast	Medium	Slow
Light	Tutuila	.12	.73	.15
	Upolu	.11	.89	0
Dark	Tutuila	.84	.14	.02
	Upolu	.67	.33	0

*Sample sizes were between 50–100 genomes; for ease of making calculations, assume that the number equals 100.

What inferences can be drawn from these data concerning (1) the genotypic constitution of populations, (2) the differences between islands, and (3) the differences between strains? Might these strains represent different species?

7. The "sign-test" was introduced in the material (item 3) at the end of chapter 3. Use it in examining the following case:

Gottlieb (1977) studied a number of allozyme loci in an obligate cross-fertilizing plant (*Stephanomeria exigua*) in an unsuccessful attempt to find genetic differences between large and small individuals. In all, 21 Hardy-Weinberg expectations were computed; the expected number of heterozygotes exceed the observed number in 18 of the 21 tests. Is this number significantly greater than might be expected by chance?

8. Carson (1969) reported that in *D. disjunata*, one of the many endemic Hawaiian *Drosophila* species, three gene arrangements (A, B, and C) are known for the fourth chromosome; their frequencies are 0.500, 0.342, and 0.158. No C/C homozygotes were found among 73 larvae examined. What proportion of samples of size 73 would be expected to lack C/C homozygotes?
Hint: see table 1.2.

9. Baker (1975) studied the allozymes at four closely linked esterase loci in *D. montana*; at each locus there is a null allele (O) and an active allele (A). Calculate the expected numbers of chromosomes bearing each of the 16 possible combinations listed below (the number of chromosomes has been adjusted in this example to 500 in order to ease the calculations somewhat):

Locus				No. of Chromosomes
1	2	3	4	
0	0	0	0	3
A	0	0	0	7
0	0	A	0	32
0	A	0	0	37
0	0	0	A	25
A	A	A	0	11
A	0	A	A	4
0	A	A	A	4
A	A	0	A	1

Locus				No. of Chromosomes
1	*2*	*3*	*4*	
A	A	A	A	0
A	A	0	0	66
0	0	A	A	96
0	A	A	0	153
A	0	0	A	30
A	0	A	0	19
0	A	0	A	12
Total				500

If observed and expected frequencies do not agree, pinpoint excesses and deficiencies.

Baker believes these four loci to be the result of two sequential duplications of a single ancestral locus; for a review of duplicated genes see MacIntyre (1976).

10. Smith and Zach (1979) measured the beak depth and tarsus length of over 100 pairs of parental song sparrows, *Melospiza melodia*, and of their offspring. The results resemble those depicted in figure 5.3; for each measurement, the value for offspring increases with that of the average of its parents. What bearing do these observations have on evolutionary theory?

11. Anderson et al. (1979) have reported that inversion frequencies of adult male *D. pseudoobscura* in Mexican populations frequently differ significantly from inversion frequencies found among larval offspring of contemporaneous females. Older reports (see Anderson et al. for references) reveal the same: captured females produce offspring that differ in gene frequency from males captured simultaneously.

Why does this discussion dwell on *males* versus *offspring* of wild-caught females? How divergent must gene frequencies of males and females be before (in respect to a given sample size) the larval genotypes differ significantly from Hardy-Weinberg expectations? What bearing might the observations of Anderson et al. have on the mating ability of males?

12. An examination of figure 5.1 reveals that a basic assumption of the Hardy-Weinberg equilibrium is the 1 : 1 segregation of two unlike alleles in the gametes of heterozygous individuals. This assumption is based on Mendel's law of segregation; the abandonment of this law leaves no precise ratio as an alternative. Often, however, organisms—even diploid organisms—behave as if they were unaware of Mendel and his experiments.

Either by means of a checkerboard or by algebraic calculations, allow the heterozygote to produce gametes in other than 1 : 1 ratios, and see how the zygotic frequencies among the progeny compare to those among parents. For a concise account of this problem, see Crow (1979).

Migration

PREVIEW: Among the assumptions underlying the Hardy-Weinberg equilibrium is that which says: no individuals enter or leave the population. This chapter examines the consequences of migration at the theoretical (algebraic) level, in reference to laboratory populations, and in respect to field observations. Migration is closely related to dispersal (chapter 2). Furthermore, because migrant individuals must enter into the breeding structure of a population in order to affect it, migration is also closely related to mating systems (chapter 7).

OUTSIDE THE LABORATORY, populations are never closed; individuals always enter and leave. Only in the large sense of a species inhabiting a remote area—such as the progenitors of Darwin's finches following their arrival on the Galapagos Islands—can immigrant individuals be neglected. In such large systems, however, the species eventually subdivides into local populations from which emigrants depart and to which immigrants come from elsewhere. The continual movement of migrant individuals genetically unites the many local populations of a species; these individuals are the links that cause all members of a species ultimately to share a common pool of genes. In the absence of migrants, carrying their genetic wares and offering these to their more sedentary cousins, even cross-fertilizing species would become differentiated into independent populations, as are self-fertilizing plants and clonal or asexual organisms.

In the present chapter, we shall examine some of the algebra of migration and consider some illustrative examples. Because migration and dispersal are intertwined, we shall touch more than once on dispersal. (Migrant individuals are those which enter a population and take part in its reproductive activities; dispersal refers to the movement of individuals, seeds, spores, or even pollen. Dispersal is a prerequisite for migration, but can occur without the dispersed individuals becoming migrants.) Finally, even though it involves selection (a topic we are not yet prepared to discuss), we shall consider the fate of immigrant genes in populations.

The Algebra of Migration

Migration, like other aspects of population genetics, can be treated formally without reference to particular organisms. Such is the nature of Figure 6.1. Two populations are shown at a time prior to the onset of migration; then, one of these populations with its new immigrants is shown after migration has stopped.

Suppose that the frequencies of the alleles A and a in population 1 are p_1 and q_1, while the corresponding frequencies in population 2 are p_2 and q_2. In each population, $p + q = 1.00$. Suppose, too, that following migration, a fraction m of population 1 are newly arrived migrants and $1 - m$ are natives. What is the frequency of A in the altered population?

An average is obtained by summing the products of the values to be averaged times the frequency with which each occurs (page 75). The final (or average) frequency of A ($= p_f$) in population 1 equals

$$mp_2 + (1 - m)p_1$$

or

$$p_f = p_1 + m(p_2 - p_1).$$

If the *change* in gene frequency is represented as the difference obtained by subtracting the old frequency from the new (a nega-

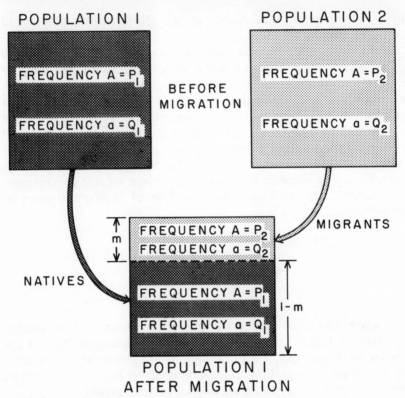

Figure 6.1 A diagrammatic representation of migration and its effect on gene frequencies. (After Wallace 1964.)

tive change if the new frequency is the smaller), then the change in the frequency of allele A, Δp, equals $m(p_2 - p_1)$.

Because $\Delta p = m(p_2 - p_1)$, the effect of migration depends upon (1) the proportion of migrants and (2) the difference in the gene frequencies of the two populations. If neither of these is zero, migration must alter gene frequencies; if both are small, the effect of migration can be very small, indeed.

The equation $p_f = p_1 + m(p_2 - p_1)$ can be used to obtain any of the four items (p_f, p_1, p_2, or m) if the other three are known. For example, p_f can be calculated if p_1, p_2, and m are known.

A more general problem, however, is that in which the three gene frequencies are known but m, the proportion of migrants, is not. In this case, the equation can be rewritten

$$m = \frac{p_f - p_1}{p_2 - p_1}.$$

We shall use the equation in this form in examining an example of migration in human populations.

Racial Intermixture

An analysis of migration, in the form of the intermixture of American Negroes and whites, has been made by Glass and Li (1953). In the simplified account given here, the Negro population is designated population 1, into which migrants move from population 2, the white population of the United States. The American blacks, then, correspond to population 1 after migration has occurred. Information on the gene frequencies in populations of African Negroes during the eighteenth and nineteenth centuries is lacking; in place of these unknown frequencies, averages of gene frequencies existing today in appropriate tribes have been substituted.

A large number of gene loci could, in theory, be used in studying the intermixture of black and white races in the United States. In practice, some are unsuitable. The *MN* blood-group system which served to illustrate the Hardy-Weinberg equilibrium (table 5.2) cannot be used because gene frequencies at this locus in the two races are too nearly alike. The frequencies of still other genes—genes that will be considered later in this chapter—are subject to modification by the differential survival of various zygotic types; these survival patterns are not the same in the United States as they are in Africa. Hence, such genes cannot be used in studying migration.

A number of genes which seem to be suitable for a study of racial intermixture in the United States are listed in Table 6.1. Four of the alleles listed (R^0, R^1, R^2, and r) belong to the Rh blood group system. Two others (I^A and I^B) are members of the ABO blood group system. The remaining allele, t, is the recessive allele of the pair involved in the ability to taste phenylthiocarbamide (PTC); tt individuals are nontasters.

The three gene frequencies—p_1, p_2, and p_f—needed to estimate m, the proportion of migrants, are given in table 6.1. Frequencies (averages obtained from three studies on different African tribes) that represent the Negro population before migration are listed under p_1; frequencies observed in studies on American blacks are listed under p_f; and frequencies representative of

Table 6.1 Estimation of the proportion of migration that has occurred from the white to the black population of the United States. The genes upon which the estimates are based are identified under "allele" and "system." The frequencies of the alleles in African blacks, American blacks, and American whites are given as p_1, p_f, and p_2. Total migration and migration per generation are listed in m and m'. (After Glass and Li 1953.)

Allele	System	p_1	p_f	p_2	m	m'
R^0	Rh	0.62	0.45	0.03	0.288	0.033
R^1	Rh	0.06	0.15	0.42	0.250	0.028
t	PTC	0.18	0.30	0.55	0.324	0.038
r	Rh	0.22	0.27	0.38	0.313	0.037
I^B	ABO	0.17	0.13	0.08	0.444	0.057
I^A	ABO	0.15	0.18	0.25	0.300	0.035
R^2	Rh	0.06	0.10	0.15	0.444	0.057

the white population of the United States are listed under p_2. The substitution of these values into the equation given above leads to an estimate of m, the fraction of immigrants. The values listed in table 6.1, ranging from 25 percent to 44 percent, suggest that the fraction of immigrants is substantial. In estimating m, a fraction is used in which $p_2 - p_1$ is the denominator; consequently, the greater this difference, the more reliable is the estimate of m. Because the first three entries in the table appear to be the most reliable, 0.30 seems to be a reasonable estimate of m.

A graphic representation of the relationship between migration and the three gene frequencies (p_1, p_2, and p_f), and which illustrates some of the data in table 6.1, is shown in figure 6.2. The relationship between gene frequencies and migration is linear; therefore, points on the two vertical axes, p_1 and p_2, can be joined by straight lines. Horizontal lines representing p_f intersect the slanted lines at points corresponding to the amount of migration that has occurred. Consequently, although the relationships between p_1 and p_2 in the cases of R^0 and t are reversed, the gene frequencies observed among American blacks lead to nearly identical estimates of m. On the other hand, T. E. Reed (1969), in a study of the Duffy blood-group system (Fy^a), has shown that the extent of migration from the white into the black population depends upon the geographic locality within the United States; a number of cities and regions are identified in figure 6.2.

Migration is generally expressed as a rate; that is, as so many migrants per generation. The data listed in table 6.1 and illustrated in figure 6.2 involve changes in gene frequencies that have occurred over three centuries. To reduce the calculated value of m to a "per generation" value (m' in table 6.1), an estimate of the number of generations is needed. Ten generations seems to be a

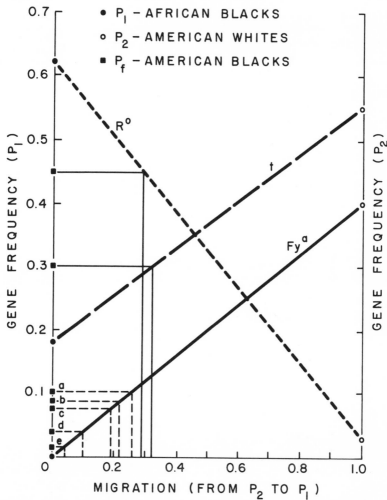

Figure 6.2 Graphic representation of the consequences of migration. Heavy lines connect the frequencies in African and white American populations of the genes R^0 (Rh blood group system), t (PTC system), and Fy^a (Duffy blood-group system). The light solid lines indicate (vertical axis) the frequencies of R^0 and t in the black population of Baltimore, and (horizontal axis) the total migration into the black population that would account for the observed frequencies. The light dashed lines indicate the frequencies of Fy^a in the black populations of (a) Detroit, Mich., (b) Oakland, Calif., (c) New York, N.Y., (d) rural Georgia, and (e) Charleston, S.C. The data on Fy^a reveal that the degree of admixture is not uniform. (After Glass and Li 1953; and Reed 1969.)

reasonable estimate for the time during which blacks and whites have intermingled within the United States.

If the proportion of natives after one generation of migration equals $1 - m'$, that remaining after two generations would be $(1 - m')^2$, and that remaining after ten generations, $(1 - m')^{10}$. Consequently,

$$(1 - m')^{10} = 1 - m,$$

or

$$10 \log (1 - m') = \log (1 - m).$$

Alternatively, this can be written

$$e^{-10m'} = 1 - m.$$

Either way, if $m = 0.30$, $m' = 0.035$; thus, it seems that 3 to 4 percent of the alleles in the American black population have come each generation (as an average) from what is known as the white population of the United States.

Migration and Dispersal

For most organisms, the dispersal of individuals is an essential part of gene migration. Human beings, through legal or social procedures, have a capacity to erect mating barriers between groups that are otherwise intermingled at work and at play. For this reason we can speak of intermixture between groups of persons as migration even though no actual travel may be involved. The religious and social castes of India (until recently, at least) represented isolated populations between which migration (intermixture) occurred at low frequencies. Ordinarily, the members of a species that inhabit a small geographic locality do interbreed with one another; immigrants are migrant individuals that have arrived from outside a locality. The arrival of immigrants depends, in turn, upon their ability to disperse from their original homeland.

To assume (as many in the past have done) that individuals who are capable of mating will in fact do so at any point within the area in which they are observed could be a grave mistake.

Crowded as sidewalks are, few human marriages are consummated on them. Similarly, many animals from insects to birds and mammals have *leks*, specific areas within which mating occurs. In some instances breeding areas correspond with feeding areas (male and female *Drosophila* are seen mating on baited food cups); in other instances (including some Drosophila species), mating occurs in areas quite distinct from feeding sites. The existence of specific breeding sites subdivides a geographical locality to an extent not recognized when it was believed that physical encounters were tantamount to mating itself.

In chapter 3, lethal chromosomes were shown to constitute a sizable fraction of all chromosomes of Drosophila populations. Random combinations of lethal chromosomes are, for the most part, not lethal to the flies that carry them; in such combinations, the lethal genes of the two chromosomes occupy different loci and, therefore, are not allelic. A few combinations of lethal chromosomes, however, do kill their carriers; in these cases, the lethal genes presumably are allelic and occupy the same locus. If the tested chromosomes are from localities hundreds of miles apart, the low frequency of allelism is taken as a measure of the number of loci at which lethal mutations occur. If lethals of independent origin can occupy the same locus, the number of loci at which lethals can occur must be limited.

When lethal chromosomes are obtained from geographically restricted areas, the frequency of allelism is greater than that observed when collection sites are more distant. The probability of allelism ascribable to the finite number of loci at which lethals can occur is present, of course, in all tests. To this low, constant frequency of allelism, however, is added a second cause of allelism—identity by common descent. In restricted localities, two lethal chromosomes may carry identical lethals because each is a direct descendant by recent replication of a single ancestral lethal gene. As a consequence, the frequency of allelism is higher among lethal chromosomes taken from a single locality than it is among those collected at remote distances.

The dependence of the frequency of allelism of lethals on distance was reported by Dobzhansky and Wright (1941) and Wright et al. (1942). These studies were based on flies (*D. pseudoobscura*) captured at various places on Mount San Jacinto in Southern California. Collections were made at three localities—Andreas Canyon, Keen Camp, and Piñon Flats—separated from one another by ten to twenty kilometers. At each locality there were two or more collecting stations—Andreas A and B, Keen A, B, C, D, and E, and Piñon A and B—separated by distances of one to two kilometers. Each collecting station consisted of a series of small baited cups placed in a circle roughly one hundred meters in diameter.

The results of the tests of lethals from Mount San Jacinto are presented in table 6.2. Their interpretation is clear; the smaller the area from which lethal chromosomes are obtained, the higher the probability (frequency) of allelism. The bottom part of table 6.2 shows that *time* can be substituted for *distance* in the preceding statement: the more nearly simultaneously two lethals are taken from a restricted locality, the more likely they are of being allelic. The observed differences in the latter case are not statistically significant; they merely agree with expectation.

The data on the dispersal of flies (chapter 2) allow us to speak more precisely about distances and the allelism of lethals, especially that portion of allelism ascribable to common descent. We saw that the "lifetime" dispersal of flies from a point of release was such that the logarithm of the number of flies recaptured at various distances decreases more or less linearly with the square root of distance (table 2.1; figure 2.7). Because lethals are carried from place to place by dispersing flies, it follows that the logarithm of the frequency of allelism (common descent, only) might also decrease more or less linearly with the square root of distances. (A recent, thorough mathematical analysis of this problem by Yokoyama [1979] reveals that the decrease may be even more rapid than the square root of distance suggests.)

Experiments to test the suspected relationship between distance and the allelism of lethals have been made using *D. melanogaster* (Wallace et al. 1966; Wallace 1966b); the results agree well

Table 6.2 The dependence of the probability of allelism of lethals upon distance and time. (After Wright et al. 1942.)

	Total Tests	Allelic Combinations	Frequency of Alleles
Within station	2,068	44	0.0213
Between stations, within locality	2,284	20	0.0088
Between localities	706	4	0.0057
Between regions	6,294	26	0.0041
Within station (simultaneous)	594	15	0.0253
Within station (different times)	1,474	29	0.0197
Between stations, within locality (simultaneous)	691	9	0.0130
Between stations, within locality (different times)	1,593	11	0.0069

with the expectations. Through the use of a modified *CIB* technique that controls both major autosomes, a number (95) of lethals were recovered from 119 tests; the frequency of lethal "genomes" (second and third chromosomes tested simultaneously) was approximately 80 percent. The flies used in this experiment were collected near Bogotá, Colombia, at four trapping sites that were spaced linearly at 30-meter intervals.

The lethals obtained at these four sites were intercrossed both within and between collecting sites. Not all possible intercrosses were made; those that were made are summarized in table 6.3. Ostensibly, the probability of allelism declines with increasing distance. Although the numbers are not large enough to give significant results with a Chi-square test, the slope of the regression of the logarithm of these frequencies on the square root of distance (see figure 6.3, solid circles) does differ significantly from zero (zero slope would mean that no relation exists between allelism and distance). The observed decrease in the frequency of allelism with increasing distance, even with distances as small as 90 meters, is probably meaningful; the data suggest that lethals spread from place to place much as the flies themselves—i.e., migrant flies do enter the breeding structure of the populations they enter.

The relation between the allelism of lethals and distance between collecting sites has been reexamined by Paik and Sung (1969), whose data are also presented in table 6.3 and figure 6.3 (open circles). These data also show a decrease in allelism over short distances (30–180 meters).

Table 6.3 Observed frequencies of allelism between lethal genomes of *D. melanogaster* collected nearly simultaneously at sites separated by various distances. (After Wallace 1966b; Paik and Sung 1969.)

Distance	Author	Number of Tests	Allelic Combinations	Frequency
0 m.	Wallace	629	29	0.0461
	Paik	3,196	112	0.0350
30 m.	Wallace	713	26	0.0365
	Paik	4,094	127	0.0310
60 m.	Wallace	586	19	0.0324
	Paik	3,509	84	0.0239
90 m.	Wallace	327	9	0.0275
	Paik	2,768	51	0.0184
120 m.	Paik	1,780	28	0.0157
150 m.	Paik	1,068	15	0.0140
180 m.	Paik	232	3	0.0129
Average	Wallace	2,255	83	0.0368
	Paik	16,647	420	0.0252

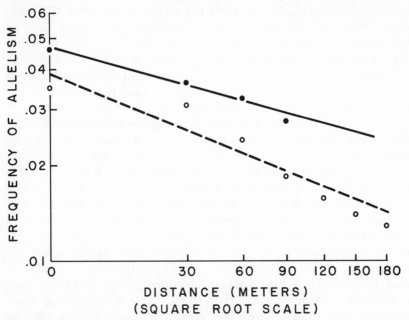

Figure 6.3 The probability that two lethal genomes are allelic as a function of the distance between the trapping sites (*D. melanogaster*) at which they were obtained. (Solid circles: after Wallace 1966b; open circles: after Paik and Sung 1969.)

The Diffusion of *Orange* in Populations of *D. pseudoobscura*

The analysis of the allelism of lethals is only one of several techniques by which the diffusion of genes through populations can be studied. The artificial introduction of a mutant gene and the analysis of its spread represent an alternative procedure. Dobzhansky and Wright (1947) carried out an experiment of this sort in *D. pseudoobscura* at Mather, Calif.

The experiments at Mather involved a study of the dispersal of flies as well as an analysis of the diffusion of the mutant gene, *orange* (eye color), within the local population. An analysis of the lifetime dispersal of some 3,800 orange-eyed flies yields results much like those presented in table 2.1. The logarithm of the total number of flies recaptured daily over a 6-day period decreases linearly with the square root of distance. The slope of the regression equals -0.109; the corresponding slope for the earlier data was -0.102. Both slopes are highly significant.

Following the analysis of the dispersal of released flies over a period of six days, approximately 25,000 additional orange-eyed flies were released at the center of the trapping field over the following twenty days (July 23 to August 11, 1945). The release was prolonged to minimize local, temporary overcrowding. Almost immediately (August 10–16), trapping of the released flies at 250-meter intervals up to 1,500 meters from the point of release was begun; the bulk of the survivors were found at the point of release. One or two orange-eyed flies were captured per day, however, even as far from the point of release as 1,500 meters. (In a recent review of release-recapture experiments of *Drosophila*, Endler [1979] has shown that a correlation exists between dispersal distances and the distances between traps that are used for recapturing the released flies; these traps are apparently among the most attractive spots in the experimental field.)

The recapture of orange-eyed flies was continued from August 22 through September 5, a period covering 11–44 days from the prolonged period of release. During this time both survivors and newly hatched orange-eyed flies were found among the mutants recovered. The genotypes of the phenotypically wildtype males were tested by mating them individually with virgin *orange* females; *orange* genes were found among the males tested. Gene frequencies for the mutant *orange* decreases less rapidly with increasing distance from the release point than do the frequencies of orange-eyed flies; the latter are, of course, determined by the square of the former.

Finally, nearly one year following the original release, several thousand flies were captured in the experimental field and examined; no orange-eyed flies were found. Again, the genotypes of the phenotypically wildtype males were determined by mating each one to *orange* females. *Orange* alleles were still present in the field, and still most common at the point of release. Despite the release of nearly 30,000 flies, however, the frequency of *orange* after one year was only 0.5 percent; before release the "control" level of *orange* in this area had been measured; it was 0.2 percent.

The migration of genes in plant populations is frequently accomplished by pollen dispersal—sometimes transported by wind, sometimes carried by insects or birds. In either case, the distribution *patterns* (not necessarily actual distances) are similar: a rapid decrease in the number of transported grains over short distances; a further, slow decrease over much longer distances. Figure 6.4 shows the flight distances of pollinators in colonies of *Liatris*

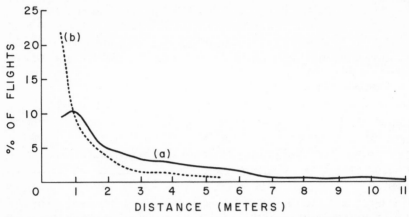

Figure 6.4 Curves illustrating the dependence of pollinator flight patterns on the density of *Liatris aspera* Michx. Plants per square meter: (a) 1.0, and (b) 11.0. (After Levin and Kerster 1969.) Reprinted by permission of the University of Chicago Press from Kerster, *Amer. Nat.* (1969), 103:61-74; © 1969 University of Chicago.

aspera of various densitites (Levin and Kerster 1969); as one might expect, the denser the stand, the shorter the average flight. Pollination by hummingbirds follows a similar pattern (Linhart 1973).

The Fate of Mutant Genes

In calculating the intermixture of Negro and white races in the United States, we tacitly assumed that the alleles followed were not subject to appreciable selection through the differential survival or fertility of their carriers. This assumption applied to alleles both within the race from which they came and within that to which they migrated.

The data on the introduction of the mutant allele *orange* into a natural population of *D. pseudoobscura* differ from those on racial intermixture in man. At least, they seem to do so. In describing racial intermixture, a model involving migrant individuals was used in analyzing the changing frequencies of genes. The changes were caused by the movement of migrant genes. In the case of *orange* in Mather, the frequency did not remain constant. It is known from matings that were observed that the mutant gene was introduced into the breeding population; nevertheless, the frequency of *orange* dropped considerably in the year between its release and the final test.

A similar elimination of a mutant gene would have been observed in the study of Negro-white intermixture had the study been based on the allele responsible for sickle cell anemia. This allele, Hb^S, is common in populations of African Negroes inhabiting malarial regions of that continent; it is rare in all other populations. Its rareness in other populations stems from the severe (often fatal) anemia suffered by homozygous individuals ($Hb^S Hb^S$). Despite its severe effect on homozygotes, the gene is common in malarial regions because its heterozygous carriers tolerate malaria better than individuals homozygous for the normal allele, Hb^A. Malaria is not a serious disease in the United States, so that Hb^S allele carried by many of the original black immigrants would have decreased in frequency because of the early death of $Hb^S Hb^S$ homozygotes. An estimate of racial intermixture based on the present frequency of Hb^S among American blacks would have differed considerably from those listed in table 6.1 The admixture of genes from the white population would have been grossly overestimated. Obviously, neutral or nearly neutral alleles are needed in studying migration between populations.

A number of experiments have been made in which mutant genes have been introduced into laboratory populations of *D. melanogaster*. Carson (1961a), for example, introduced single females heterozygous for the recessive mutations *sepia*, *rough*, and *spineless* (and for the corresponding wildtype alleles) into populations that (1) were homozygous wildtype at these three loci or (2) were homozygous for the three mutants. Rapid changes took place in gene frequencies at all three loci following both types of "contamination." Each of the three mutations, after nearly two years following both types of introductions (single mutant into wildtype; single wildtype into mutant), reached similar final frequencies. Both types of contaminated populations approached what seemed to be the same equilibrium frequencies at each locus.

A *sepia* allele recovered from wildtype *D. melanogaster* captured in North Carolina was introduced (2 *sepia* pairs plus 4 wildtype pairs) into populations of various wildtype strains; its average fate in each case was clearly dependent upon the strain (Wallace 1966c). The wild strains of this study were originally from Bogotá, Barcelona, California, and North Carolina. The frequencies of *sepia* homozygotes and of the *sepia* allele (gene frequency) after one year in these populations are listed in table 6.4.

Table 6.4 Observed frequency of *sepia* homozygotes and the calculated frequency of the *sepia* allele in four "populations" involving wildtype flies of different geographic origins. Each "population" consisted of thirty cultures that were maintained by the mass transfer of adults each generation. The observations reported here were made one year after the cultures were started. (Wallace 1966c.)

Source of Wildtype	Frequency of sepia Homozygotes Observed, %	Number of Individuals Examined	Calculated Frequency of sepia Genes, %
North Carolina	0	7,731	0.4
Bogotá	16.5	8,766	37.5
Barcelona	2.8	8,925	20.2
California	38.5	8,380	62.2

That the final frequency of *sepia* in each instance depends upon the source of the *wildtype* flies is quite clear.

The *sepia* mutation also provides a means for following the vagaries of a migrant allele arriving in seemingly similar populations. Fifty populations of wildtype *D. melanogaster* from Riverside, Calif., were started in the laboratory; in each generation all the progeny flies in each old culture were transferred as parents to a new culture. By a random process one or more *sepia* flies were allowed to enter a few wildtype cultures each generation; the majority of cultures at any transfer received no mutant migrants. The results of these experiments are shown in table 6.5 (Wallace 1979a). Among the ten populations chosen here as illustrative examples, the actual number of migrant flies ranged from 2 to 11; the proportion of sepia-eyed flies (i.e., homozygotes) after nearly two years ranged from 0 percent to nearly 90 percent. The corre-

Table 6.5 Variation in the fate of the *sepia* mutant allele when introduced into various laboratory populations of wildtype *D. melanogaster* from Riverside, Calif. The table lists events that occurred in 10 of 50 populations that were studied. Time of arrival of "first immigrant" is given in terms of eight 4-generation intervals into which the study was divided; "total immigrants" are the total number of *sepia* migrants entering the population during all eight intervals; "final frequency" is the frequency of *sepia* homozygotes in the last (34th) generation. The total number of *sepia* flies immigrating into any population is less than 1/1000th of the total flies of that population. (Wallace 1979a)

Population	First Immigrant	Total Immigrants	Final Frequency (%)
1	1	2	0
2	1	6	9.5
4	2	9	10.7
8	1	9	12.7
12	1	11	55.5
13	2	6	38.0
15	1	6	87.5
29	3	4	0
31	1	9	50.0
36	1	9	2.2

lation between the number of migrants and the final frequency of sepia-eyed flies is far from perfect: among the four populations that received nine migrants, *sepia* flies had final frequencies ranging from 2 percent to 50 percent; the two populations with six migrants had final frequencies of about 10 percent and 90 percent. Notice the contrast between the release of *orange* mutants into wild populations of *D. pseudoobscura* and the introduction of *sepia* mutants into laboratory populations: carriers of the first mutant gene were eliminated, those of the second were retained; the answer to such discrepancies is to be found in chapters 10 and 11—chapters that deal with selection.

Clines and Introgressions

Migration should not be thought of as a one-time event; on the contrary, nearby populations continually share genes by means of wandering individuals. Such constant sharing can be expected to mold the patterns of gene frequencies that characterize the assemblage of local populations that occupies a geographic region. Earlier (page 124), we saw that if both m and $(p_2 - p_1)$ were small, the effect of migration on gene frequency could be small, indeed. That statement can be extended somewhat: neighboring populations that exchange the largest proportions of migrant individuals are those that usually differ least in gene frequencies; populations that differ considerably in gene frequency are usually so remote that they exchange virtually no migrants. Consequently, if for some reason or another, populations occupying two geographic regions differ considerably in gene frequency (say 100 percent A in one, 100 percent a in the other), intervening populations should, through an exchange of migrant individuals, exhibit a gradient (or *cline*) of intermediate frequencies.

Endler (1973) has studied the effect of gene flow on population differentiation in laboratory cultures of *D. melanogaster*; some of his results have been summarized in figure 6.5. The heavy line in the figure represents the distribution of frequencies of the mutant, *Bar*, in a series of cultures of *D. melanogaster*; this distribution was maintained by applying selection of graded intensities against the wildtype (non-*Bar*) flies in these cultures. The dashed line represents the frequencies of the mutant *Bar* in cul-

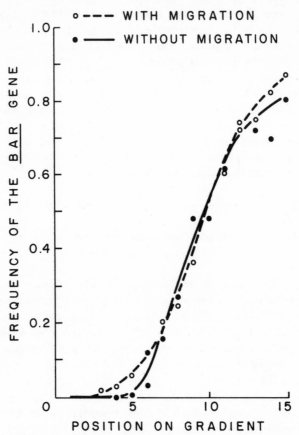

Figure 6.5 Curves illustrating the extremely slight effect of migration between neighboring populations on an artificially generated cline in a series of laboratory populations of *D. melanogaster*. (After Endler.) Reproduced with permission: from *Science* (1973), 179:243–50; © 1973 American Association for the Advancement of Science.

tures maintained as the earlier ones but in which each culture received 40 percent migrant individuals each generation (20 percent from the culture on one side, 20 percent from the one on the other). Clearly, the migrant individuals have had only a slight effect on the overall pattern of frequency distribution. The explanation lies in the cancelling effect of the two sets of migrants. Even though the proportions were large (20 percent from one culture, 20 percent from another), the *average* frequency of *Bar* in the two sets of migrants was nearly the same as that in the recipient culture; consequently, the migrants changed the recipient population surprisingly little.

Migration need not be limited to an exchange of individuals between the local populations of one species; it may lead to the introduction of genes from one species to another—interspecific introgression. A species is defined here as a group of individuals that for one or more reasons (excluding spatial isolation) does not regularly exchange genes with any other similar group.

That different species do not exchange genes is, in most cases, a true description of events in nature. Nevertheless, because organisms evolve, because isolating mechanisms also evolve, and because evolution is a gradual process, exceptions are occasionally found. Such exceptions, together with methods by which they can be studied, have been described by Anderson (1949).

An excellent example of introgression resulting in the formation of a cline involves two species of towhee in Mexico, *Piplio erythrophthalmus* and *P. ocai* (see Sibley 1954). Members of these two species differ in plumage color in six areas of the body:

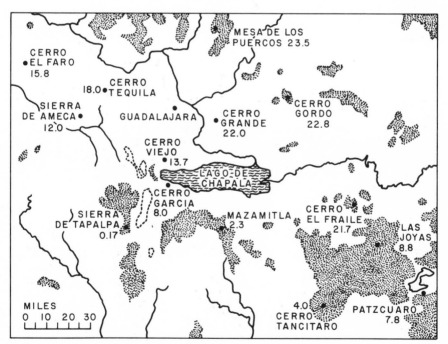

Figure 6.6 Detailed map of southwestern Mexico showing (shaded areas) regions favorable for towhees. Numbers are hybrid indices as explained in the text. (After Sibley 1954.)

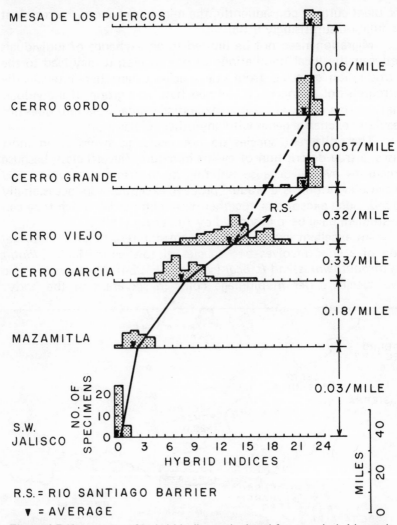

Figure 6.7 Histograms of hybrid indices calculated for some hybrid populations of towhees in Mexico. (After Sibley 1954.)

pileum, back- and wingspots, back, throat, flank, and tailspots. For each of these six areas, the color usually found in *P. ocai* was scored 0, while that which is characteristic of *P. erythrophthalmus* was scored 4. Intermediate colors were scored 2, while those displaced one way or the other were scored 1 or 3. The result is a scale extending from 0 to 24; members of *P. ocai* fall near the zero end, those of *P. erythrophthalmus* fall near 24. Individuals carrying mixtures of genes from the two species should tend to fall between the two extremes.

One area in southwestern Mexico studied by Sibley is shown in figure 6.6; a series of histograms representing the samples of towhees taken at seven sites within this area is shown in figure 6.7. The histograms show that in a number of localities the towhee populations have intermediate indices and, in addition, are highly variable. The genes of the two species are intermixed in these localities. Outside the narrow zone in which the two species meet, the "hybrid indices" take on values of 0 and 24, those of *P. ocai* and *P. erythrophthalmus*, respectively.

(The status of these two towhees as separate species has been questioned; experts will have to settle that matter. Whatever the outcome, the cline resulting from gene exchange through migrant individuals remains, and it is remarkably similar to those shown in figure 6.5.)

Concluding Remarks

A discussion of migration is a discussion of genes and not of individuals, even though genes are transported by individuals. A movement of individuals is not the same as that of genes. At the beginning and end of each school year and at each of the major intervening holidays, thousands of young adults of both sexes stream across the United States in a most complex pattern, clogging airports as they return to their campuses or travel to their favorite mountain or beach resorts. Despite the tremendous economic impact such migration has on local populations, the genetic consequences are surprisingly modest.

In closing this chapter, it may be worthwhile to refer once more to the results obtained by Carson (1961a) and those described in table 6.5, and to consider the bearing these results have in reference to the "bed-of-nails" concept of population structure illustrated in figure 2.12. Local populations, according to that figure and to the arguments upon which it was based, are more restricted and more inbred than most persons would suspect. Nevertheless, because of the common shape of dispersal patterns (the wide-brimmed witch's hat) migrant individuals, though low in number, do at times travel enormous distances. The local populations may resemble the wildtype populations into which only one or two foreign genomes were introduced (*sepia* in Table 6.5; *sepia rough spineless* in Carson's experiments): very few migrants may enter, but the genes they carry may at times increase in frequency with explosive speed.

FOR YOUR EXTRA ATTENTION

1. To see how small an effect migration might have on the gene frequency of a local population, consider the following example which involves the *Mdh*-2 and *Me* loci in *D. subobscura* populations at seven localities distributed over a 3,170 kilometer transect (see Pinsker and Sperlich 1979). What is the average change in gene frequency per kilometer?

Locality	Distance	Mdh − 2 (.96)	Me (106)
Sunne	0	.975	.972
Tubingen	1,280	.973	.935
Zernez	1,520	.975	.937
Formia	2,180	.959	.936
Ponza	2,250	.965	.931
Cinizi	2,860	.916	.923
Bizorte	3,170	.915	.916

Hint: With a pocket calculator, calculate the slope of the regression of frequency (Y) on distance (X). Otherwise, plot the data and calculate the slope of what appears to be the line of best fit.
Recalling that migrants would arrive in equal numbers from both sides of any population, what is the (approximate) net effect on gene frequency of migration within this cline?

2. Halkka et al. (1975) subjected populations of *Philaenus spumarius* (the spittlebug) displaying stable color polymorphisms to the introduction of alien phenotypes. In an exchange experiment between populations of different islands $\frac{2}{3}$ of population A was taken to P, from which $\frac{2}{3}$ of the local population was taken to A. The immediate effect of this transfer was that each population (A and P) took on the characteristics of the other. After four generations the pre-transfer frequencies were almost completely restored.

How can these changes in gene frequency be explained?

3. What do the following observations on *D. nigrospiracula* (Johnston and Heed 1976) suggest in respect to the migration of genes in *Drosophila* populations?

Newly arrived males (marked so they could be recognized) courted as many as 10–20 females while searching for a feeding site. Newly arrived females (similarly marked) were met by 5 to 10 courting males. Twenty mating pairs were captured: 4 of the 20 females were migrants (migrant flies made up only 7 percent of the population); 2 of the mating males were migrants.

Chapter

7

Patterns of Mating

PREVIEW: The Hardy-Weinberg equilibrium is based on a number of assumptions, including that of random mating. The patterns discussed in this chapter represent, for the most part, deviations from random-ness: like preferring like, like avoiding like, self- versus cross-fertilization, and other non-random patterns. Migrant individuals are strangers; consequently, patterns of mating are not unrelated to the genetic consequences of migration (chapter 6).

IN CALCULATING THE proportions of individuals of various genotypes expected under the Hardy-Weinberg equilibrium, random mating between individuals of the different genotypes was taken for granted; the assumption of random mating is implicit in the use of a "checkerboard" (figure 5.1) as a means for illustrating probabilities. In the following pages we shall examine this assumption more closely. The results of our efforts will disappoint those seeking neat generalizations: matings may occur at random in some respects, but not in others. In those instances when matings do not occur at random, like may prefer like (positive assortative mating), unlikes may prefer each other (negative assortative mating), rare phenotypes may have an advantage, or they may not. Furthermore, as both field data and laboratory data suggest, mating preferences are subject to change.

Much of the interest in mating behavior is centered on the reproductive isolation that accompanies speciation, not on the validity of the Hardy-Weinberg equilibrium. Nevertheless, mating behavior does affect the genetic composition of populations; therefore, we shall take this opportunity to discuss mating behavior, the effect of assortative mating (positive and negative) on gene and zygotic frequencies in populations, and the difficulty in separating these effects from others caused by natural selection. The discussion will be based on empirical observations; the algebra of assortative matings will not be presented (see Spiess 1977: 294ff).

The Problems of Mating: Introduction

Unthinking persons accustomed to modern societies may realize neither that lower organisms are confronted with problems nor that they have by and large found workable solutions to them. These persons have a great deal of leisure time; lower organisms do not. Most persons get their food through a system of monetary exchange; animals get their food through their own, immediate efforts. Human beings have means of communication unrivaled in complexity; most other organisms rely on sounds and smells that are effective only over relatively short distances.

The main problems in the life of most plants and animals are getting food (making it, in the case of green plants) and reproducing. Reproduction requires that a suitable mate be found, and that unsuitable ones be avoided. In this sense, a suitable mate is an acceptable member of one's own species who is of the opposite sex; an unsuitable mate is anything else. The bulk of all natural sounds—insect, bird, amphibian, and mammalian—which we hear are made in an effort to bring mates together, and to keep them together.

Even when a pair of individuals have met, mating is deferred until a courtship ritual has been performed. The courtship largely assures that the mating individuals are of the same species; otherwise, mating may lead to sterile offspring, or none at all. Flies of several *Drosophila* species may descend on the same fermenting fruit; courtships between the males and females of these different species must be terminated with a minimum of wasted time; those between members of the same species must terminate in successful matings. A species of insect that suffers a daily 10 percent mortality, and which has only one or two hours each day for moving about, must save its minutes; it has very little time to spare.

In many respects, the courtship of lower organisms is as complex and as stereotyped as the combination for the lock of a safe. Courtship begins with, let us say, action *A* on the part of the male, the female responds with *B,* whereupon the male is stimulated to do *C,* the female's response is then *D,* and so it goes until mating occurs. Averhoff and Richardson (1974) have shown that the sterility commonly encountered upon inbreeding (brother-sister matings) *D. melanogaster* for a number of generations need not be a functional sterility of either sex; rather, it may be caused by a failure of each sex to be stimulated to courtship by the other. Airborn odors drawn from cultures of other flies are enough to cause such inbred brothers and sisters to investigate and court one another, and eventually to mate.

A successful courtship requires not only that the proper signals be generated but that they be generated in the proper sequence. Different species of ducks use largely the same set of signals but each species has its own sequence according to which the signals (and responses) must be given (Lorenz 1958).

How, if life is so complex, can matings ever occur at random within any population? The fact is that they do not occur at random. They may, however, occur at random *in respect to* a given pair of alleles or a given phenotype. Human beings do not mate at random; this statement is obviously true in respect to phenotypic traits (skin pigmentation, for example) commonly associated with racial designations; it is less obvious but true nevertheless in respect to height; on the other hand, marriages are seemingly contracted at random in respect to physiological or biochemical traits that can be detected only by sophisticated techniques or elaborate physical examinations.

The Consequences of Assortative Matings

Matings can deviate in two obvious ways from randomness—mating individuals may be more similar (positive assortative mating) on the average than are corresponding pairs chosen by some random process, or they may be less similar (negative assortative matings). If the similarity in question has a reasonably simple genetic basis, positive assortative mating promotes homozygosity while negative assortative mating (to the extent that it involves matings of dissimilar homozygotes) promotes heterozygosity.

Assortative mating (positive or negative) alone has no effect on gene frequencies, only on the association of alleles in the formation of zygotes. Assortative mating may, however, be important in connection with natural selection. If we postulate that *AA* individuals of either sex prefer mates of type *aa*, and if the frequencies of these two genotypes are not identical, it may be necessary to admit as well that some individuals tend to remain without mates. Negative assortative mating, in other words, may imply differential fertility—a form of natural selection.

Similar problems involving natural selection arise in the case of positive assortative mating. The mating of similar homozygotes tends to increase their frequency at the expense of heterozygotes, much as does inbreeding. The fitness of one homozygote is often much lower than that of the other. Consequently, positive assortative mating tends to exaggerate the frequency with which these homozygotes are formed and, thus, the frequency with which they are exposed to the action of selection.

Random and Assortative Mating in Man

Marriages in man are or are not entered into at random depending on whether the phenotypic traits involved are obvious to all or are cryptic. An earlier statement was more precise: marriages are not contracted at random; in respect to certain cryptic traits, however, they appear to occur in a random fashion. The extreme case involves those characteristics which (for wrong reasons) have been associated with racial differences: inter-racial marriages, though no longer uncommon, are still a minority of all marriages.

In contrast to externally visible characteristics, genes whose effects can be revealed only by laboratory tests usually have no discernible influence on marriage patterns. Parental combinations in respect to the MN blood group system were listed in table 5.2; the numbers of marriage combinations observed, together with the corresponding expected numbers, have been listed in table 7.1. The observed deviations are no larger than those expected by chance alone.

Table 7.1 Comparison of the observed and expected number of marriages between couples of M, N, and MN blood-group phenotypes, and of 1-1, 1-2 and 2-2 haptoglobin phenotypes. These data offer no evidence that marriages are contracted other than randomly in respect to these two phenotypic classes.

M-N Blood Groups			Haptoglobins		
Mating	Observed	Expected	Mating	Observed	Expected
M × M	119	111.5	1-1 × 1-1	6	4.4
M × N	142	141.0	1-1 × 1-2	21	28.8
N × N	33	44.6	1-1 × 2-2	28	25.0
MN × M	341	355.3	1-2 × 1-2	48	47.4
MN × N	250	224.6	1-2 × 2-2	91	82.4
MN × MN	275	283.1	2-2 × 2-2	30	35.8
Total	1,160	1,160.1		224	223.8

Another cryptic trait for which familial data including parental genotypes are available is that concerned with the electrophoretic migration of haptoglobins (see table 7.1). Two forms of haptoglobin (a blood-serum protein) are recognized in this study; each form is the product of one of the two allelic forms of the responsible gene. Thus, individuals identified as 1-1, 1-2, and 2-2 are genotypically Hp^1Hp^1, $Hp^1 Hp^2$, and Hp^2Hp^2, where Hp^1 and Hp^2 are codominant alleles at the Hp locus.

The data listed in table 7.1 suggest that marriages are contracted without the aid of high-voltage electrophoretic equipment; each of the parental combinations corresponds in number to that predicted from gene and zygotic frequencies under random mating.

Between the randomness of marriages in respect to cryptic phenotypic traits such as blood type and blood-serum proteins and the strong correlation between mates in respect to racially associated phenotypic traits, lies a series of traits for which marriage partners show intermediate correlations. Stature is an obvious example. Pearson and Lee (1903) studied the correlation between husband and wife in respect to stature (height), span (distance from fingertip to fingertip with arms outspread), and length of forearm (tip of center finger to point of elbow). The heights of husbands and wives are correlated; the correlation coefficient is 0.28. Span and length of forearm are, of course, correlated with height; consequently, that these dimensions show a (significantly lower) correlation between husbands and wives of 0.20 need not be surprising. Nor is it surprising that potential marriage partners size up one another more by height than by span or length of forearm.

Random and Non-random Mating in Insects

In insects, as in man, matings do not occur at random but may appear to do so in respect to certain traits.

Table 7.2 summarizes a considerable amount of data for *D. melanogaster* in which the "trait" being studied was the origin of the tested flies; these origins consisted of one or the other of five laboratory populations which has been isolated from one another for about two years. In each test, males from one population were exposed to females from the same population and from a different one; the basic test consisted of 10 males (let's say, of type *A*) and 10 females of *A* plus 10 females of *B*. These flies carried no obvious mutations; each would be considered wildtype. Summarized through all 1,964 females examined for the presence of sperm, it appears that males inseminated as many females from other populations when confronted with a choice as they did females of their own population. The probability of seeing deviations as high or higher than those actually observed exceeds 50 percent; matings in these tests appear to take place at random in respect to the origin (laboratory population) of these flies.

Data such as those summarized in table 7.2 can be condensed still further into a single number: an isolation index. A commonly used method for calculating an isolation index (Stalker 1942) is to divide the *difference* between the frequencies of *homogamic* and *heterogamic* matings (that is, matings between *like* males and females, and matings between *unlike* males and females) by the *sum* of homogamic and heterogamic matings. An isolation index com-

Table 7.2 Summarized data obtained from multiple-choice tests of sexual preferences among flies from different laboratory populations of *D. melanogaster*. Each test vial contained 10 males of population *x*, 10 females of population *x*, and ten females of population *y*. These data fail to suggest that males of any one population "prefer" females of that population to those of another, despite 70 or more generations of separation. Homogamic matings are those involving males and females of the same population; heterogamic are those involving males and females of different populations. (After Wallace 1954a.)

	Females		Total
	Fertilized	Unfertilized	
Homogamic	481	499	980
Heterogamic	465	519	984
Totals	946	1,018	1,964
$p > 0.50$			

puted in this way will have values lying between + 1.00 (only homogamic matings) and −1.00 (only heterogamic matings); if the flies exhibit no preference, the expected value is zero because homo- and heterogamic matings would be equally frequent. In the case of the laboratory populations, the frequency of homogamic matings equaled 0.491 and that of heterogamic ones, 0.473. Thus, the isolation index equals (0.491 − 0.473)/(0.491 + 0.473) or 0.019, a value not significantly different from zero.

That flies from laboratory populations mate among themselves in a random fashion when given an opportunity does not imply that mating preferences (or the lack of such preferences) are unchanging. *Drosophila paulistorum* is a complex species composed of so-called "semispecies", many persons believe the latter deserve species rank. Be that as it may, the geographical distributions of these semispecies are such that many overlap one another. Thus, as is shown in table 7.3, various semispecies can be captured in the same area *(sympatric)* or either one can be captured where the second member of the pair does not exist *(allopatric)*. If tests of mating preferences are made so that the isolation indices for sympatric and allopatric strains of the same semi-species pair can be compared, one finds (as shown in table 7.3) that sympatric strains avoid one another to a greater extent (higher isolation indices) than do strains which have never encountered one another before. Mating preference, as we suggested immediately above, is an aspect of the phenotype that is subject to change.

Because the discussion has become involved with *origins* of the mating individuals, and the patterns of mating these individuals exhibit, evidence bearing on the *change* in mating behavior mentioned above might be presented here. Such evidence fre-

Table 7.3 Numbers of matings observed and isolation indexes calculated for crosses within and between sympatric and allopatric strains of the "semispecies" of *D. paulistorum*. The total number of matings observed was 1,695. (After Ehrman 1965.)

Semispecies Pair	Sympatric		Allopatric	
	Matings	Isolation Index	Matings	Isolation Index
Amazonian × Andean	108	0.86	100	0.66
Amazonian × Guianan	104	0.94	109	0.76
Amazonian × Orinocan	106	0.75	124	0.61
Andean × Guianan	109	0.96	102	0.74
Orinocan × Andean	100	0.94	111	0.46
Orinocan × Guianan	104	0.85	100	0.72
Centro-American × Amazonian	102	0.68	103	0.71
Centro-American × Orinocan	110	0.85	103	0.73

quently exists as a displacement of one or more components of
the reproductive process in areas where two potentially interfering
forms coexist—a phenomenon known as "character displacement"
(Brown and Wilson 1956). These are forms which, if they were to
interbreed, would produce at least partially sterile or inviable hybrids. In table 7.3, we saw that sympatric strains of the species of
D. paulistorium exhibit a greater aversion to one another than do
allopatric strains of the same pairs of semispecies; the nature of
that aversion is unknown. In figure 7.1, the style and anther
lengths of two species of nightshade (*Solanum grayi* and *S.
lumholzianum*) are depicted in areas where the two species live
separately, and in those where they coexist at close quarters. In
the latter area, the lengths of the style and anthers in *S.
grayi* change abruptly; because of the differences in the area of
overlap, pollination of the two species is carried out by different
sets of insects.

    Still another device by which matings between "wrong"

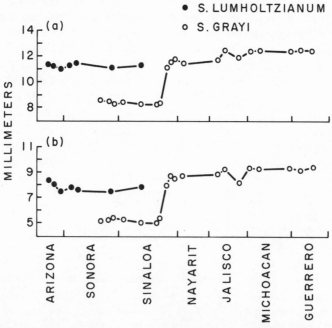

Figure 7.1 The abrupt change (= character displacement) of both (a) style and
(b) anther length of *Solanum grayi* flowers in areas where the species coexists
with *S. lumholzianum*. The size difference prevents the two species from
utilizing the same pollinator insects. (After Levin 1978a.) Reproduced with
permission: from *Evolutionary Biology*, vol. 11; © 1978 Plenum Publishing
Corp.)

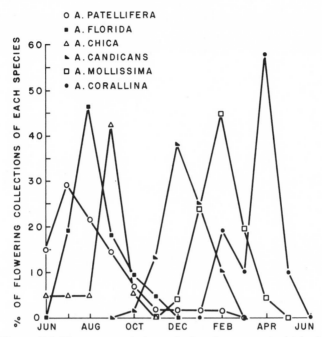

Figure 7.2 The staggered times during the year at which six species of *Arrabidaea* reach their peak flowering periods in Central America. (After Gentry 1974; see Levin 1978a.)

forms may be prevented, lies in the subdivision of time. An example, illustrating how the year in the tropics is subdivided by six species of one genus of plants (*Arrabidaea*), is shown in figure 7.2. Here, as in cases that will reappear later in other contexts (see page 263 and 613), the peak flowering times of the different species fall during different times of the year—actually, at rather regular intervals.

This brief digression can be terminated by returning once more to (genetically) less complex situations. For example, Burns (1966) has reported on the mating behavior of a butterfly species, *Papilio glacus*. In this species, all males are yellow (light-colored); females are of two sorts: a light, non-mimetic form resembling the males, and a dark form that mimics a second butterfly species, *Battus philenor*. The history of a female's mating behavior can be reconstructed from the number of spermatophores she carries; each one presumably represents a single mating.

Table 7.4 lists the number of spermatophores found per female for both *B. philenor* and *P. glacus*; the average number of spermatophores per female seems to be the same for the two species: 1.73 (= 57/33) and 1.74 (= 197/113). Within *P. glacus*,

Table 7.4 Number of spermatophores found in light and dark forms of *Papilio glaucus* and in *Battus philenor*, an unpalatable model of the dark *Papilio*. The observed distribution of spermatophores has been compared in each instance with that expected on the basis of the Poisson distribution. (After Burns 1966.)

| | Number of Spermatophores per Female | | | | | | |
	0	1	2	3	4	5	Total Females
Mt. Lake, Va.							
B. philenor	0	17	11	3	1	1	33
$\bar{x} = 1.73$	5.9	10.3	8.9	5.1	2.2	0.8	Poisson
P. glaucus (dark)	0	33	30	8	0	1	72
$\bar{x} = 1.69$	13.0	21.9	18.5	10.4	4.4	1.5	Poisson
P. glaucus (light)	0	6	2	2	1	1	12
$\bar{x} = 2.08$	1.5	3.1	3.2	2.3	1.2	0.5	Poisson
Baltimore Co., Md.							
P. glaucus (dark)	0	8	3	2	0	0	13
$\bar{x} = 1.54$	2.7	4.2	3.2	1.7	0.6	0.2	Poisson
P. glaucus (light)	0	4	10	2	0	0	16
$\bar{x} = 1.88$	2.4	4.5	4.2	2.7	1.3	0.5	Poisson

however, light-colored females seem to carry more spermatophores than dark females do: 1.96 (= 55/28) and 1.67 (= 142/85).

These data are interesting not only in reference to the apparent "preference" of male *P. glacus* for females of the same phenotype but also in respect to mating behavior generally. The distributions of spermatophores among females are listed in table 7.4, together with the distributions that would be expected on the basis of the Poisson distribution. The latter distribution would be generated if each brief, chance encounter of a male and female butterfly was accompanied by a certain, constant probability of mating success. This, however, is not the case. According to the Poisson distributions, 25 females among the 146 should have been virgin (no spermatophores); not one was observed. Furthermore, 35 females should have carried 3 or more spermatophores; only 22 carried this many. Consequently, we can conclude that encounters between males and virgin females are accompanied by a relatively high probability of mating success, whereas those between males and females that have already mated two or more times tend not to result in still another mating.

Rare Genotype Effects

Patterns of mating have been shown in numerous experiments to be dependent upon the relative frequencies of the genotypes of mating individuals. Petit (1958) and, especially, Ehrman (1966, 1967, 1970) have shown in *Drosophila* that rare genotypes (particularly, rare males) frequently have an advantage over their more common colleagues. Table 7.5 presents data emphasizing that both rare males and rare females have such an advantage: when either genotype of either sex is rare, the proportions of matings involving that genotype are considerably higher than are the proportions of that genotype among the competing flies. Rare types have disproportionate success at mating. Ehrman and Probber (1978) have also shown that the advantage of rare flies exists as well when three genotypes compete simultaneously; when only one of three genotypes is common, that genotype is relatively unsuccessful in mating. In this same report, these authors describe the results obtained by releasing virgin males and females in a large laboratory room; recapturing the females; and through progeny tests, determining the genotype of the male with which the fertilized females had mated. The advantage (as in the case of experiments performed in small observation changers) lay with rare males.

A mating advantage that resides with rare males has important consequences for the genetics of populations. It is a behavior, for example, that promotes outbreeding; thus, it can be regarded as negative assortative mating. It is also a behavior which, if it were to extend to single gene loci, would delay the possible loss of rare alleles from populations; the carriers of such alleles would be preferred as mates, thereby making the alleles common once more. Because of these implications, it is important to point out that not all species of *Drosophila* have exibited the so-called "rare male" effect. Furthermore, as shown in table 7.6, rare

Table 7.5 The effect of rareness on the mating success of *AR/AR* and *CH/CH* individuals of *D. pseudoobscura*. The asterisk (*) calls attention to the high ratio of observed to expected numbers of matings for both males and females of rare genotypes. (After Ehrman et al. 1965.)

Number of Pairs			Females		Males	
AR/AR	*CH/CH*	*Number of Matings*	*AR/AR*	*CH/CH*	*AR/AR*	*CH/CH*
12	12	265	137	128	131	134
		Expected	132.5	132.5	132.5	132.5
		Obs./Exp.	1.03	0.97	0.99	1.01
20	5	207	138	69	136	71
		Expected	165.6	41.4	165.6	41.4
		Obs./Exp.	0.83	1.67*	0.82	1.72*
5	20	209	69	140	105	104
		Expected	41.8	167.2	41.8	167.2
		Obs./Exp.	1.65*	0.84	2.52*	0.62

Table 7.6 An example of rare genotype *disadvantage* in *Phlox drummondii* Hook. Nana and Twinkle phenotypes result from a single gene difference. (After Levin 1972.)

Ratio Nana : Twinkle	Number of Progeny Scored	Percent Heterozygotes (Nana progeny)	Percent Outcrossing
9:1	1,432	.0428	42.8[a]
3:1	912	.1426	57.0[b]
1:1	1,085	.2861	57.2[b]
1:3	628	.4633	61.8[b]
1:9	418	.4447	49.4[a]

[a]Both significantly lower than the proportions of outcrossing characteristic of intermediate ratios (b).

[b]Proportions not different from one another; higher than 50 percent because *Phlox* tends to outcross.

genotypes under some situations are at a *dis*advantage (Levin 1972); pollinators tend to maintain a consistent search pattern, thus paying even fewer visits to rare floral types than would be expected by chance.

Correlations (Positive and Negative) among Mating Couples

Marriages among persons and matings among other organisms are not entered into by chance; both positive and negative correlations can be identified. The old adage says, "opposites attract," but the correlation between the height of marriage partners approaches 0.30.

Parsons (1965) has shown that *Drosophila* are also "aware" of size during courtship and mating. Males and virgin females that were grown under two levels of larval crowding were collected. After aging several days, individuals of the two sexes were placed together so mating could occur. Mating pairs were drawn from the mating chamber, pair by pair, by means of an aspirator; each pair was stored separately. When approximately one-half of the flies had mated, the experiment was stopped.

The two members of each pair were examined under a microscope and the number of sternopleural bristles on each was recorded. Unmated flies that remained at the end of the experiment were arbitrarily paired by the observer, and the sternopleural bristles of these artificial pairs were also recorded. The results are shown in table 7.7.

The table reveals a number of facts: The average number of

Table 7.7 The correlation between numbers of sternopleural bristles on male and female members of single pairs of *D. melanogaster.* (After Parsons 1965.)

	Number of Pairs	Average Number		Correlation Coefficient	p
		Females	Males		
Low larval competition					
Mated	212	23.9	22.9	0.21	<0.01
Unmated	247	23.9	22.9	−0.04	>0.50
High larval competition					
Mated	568	20.2	19.8	0.11	<0.01
Unmated	610	20.1	19.7	−0.05	>0.20
Mixed (high and low)					
Mated	172	22.0	21.2	0.20	<0.01
Unmated	239	21.6	21.0	−0.01	>0.90

bristles on flies reared under uncrowded conditions is larger than that on those reared under crowded ones. There is no correlation between the bristle numbers on unmated flies that were arbitrarily paired by the experimenter. In the case of the mated flies, however, there is a positive and a statistically significant correlation (0.11–0.21) in each of the three experiments that were carried out. This represents an assortative mating between flies, probably based on size rather than bristle number, that is comparable to that seen among married persons.

Averhoff and Richardson (1974) found that sterility of inbred cultures need not reflect a functional sterility of either the inbred males or females but, rather, a mere disinterest of inbred brothers and sisters in one another (page 145). They carried out a test on the relative mating preferences of brother and sisters of inbred lines for one another and for comparable flies of other, unrelated inbred lines. As shown in table 7.8, during the early generations, matings occured within and between lines at random; the average proportion of between-line matings through the

Table 7.8 Results of mate-choice trials conducted with nine inbred lines of *D. melanogaster* established from recently collected wildtype flies. (After Averhoff and Richardson 1974.)

Generation	Trials	Possible Matings	Matings Observed	Number (and %) of Matings between Lines
P1	19	189	144	68 (47%)
F1	10	100	86	49 (57%)
F2	20	200	148	65 (44%)
F3	7	70	55	28 (51%)
F8	13	130	70	53 (76%)
F10	16	160	100	65 (65%)
F12[a]	22	220	147	95 (65%)

[a]Sterility of the inbred lines caused the termination of the experiment before F14.

F_3 generation is almost exactly 50 percent. During the last three generations tested, however, nearly 70 percent of all matings were *between* lines. And, by generation 14, the inbred flies no longer produced offspring. This type of behavior, characterized by a pronounced negative correlation between mating pairs, would clearly promote outcrossing in *Drosophila* populations. (*D. pseudoobscura* does not show a similar preference for inter-strain mating after 12 generations of inbreeding according to Powell and Morton 1979).

Sheppard (1952) has described a mating preference in the moth, *Panaxia dominula*. This moth has three phenotypes that correspond to the two homozgotes and the heterozygotes of a single pair of alleles: *dominula* (*DD*), *medionigra* (*Dd*), and *bimaculata* (*dd*); brief reference to these genotypes were made on page 112 in regard to the accuracy of their classification. In a total of 150 successful tests involving either one male and two dissimilar females or one female and two dissimilar males (but where one male and one female were of the same genotype), Sheppard found that 97 of the 150 matings were between unlike individuals (table 7.9). This deviation from random expectation is highly significant.

In concluding this section, and as a final reminder that positive correlations also exist, data obtained by Levin (1978a: page 223) in respect to the stature of plants and the flight behavior of insect pollinators are presented in table 7.10. Foraging pollinators tend to remain at one height as they move from flower to flower; abrupt rises in flight level presumably require the expenditure of considerable energy. Thus, one would expect that flowers visited on successive segments of an insect forager's path would be those whose heights are correlated. And, indeed, that is what the data in table 7.10 show: when the difference in floral

Table 7.9 Test for randomness of matings between the *dominula* (*d*), *medionigra* (*m*), and *bimaculata* (*b*) forms of *Panaxia dominula*. Each trial involved three individuals, two of one sex and one of the other. (After Sheppard 1952.)

Males		Females		Like	Unlike	Total
d	—	d	m	8	20	28
m	—	d	m	12	14	26
d	m	m	—	13	14	27
d	m	d	—	11	22	33
m	—	m	b	2	0	2
b	—	m	b	0	1	1
m	b	b	—	3	15	18
m	b	m	—	2	10	12
d	—	d	b	2	1	3
	Total			53	97	150

Table 7.10   The effect of flight patterns of insect pollinators on the correlation between the statures of successively visited plants (*Liatris salicaria* and *L. alatum*). The heights of plants in this experiment were controlled experimentally; normally, *L. salicaria* would bear blossoms 2 to 6 feet higher than *L. alatum*. Numbers in the table are correlation coefficients. (After Levin 1978a.)

		Spacing		
	Difference in Stature	*2'*	*4'*	*8'*
	6"	.40**	.14*	-.05
L. salicaria	10"	.72**	.59**	.06
	20"	.80**	.72**	.31**
	6"	.29**	.15*	.02
L. alatum	10"	.68**	.62**	.08
	20"	.78**	.69**	.29**
	0"	.14	.06	.02
Both species	6"	.54**	.22*	.06
	10"	.74**	.68**	.14*
	20"	.83**	.75**	.38**

*Significant at the 5 percent level; **, at the 1 percent level.

level was as great as twenty inches, there was a strong tendency for the pollinating insects to remain at the same level even when the individual flowers were eight feet apart. The closer the spacing between flowers, the higher the correlation between heights of flowers visited. At distances of two feet, even a six-inch change in altitude is most often avoided. Here we see a correlation between heights of parental plants that arises from the economics of insect flight.

Summary

Mating preferences can scarcely be discussed without reference to natural selection. Positive assortative mating, at the genic level, promotes homozygosis and, thus, exposes what are frequently dissimilar phenotypes to selective forces such as differential survival or reproduction. Negative assortative mating may leave certain individuals without mates, thus rendering them sterile.

    Matings that do not occur at random in respect to certain genotypic or phenotypic attributes have no apparent bias toward positive or negative assortative mating. One might guess that what passes for random mating is really the cancellation of many opposing tendencies; every individual, in terms of his genotype, is a rare individual—unique, in fact. Thus, because all possibilities regarding mating preferences exist, populations seem poised, ready to respond to any demand on mating behavior that conditions might dictate. And, in passing, we did note that *character displacement* represents one such response.

FOR YOUR EXTRA ATTENTION

1. Extract as much information as possible, including mating patterns, from the following data (Berberović 1969). These data were obtained from tests of 616 women who were bearing children of disputed parentage; consequently, the number of offspring also equals 616.

		Offspring			
Mother	Father	M	MN	N	Total
M	M	36	–	–	36
M	N	–	17	–	17
M	MN	32	29	–	61
N	M	–	28	–	28
N	N	–	–	15	15
N	MN	–	32	20	52
MN	M	29	48	–	77
MN	N	–	38	16	54
MN	MN	31	214	31	276
Total		128	406	82	616

2. Suppose that no bias, for or against, exists in respect to the marriage of first cousins. Let b equal the number of children surviving to adulthood in each family. If the proportion of cousin marriages equals c, show that the number of potential marriage partners available to any person equals $n = 2b(b - 1)/c$.

How many potential marriage partners are available to a man or woman living in a community where $c = 0.25$ percent and where $b = 2$?

3. Dobzhansky and Koller (1938) studied the mating preferences of flies (*D. miranda*) captured at two localities: Olympic (Washington) and Whitney (California). Here I shall identify the males from the two localities only as A and B, and the females as C and D. The results of this study (which involved nearly 2,250 females) are summarized as follows (percent females mated within 4 days):

	Females	
Males	C	D
A	73%	88%
B	84%	78%

Which males and females came from the same locality? On what do you base your decision?

4. Muggleton (1979) has reported on the mating patterns of melanic and non-melanic forms of the ladybird beetle, *Adalia bipunctata*. Two sets of data are given below:

Sample	Beetles Captured as Mating Pairs		Beetles Captured Singly	
	Non-melanic	Melanic	Non-melanic	Melanic
A	536	44	1597	88
B	156	44	536	67

Do the mating beetles represent a random sample of *all* beetles in a locality? Do these data differentiate preference from, let's say, duration of coupling?

5. The following data are taken from a report (Fontdevila and Mendez 1979) on frequency-dependent selection in *D. pseudoobscura*. The two alleles (104 and 100) are allozyme variants of the esterase-5 locus; only homozygous flies were used in this experiment.

Input Ratios		Mated Females		Mated Males	
104	100	104	100	104	100
32	8	276	100	295	81
20	20	183	200	180	203
8	32	94	281	80	295

Identify the evidence for frequency dependence in these data.

Inbreeding and Chance Events

PREVIEW: In certain respects, this chapter extends the consequences of randomness *versus* non-randomness of mating patterns discussed in the previous one. Because individuals of a species are generally distributed over a large area, and (especially) because these individuals occur spatially in partially isolated patches, mating cannot be random over the entire species range. The limited dispersal ranges of individuals and their gametes impose restrictions on the numbers of interbreeding individuals. Consequently, the gene frequencies in these local populations are subject to chance fluctuations through sampling error. The Hardy-Weinberg equilibrium was based, of course, on the assumption that such chance fluctuations do not occur.

THE HARDY-WEINBERG equilibrium, as we have stressed in each recent chapter, is based upon a number of assumptions, including an absence of chance fluctuations in gene frequencies. This assumption would be met if populations contained infinite numbers of individuals, but real populations do not. The present chapter concerns the consequences of finite population size on the genetics of populations.

No matter what plan might be followed in organizing this chapter, it would still consist of an interwoven account of (seemingly) loosely related topics. I suggest, then, that the following statement by R. A. Fisher (1958:9–10) be kept in mind as the focal point for the several topics that are to be discussed; each topic in its own way brings us back to this common point: "In a population breeding at random in which two alternate alleles . . . exist in the ratio p to q, the three genotypes will occur in the ratio $p^2 : 2pq : q^2$, and these assure that their characteristics will be represented in fixed proportions of the population . . . provided that the ratio $p:q$ remains unchanged. This ratio will indeed be liable to slight changes; first by chance survival . . . ; and secondly by selective survival. . . . The effect of chance survival is easily susceptible to calculation, and it appears . . . that *in the population of n individuals breeding at random the variance will be halved by this cause acting alone in 1.4 n generations. . . .* [It] will be seen that this cause of diminution of hereditary variance is exceedingly minute, when compared with the rate of halving in one or two generations by blending inheritance" (Italics added).

As the central task of this chapter, I suggest that we gain an understanding of these few sentences. In particular, I suggest that we discover why Fisher says (with great precision) that hereditary variance will be halved in 1.4 n generations. I also suggest that we discover the multitude of seemingly diverse events that lead us repeatedly to this same conclusion.

The Isolated Population

Assume that an isolated population of diploid plants consists of N individuals. Assume, too, that at the a locus, every one of the $2N$ alleles has been identified: a_1 and a_2 of plant #1, a_3 and a_4 of plant #2, . . . , a_{2i-1} and a_{2i} of plant #i, . . . , a_{2N-1} and a_{2N} of plant #N, as shown in Figure 8.1. Finally, assume that at one moment the pollen grains formed by these plants rise in a cloud, become thoroughly mixed, and then rain back down onto the plants where fertilization occurs—either self- or cross-fertilization as chance decrees.

A pollen grain carrying the allele a_i, for example, has one chance in N of landing on the plant (plant #i) from which it came; having landed on that plant, it has one chance in two of fertilizing an egg carrying the same rather than the other allele. Upon com-

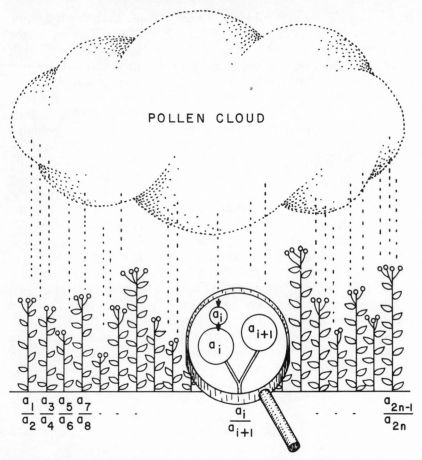

Figure 8.1 A diagram representing a small population of an open-pollinated plant species, and illustrating the probability that an egg cell will be fertilized by a pollen grain (1) from the same plant and (2) carrying the identical allele. If there are N plants in the population, this probability equals $(1/2N)^2 \times 2N$, or $1/2N$.

bining these two probabilities, we see that the overall probability that two uniting gametes carry alleles which are identical by descent (both formed by the same individual) equals $1/2N$. The probability that an individual, after a single generation, will carry two alleles identical by descent equals $1/2N$; this probability which increases with time is called F, the inbreeding coefficient.

If, after one generation, the probability that an individual carries two alleles at a given locus that are identical by descent equals $1/2N$, the probability that it does *not* carry identical alleles is obviously $1 - 1/2N$. The probability that an individual does *not* carry alleles identical after 2, 3, or t generations equals $(1 - 1/2N)^2$, $(1 - 1/2N)^3$, or $(1 - 1/2N)^t$. Consequently, the probability that an individual, after a lapse of t generations, *does* carry two alleles that are identical by descent equals $F_t = 1 - (1 - 1/2N)^t$. This can also be written as

$$F_t = 1 - e^{-t/2N}.$$

The proportion of individuals carrying two alleles that are identical by descent increases as the number of generations increases. Eventually, this proportion approaches 1.00 (as the number of elapsed generations becomes extremely large). That F should approach 1.00 suggests that the inbreeding coefficient has meaning only in reference to a stated number of elapsed generations. Furthermore, that F approaches 1.00 as the number of generations increases, reminds us that *all* genes are presumably descended from an original one: genomes have grown through gene and chromosome duplication and corresponding alleles in different forms of life have evolved, but all genes of all living things are related by descent through geologic time.

The proportion of individuals that are *not* homozygous by decent equals $1 - F_t$ or $e^{-t/2N}$. These individuals are the reservoir of remaining genetic variation; the others, being homozygous, are invariant. Thus, we can calculate the time required for a population to lose one half its original, heritable variation—that is, for the level of heterozygosity to drop from its original level (say, H_0) by one-half (to $0.5\,H_0$):

$$H_0 e^{-t/2N} = 0.5\,H_0$$
$$e^{-t/2N} = 0.5$$
$$-t/2N = \ln 0.5 = -0.693$$
$$t = 1.39\,N.$$

Here, then, is the basis for Fisher's statement that one-half of hereditary variation is lost in $1.4\,N$ generations. Contrary to expectations that might have been encouraged by the original account of the Hardy-Weinberg equilibrium, genetic variation remains constant only in hypothetical populations of infinite size; real populations lose their variation at a steady rate. This rate, however, is a function of population size, N: only in a "population" consisting of a single, self-fertilizing individual will hereditary variation be lost at a rate comparable (when $N = 1$, $1/2N =$ 1/2) to that at which a population of any size would lose it under blending inheritance (page 104).

Inbreeding and the Hardy-Weinberg Equilibrium

The Hardy-Weinberg equilibirum was derived in chapter 5 by means of a checkerboard (figure 5.1). However, we have now learned that a number of individuals in a population are homozygous because the alleles they carry are identical by descent. If their proportion equals F, the Hardy-Weinberg equilibrium applies only to the remainder $(1 - F)$ of the population. The consequence, as shown in figure 8.2, is a population divided: the non-inbred portion possesses AA, Aa, and aa individuals in the proportions p^2, $2pq$, and q^2; the inbred portion contains only AA and aa individuals at frequencies p and q. Overall, inbreeding increases the frequency of homozygotes at the expense of the heterozygotes.

The frequency of AA individuals in the population illustrated in figure 8.2 equals $(1 - F)p^2 + Fp$. This also equals $p^2 + F(p - p^2)$, or $p^2 + Fpq$. The frequency of aa individuals equals $(1 - F)q^2 + Fq$ or, after some manipulations, $q^2 + Fpq$. The frequency of Aa heterozygotes equals $2pq - 2Fpq$. The sum of all three frequencies still equals 1.00.

In the previous section we learned that $F_t = 1 - (1 - 1/2N)^t$. This value for F can be substituted into the modified Hardy-Weinberg equilibrium developed here to yield:

Genotype	Frequency
AA	$p^2 + pq \left[1 - \left(1 - \dfrac{1}{2N} \right)^t \right]$
Aa	$2pq - 2pq \left[1 - \left(1 - \dfrac{1}{2N} \right)^t \right]$
aa	$q^2 + pq \left[1 - \left(1 - \dfrac{1}{2N} \right)^t \right].$

As t becomes large, $(1 - 1/2N)^t$ approaches zero, and the frequency of the two homozygotes approach $p^2 + pq$ and $q^2 + pq$; these two terms as we learned in chapter 4 equal p and q. The

	1 − F	F
A A	p^2	p
A a	$2pq$	NONE
a a	q^2	q

Figure 8.2 The subdivision of a population into two portions: one within which the Hardy-Weinberg equilibrium applies $(1 - F)$, and the other which, because it is inbred, contains only homozygous individuals (F).

frequency of heterozygotes approaches $2pq - 2pq$, or zero. When inbreeding is complete, no heterozygotes exist; and the proportions of homozygotes equal the gene frequencies, p and q, themselves. Once more we see the importance for specifying t, the number of generations that have elapsed; otherwise, all present-day populations are homozygous by *descent* (although not necessarily in *content* because mutational changes may have occurred during the course of descent).

This and the previous section have dealt with a population in isolation; we shall leave this topic for the moment to consider a *number* of isolated (or partially isolated) populations.

Wahlund's Principle

The Hardy-Weinberg equilibrium is not a *stable* equilibrium. Gene frequencies *tend* to remain at their present values but, once changed, they have no tendency to revert to an earlier value. The Hardy-Weinberg equilibrium has no memory; it is an indifferent equilibrium.

Local populations that together constitute a much larger one should, by chance alone, come to differ somewhat in the frequencies of various alleles. This is so even though all local populations may trace back to a common source, thus having identical initial frequencies. Indeed, a good deal of the effort invested in the studies described in chapters 3 and 4 was expended in demonstrating that populations of the same species do, in fact, differ both in space and in time.

In this section, we shall compare the gene and zygotic frequencies found in a number of small populations with the gene frequency and the frequencies of zygotes that would be seen if the small populations were consolidated into a larger one. This problem was touched upon earlier (page 113) when the danger of pooling data was described. Even now, we shall not give an extensive treatment: Li (1955), Crow and Kimura (1970), and Spiess (1977) give more thorough accounts.

Suppose that within a circumscribed geographical region a species exists as n populations of equal size. Within these populations, the frequencies of A and a are $p_1 : q_1, p_2 : q_2, p_3 : q_3, \ldots,$ $p_n : q_n$, where $p_i + q_i$ equals 1.00 (see table 8.1). The zygotic frequencies of AA, Aa, and aa in these populations will be $p_1^2 : 2p_1q_1 : q_1^2$; $p_2^2 : 2p_2q_2 : q_2^2$; ...; and $p_n^2 : 2p_nq_n : q_n^2$. The zygotic distributions that are expected under the Hardy-Weinberg equilibrium are expected *within* local populations; these expected frequencies are determined by local gene frequencies.

If these local populations were to be combined into a single

Table 8.1 Calculations showing the relation between observed and expected frequencies of homozygotes and heterozygotes when data from several populations are accidentally (or otherwise) pooled.

Population	Gene Frequency		Zygotic Frequency		
	A	a	AA	Aa	aa
1	p_1	q_1	p_1^2	$2p_1q_1$	q_1^2
2	p_2	q_2	p_2^2	$2p_2q_2$	q_2^2
3	p_3	q_3	p_3^2	$2p_3q_3$	q_3^2
.	.	.	.	.	.
.	.	.	.	.	.
i	p_i	q_i	p_i^2	$2p_iq_i$	q_i^2
.	.	.	.	.	.
.	.	.	.	.	.
n	p_n	q_n	p_n^2	$2p_nq_n$	q_n^2
	$\bar{p}$	$\bar{q}$	$\overline{(p_i^2)}$	$\overline{(2p_iq_i)}$	$\overline{(q_i^2)}$
Average	or	or	or	or	or
	$\Sigma p_i/n$	$\Sigma q_i/n$	$\Sigma p_i^2/n$	$\Sigma 2p_iq_i/n$	$\Sigma q_i^2/n$

$$\sigma_p^2 = \Sigma p_i^2/n - \bar{p}^2 = \sigma_q^2$$

$$\overline{(p_i^2)} = \Sigma p_i^2/n = \bar{p}^2 + \sigma_p^2$$

$$\overline{(q_i^2)} = \Sigma q_i^2/n = \bar{q}^2 + \sigma_p^2$$

$$\overline{(2p_iq_i)} = 2\bar{p}\bar{q} - 2\sigma_p^2$$

large population (as an unobservant collector might do with specimens brought back from a wide-ranging field trip; or as might be
the case when highways and automobiles connect once isolated
villages in a mountainous nation), the frequencies A and a would
be $\bar{p}$ and $\bar{q}$, the averages of the local frequencies ($\bar{p} = \Sigma p_i/n$;
$\bar{q} = \Sigma q_i/n$). The zygotic frequencies expected in the pooled material would be $\bar{p}^2 : 2\bar{p}\bar{q} : \bar{q}^2$.

The expected zygotic frequencies are now to be compared
with the observed ones. The latter, however, consist of n observations each of which might be represented as $p_i^2 : 2p_i q_i : q_i^2$. Pooled
over all n local populations, the observed zygotic proportions
become

$$\overline{(p_i^2)} = \Sigma p_i^2/n$$
$$\overline{(2p_i q_i)} = \Sigma 2p_i q_i/n$$
$$\overline{(q_i^2)} = \Sigma q_i^2/n.$$

Now, it happens (see Boxed Essay, chapter 3) that the variance of a series of observations is frequently calculated not as the
average deviation squared ($\sigma_X^2 = \sum_{i=1}^{n} (X_i - \overline{X}^2)$) but by the mathematically equivalent, computational short cut ($\sigma_X^2 = \Sigma X_i^2/n - \overline{X}^2$).
The latter can be rearranged

$$\Sigma X_i^2/n = \overline{X}^2 + \sigma_X^2.$$

upon substituting gene frequencies for X, we see that

$$\Sigma p_i^2/n = \bar{p}^2 + \sigma_p^2$$
$$\Sigma q_i^2/n = \bar{q}^2 + \sigma_q^2 = \bar{q}^2 + \sigma_p^2.$$

That is, the frequencies of homozygous AA and aa individuals
are larger than expectations based on the square of the corresponding average frequencies, $\bar{p}$ and $\bar{q}$, by an amount equal to the variance of p among local populations. Because q equals $1 - p$, $\sigma_q^2 =
\sigma_p^2$. The frequency of heterozygotes observed in the consolidated
data is smaller than the value calculated as $2\bar{p}\bar{q}$ by twice the variance of p.

The results of these calculations (calculations first carried
out by Wahlund 1928) can be summarized by stating that the subdivision of a large population into a series of smaller ones results in
an increase in the overall proportion of homozygotes at the expense of heterozygotes. The excess homozygotes equals the vari-

ance of p (for a two-allele system) among local (or sub-) populations. Conversely, the results can be restated as follows: if for any reason a once-subdivided population is permitted to mate more nearly at random across earlier boundaries, fewer homozygotes will be found after the breakdown of these barriers than before. In at least one sense of the word, the inbreeding that is characteristic of local populations will have been reduced.

The Subdivision of a Large Population

The local populations that were discussed above under Wahlund's principle, represented an already-existing situation, one that might reflect either the subdivision of a snail population in a city park or the isolation of remote villages in the Alpine region of Yugoslavia. In the present section, we shall consider a theoretical case in which a large population is subdivided at a given moment into numerous smaller ones; we shall then reconstruct events as they might occur, generation by generation. These events should illustrate the steps which must occur (although not so neatly) during the subdivision of natural populations. The account that we shall give closely follows one presented by Falconer (1960).

Imagine a large population in which the alleles A and a have frequencies p_0 and q_0 (figure 8.3). Imagine, too, that the members of this population are arbitrarily apportioned to n smaller populations, each of size N. Because of sampling error, the frequency of A will not be identical in all these smaller populations; rather, these populations will have gene frequencies equal to $p_1, p_2, p_3, p_4, \ldots$; the average frequency, $\bar{p}$, however, will equal p_0.

The expected distribution of the gene frequencies ($p_1, p_2, p_3, \ldots$) among the small populations about their mean, $\bar{p}$ ($= p_0$), is given by the binomial expansion $(p_0 + q_0)^{2N}$. Because the N individuals of each population carry $2N$ alleles at each locus, and because the frequency of A in the large population being subdivided was p_0, then the probability that all alleles in any one population are A would equal p_0^{2N}. Because there are $2N$ different ways in which only a single a allele could occur in any one population of size N, the probability in this case would equal $2Np_0^{2N-1}q_0$. Such calculations are not difficult when the exponent is small; for example, in elementary genetics each of us has at some time calculated the probability that a family of three children will consist

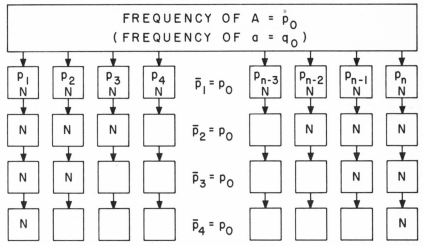

Figure 8.3 A diagram illustrating the subdivision of an initially large population in which the frequency of A was p_0, into n smaller populations each of size N. The latter are then maintained at this smaller size. In the absence of selection, the average frequency of A $(\bar{p})$ remains p_0; the variance of p, however, grows as described in the text. (Some boxes have been left empty in order to avoid cluttering the diagram.)

of three daughters, two daughters and one son, one daughter and two sons, or three sons (Answer: $\frac{1}{8} : \frac{3}{8} : \frac{3}{8} : \frac{1}{8}$). The calculations become exceedingly difficult when the exponent is large. The problem is considerably simplified by the knowledge that the binomial variance of p (σ_p^2) among the small populations once they have been set up is $pq/2N$.

Suppose now that each small population perpetuates itself through time, while maintaining a constant population size, N (see figure 8.3). First, what might we expect in respect to the mean frequency, $\bar{p}$, each generation? Because no provision has been made allowing for a systematic change in $\bar{p}$, we must assume that it remains constant (or as nearly so as sampling errors will permit); the expectation for $\bar{p}$ in each succeeding generation is the original frequency of A itself, p_0.

What do we expect concerning the variance of p about its mean? We saw that binomial variance for a sample size of $2N$ (the number of alleles carried by N diploid individuals) equals $pq/2N$. We also know that variances are additive; consequently, we can guess (and be approximately correct) that in the first several generations the variance around the mean will grow: $pq/2N$, $2pq/2N$, $3pq/2N$, ...

The variance cannot grow forever, however. At some point,

as the result of chance fluctuations in gene frequency, each population will have lost either the A allele or its alternative, a. Because the average frequency remains constant, we would expect to find when all populations have become inhabited by AA or aa individuals only, that there be p populations of AA individuals and q populations of aa individuals. This represents the largest possible variance among these populations; its value can be calculated rather easily:

Proportion of Populations (=f)	Frequency of A within Population	$(p - \bar{p})^2$	$f(p - \bar{p})^2$
p	1.00	$(1 - p)^2$	$p(1 - p)^2$
q	0	$(0 - p)^2$	qp^2

Mean $= (p \times 1.00) + (q \times 0) = \bar{p} = p$

Variance $= pq^2 + qp^2 = pq(q + p) = pq$ [Recall that $(1 - p)^2 = q^2$.]

When all populations have attained gene frequencies of either 0 or 1.00 (frequencies at which *within* population variance equals zero), the variance *among* populations has become pq; it can become no larger. Here is the limit that is placed on our earlier approximation in which we estimated the variance among populations as the number of generations times $pq/2N$; the approximation fails long before the number of generations approaches $2N$ and would give absurd results for generations exceeding $2N$.

It is not possible to derive here the mathematical expression describing the gradual increase in variance from zero (original large population), to pq/N (first array of small populations) to pq (when all populations have become fixed at 0 percent and 100 percent A). The equation that describes this increase in

$$\sigma_p^2 = pq \left[1 - \left(1 - \frac{1}{2N} \right)^t \right]$$

where t is the number of generations elapsed since subdividing the large population. When t equals 1, for example, $\sigma_p^2 = pq/2N$ as we saw above.

Returning to the calculations made earlier in illustrating Wahlund's principle, we saw that upon averaging through what were really isolated or partially isolated local populations, homozygotes exceeded expectations, while heterozygotes fell short of their expected frequencies. The discrepancies involved the variance of p about the average, $\bar{p}$, of the lumped data. These reuslts can now be rewritten

Genotype	Frequency
AA	$\bar{p}^2 + \sigma_p^2 = p^2 + pq \left[1 - \left(1 - \dfrac{1}{2N} \right)^t \right]$
Aa	$2\bar{p}\bar{q} - 2\sigma_p^2 = 2pq - 2pq \left[1 - \left(1 - \dfrac{1}{2N} \right)^t \right]$
aa	$\bar{q}^2 + \sigma_p^2 = q^2 + pq \left[1 - \left(1 - \dfrac{1}{2N} \right)^t \right].$

Written in this way, the zygotic frequencies seen in isolated populations take on precisely the values which were calculated earlier under inbreeding (see figures 8.1 and 8.2). The divergence of populations in isolation that is governed by chance events results in precisely the same excess of homozygotes as does the chance union of gametes carrying alleles that are identical by common descent within a single isolated population. Both events revolve about the population size, N, the numbers of alleles, $2N$, and time, t, in generations. The one, inbreeding, gives rise to two types of individuals, *AA* and *aa,* in one population with the exclusion of heterozygotes; the other, the divergence of isolated populations, generates two types of populations, those consisting only of *AA* individuals or, alternatively, of *aa* individuals at the expense of segregating populations of intermediate frequencies. The rules are the same whether, as in the account of inbreeding, we insist that p (and q) remain constant while repackaging individuals at the expense of heterozygotes or, as in the case of isolated populations, we let the individual p's and q's vary, and repackage populations at the expense of segregating ones.

Chance Events: Mechanical Simulation

At this moment, we have seen that the random union of gametes in a finite population leads to an increase in F, the inbreeding coefficient, which reflects in turn an increase in the proportion of individuals homozygous for alleles that are identical by descent from a common, ancestral allele. We have also seen that the subdivision of a large population into smaller ones leads to a divergence of gene frequencies in these populations. In the latter case, genetic variation which at the outset was present as segregating alleles *within* a single population became, through the operation of chance events, genetic variation *among* dissimilar populations. In this section, the process of population differentiation will be examined by means of mechanical or computer models.

The operation of chance in determining gene frequencies in populations will first be illustrated by means of numerous red and white beans and nine one-pound coffee tins. The model involves samples of ten beans, some of which are red and some white. Excluding the terminal frequencies—100 percent red and 100 percent white—only nine possible frequencies of red beans exist in a sample of 10: 10, 20, 30, 40, 50, 60, 70, 80, and 90 percent. In preparation for the experiment, then, nine tins, clearly labeled, are set up in which there are a total of some 250–300 beans, but in which the frequency of red ones is precisely 10 percent, 20 percent, and so forth.

The nine containers are for convenience; they make the experiment go faster than it otherwise would. The experiment consists of games. Each game begins by drawing 10 beans from the container labeled 50 percent (see table 8.2). Among the beans of the first draw, as shown in game #1 of the table, there may be

Table 8.2 Results of seven games played on the random drift machine as described in the text; numbers in the body of the table are the numbers of red beans in drawings of ten beans each.

Draws	Games						
	1	2	3	4	5	6	7
0	5	5	5	5	5	5	5
1	4	4	4	8	6	6	5
2	7	3	6	8	6	6	4
3	5	4	9	8	7	5	3
4	7	1	9	7	6	3	4
5	6	1	9	7	6	3	4
6	7	0	9	9	7	2	3
7	6	—	7	10	6	3	3
8	8	—	7	—	6	4	2
9	9	—	4	—	3	6	3
10	9	—	5	—	4	7	1
11	10	—	5	—	4	7	2
12	—	—	6	—	0	6	0
13	—	—	7	—	—	7	—
14	—	—	10	—	—	7	—
15	—	—	—	—	—	7	—
16	—	—	—	—	—	8	—
17	—	—	—	—	—	9	—
18	—	—	—	—	—	10	—
19	—	—	—	—	—	—	—

only 4 (40 percent red ones). The ten beans are returned to the 50 percent container (so that the frequencies of red and white beans in each coffee tin remains constant), and a second draw is made, this time from the container marked 40 percent. In the first game listed in table 8.2, the second draw contained 7 (70 percent) red beans. The 10 beans were returned to the tin from which they had been taken, and a third draw was made from the tin labeled 70 percent. This draw contained 5 red beans which dictated that the next one would once more be from the tin labeled 50 percent. The game contained until the draw (the eleventh in the game we are following) contained only red beans.

The steps in this game have their biological counterparts. A drawing of 10 beans represents a sample of $2N$ gametes of N (= 5) diploid individuals (the game ignores sex). The 10 beans yield gene frequencies: p red beans, q white ones. The five individuals produce many gametes (the 250–300 beans in the labeled tins) among which gene frequencies are the same as in the sample of 10. From these many gametes, 10 are drawn at random; these represent the $2N$ genes carried by 5 (= N) individuals of the next generation. A game represents a population; each draw represents a generation. Drawings continue until the population reaches fixation at 0 percent red or 100 percent red—all white or all red beans. In the absence of mutation, there is no return from fixation at one or the other terminal frequencies. Although the 0's and 10's are not listed repeatedly following the drawing on which they are attained, it is understood that these populations continue with those extreme frequencies.

Data of the sort tabulated in table 8.2 have been summarized for 100 games in table 8.3. The first draw was always made

Table 8.3 Results of 100 games played on the random drift machine. Entries are the number of games in which various numbers of red beans were found remaining after the stated number of draws.

					Number of Reds					
Draws	0	1	2	3	4	5	6	7	8	9	10
0	—	—	—	—	—	100	—	—	—	—	—
1	0	2	4	11	17	24	24	11	4	2	1
3	3	1	9	11	13	17	11	14	7	9	5
5	9	7	6	8	11	6	10	11	10	12	10
10	28	4	4	2	5	5	3	4	4	8	33
15	36	2	1	2	3	2	4	2	2	0	46
20	37	0	2	3	0	3	0	3	1	2	49
25	42	0	0	0	1	0	3	0	1	0	53
30	42	0	0	0	1	0	1	0	0	1	55
35	43	0	0	0	0	0	0	0	0	0	57

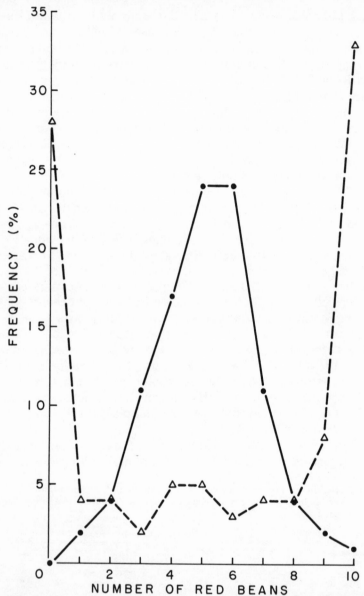

Figure 8.4 The results of 100 games consisting of draws of 10 beans from containers with 10, 20, . . . , 90 percent red beans. The first draw was always from the container with 50 percent red beans; subsequent draws were from containers whose contents corresponded to the results of the previous draw. The bell-shaped curve shows the results of the first draw; the ∪-shaped curve shows those of the tenth drawing.

from the tin in which the proportions of red and white beans were 50:50. The results of draw #1, consequently, should fit the binomial distribution corresponding to $(\frac{1}{2} + \frac{1}{2})^{10}$. Rounded to the nearest whole draw, the numbers of draws of 100 that are expected to contain 0, 1, 2, 3, . . . , 10 red beans are 0, 1, 4, 12, 21, 25, 21, 12, 4, 1, and 0. A comparison of these expectations with the observed numbers listed in table 8.3 reveals that the fit is excellent.

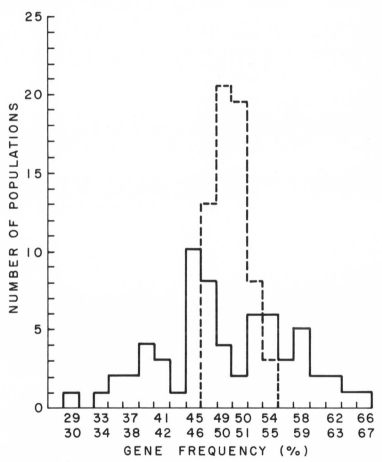

Figure 8.5 Results of a computer-simulated drawing comparable to that in figure 8-4, but on a scale that could not be tested by manual draws of beans from cans. Population size N = 250. Number of populations studied, 64. Initial gene frequency (generation 0), 0.50. Dashed line, distribution of gene frequencies in generation 1; solid line, distribution in generation 29. (Computer data compiled by Paul Miller.)

As the drawings continue, the number of populations (= games) containing intermediate numbers (4, 5, and 6) of red beans becomes smaller, whereas, correspondingly greater numbers contain few (1, 2, 3) or many (7, 8, 9) red ones. By the tenth draw (table 8.3 and figure 8.4) the distribution of populations with various numbers of red beans, rather than being bell-shaped as it was after the first draw, is U-shaped; many populations have already become fixed at 0 or 10 red beans, the others are now uniformly distributed over intermediate values. By generation 35, all populations are fixed at the terminal frequencies, 0 percent and 100 percent red. The number of populations fixed at these two extremes should be equal and, in agreement with expectation, the observed numbers, 43:57, do not differ significantly from the expected ones, 50:50.

Figure 8.5 shows the results of a computer-simulated drawing involving a hypothetical population of 250 individuals ($2N = 500$); distributions of gene frequencies are given for generations 1 and 29 (starting gene frequencies were 50:50 as in the bean-drawing described above). Here we can see the stabilizing influence of population size on gene frequencies. During 29 generations, the frequency distribution has spread (its variance has increased) but the distribution is still bell-shaped, and no population has yet approached the limiting gene frequencies of 0 percent and 100 percent.

Chance Events: Experimental Studies

The outcome of experiments done by computer or using coffee tins can be verified as well by experiments using laboratory organisms. One such experiment (Buri 1956) has been thoroughly analyzed by Falconer (1960). Here we shall present the results of one of three experiments described by Kerr and Wright (1954) in successive issues of the journal, *Evolution.*

Kerr and Wright studied ninety-six lines (a "line" corresponds to a "game" or a "population") each of which contained at the outset two alleles, *forked* (*f*) (a recessive, sex-linked mutation affecting bristle shape) and its wild-type allele (+) in equal frequencies. The zygotic constitution of the initial generation in each line was: 1 *f/f* : 2 *f/*+ : 1 +/+ (females) and 2 *f/Y* : 2 +/Y (males).

The initial parents were eight in number; in each succeeding generation, eight flies (4 females; 4 males) were chosen at random to be the parents of the next generation. The frequencies of *f* and + fluctuated from generation to generation. If at any time all eight flies chosen to be parents were *forked* (*f/f* and *f/Y*), that population was fixed at 100 percent *forked*. If, on the other hand, all eight were wildtype in appearance, and if there were no *forked* flies among their progeny, the population was assumed to be fixed at 0 percent *forked*. Gene frequencies were not determined for segregating populations because +/+ and +/*f* females are both wildtype in appearance.

The results of this experiment are summarized in table 8.4 and in figure 8.6. The table lists the numbers of populations that had become fixed during the 16 generations of the experiment as well as the number still segregating both + and *f* alleles.

In figure 8.6 (dotted line) the proportions of segregating populations remaining in successive generations are shown in a semilog plot. From the third until the sixteenth (last) generation, the proportion of segregating populations decreases linearly. On theoretical grounds, the *loss* of segregating populations is expected to approach a constant equal to $1/2N$ ($1/1.5N$ in the case of sex-linked alleles because males are hemizygous). The observed proportion of segregating populations *remaining* each generation from generations 3 to 16 equals 0.911; the expected proportion equals

Table 8.4 Genetic drift in small laboratory populations of *D. melanogaster*. For each generation are listed the number of populations that contained only wildtype or mutant (*forked*) alleles or that were still segregating. (After Kerr and Wright 1954.)

Generation	Wild	Unfixed	Forked
0	0	96	0
1	1	94	1
2	1	92	3
3	2	87	7
4	7	79	10
5	10	70	16
6	11	66	19
7	16	59	21
8	17	56	23
9	20	52	24
10	24	47	25
11	29	39	28
12	31	37	28
13	34	34	28
14	37	30	29
15	38	29	29
16	41	26	29

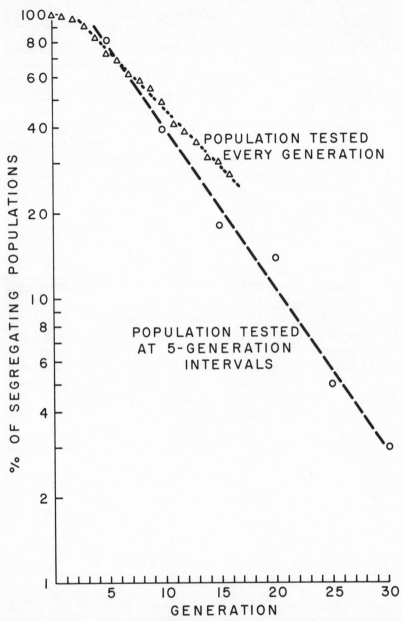

Figure 8.6 Semilog plot of the decline in the proportions of segregating (i.e., unfixed) populations of tables 8.3 (dashed line and circles) and 8.4 (dotted line and triangles). Beyond generation 5 the loss of unfixed "populations" in table 8.3 becomes exponential with the ratios of remaining proportions in successive generations being 0.877; beyond generation 3 the loss of unfixed populations in table 8.4 also becomes exponential with the corresponding ratio being 0.911. These values very nearly correspond to 0.900 and 0.917, the expected values as explained in the text.

$\frac{11}{12}$ or 0.917. Figure 8.6 (dashed line) also shows the loss of segregating populations which can be calculated from table 8.3. In this case, the proportion of segregating populations remaining each generation from generation 5 to 30 was 0.877; the expected proportion ($\frac{9}{10}$) equals 0.900. In each case, the loss of segregating populations corresponds to that expected on the basis of the number of sampled genes.

The Chance Loss of Lethal Genes

As we have seen throughout this chapter, chance events affect gene frequencies and the frequencies of zygotes within populations. These events also affect *numbers* of genes in ways that might not be obvious from discussions devoted to *frequencies.* Suppose, for example, one imagines a large number of populations each containing one and only one allele, *a*. This one allele, of course, is carried by a heterozygous, *A/a*, individual. What is the probability that the *a* allele will be present in a given population in the following generation? Provided that the number of progeny per parental pair is a random number (with an average value of 2), the distribution of *a* in the next generation equals that of a Poisson distribution with $m = 1$: 37 percent of all populations will have lost *a*, 37 percent will have retained one *a* allele, $18\frac{1}{2}$ percent will have two, some 6 percent will have three, $1\frac{1}{2}$ percent will have four, and still fewer will have five or more *a* alleles. This pattern of loss and gain is affected only slightly by any small effect *a* might have on the survival or fertility of its heterozygous carriers.

Table 8.5 presents the results of calculations similar to the above but which have been carried through eight generations. (Sewall Wright once presented me with a corresponding table carried through 10 generations. That handwritten table was eventually lost. I never appreciated, until I generated by hand the numbers in table 8.5, the effort that Dr. Wright expended in providing me with the original data, an effort typical of the aid Dr. Wright extended to his less mathematically inclined colleagues.) Imagine a very large population in which *1* recessive lethal allele is found at each of 1,000 autosomal loci. The population is so large that, even though a given lethal may eventually be represented several times, elimination by homozygosis does not occur. The question is: What is the fate of these lethals in successive

Table 8.5 The chance fate of 1,000 unique recessive lethal alleles in a very large population where, even though there may eventually be more than one copy of some alleles,
homozygosis does not occur. (The sums of the columns do not always equal 1,000 because of rounding errors in the computations.)

Number of copies of various lethals remaining in population	Generation					
	0	1	2	4	6	8
0	—	368	531	690	786	842
1	1,000	368	195	84	48	31
2	—	184	134	72	44	28
3	—	61	72	52	34	23
4	—	15	37	37	27	20
5	—	3	17	22	19	15
6	—	1	7	15	13	10
7	—	—	3	10	10	9
8	—	—	—	8	8	8
9	—	—	—	4	5	6
10	—	—	—	1	4	5
11	—	—	—	—	1	2
12	—	—	—	—	—	1
13	—	—	—	—	—	—
Total	1,000	1,000	996	995	999	1,000

generations? Because no procedure for eliminating them has been
provided, we must assume that the total number of lethal alleles
in this population remains approximately 1,000.

As shown in table 8.5, in the first generation, about 37 percent of the lethals are lost and the rest (1,000 of them in all) are
represented once, twice, three, or more times. In going from generation 1 to generation 2, the lost alleles remain lost, 37 percent of
the 368 that are present once are now lost, 18.5 percent of the
184 that are present twice become lost, 6 percent of the 61
present three times, become lost, and so forth. Calculations such
as these lead to the distribution shown under generation 2. In each
succeeding generation, more and more of the original 1,000 lethals
are accidentally lost. Because in the absence of selection the total
number of lethal alleles remains 1,000, the few remaining lethals
become (accidentally) more and more common. The outcome is
plain: within very few generations most of the original lethals are
lost; the few that remain, however, are surprisingly common.

The information contained in table 8.5 is extremely important in respect to experimental data such as those presented in
table 8.6. The experimental data concern the frequencies of individual lethal-bearing chromosomes in once-irradiated populations
of *D. melanogaster* ($N \sim 5000$) 45 generations after exposure to
irradiation had ceased. Most lethals found in the experimental

Table 8.6 Numbers of lethals found once, twice, or three or more times in irradiated populations of *D. melanogaster* some 45 generations after irradiation had ceased. (After Salceda 1967.)

Population	Times Found									
	1	2	3	4	5	6	7	8	9	10
C	6	12	2	1	1	—	—	—	1	1
D	10	7	7	1	2	1	1	—	2	2
Total	16	19	9	2	3	1	1	—	3	3

populations occurred once or twice; six occurred nine or ten times each. Were the latter lethals of a sort unlike the others which remained at low frequency? Perhaps, but not necessarily. The calculations presented in table 8.5 show that by chance alone, some lethals are *expected* to attain high frequencies in populations.

Chance Events and Gene-Frequency Surface

In this, the concluding section on chance events, I want to shift the emphasis from what was a consideration of essentially independent populations to a surface of interconnected ones—a surface in which the sharing of gene pools between overlapping populations (migration) interacts with chance events. The topic of discussion is a surface generated by computer-simulation (figure 8.7). The entire surface consists of 10,000 individuals in a 100 X 100 array. Chance events occur in neighborhoods of size 25 (5 X 5) and 9 (3 X 3). The neighborhoods work their way back and forth across the entire surface so that nearby neighborhoods share chance events in a manner corresponding to a two-dimensional moving average.

The figure illustrates that the surface of gene frequencies is smoother and more regular with a neighborhood size 25, than it is with one of 9. That should not be surprising: if neighborhoods had been reduced to 1, each one would have been independent; if it had been enlarged to 10,000, no irregularities would have occurred because all individuals would have contributed to one calculated gene frequency.

What may be surprising in figure 8.7 is the persistence of certain irregularities in the surface over as many as 60 or 80 generations. Persons often overlook the inertia that accompanies cumulative chance events. If, for example, one were to toss a coin every

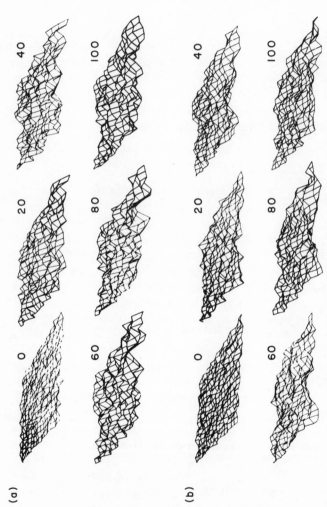

Figure 8.7 A perspective view of surfaces of computer-simulated gene frequencies for various generations over a 100 × 100 array of individuals. (a) The surface generated when the neighborhood size is 9; (b) the corresponding (but less irregular) surfaces generated when the neighborhood size is 25. Surfaces are shown for every 20 generations. (After Rohlf and Schnell 1971. Reprinted by permission of the University of Chicago Press from Rohlf and Schnell *Amer. Nat.* (1971), 105:295-324); © 1971 University of Chicago.

second for an entire year and to note at the end of that time which side (heads or tails) had predominated *by chance*, in one case out of five one would find upon examining the entire record that the side which predominated at the end of the year had predominated for approximately 356 of the 365 days. That is, the inertia accompanying an excess of heads, lets's say, gained during the first days of tossing, is not likely to be overcome by isolated chance events occurring in the subsequent tosses, even in the 30 million that follow. In a similar way, once a pronounced irregularity has appeared in the gene-frequency landscape as shown in figure 8.7, future chance events are quite ineffective at erasing it.

Summary

The material covered in this chapter was discussed in some detail, but with no pretence at mathematical expertise. We have covered a number of topics; some of which proved to be quantitatively interrelated. The effect of inbreeding on the loss of hereditary variation is clear: the identity of alleles by descent does, indeed, lead to a loss of heterozygosity. I will warn the reader, however, that the interrelations between population size, the subdivision of populations, chance fluctuations in gene frequencies, interpopulation variance in respect to gene frequencies, and the loss of heterozygotes through inbreeding within individual populations are subtle—and elusive.

FOR YOUR EXTRA ATTENTION

1. Rich et al. (1979) followed the fate of the wildtype (b^+) and black (b) alleles in 12 populations of flour beetles (*Tribolium castaneum*) of each of four sizes (10, 20, 50, and 100 individuals) for 20 generations. A summary of their data is tabulated below; both the mean frequency of the wildtype gene and standard deviation among the 12 populations are given:

Generation	Population Size							
	10		20		50		100	
	$\bar{X}$	σ	$\bar{X}$	σ	$\bar{X}$	σ	$\bar{X}$	σ
0	.50	0	.50	0	.50	0	.50	0
5	.64	.18	.62	.14	.61	.05	.64	.05
10	.68	.26	.65	.20	.70	.09	.74	.04
15	.77	.28	.76	.21	.77	.12	.79	.05
20	.85	.28	.78	.24	.81	.11	.87	.04

Which reflects the role of population size: the mean or the standard deviation?

2. Relate the following observation (Levin 1978b) to material covered in this and the previous chapter: Crosses between plants (*Phlox*) separated by two meters or less yield a larger percentage of aborted embryos than crosses between plants more distantly spaced.

Compare the results of these crosses with those summarized in figure 6.3.

3. Connor and Bellucci (1979) perpetuated 10 lines of wild-caught mice by brother-sister matings for 20 generations.

What is the genetic consequence of brother-sister matings? Can you quantify this expectation, recalling that $N = 2$? What should be the effect of inbreeding such as this on allozyme loci that were polymorphic in the original population? Or, on the success of inter-sib skin grafts?

Compare your expectations with these observations:

1. Inbreeding depression was discontinuous (irregular).
2. Skin grafts failed regularly in four of the five lines (the other five were lost during the inbreeding regime).
3. Surviving lines were generally heterozygous for allozyme loci that were polymorphic among the parental mice.

Explanations for discrepancies between expectations and observations will be encountered in subsequent chapters.

Mutation

PREVIEW: The Hardy-Weinberg equilibrium does not provide for the conversion of one allele (say, *A*) into another (*a*, for example); it assumes that there is no mutation. The present chapter presents a brief account of the role of mutation, the molecular basis of mutation, an outline of the phenotypic effects of mutations, a description of the procedures by which mutations are detected and mutation rates measured in various organisms, and a theoretical analysis of the consequences of mutation on gene frequencies in populations.

The Role of Mutation

MUTATION IS THE ultimate source of new variation. Whenever we mention two alleles, *A* and *a*, we assume that one of these alleles has arisen by mutation from the other or that each has arisen from an earlier common form; different allelic states of a gene arise only as a result of mutational changes. Within a local population, migration acts as an alternative, immediate (proximate) source of variation. Without mutation somewhere in the past, however, natives and immigrants would be identical. Similarly, recombination can give rise to enormous arrays of new combinations of genes once allelic differences exist at a number of gene loci. Without the initial mutational divergence at individual loci, though, there would be nothing different to recombine. Migration and recombination are powerful secondary generators of hereditary variation, but mutation remains as the sole ultimate source.

Mutation plays a special role in population genetics as the spoiler of genetic fixation. It provides a means by which alleles, once lost, can be regained. In chapter 8 we saw that no matter how large the population, there is always a chance that an allele will be fixed within or lost from that population. In discussing genetic drift, these limiting frequencies—0 percent and 100 percent—were treated as terminal frequencies; games in which the frequency of red beans reached either one were terminated. In natural populations these terminal states are not foolproof traps. A population that contains only A can regain a by mutation; similarly, one containing only a can regain A. The rate of fixation and loss of genes from a large population is very small (the loss of segregating populations occurs at a rate equal to $1/2N$), as we have seen; only a small rate of mutation is required to counterbalance this loss of genetic variation.

Some Molecular Aspects of Mutation

In the earier version of this text, the following quaint sentence appears: "Molecular evolution at the moment does not greatly concern the population geneticist" (Wallace 1968a:131). Since that sentence was written, many have concluded that molecular evolution *is* population genetics. My view is not so extreme, but during the intervening years, I have realized that an understanding of the types of mutations and of mutational events which are possible (and do occur) at the molecular level is important. Only when these events are understood, can symbols be devised and safely manipulated in population models, because behind each symbol is a tacit notion of the physical system itself. Without a knowledge of the physical system, symbols become empty; like commercial television programs, they become associated with concepts corresponding to the lowest denominator common to those who use them. For example, gene control during the development of a higher organism may involve the rearrangement of DNA, the consequent sequestering of some genes (so that they cannot be transcribed), and the exposure of others to the transcribing mechanism. Such a system of control can be discussed in terms of somatic mutation (for example, Becker 1957) or in terms of normally occurring mosaic individuals (Stern 1968); furthermore, a symbol such as a_1 can be defined or described in such a way as to include mosaicism of its carriers as one of its properties. Such properties will not normally be read into such symbols, however, unless those who devise and use them are aware of the molecular mechanisms for which the symbols are a shorthand notation. At the moment the three notations—*AA, Aa*, and *aa*—convey a sense only of the stability of the genes for which the letters, *A* and *a* stand, and convey no sense of the machinations these same genes go through in affecting the development and survival of their carriers.

Structural genes

In figure 9.1, a brief catalogue of mutations occurring in structural genes (those that specify the amino-acid composition of specific proteins) is presented. These mutations are caused by the substitution of one nitrogenous base for another in the DNA molecule, or from the addition or loss of bases from DNA. The figure identifies the alteration in the DNA molecule, the composition of messenger RNA that is transcribed from the (bottom) strand of the DNA molecule, and the sequence of amino acids that is assembled under the direction of the mRNA molecule. The genetic code is not reproduced here; elementary texts such as Watson (1976) and Lehninger (1970) present not only the familiar table of codons but also additional important details.

Three mutations that occur through the substitution of one base pair for another are shown in figure 9.1; these are missense mutations. Transitional mutations are those in which a purine replaces a purine, or a pyrimidine a pyrimidine. The first substitution of this sort shown in the figure does not alter the structure of the polypeptide chain; the missense mutation is transitional, but it is also synonymous. Nei (1975:24-25) has estimated that nearly 25 percent (0.23-0.24) of all base substitutions result in synonymous mutations; these mutations do not alter protein structure. [They may, as Richmond (1970) has emphasized, require the use of different transfer RNAs in translating codons into amino acids.]

The replacement of a purine by a pyrimidine results in a

```
        DNA                        mRNA                POLYPEPTIDE

NON-MUTATED
A-T-T-C-T-G-A-C-T-T-G-C-
T-A-A-G-A-C-T-G-A-A-C-G-    A-U-U-C-U-G-A-C-U-U-G-C-   Ile-Leu-Thr-Cys-


MISSENSE MUTATION: TRANSITIONAL, SYNONYMOUS
A-T-T-T-T-G-A-C-T-T-G-C-
T-A-A-A-A-C-T-G-A-A-C-G-    A-U-U-U-U-G-A-C-U-U-G-C-   Ile-Leu-Thr-Cys-


MISSENSE MUTATION: TRANSITIONAL, NON-SYNONYMOUS
A-T-T-C-T-G-G-C-T-T-G-C-
T-A-A-G-A-C-C-G-A-A-C-G-    A-U-U-C-U-G-G-C-U-U-G-C-   Ile-Leu-Ala-Cys-


MISSENSE MUTATION: TRANSVERSION
A-T-T-C-T-G-A-C-T-G-G-C-
T-A-A-G-A-C-T-G-A-C-C-G-    A-U-U-C-U-G-A-C-U-G-G-C-   Ile-Leu-Thr-Gly-


NONSENSE MUTATION
A-T-T-C-T-G-A-C-T-T-G-A
T-A-A-G-A-C-T-G-A-A-C-T     A-U-U-C-U-G-A-C-U-U-G-A    Ile-Leu-Thr STOP


FRAMESHIFT MUTATION
G-A-T-T-C-T-G-A-C-T-T-G-C-
C-T-A-A-G-A-C-T-G-A-A-C-T-  G-A-U-U-C-U-G-A-C-U-U-G-C-  Asp-Ser-Asp-Leu
```

Figure 9.1 A brief catalogue of gene mutations caused by a gain (or loss) or substitution of base pairs in DNA. Each example shows (1) the alteration in the DNA, (2) the modified mRNA, and (3) the (generally) altered sequence of amino acids that is finally assembled. The genetic code can be found in most modern textbooks on general genetics; it might be recalled here, however, that there are three translation-terminating codons: UAA, UAG, and UGA. Naturally occurring allozyme variation consists primarily of nonsynonymous missense mutations.

transversion. Because of the nature of the genetic code, transversions result in fewer synonymous mutations than do transitional mutations.

The last two mutational events illustrated in figure 9.1 are mutations to nonsense (or terminating) codons, codons that bring the growing polypeptide chain to an end, and frameshift mutations. Frameshift mutations result in a misreading of the genetic message carried by the mRNA molecule; consequently, such mutations can be caused by the addition or deletion of one or two base pairs. The addition or deletion of three base pairs, of course, either adds or subtracts one amino acid from the protein molecule while leaving the remainder unchanged. Among the 64 possible codons, three are termination signals. The misreading of the mRNA molecule carries with it a 5 percent chance of encountering a termination signal at each newly generated codon; frameshift mutations, consequently, result ("downstream" from their point of occurrence) in synthesis of rather short, more or less randomly constructed sequences of amino acids.

The mutations described in figure 9.1 are those that affect the amino-acid composition of protein molecules. Some of the mutations are synonymous, and fail to result in an alteration of the amino-acid sequence; about one quarter of all base pair substitutions are of that sort. Furthermore, amino acid substitutions are not always revealed by electrophoresis; the migration of a protein molecule in an electrical field is altered only if the substitution of one amino acid for another causes a change in the electrical charge of the molecule (chapter 4).

Of the 20 amino acids, 2 are positively charged, 2 are negatively charged, and 16 are neutral. The probability that the substitution of one amino acid for another will result in a charge change is about 36 percent, as we saw on page 91. Nei (1975:25) presents more precise calculations based on changes in DNA itself as well as the results of empirical studies: approximately one-third of all amino acid substitutions result in a changed electrical charge on the protein involved, a change that is detectable by electrophoresis.

Regulatory genes

Following the classic studies of Monod and Jacob (1961), geneticists have recognized that structural genes are themselves under genetic control, and that mutations affecting the presence or absence of functional enzymes may be regulatory mutations rather than mutations that alter the enzyme's molecular structure. The control of the *lac*-operon of *E. coli* is shown in figure 9.2, following an account given by Dickson, et al. (1975). The chemical structure of three enzymes (β-galactosidase, a permease, and a transacetylase) are specified by three adjacent structural genes (*z, y,* and *a*); the first two of these enzymes are needed for the utilization of the sugar, lactose. These three genes cannot be transcribed unless region *O*, the operator, is "open"; normally, the

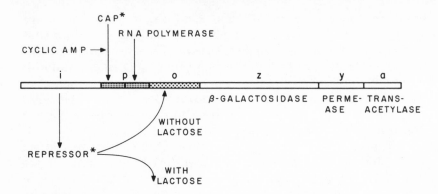

*REGULATOR PROTEINS

Figure 9.2 Some of the regulative processes controlling the normal functioning of the *lac* region in *E. coli*. The genes, *z*, *y*, and *a*, specify the primary structure of three enzymes, two of which are required for the metabolism of lactose. The control of these genes and the effects various mutations have on this system are described in the text. (After Dickson et al. 1975.) Reproduced with permission: from *Science* (1975) 187:27–35; © 1975 American Association for the Advancement of Science.

operator is "closed" because of the attachment of a repressor protein that is made under the direction of a repressor gene (*i*). Lactose (and certain other chemicals) binds to the repressor protein, alters its shape, and prevents it from binding with the operator region; thus, lactose can turn on the genetic and enzymatic machinery necessary for its own utilization.

Bacteriologists have long known that cultures of *E. coli* that are grown in the presence of the two sugars, glucose and lactose, will exhaust the glucose from the medium, pause momentarily, and then utilize the lactose. Thus, the lactose-utilizing machinery need not function even in the presence of lactose if glucose is also present. This additional control is effected through a catabolite gene-activator protein (CAP) whose presence is needed in order that RNA polymerase molecules be able to attach themselves to the *E. coli* chromosome at the promoter (*p*) region; such attachment is necessary for the transcription of the three structural genes *z, y,* and *a*. The activator protein (CAP) in turn requires cyclic adenosine monophosphate (cyclic AMP), but glucose reduces the amount of cyclic AMP present in the cell. Consequently, even though the presence of lactose may "open" the operator, transcription of the three structural genes of the *lac*-operon does not occur until glucose has been removed from the growth medium.

The model of gene control exemplified by the *lac*-operon (models represented by other loci differ in many details) identifies a number of sites at which mutations affecting the proper functioning of the operon might occur. Mutations within region *0* might affect the recognition and binding of the repressor protein. Mutations within the gene *i* might affect the repressor protein so that it fails to recognize the operator region of the chromosome, recognizes it but falls off spontaneously, or fails to recognize lactose when that sugar is present. Similar types of changes might occur in the gene-activating protein (CAP) or in proteins mediating the effect of glucose on cyclic AMP. Mutations in the promoter region (*p*) could affect the attachment of CAP, the attachment of RNA polymerase, or both. Finally, although the effect would not be limited to the *lac*-operon alone, mutations in the locus responsible for specifying the RNA polymerase molecule would also affect the accuracy with which mRNA is transcribed from the bacterial chromosome.

Genetic control systems with their numerous feedback loops are capable of an infinite variety of normal patterns; as more information is gained by microbial geneticists concerning the working of different systems, the less likely it seems that any one has provided a master plan for all others (Stanley A. Zahler, personal communication). The extent, then, that these complex systems (and even the simplest-appearing ones *are* complex) can be modified by mutational events is beyond comprehension. A number of persons (Wallace 1963a; Zuckerkandl 1963; Wallace and Kass 1974; Britten and Davidson 1969; Lerner et al. 1964; Wilson 1976) have emphasized the possible importance of regulatory mutations in populations and their evolution.

Insertional elements

Still another mechanism for gene control is illustrated in figure 9.3, a mechanism that frequently masquerades as gene mutation (Starlinger and Saedler 1972). Organisms are often characterized by seemingly mutually exclusive, alternative states (the *a* and α mating types of yeast are an example) which may "mutate" at rare intervals from one to the other; nevertheless, individuals do not normally exhibit both states: the gain of one is accompanied

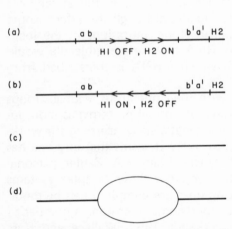

Figure 9.3 A model illustrating the "flip-flop" control of flagellar protein genes in *Salmonella*. The gene H2, lies adjacent to a segment of DNA that can take either of two orientations (*a* and *b*). Each orientation can at rare intervals change to the other, possibly through intrastrand recombination between homologous segments having reverse orientations (*ab–b'a'*) as illustrated in the text. A test of this model consists of melting DNA fragments including this region and allowing the single strands to reanneal. Fragments containing DNA corresponding to either *a* or *b* reanneal to give intact double-stranded DNA as shown in *c*. A mixture of fragments corresponding to both *a* and *b*, however, upon reannealing produces not only intact double-stranded DNA as in *c* but also strands (*d*) that exhibit single-stranded "bubbles." The latter reveal the position and length of the inverted segment of DNA. (After Zieg et al. 1977.)

by the concomitant loss of the other. What may prove to be a general mechanism for genetically controlled switching from one state to another has been described in *Salmonella*; a few details are shown in figure 9.3. Although they are not closely linked, the transcription of two gene loci is controlled by a stretch of DNA that is capable of reversing its orientation within the *Salmonella* chromosome. This "flip-flop" portion allows transcription in only one direction. Only one of the two loci can be transcribed at one time; when the one is "on," the other is "off." A structure of DNA that can permit the rotation of an intervening segment is that of a separated reverse-repeat:

$$-a\,b\,c\,d\,.\,.\,A\,B\,C\,D\,.\,.\,d'\,c'\,b'\,a'-$$

DNA with this physical structure can pair on itself and recombine

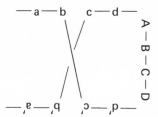

thus, generating the following:

$$-\text{a b c d} \ldots \text{d' c' b' a'}-.$$

The flip-flop control of gene action described in figure 9.3 is important because the physical structure is similar to that which underlies the creation of the extrachromosomal elements which are responsible for many insertional mutations in bacteria, and for the insertion of viral DNA into, and their excision from, bacterial chromosomes. Furthermore, it is suspected that the same geometrical properties of DNA are responsible for the transposable elements in higher organisms which, at least when things go wrong,

LICUALA PELTATA

CARYOTA MITIS

Figure 9.4 The control of leaf shapes in palm trees. The individual leaflets of *L. peltata* are virtually identical in length, thus forming circular fronds. Analogous segments of the individual leaflets of *C. mitis* show nearly random patterns of growth. Three points can be made in reference to the leaves of *C. mitis*: (1) leaves of random shape provide protection against insects that search according to an established pattern or avoid seemingly damaged leaves, (2) randomization, like many evolutionarily important traits, is more a matter of the control of gene action than one of new enzymes and new metabolic pathways, and (3) transposable control elements possess properties that can easily account for patterns, including random ones.

given rise to unstable, mutable loci (see McClintock 1956; Green 1969). The contrast in the regularity and irregularity of the leaves of different species of palms shown in figure 9.4 is only one of many possible examples in which the individual's phenotype reflects a control of genes during development that might be regarded as a controlled series of "mutational" changes.

The Phenotypic Consequences of Mutation

Types of mutations
The types of mutations that are detected and studied vary with the experimental organism and the interests of the experimenter. The genetic control of metabolic pathways is still the main interest of microbial geneticists: the biochemical mutant is still the tool, although recent work centers on its control (see Hicks et al. 1979). Higher organisms do not live on chemically defined media and so (with the notable exception of those studying the inborn errors of human metabolism) most workers deal with morphological mutants. Population geneticists deal largely with mutant alleles that flow in and out of populations. For the most part these have been viability mutations—mutations that affect the survival and reproduction of their carriers. Electrophoretic procedures have allowed the separation of *detection* of a mutant allele from its *effect,* and with this separation has come the suggestion that many protein alterations have no phenotypic effect on their carriers, that the various alleles are functionally equivalent, and, hence, that they are neutral in respect to natural selection.

A single amino acid substitution
Mutational changes are ultimately changes in the composition of DNA, the smallest of which might be a transitional change at the site of a single base pair. Britten and Davidson (1969) have pointed out that such a change in a region that is responsible for initiating transcription may stop transcription completely: the least conceivable genetic change, therefore, may have a considerable effect. Miozzari and Yanofsky (1978) have reported that a single base-pair substitution in the control region of the *trp* locus in *Shigella dysenteriae* seems to be responsible for the 90 percent reduction in the promoter activity in this bacterium relative to

the activity which is characteristic of *Escherichia coli*. Single base-pair substitutions, consequently, can be important in respect to certain evolutionary and adaptive changes.

The substitution of one amino acid for another in a protein molecule may have no effect on the organism; indeed, as we have seen, it may even have no detectable effect on the electrophoretic mobility of the molecule. On the other hand, such a simple change may have an enormous, even fatal, effect. A mutation at the locus responsible for specifying the amino acid sequence in the β chain of human hemoglobin results in the substitution of one amino acid (valine) for another (glutamic acid) in the sixth of 146 positions. The result of this substitution is an inherited abnormal hemoglobin: hemoglobin S, or sickle-cell hemoglobin. The term "sickle cell" refers to the bizarre shapes taken on by the red blood cells of homozygous carriers of the mutant gene ($Hb^S Hb^S$), and by those of heterozygotes ($Hb^A Hb^S$) when the oxygen tension of the blood is lowered. Figure 9.5a summarizes the cascading (or pleiotropic) effects of this single amino acid substitution. The sometimes-terminal characteristics for the sickle-cell homozygote are heart and kidney failure, and pneumonia. Figure 9.5b shows a similar table of pleiotropic effects stemming from a mutation causing abnormal cartilage in the rat. In this case the initial lesion is not known; it may or may not be a simple molecular change in collagen. However, the ramifications that arise from the initial anomaly stress once more the complications accompanying genetic disorders in higher organisms. In a sense, the greater the network of pleiotropic effects such as those shown in figure 9.5, the more difficult it becomes to imagine an unimportant genetic change.

Returning to single amino acid substitutions once more, hemoglobin variants in human beings can be used to illustrate changes that seem to have relatively minor effects, if any, on their carriers. Because these variants will be used in measuring mutation rates (page 206), we shall simply say here that approximately 100 abnormal human hemoglobins are known in all, and that the bulk of the amino acid substitutions—as one might expect—are those that can be explained by single base alterations in individual codons.

Visible mutations

Depending upon their experimental organisms, geneticists have developed a number of techniques for scoring the rate at which gross morphological (visible) mutations arise.

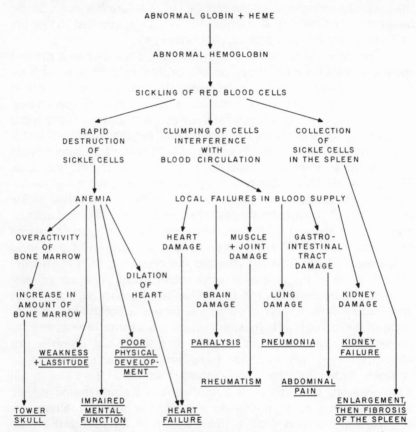

Figure 9.5a A chart showing the physiological, developmental, and anatomical pathologies that follow as consequences of homozygosis for the sickling gene, a mutant gene that alters one of 287 amino acids in the hemoglobin molecule. Reprinted by permission of the University of Chicago Press from Neel and Schull, *Human Heredity*; © 1954 University of Chicago.

The multiple-locus technique of mouse geneticists has proved to be an excellent way to study mutations in mammals. Large numbers of females homozygous for seven recessive mutations, each of which causes a distinct morphological change in the mouse, are crossed with males that are homozygous for the wildtype allele at each locus. Because the wildtype alleles are dominant, most F_1 offspring are wildtype in appearance; a few are not. Table 9.1 lists data on spontaneous mutations obtained in a study by W. L. Russell of more than one-half million progeny mice. Thirty-two mutations of all sorts among this total yields a mutation rate per locus of 0.84×10^{-5} in the case of spermatogonia;

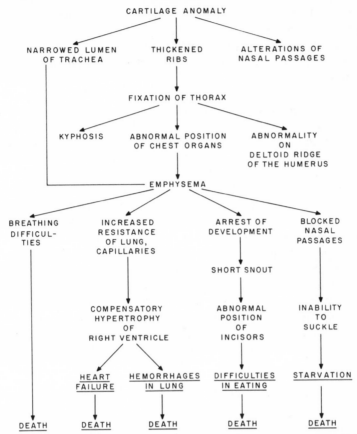

Figure 9.5b The cascade of developmental abnormalities, all leading to the individual's death, caused by an inherited abnormality of cartilage in the rat, *Rattus norvegicus*. (After Grüneberg; from Sinnott et al. 1958.)

Table 9.1 Spontaneous mutations at seven gene loci in the mouse *Mus musculus*. (After Russell; U.N. Sci. Comm. 1962:106.)

Spermatogonia	
Number of offspring	544,897
Number of mutations	32
Mutation/locus/gamete	0.84×10^{-5}
Oocytes	
Number of offspring	98,828
Number of mutations	1
Mutation/locus/gamete	0.14×10^{-5}

the rate (0.14×10^{-5}) obtained in a more modest study of oogonia appears to be significantly smaller. The term "rate," as it is used here, is really a "proportion"; the test measures the *proportion* of supposedly wildtype gametes that carry an unexpected mutant allele.

Drosophila geneticists have long used special techniques for studying the occurrence of mutations; one of these is based on the abnormal inheritance patterns resulting from the attachment of two X-chromosomes to a single centromere. An attached-X female carrying a normal Y-chromosome (XXY female) transmits her two X-chromosomes to her daughters (who are then also XXY because of the Y chromosome received from the father), and her Y-chromosome to her sons (each of whom receives his X-chromosome from his father). Hence, a wildtype male should produce wildtype sons; sex-linked visible mutations that occur, however, during the lifetime (including spermatogenesis) of the father will give rise to mutant sons if he is mated with an attached-X female. Examples of the use of this technique are given in table 9.2. Timofeeff-Ressovsky (1932) determined that the mutation rates from wildtype to a mutant allele at the *white* locus for two dif-

Table 9.2 (a) Mutations from wildtype to mutant alleles at the *white* locus in two geographic strains, American (A) and Russian (R), of *D. melanogaster*. (After Timofeeff-Ressovsky 1932.) (b) EMS-induced reverse mutations at the *white* locus of *D. melanogaster*. (After Banjeree et al. 1978.)

(a) Allele	Background	Number of Tests	Number of Mutations	Frequency
w^A	American	31,000	27	0.00087
w^A	Russian	28,200	28	0.00100
Total		59,200	55	av. 0.00093

41 of 55 mutations were *white*

w^R	Russian	49,200	26	0.00053
w^R	American	26,100	14	0.00054
Total		75,300	40	av. 0.00053

19 of 40 mutations were *white*

difference: 0.00040 ± 0.00013

(b) Mutant	Control			EMS Treated		
	Total number of chromosomes observed	Revertants	Frequency in 10^6 gametes	Total number of chromosomes observed	Revertants	Frequency of revertants in 10^6 gametes
w^{sp}	98,204	1	10.19	97,692	3	30.71
w^e	113,218	–	–	108,270	2	18.48
w^a	106,240	1	9.42	100,432	2	19.92
w^{col}	116,842	–	–	104,694	1	9.56
w^{Bwx}	121,305	–	–	118,032	–	–

ferent (Russian and American) wildtype alleles differed (table
9.2a), and that the difference was to be ascribed to the locus, not
to the remainder of the genotype. Table 9.2b, involving even larger
numbers of tested progeny, shows that a chemical mutagen (*E*thyl
*M*ethane *S*ulfonate or EMS) increases the frequency with which
various mutant alleles at the *white* locus in *D. melanogaster* revert
to wildtype. Because this mutagen usually causes base changes in
DNA rather than gross rearrangements, the results suggest that
these mutants themselves were the result of purine and pyrimidine
substitutions in the fly's DNA.

Viability mutations

The study of visible, morphological mutations suffers from two
serious flaws. One, as pointed out by Lewontin (1974a:36), gene
loci at which visible mutations arise seem to be a minority among
loci generally; perhaps they represent a minority of changes occurring at any one locus as well, in the sense that they represent
"leaky" mutations whose low level of activity permits the survival
of the affected individual, albeit in a deformed state. A second,
and in my opinion a more serious flaw, is that the scoring of
visible mutations is largely subjective: not all persons are equally
adroit at detecting minor but consistent deviations from that *array*
of phenotypes which is considered "normal." A crumpled wing
scored by one investigator is unnoticed by another; subtle differences in eye color represent striking changes to one person but
are invisible to the next. The results obtained by different persons
under such circumstances cannot be compared easily.

The modified *CIB* techniques described in chapter 3 largely
remove the idiosyncracies of the investigator from the experimental results themselves. In table 9.3, for instance, are listed

Table 9.3 Mutation rates for the second and third chromosomes in the two sexes in
D. melanogaster. (After Wallace 1968b.)

Chromosome	Sex	Lethals	Total	u
II	Male	22	3,454	.0064
II	Female	20	2,337	.0086
III	Male	16	2,797	.0057
III	Female	22	3,067	.0072
Males		38	6,251	.0061
Females		42	5,404	.0078
Chromosome II		42	5,791	.0073
Chromosome III		38	5,864	.0065
Grand total		80	11,655	.0069

the proportions of second and third chromosomes, recovered from lethal-free males and females, that proved to be lethal when homozygous. The data suggest that there are no marked differences in the proportions of newly arisen lethals among the gametes of the two sexes, nor are there large differences between the two chromosomes. Because each lethal was identified by the absence of an entire expected *class* of flies, no subtlety in scoring was involved. The results of a large number of second-chromosome tests agree in estimating the lethal mutation rate for that chromosome as about 0.005. A test, completely unlike that described in table 9.3, is shown in table 9.4; here the mutation rate per chomosome-generation is estimated to be 0.0044. There is reason in this case to believe that this figure may be biased downward by underestimating the length of one generation.

Because viability mutations are *defined* (not just *identified*) in terms of deviations from expected proportions of flies of different genotypes (see page 68), modified *ClB* techniques can be used to detect the origins of mutations with less than lethal effects. Table 9.5 lists results obtained by Timofeef-Ressovsky (1935) following the exposure of *D. melanogaster* males to X-rays; clearly, cultures with intermediate frequencies of males carrying irradiated X-chromosomes are proportionately more numerous in the experimental (X-ray) than in the control tests (see figure 9.6).

Table 9.4 Estimating the mutation rate of spontaneous lethals and semilethals from the accumulation of these mutations in an initially lethal-free population of *D. melanogaster*. A "generation" for this and other laboratory populations maintained at Cold Spring Harbor is really a 2-week interval between egg samples taken from the population cages; the actual generation time would be somewhat longer. Hence, the estimated mutation rate is probably somewhat low.

Generation	Tests	Lethal + Semilethal	Frequency, %
1	133	1	0.8
2	52	0	0
3	183	4	2.2
5	212	3	1.4
7	263	15	5.7
9	285	10	3.5
11	283	16	5.7
13	386	24	6.2
15	377	23	6.1
17	408	33	8.1
19	409	26	6.4
22	289	26	9.0
	3,280	181	

NOTE: Total chromosome generations tested: 41,103

$$u = 0.0044$$

Table 9.5 Detection of viability mutations, including lethals, induced on *X* chromosomes of *D. melanogaster* following an exposure to *X* radiation. The classification of viabilities is based on the ratio of males hemizygous for an irradiated chromosome to females appearing in the same culture and carrying the same chromosome in heterozygous condition. (After Timofeeff-Ressovsky 1935.)

Males/Females	Control No. of Cultures	Control Frequency	Irradiated No. of Cultures	Irradiated Frequency
0	1	0.001	107	0.123
0.05	—	—	7	0.008
0.15	—	—	10	0.012
0.25	—	—	7	0.008
0.35	—	—	15	0.017
0.45	2	0.002	31	0.036
0.55	—	—	53	0.061
0.65	—	—	49	0.056
0.75	9	0.011	53	0.061
0.85	46	0.055	58	0.067
0.95	655	0.783	383	0.441
1.05	110	0.131	84	0.097
1.15	14	0.017	11	0.013
Total	837	1.000	868	1.000

Proof that the chromosomes resulting in intermediate frequencies of their male carriers had suffered permanent alterations and that the results were not caused by transient events, or even statistical accidents, was provided (see table 9.6) through the use of appropriate re-test cultures. The diminished viabilities that characterized a number of irradiated chromosomes proved in these tests to be heritable.

A great deal of information concerning the relative rates of

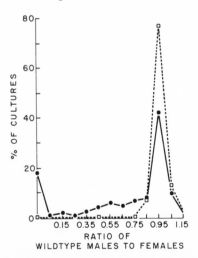

Figure 9.6 Viability mutations induced on the *X* chromosome of *D. melanogaster* by a large (approximately 6,000 r) does of X radiation. Dashed line: control; solid line: irradiated. (After Timofeeff-Ressovsky 1935.)

Table 9.6 Tests showing that nonlethal viability mutations are heritable. As in Table 9.5, these proportions (%) represent the ratio of males hemizygous for an irradiated chromosome to females appearing in the same culture and carrying the same chromosome in heterozygous conditions. (After Timofeeff-Ressovsky 1935.)

| | | Re-test Cultures | | | | | | | | | | | Number |
Culture	Original	1	2	3	4	5	6	7	8	9	10	Average	of Flies
R12	96.8	89	96	99	92	85	96	101	94	88	97	93.7	2,247
R73	94.3	96	97	91	102	93	104	90	92	95	94	95.4	2,374
R161	98.2	90	98	96	93	98	99	97	100	95	95	96.1	2,014
R253	97.4	106	97	90	96	95	91	93	92	98	90	94.8	2,116
R41	65.1	61	65	69	60	57	60	56	63	54	62	60.7	1,818
R69	51.8	49	55	60	54	62	52	51	59	56	50	54.8	1,619
R132	66.9	71	64	78	73	68	72	69	77	70	75	71.7	2,189
R214	77.2	72	80	89	78	83	87	85	76	82	88	82.0	2,260
R307	60.3	68	57	63	70	65	59	66	64	60	67	63.9	1,754
R341	63.5	72	64	73	59	66	75	78	63	70	72	68.2	2,410
R379	75.1	75	83	91	67	86	76	79	81	77	–	79.4	1,733
R422	58.2	58	69	57	60	67	62	65	54	63	61	61.6	1,895

mutations of various sorts that are induced by the exposure of *D. melanogaster* to low levels of X radiation were presented (figure 9.7) in a whimsical form by H. J. Muller (1950a). There is no reason to believe that the relative proportions of the various sorts of mutations depicted in this figure are grossly distorted by exposure to radiation; the greatest uncertainty lies with the estimation of subtle viability modifiers. The more subtle these mutations are, the more difficult their rate of origin is to measure; the Boxed Essay at the end of this chapter outlines a procedure used by Bateman (1959) and Mukai (1964) which suggests that the total mutation rate for very slight viability modifiers may be considerably higher than Muller suspected.

Mutations: Their Interactions

A discussion of mutations, especially those with a (sometimes) slight effect on viability, which leaves an impression that the action of mutant genes is fixed and incapable of modification would be incomplete at best. The effects of mutations at the level of the survival of their carriers (even if not at the level of primary gene products) is subject to considerable modification. Table 9.7 and figure 9.8 illustrate just how great such modification can be. The data concern nine chromosomes of *D. pseudoobscura* that were obtained (using a modified *ClB* technique) from three popu-

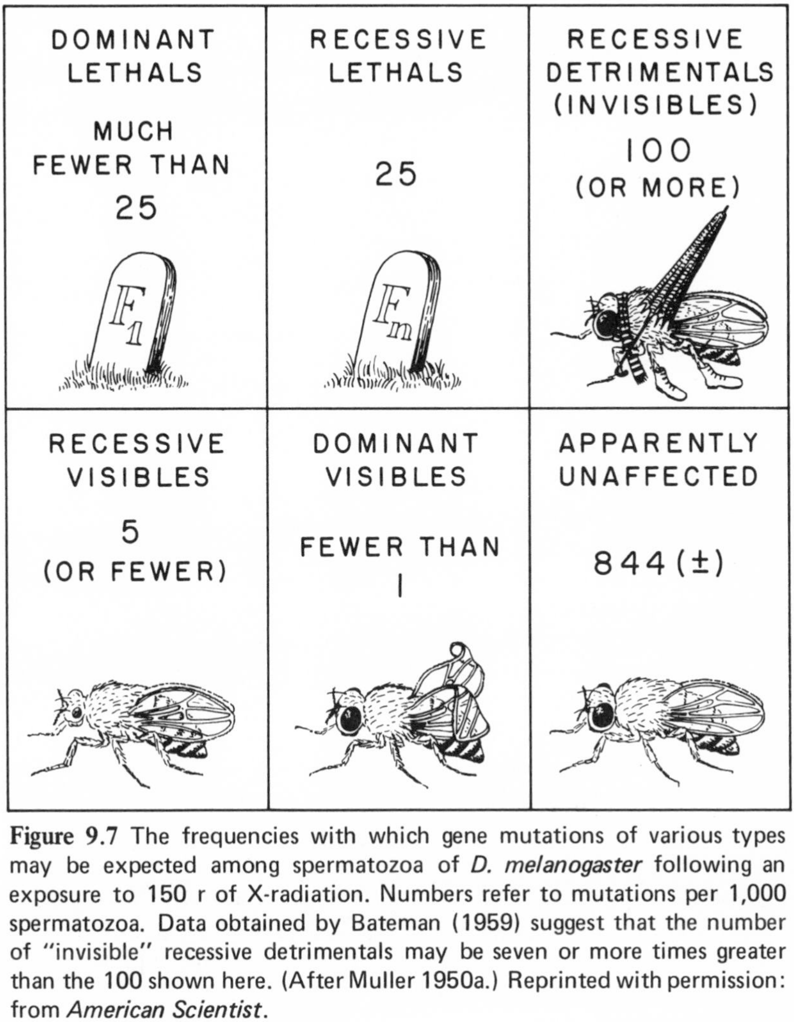

DOMINANT LETHALS MUCH FEWER THAN 25	RECESSIVE LETHALS 25	RECESSIVE DETRIMENTALS (INVISIBLES) IOO (OR MORE)
RECESSIVE VISIBLES 5 (OR FEWER)	DOMINANT VISIBLES FEWER THAN I	APPARENTLY UNAFFECTED 844 (±)

Figure 9.7 The frequencies with which gene mutations of various types may be expected among spermatozoa of *D. melanogaster* following an exposure to 150 r of X-radiation. Numbers refer to mutations per 1,000 spermatozoa. Data obtained by Bateman (1959) suggest that the number of "invisible" recessive detrimentals may be seven or more times greater than the 100 shown here. (After Muller 1950a.) Reprinted with permission: from *American Scientist*.

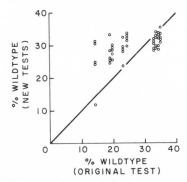

Figure 9.8 Comparison of the viability of nine different chromosomes of *D. pseudo-obscura* in the original tests and in novel genetic backgrounds; see table 9.7 for the data on which this figure is based. (Data from Dobzhansky and Spassky 1944.)

Table 9.7 Effect of background genotype on the viability of flies (*D. pseudoobscura*) homozygous for various wildtype second and fourth chromosomes. The background genotypes consist of wildtype chromosomes from various geographic regions. The numbers in this table are the frequencies (%) of homozygous wildtype flies in *ClB*-like test cultures where the expected frequency of these flies is 33.3 percent. (After Dobzhansky and Spassky 1944.)

Chromosome	Original	Washington	Colorado	California	Mexico	Guatemala	Average
AA1033	34.6	32.1	33.9	33.2	28.7	29.9	31.5
AA1015	14.2	12.8	31.2	30.8	24.7	24.3	24.8
AA1178	34.8	35.0	31.0	33.6	31.9	32.1	32.7
KA667	33.6	34.0	29.9	32.9	29.8	33.4	32.0
KD745	24.4	33.6	31.6	32.7	29.9	32.3	32.0
AA955	32.8	28.4	31.8	31.1	30.5	28.5	30.1
AA1035	23.2	28.0	32.5	23.4	28.4	29.4	28.7
PA851	19.0	25.9	25.7	33.0	28.2	24.5	27.5
PA998	19.4	29.3	27.2	28.3	29.8	25.3	28.0
Average	26.2	28.8	30.5	31.0	29.1	28.9	29.7

lations on Mount San Jacinto (near Palm Springs), California. The effects of these chromosomes on their homozygous carriers were studied (these are the *original* tests of both the table and the figure). By repeated backcrosses, these same chromosomes were transferred into genetic backgrounds whose geographic origins were scattered from the State of Washington to Guatemala. Once in these foreign backgrounds, the nine chromosomes were once again tested in homozygous condition. On the average, the chromosomes that resulted in poor viability in the original tests did so in the later ones as well, but the correlation is exceedingly poor. Examples of striking individual exceptions can be identified in table 9.7: Chromosomes AA1015 produced 12.8 percent wildtype homozygotes in the Washington background, but about 31 percent in those of Colorado and California. An examination of the distribution of individual points in figure 9.8 strongly suggests that the original classification of chromosomes as those causing low or high viability of homozygotes depended strongly upon specific genetic interactions that were characteristic of the early tests; these strong interactions were disrupted by the substitution of foreign genetic backgrounds. Unlike the more common adage, here we see that the poor become richer and the rich, poorer.

Dobzhansky and Spassky (1944) also demonstrated that the observed viabilities of individuals homozygous for chromosomes obtained from wild populations of *D. pseudoobscura* depended on the temperature at which the test cultures were raised (table 9.8). Temperature-sensitive lethals were not at all infrequent within their tested material; ironically, concentration on the study of gene-environment interactions obscured the potential value of

Table 9.8 Gene-environment interaction: The viability of wildtype *D. pseudoobscura* homozygous for various second chromosomes and tested at three different temperatures. Notice the two temperature-sensitive lethals (AA966 and 1015) included in this tabulation. As in the previous two tables, the numbers in this table are the percentages of wild-type homozygotes in *ClB*-like test cultures where the expected frequency of these flies is 33.3 percent. (After Dobzhansky and Spassky 1944.)

	Temperature, $^\circ C$		
Chromosome	16.5	21	25.5
AA958	0	13.1	1.0
AA966	26.0	12.6	2.4
AA995	31.7	27.2	22.7
AA1003	31.6	34.6	26.9
AA1015	29.7	14.2	0
PA736	27.8	18.4	9.0
PA748	23.0	13.6	9.9
PA784	26.9	21.4	20.5
PA841	32.2	26.7	11.5
PA858	31.5	37.0	30.0

such mutations for studies other than those dealing with natural populations (see, for example, Suzuki 1975).

An opportunity to refer once more to sickle-cell hemoglobin arises in conjunction with a discussion of the modification of gene action. Figure 9.9 shows the percent of abnormal hemoglobin in the red blood cells of persons heterozygous for the sickle-cell gene ($Hb^A Hb^S$). Although these persons carry one allele of each sort, and although each allele specifies one type of hemoglobin, these individuals do not have 50 percent abnormal hemoglobin: they have less. In addition, there appear to be concentrations of persons at two, maybe three, levels of abnormal hemoglobin. The point is a simple one that needs no belaboring: the variation between families exceeds that within families; therefore, it seems that the amount of abnormal hemoglobin found in the cells of heterozygous cariers may be subject to genetic control.

Mutation Rate: Abnormal Hemoglobins

Human genetics, once a stepchild of the science of genetics, has progressed enormously during the past three decades. Haldane once lamented that his personal fame rested on some rather routine calculations concerning enzyme kinetics while his estimate of human mutation rates (Haldane 1932:57) was the first such

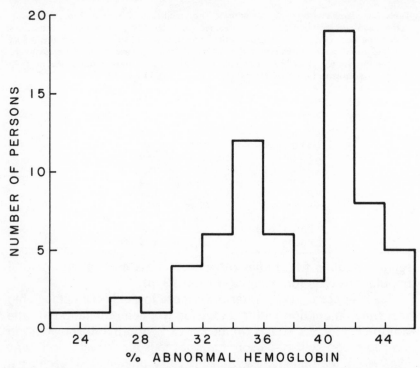

Figure 9.9 Histogram showing the number of sickle-cell heterozygotes possessing various proportions of abnormal hemoglobin in their red blood cells. An accompanying statistical test showed that the variation between members of the same family is smaller than that between different families; in the absence of environmental correlations, genetic modification of the relative proportions of the two hemoglobins in heterozygous individuals seem to be the most plausible explanation. (After Neel et al. 1951.) Reproduced with permission: from *Journal of Clinical Investigation* (1951), 30:1120-24.

estimate ever made. An increase in research in public health, the development of computers and the computerization of enormous collections of data, and the advances in biochemical procedures (including those of electrophoresis) and their widespread use in medicine have all combined to make human biochemical and molecular genetics an extremely advanced science. For this reason, the calculations of mutation rate will be made using abnormal hemoglobins in man; I shall follow the excellent account presented by Nei (1975:29-30). The arithmetic can be organized as follows:

    1. A study of 320,000 persons revealed that 62 persons were

heterozygous for a total of 44 electrophoretic hemoglobin variants.

2. Because only one-third of all amino acid substitutions are detected by electrophoresis, the frequency of hemoglobin variants in this sample is estimated as

$$(62 \times 3) \div (2 \times 320{,}000) = 186 \div 640{,}000 = 3 \times 10^{-4}$$

3. Parents of 18 of the observed heterozygotes were examined in respect to their hemoglobins; both parents in 2 instances proved to be homozygous normal. The fraction of new mutations, consequently, is 1/9; thus the mutation rate to abnormal hemoglobin is estimated as $3 \times 10^{-4} \times 1/9 = 3.3 \times 10^{-5}$. (That the proportion of new mutations is one-ninth of the total frequency of abnormal hemoglobins is important, and will be mentioned again on page 208.)

4. The total number of amino acids in the combined α and β chains of the hemoglobin molecule equals $141 + 146$, or 287. Consequently, the detectable mutation rate per codon equals $3.3 \times 10^{-5} \div 287$, or about 1.1×10^{-7}. Because 25 percent of base-pair substitutions leave the amino acid sequence unchanged (synonymous mutations), the mutation rate per codon is estimated as

$$1.1 \times 10^{-7} \times 4/3 = 1.47 \times 10^{-7}$$

5. Because there are three bases per codon, the mutation rate per base pair is about 5×10^{-8}.

6. If one assumes that 50 cell generations occur during the early development of a human being before gametes are formed, the mutation rate per base pair per cell division can be estimated at 10^{-9}, a value not far from one estimated by Watson (1976:254) on theoretical grounds.

Despite the widespread use of electrophoretic variants in the study of Drosophila populations, relatively few studies have been carried out on their rates of mutation in these flies. One such study was carried out by Tobari and Kojima (1972) using ten allozyme systems in *D. melanogaster*; with three mutations recovered in all (two of these exhibited complex inheritance patterns), they estimated the mutation rate per locus to be 4.5×10^{-6}. The number of mutations recovered in all does not warrant an estimate any more precise than 10^{-6} (see Voelker, et al., 1980).

Mutations: Algebraic

A consideration of the quantitative aspects of mutations and their role in the genetics of populations will be limited here to (1) the loss of a gene by forward mutation and (2) the equilibrium that would arise through the opposition of forward and reverse mutations. Because mutant genes frequently affect the health and reproductive abilities of their carriers, mutation will be considered later (chapter 10) in conjunction with selection.

Forward mutation

Assume that an allele A mutates to a at a rate u per generation. Assume, too, that the frequency of A in the population is p while that of a is q. Following mutation, the frequency of A is reduced to $p(1 - u)$. If there is an equilibrium frequency it must be such that

$$p = p(1 - u).$$

This equation can be satisfied, however, by only the two trivial conditions: $p = 0$, or $u = 0$. Mutation of A to a, if unopposed, will remove A from the population.

The loss of A by forward mutation alone is not necessarily a rapid one. The frequency of A in any generation, t, equals

$$p_t = p_0 (1 - u)^t = p_0 e^{-ut}.$$

Thus, the original frequency of A will be reduced by half when

$$e^{-ut} = 0.5$$
$$ut = - \ln 0.5 = 0.7$$
$$t = 0.7/u.$$

If u equals 10^{-6}, $0.7/u$ equals 700,000 generations. Loss through unopposed mutation is a slow process.

When the number of generations is small, $(1 - u)^t$ equals approximately $1 - tu$. If the loss of A amounts to tu, then the gain of a must also equal tu. The term q/u bears on the relationship between the frequency of abnormal hemoglobins in human populations and the estimated mutation rate reported on page 207. It appears that the frequency of abnormal hemoglobins in the sampled population was only nine times greater than the mutation rate. Ignoring for the moment that this might represent an equilibrium between mutation and selection (a matter subsequently discussed on page 269), it would appear that abnormal hemoglobins have been accumulating in human populations for only nine generations, some 200 years. Thus, although King George III of Britain may have suffered from porphyria, a hereditary metabolic disorder involving the breakdown of discarded hemoglobin, he and his contemporaries had not yet begun accumulating abnormal hemoglobins. Such are the absurd conclusions which are reached when the assumption of equilibrium conditions is abandoned.

Forward and reverse mutations

Suppose that the frequencies of A and a are p and q. Suppose, too, that A mutates to a at a rate u per generation while a mutates to A at a rate v. At equilibrium

$$up = vq.$$

Recalling that $q = 1 - p$, the frequency of A at equilibrium, $\hat{p} = v/(u + v)$ and $\hat{q} = u/(u + v)$ ($\hat{p}$ and $\hat{q}$ are read p-hat and q-hat).

This is the first *stable* equilibrium we have encountered. The Hardy-Weinberg equilibrium is an indifferent equilibrium; however often the value of p is disturbed, it never tends to return to an original value. Not so with the frequencies, $\hat{p}$ and $\hat{q}$, described here. A population consisting of only AA individuals will accumulate mutations until the frequency of a approaches $u/(u + v)$. Conversely, a population consisting originally of aa individuals will accumulate A mutations until their frequency equals $v/(u + v)$.

Figure 9.10 illustrates the establishment of an equilibrium between forward and reverse mutation in geometric rather than algebraic terms. When the frequency of A equals 1.00, it is lost at a rate u whereas the nonexistent a cannot be lost. When the frequency of a equals 1.00, it is lost at a rate v but the now nonexistent A cannot be lost. The sloping lines connecting these extremes intersect where $u/h = 1/p$ and $v/h = 1/(1 - p)$ (h equals the height of point of intersection, p equals the frequency of A). Solving this relationship yields $p = v/(u + v)$ as before.

Although forward and back mutation may give rise to a stable equilibrium, the mutational forces that tend to restore this

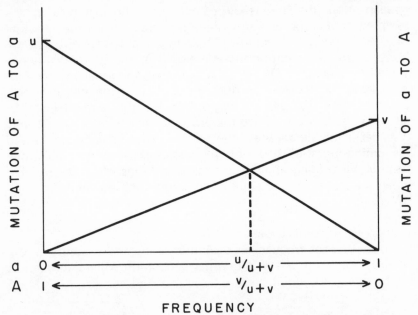

Figure 9.10 A geometrical representation of the equilibrium established between forward and reverse mutation. The equilibrium frequencies of the two alleles (A and a) can be obtained by noting the following two proportionalities: $u/h = 1/p$ and $v/h = 1/(1 - p)$, where h is the length of the dashed line and p = frequency of A. This representation is analogous to that used by MacArthur and Wilson (1967) in accounting for the number of species found on oceanic islands (see figure 19.3).

equilibrium are extremely weak. Once p has been displaced from $\hat{p}$, scarcely any tendency to return once again exists (see Haldane 1964). As a rule, chance events are greater at near-equilibrium frequencies than are the net mutational changes.

BOXED ESSAY

Bateman (1959) and, even more extensively, Mukai (1964) have estimated the rates of mutation to alleles that have extremely small, deleterious effects on the viability of their carriers. The mathematical procedures used in these studies are described here:

Experimental data (obtained through modified *CIB* techniques) provide empirical estimates of (1) control viability (v_0), (2) mean viability following mutation ($\bar{v}$), and (3) genetic variance in viability among parallel cultures (σ_g^2).

The theoretical analysis proceeds as follows:

- Let a be the effect of a single mutation on viability; then the average effect will be $\bar{a}$, and the variance in effect will equal σ_a^2.
- Let p be the average number of mutations per chromosome; then the mean depression of viability per chromosome will equal $p\bar{a}$.

If mutations are distributed among chromosomes according to the Poisson distribution, the following will be true:

(a)

	\multicolumn{5}{c}{*Mutations/chromosome*}		
	0	*1*	*2*	*3*	*4*	•	•
Frequency	$f(0)$	$f(1)$	$f(2)$	$f(3)$	$f(4)$	•	•
Depression	0	$\bar{a}$	$2\bar{a}$	$3\bar{a}$	$4\bar{a}$	•	•
or	$\bar{a}[$ 0	1	2	3	4	•	•]

Because this is a Poisson distribution, the variance of the numbers within the brackets equals p. Therefore, the variance *among* chromosomes with different numbers of mutations equals $\bar{a}^2 p$. (Multiplying each observation by a constant will multiply the mean by the same constant, but will multiply the variance by the *square* of the constant; see Dixon and Massey 1951:23.)

(b)

	\multicolumn{4}{c}{*Mutations/chromosome*}		
	0	*1*	*2*	*3*	•	•
Frequency	$f(0)$	$f(1)$	$f(2)$	$f(3)$	•	•
Variance *within* each	$0[f(0)]$	$\sigma_a^2 [f(1)]$	$2\sigma_a^2 [f(2)]$	$3\sigma_a^2 [f(3)]$	•	•]
or	$\sigma_a^2 [0f(0)$	$1f(1)$	$2f(2)$	$3f(3)$]	•	•]

Now, the sum of $0f(0) + 1f(1) + 2f(2) + \ldots = p$. Therefore, within group variance equals $\sigma_a^2 p$. Because variances are additive, the total genetic variance equals $\bar{a}^2 p + \sigma_a^2 p$, or $p(\bar{a}^2 + \sigma_a^2)$.

The theoretical analysis can now be combined with the empirical data:

$$\bar{a}p = v_0 - \bar{v}$$

and

$$(\bar{a}^2 + \sigma_a^2)p = \sigma_g^2.$$

Let

$$A = v_0 - \bar{v} \quad \text{and} \quad B = \sigma_g^2.$$

Then

$$\bar{a}p = A$$

and

$$(\bar{a}^2 + \sigma_a^2)p = B.$$

Consequently,

$$(\bar{a})^2 A/\bar{a} + \sigma_a^2 (A/\bar{a}) = B.$$

This can be rewritten as

$$\bar{a}^2 A - \bar{a}B + \sigma_a^2 A = 0.$$

If an equation has the form $ax^2 + bx + c = 0$,

$$x = \frac{-b \pm \sqrt{b^2 - 4ac}}{2a}.$$

Furthermore, if x is a real number,

$$\sqrt{b^2 - 4ac} \geqslant 0$$

Substitution shows that

$$0 \leqslant \sigma_a^2 \leqslant B^2/4A^2$$

Solving further we find that if $\sigma_a^2 = 0$, then $B = \bar{a}^2 p$ and that $\bar{a} = B/A$ and $p = A^2/B$.

Mukai found that A $(= v_0 - \bar{v})$ equaled $0.1261 = \bar{a}p$, and that the increase in variance per generation equaled 0.1127 $[= p(\bar{a}^2 + \sigma_a^2)] = B$. Thus, $\bar{a} = 0.1127/0.1261$ or 0.89 and $p = 0.0159/0.1127$ or 0.14. For *D. melanogaster*, then, the data suggest that as many as one second chromosome in seven may contain a new, somewhat deleterious mutation each generation. If all of this fly's chromosomes were to have a similar mutation rate, about 35 percent of all gametes produced by these flies would carry *newly arisen mutations*.

FOR YOUR EXTRA ATTENTION

1. Imagine that at each of 10 loci in a bacterium the wildtype allele (+) mutates to a mutant allele (*m*) nine times as frequently as the mutant allele mutates back to wildtype. What are the expected frequencies of mutant alleles at each locus? What is the expected frequency of individuals possessing wildtype alleles at all ten loci? Anticipating topics to be discussed in subsequent chapters, can you imagine how "wild" individuals come normally to possess wildtype alleles?

2. On several occasions we shall speak of the genetic bases of certain traits in terms of what is "necessary" and what is "sufficient" for the expression of a certain phenotype. Two sentences from Haldane (1932) can be considered from the same point of view: "Genetics can give us an explanation of why two fairly similar organisms, say a black and a white cat, are different. It can give us much less information as to why they are alike."

3. The average incidence of Down's syndrome at birth for mothers of ages 17–35 is about 1 per 1,500 births. Down's syndrome results from trisomy of one small chromosome; presumably an equally frequent chromosomal misdivision of this chromosome causes the death of affected monosomic embryos (they would carry only one of these chromosomes instead of the normal pair). Assuming that the accidents which befall this one pair of chromosomes during meiosis affect all other members of the 22 autosomal pairs, but that in the remaining cases both trisomy *and* monosomy are lethal, estimate the proportion of embryos that are eliminated by meiotic accidents (nondisjunction).
Hint: The probability of surviving accidents involving one of the 22 autosomal pairs seems to be 749/750.
Later in the mother's life, the incidence of Down's syndrome at birth rises to 1 per 50 births. What proportion of all zygotes at that age successfully avoid the difficulties of trisomy and monosomy.

4. The spontaneous mutation rate for sex-linked lethals in *D. melanogaster* is about 0.003. The frequency of sex-linked lethals recovered from male flies exposed to 1,000 r of X-radiation is about 0.03. The exposure of living things to natural radiation (cosmic rays, gamma and other rays from disintegrating atoms) is about 0.15 r per year. Assuming that a fly's life is two weeks, are the spontaneous mutations likely to be radiation induced?
Hint: Radiation-induced mutations increase linearly with dose over all levels of radiation that have been tested experimentally.

10

Selection I

PREVIEW: Because it is the guiding force in determining the direction of genetic change in both natural and domesticated populations, selection will be discussed in this and the following chapters. The present chapter provides illustrative examples of selection, presents algebraic accounts of various types of selection acting on alleles at a single gene locus, and concludes with a general discussion of the consequences of selection as illustrated by in vitro studies of replicating molecules.

MODERN BIOLOGY EMBRACES two dogmas: Darwin's evolution through natural selection, and the central dogma of molecular genetics. Both have undergone modifications since first being enunciated; both remain, nevertheless, virtually unaltered. Reverse transcriptase and error-prone polymerases introduce some ambiguity into the central dogma, but these exceptions serve only to emphasize the overall importance of the dogma in describing genetic phenomena. Random, chance events thwart the operation of natural selection, but these, too, serve primarily to stress the overall importance of natural selection in adjusting living organisms to the world in which they live.

The logic which ascribes the adaptation of organisms to their environments rests upon three, and only three, premises: (1) that the individuals of a population exhibit hereditary variation, (2) that some individuals are genetically better equipped to survive than are others, and (3) that there are generally more zygotes in a population initially than can survive to adulthood. The consequences of this logic are (1) that elimination of individuals occurs, (2) that this elimination is non-random in respect to hereditary variation, (3) that the characteristics which favor (*not* guarantee!) the survival and reproduction of some individuals will be transmitted to their offspring, and (4) that the number of these offspring will once more exceed the number that can survive and reproduce in the local environment. In short, *the characteristics of a population are determined by its breeding members.* This account describes the logic of evolution by natural selection. It also explains the logic of those solutions which living things (even plants) arrive at in solving problems of survival. This last point should not be overlooked: because its structure allows DNA both to reproduce itself accurately and to specify what (and when) enzymes and other proteins will be synthesized, DNA is capable of arriving at logical solutions to even complex problems. The solutions may take longer in being expressed than are those of the brain, but tardiness should not detract from their logic.

The above points are stressed here because, perhaps as a reaction to a perceived intrusion of "biological determinism" in the social sciences, a reaction to the role of natural selection in evolution has arisen. Pan-selectionism is attacked as a point of view which, oftentimes by tortuous means, can explain everything. While the criticisms may in some cases be well taken (see, for example, Shapiro 1978) nothing is gained by embracing random events as an acceptable alternative. Every statistical test asks, "What is the probability of observing deviations as large as this or larger by chance alone?" Every experimenter stands ready to admit that chance events offer a reasonable explanation for his most cherished observations. Chance events can also explain all observations. The problem facing evolutionists is not so much the construction of fanciful ad hoc explanations for isolated observations as it is the erection of falsifiable theories making sense of numerous—sometimes seemingly unrelated—observations.

How, and what, does natural selection select? First, selection need not be all or none; in time slight differences can effectually alter the composition of a population. Second, it is the individual that will or will not leave a full quota of offspring; the bulk of all selection operates on differences between individuals. Third, the phenotype, not the genotype, exposes the individual to hunger, predation, accident, and the multitude of other challenges posed by the surrounding environment. Fourth, if the population is to undergo a genetic alteration as the result of selection, certain genotypes must be differentially associated with the selected (or rejected) phenotypes; the effect is a change in the frequencies of various alleles in the population. Fifth, if a change in gene frequency is regarded as the ultimate effect of natural selection, then it must be remembered that devices exist by means of which certain alleles do, or could, increase their frequencies at the expense of homologous alleles (Dawkins 1976; Crow 1979). In the sections that follow, we shall first illustrate the action of selection on genetically dissimilar members of various species of organisms (including man), and then provide an algebraic basis for understanding the outcome of selective processes. The chapter ends with an account of Darwinian selection among self-replicating molecules.

Natural Selection

Survival and reproductive success are the ingredients of natural selection. Selection does not *cause* differential survival and reproduction; selection *is* differential survival and reproduction.

A dominant gene, *D,* is responsible for abnormal human growth: *Dd* individuals are chondrodystrophic dwarfs, men and women who are characterized by short muscular limbs, but nearly normal torsos. Normal individuals have the genotype, *dd*. Because the gene *D* is rare, relatively few marriages between *Dd* individuals, husband and wife, are known; there is reason to suspect, however, that *DD* individuals are fetal monsters dying before or at birth.

Table 10.1 shows the number of children born to *Dd* (dwarfs) and to *dd* (sibs of the dwarfs) persons. The average number of children born to a normal sib in this study was 1.27; the average number born to dwarf members of the same families was 0.25. For every child left as a descendant by a normal indi-

Table 10.1 Calculation of the relative reproductive success of chondrodystrophic dwarfs based on the number of progeny produced by dwarfs, and by their nondwarf sibs.

Parents	Number	Number of Progeny	Progeny/Parent
Dwarf	108	27	0.25
Normal	457	582	1.27

$(27/108) \div (582/457) = 0.196$

vidual, a dwarf would leave only 0.20 child. Thus, in respect to reproductive success, dwarfs are only one-fifth as successful as normal individuals.

Table 10.1 also suggests that the survival of dwarfs is less than that of normal individuals, possibly explaining in part the lowered reproductive success of dwarf individuals. The most common marriage that is capable of producing families segregating both dwarf and normal sibs is that involving one dwarf (*Dd*) and one normal (*dd*) partner. From such marriages, dwarfs and normal sibs are expected in equal frequencies. The data given in table 10.1 suggest, however, that the proportion of dwarfs is much less than the expected 50 percent.

The fitness of an individual is defined as the overall relative survival and reproductive success enjoyed by that individual. Normally, because individuals of any sort are subjected to debilitating accidents, a fitness value is assigned to the average value exhibited by individuals of a certain genotype (or, in some instances, phenotype). The average, all-inclusive fitness which incorporates every facet of the survival and reproductive success of an individual (or group of similar individuals) is known as the Darwinian fitness of individuals of that genotype.

Unfortunately, the identification and measurement of every facet of one's behavior in respect to reproduction is a heroic, if not an impossible, task. Consequently, experimentalists usually settle on measurements of those components of fitness that can be measured relatively easily. Understandably, components of fitness are sometimes discussed as though they were fitnesses themselves, but they are not. For example, the estimates of relative "viability" obtained through the use of modified *CIB* techniques are neither measures of viability (speed of development is included; survival of adults is not) nor of fitness (because no aspect of reproductive ability is included in the data); nevertheless, the results of such tests are used as an indication of the relative fitnesses of the flies involved. Similarly, the relative reproductive success of dwarfs has been calculated as 0.20, but we have reason to believe that this estimate includes survival of the progeny as well as the reproductive behavior of the parent.

The Darwinian fitnesses of individuals of different genotypes determine the fates of alleles within populations. Favored genotypes increase the frequencies of the alleles they carry; disfavored ones decrease the frequencies of their alleles. Some patterns of selection leave gene frequencies unchanged (when the favor

Table 10.2 Selective changes in the frequencies of alleles at eight gene loci in the barnacle, *Balanus amphitrite*, caused by the warming of seawater as it passes through the cooling systems of electrical power stations at Haifa, Israel. Samples of barnacles were obtained by immersing 25 cm X 25 cm plastic plates 100 cm below the surface of the water entering ($24°$–$25°$C) and leaving ($33°$–$38°$C) the cooling systems. All sample sizes were approximately 250 individuals. Although seven of the eight alleles change significantly in frequency as a result of the temperature change, the evidence does not prove that these alleles, rather than others at nearby loci, are the basis for the selective change. (After Nevo et al. 1977.)

	Frequency	
Locus (allele)	Intake	Outlet
Pgi(S)	.014	.033
Me(M+)	.334	.510
Ao-3(F)	.011	.133
Mdh-2(F)	.010	.073
Acph-1(M)	.893	.964
Est-3(F)	.238	.451
Est-4(F)	.017	.067
Est-8(F)	.014	.104

bestowed on an allele via one genotype is cancelled by the disfavor arising via another one); others do not. Table 10.2 shows the effect of water temperature on the frequency of eight electrophoretic alleles in a barnacle living near an electrical power station. Seven of the eight alleles (the *Pgi* locus was the exception) show highly significant changes in frequency as a result of the $10°$C change in temperature.

The term "protective coloration" implies that individuals of certain colors are inconspicuous and, as a result, are overlooked by potential predators; other individuals, less perfect in their camouflage, do not escape and are eaten. This was a hypothesis that, in its day, had to be tested. One of the early tests (Cesnola 1904) involved green and brown forms of the praying mantis (*Mantis religiosa*). Cesnola tethered green and brown individuals on patches of green and brown grass. Within nineteen days on green grass, birds had eaten 35 of 45 brown insects, but none of the 20 green ones. Conversely, on brown grass, all 25 greens were eaten within eleven days, but none of the 20 browns even by the nineteenth day. On each background, enormous differences in relative survival of the contrasting color morphs exist. The experiment does not tell us whether the green and brown mantises seek out appropriate backgrounds, nor does it tell us the overall fitness of these forms in a patchy—green and brown—environment. It does tell us that under certain circumstances the possession of either color is fatal; consequently, the intensity of natural selection for color could be high, indeed.

Table 10.3 Evidence for the protective value of changeable coloration in the fish *Gambusia patruelis* when exposed to predation by penguins. (After Sumner 1935.)

Tank	Fish	Eaten	Not Eaten	Total
Gray	Gray	176	352	528
	Black	278	250	528
		454	602	1,056
Black	Gray	217	118	335
	Black	78	257	335
		295	375	670

Table 10.3 presents data on the protective coloration of fish. In this case, the fish can adjust their body shade to more or less match their background. By allowing fish (*Gambusia patruelis*) to match their colors to those of light and dark tanks over a period of days, Sumner was then able to transfer about half of those in either tank to the other one and then, almost immediately to expose them to predation. The results obtained are highly significant: in gray tanks, blacks are preferentially eaten by penguins and gray fish are overlooked; in dark tanks, gray fish are preferentially eaten and blacks are overlooked. Here, then, we can identify predation as a selective agent conferring an advantage to genotypes that enable fish to match their body colors to those of their surroundings.

The drastic increase in the melanic forms of numerous species of moths in industrialized Europe and North America is a matter of record. The first occurrences and subsequent increases can be noted in museum collections; present-day frequencies can be verified by any collector. A plausible explanation for the change in coloration (see Haldane 1924) awaited the release-recapture experiments of Kettlewell using *Biston betularia*; some of his data (Kettlewell 1973) are listed in table 10.4. In the heavily industrialized region near Birmingham, where the trunks of trees are blackened with soot and where the lichens normally found on tree

Table 10.4 Release-recapture data on the *typica* (light) and *carbonaria* (melanic) forms of the pepper moth, *Biston betularia*, at two localities in Great Britain: Birmingham (heavily industrialized) and Dorset (a relatively unpolluted rural area). (After Kettlewell 1973.)

Locality	Type	Recaptured	Lost	Total released
Birmingham	Carbonaria	123	324	447
	Typica	18	119	137
	Total	141	443	584
Dorset	Carbonaria	30	443	473
	Typica	62	434	496
	Total	92	877	969

trunks in Britain (and on which the moths would normally rest) are missing, nearly 30 percent of the dark (*carbonaria*) moths that were released were recaptured; only 13 percent of the released light forms were found again. On the contrary, in the nonindustrialized region in Dorset, where trees still maintain their lichen covers, only 6 percent of the released dark forms were recovered whereas 12 percent of the light ones were. These data, together with the observations (1) that when resting these moths normally orient themselves so that their color patterns match those of the surrounding lichens and (2) that birds have been seen taking conspicuous individuals from both lichen-covered and soot-covered trees, adequately account for the prevalence of industrial melanisms in this and many other species of moths.

For reasons that will become apparent in later chapters, evidence bearing on the differential survival or reproduction of allozyme (= electrophoretic) variants are of special interest to modern population geneticists. One example already mentioned concerned the frequencies of allozyme alleles before and after exposure of barnacles to heated water (table 10.2). Figure 10.1 and table 10.5 illustrate the effect of selection within a laboratory population of *D. melanogaster* on the frequencies of different alleles at the *amylase* locus. Figure 10.1 represents a starch-gel electrophoretic analysis of four alleles at the *amylase* (*amy*) locus. The locus appears to be a compound one consisting of tightly linked subunits; each of three of the alleles shown produces two enzymes with clearly different mobility patterns. One allele (amy^1) produces only one band, but one that is common to two others ($amy^{1,2}$, $amy^{1,3}$).

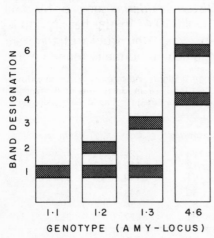

Figure 10.1 Electrophoretic patterns generated by allozyme variation at the *amylase* (*amy*) locus in *D. melanogaster*. An examination of this figure reveals why $amy^{1,3}/amy^{1,3}$ and $amy^{1,3}/amy^1$ flies have not been listed separately in table 10.5. The same explanation applies to $amy^{1,2}/amy^{1,2}$ and $amy^{1,2}/amy^1$.

Table 10.5 Changes that occurred in the frequencies of various *amylase* (*amy*) alleles within eight months after a sample of flies from a two-year-old laboratory population of *D. melanogaster* was transferred from the previous sucrose-containing medium to one made with cornmeal. (After Jong et al. 1972.)

Genotypes of Sampled Flies	Sucrose Population	Cornmeal Population
$amy^{4,6}/amy^{4,6}$	—	0.122
$amy^{1,3}/amy^{1,3}$		
$amy^{1,3}/amy^{1}$	0.444	0.169
amy^{1}/amy^{1}	0.556	0.272
$amy^{1,2}/amy^{1,2}$		
$amy^{1,2}/amy^{1}$	—	—
$amy^{1,3}/amy^{4,6}$	—	0.091
$amy^{1,2}/amy^{4,6}$	—	—
$amy^{1}/amy^{4,6}$	—	0.347
$amy^{1,3}/amy^{1,2}$	—	—
Numbers of flies tested	275	254

NOTE: — = Frequency so low that genotype did not appear in sample; possibly higher eight months earlier.

A population of *D. melanogaster* that had been maintained for two years on sucrose-containing food was subdivided by transferring a large sample of flies to a second cage in which the medium contained cornmeal (i.e., starch rather than sucrose) as the main carbon source. Eight months after the original sucrose-population was divided, between 250 and 300 flies of both the sucrose and cornmeal populations were tested in respect to amylase allozymes. The results (table 10.5) show that the allele *amy* ^{4,6} is considerably more frequent in the population maintained on cornmeal than in the one containing sucrose. Neither $amy^{4,6}$ / $amy^{4,6}$ nor any of the possible $amy^{4,6}$ -carrying heterozygotes were recovered in the sucrose population. It appears from these data that flies (probably larvae) carrying the $amy^{4,6}$ allele survive on the cornmeal diet better than larvae carrying either amy^{1} or *amy* ^{1,3}; during the original two years while kept on sucrose, the allele $amy^{4,6}$ was reduced to a very low frequency.

In lieu of information on the survival of individuals, experimenters at times substitute laboratory (in vitro) tests of allozyme activities in an attempt to discern differences that might cause differential survival of individuals possessing different genotypes. Figure 10.2 illustrates the results of one such test. In this figure are seen the relative activities of esterases that are characteristic of the different genotypes of a fish (*Catostomus clarkii*). The activities of the allozymes produced by the two types of homozygous individuals differ: one is most active at low temperatures, the other at high ones. The enzyme that is characteristic of heterozygous individuals exhibits its greatest relative activity at intermediate temperatures. If high activity in vitro were synonymous with survival in the field (an unproven possibility), the data presented in figure 10.2 would suggest that heterozygous individuals in respect to this *esterase* locus would have a selective advantage

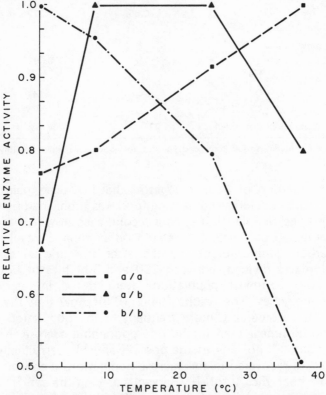

Figure 10.2 The relative in vitro activities of the esterases obtained from *Catostomus clarkii* (a fish) of three genotypes— *Es*-I$^{a/a}$, *Es*-I$^{a/b}$ and *Es*-I$^{b/b}$—at four different temperatures. The enzyme with the highest activity at a given temperature is assigned the value 1.00. Note that, if enzyme activity were directly related to overall fitness, heterozygotes would have selective advantage over a 20° range of temperatures. (After Koehn 1969.) Reproduced with permission: from *Science* (1969), 163:943-44; © 1969 American Association for the Advancement of Science.

over much of the tested temperature range. There is no reason, however, to believe that the relation between activity of an enzyme under laboratory conditions and the survival in the field of individuals carrying this enzyme are absolutely correlated.

Modern society has, for a multitude of reasons, come to rely heavily on chemicals as a means for controlling pests—both plant and animal—that seemingly threaten people and their resources. The discovery of antibotics not only led to seemingly miraculous cures of life-threatening infections but also eliminated the almost-daily manual sterilization of the walls and floors of operating rooms and clinics by means of phenolic scrubbings. The immediate result in hospital administration was a reduction in labor costs; the result in respect to medical genetics was a tremendous pressure for microorganisms to develop and, it turns out, to share with one another, genetic mechanisms for achieving resistance to these modern miracle drugs (many of which have been the bane—literally—of soil bacteria since time immemorial). Similarly, modern agricultural practice and the enormous expanses of single crops (monocultures) which it encourages have encouraged insect and other arthropod pests; these have been fought with a variety of lethal chemicals. Rodents have also been attacked more and more by chemical means of which, until recently, anticoagulants have been particularly effective.

In each of the listed instances, the man-inspired attack has represented a (relatively) new environmental challenge to the pest organism. Each attack has set in motion a series of events amounting to natural selection within the organism under attack, an onslaught in which certain genotypes have proven to have higher fitness than others. Penicillin-resistant bacteria are today extremely common; pencillin-resistant gonococcus is a serious threat in certain strata of society. Mice and rats resistant to warfarin, a chemical commonly used in controlling these rodents, have appeared in many localities.

Figure 10.3 illustrates the historical development of insecticides in Denmark, their introduction into general use, the origins of resistant strains of houseflies, and the termination of the usefulness of each insecticide. Within a period of 25 years, the Danes have used and, for the most part, have abandoned a dozen and a half chemicals in an attempt to control houseflies. The figure contains no evidence that the Danes are winning the contest with these insects; natural selection of heritable variation among houseflies constitutes formidable opposition.

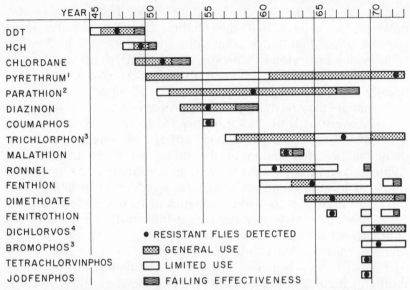

Figure 10.3 Countrywide use of insecticides for housefly control on Danish farms from 1945 to 1972, and the development of insecticide resistance by the fly population. Stippled portions of the bars indicate that the insecticide was in general use; the appearance of resistant flies is shown by a large dot; a period of limited use is shown by open bars; and a period of acknowledged failing effectiveness is shown by horizontal lines. The economic usefulness of an insecticide is governed by its own nature and by the availability of promising substitutes. The superscripts indicate the means by which the insecticides are applied: 1, space spray; 2, impregnated strips; 3, bait, 4, vapor; otherwise, residual spray. (After Keiding 1977.)

The Algebra of Selection

Recessive alleles

The effect of natural selection through differential mortality or fertility of individuals of different genotypes can be calculated quite easily. Suppose that the alleles, A and a, have frequencies, p and q ($p + q = 1.00$), and that at the moment of fertilization the genotypes AA, Aa, and aa have frequencies, p^2, $2pq$, and q^2. Suppose, too, that between the formation of zygotes at fertilization and the attainment of sexual maturity, only $(1 - s)$ aa individual survives for each AA or Aa individual. This supposition does not imply that all or even most AA and Aa individuals survive; on the contrary, mortality in the population may be extremely heavy.

The supposition says only that for each AA or Aa survivor, less than one, $1 - s$, aa survives. The quantity s is known as the selection coefficient; $1 - s$ is the adaptive value or Darwinian fitness (actually, as stated here, the component of Darwinian fitness based on survival) of aa individuals.

The description given in the preceding paragraph can be summarized algebraically as follows:

Genotype	AA	Aa	aa
Darwinian fitness (w)	1	1	$1 - s$
Starting frequency	p^2	$2pq$	q^2
Final frequency	p^2	$2pq$	$q^2 - sq^2$
Average fitness $(\bar{w})$		$1 - sq^2$	

Frequency of a after selection:

$$\frac{q - sq^2}{1 - sq^2} \quad \text{or} \quad \frac{q(1 - sq)}{1 - sq^2}$$

Change in the frequency of a (Δq):

$$\frac{q - sq^2}{1 - sq^2} - q = \frac{q - sq^2 - q + sq^3}{1 - sq^2}$$

$$\Delta q = \frac{-sq^2 (1 - q)}{1 - sq^2}.$$

If q is small,

$$\Delta q = -sq^2.$$

If aa individuals fail to survive, (that is, $s = 1$), $q(1 - sq)/1 - sq^2$ becomes

$$\frac{q(1 - q)}{(1 + q)(1 - q)}, \quad \text{or} \quad \frac{q}{1 + q}.$$

If $q = 1/n$, $q/(1 + q)$ equals $1/(n + 1)$. Thus, if the original frequency of a recessive lethal in a population is $\frac{1}{2}$, for example, the frequencies expected in the following generations are $\frac{1}{3}$, $\frac{1}{4}$, $\frac{1}{5}$, $\frac{1}{6}$, and so forth. This series reveals that although the frequency of a recessive lethal may be rapidly reduced if it is large (only 2 generations are required to lower 0.50 by half), the rate of elimination becomes extremely slow when the frequency of a lethal is low (1,000 generations are needed to reduce 0.0010 to 0.0005).

The above calculations demonstrate the inefficiency of negative eugenic measures proposed (and at times carried out) so vigorously in the early 1900s. The proposed cure-all for genetic ills was sterilization. The ills against which the measures were proposed and which did have a genetic basis (many of the syndromes under attack were the result of the affected individual's environment) were, when considered separately, rare abnormalities. Furthermore, the diseases themselves severely handicapped those who were afflicted. Consequently, sterilization merely reduced an already severely impaired reproductive ability. The effect of sterilization on the elimination of the genes involved was negligible; only the most fanatic or the least informed person could detect any value in these early sterilization programs.

Incompletely recessive alleles

The algebra dealing with the elimination of recessive alleles from populations can be extended rather easily to cover the elimination of incompletely recessive ones as well. A quantity, h, is introduced into the algebraic notation; this quantity measures the extent to which s (the amount by which fitness is lowered in aa individuals) is expressed in heterozygous (Aa) individuals. Thus, h is a measure of dominance which may take on any value between 0 and 1.00. Again, the algebra can be summarized in the following form:

Genotype	AA	Aa	aa
Darwinian fitness (w)	1	$1 - hs$	$1 - s$
Original frequency	p^2	$2pq$	q^2
Final frequency	p^2	$2pq - 2hspq$	$q^2 - sq^2$
Average fitness ($\overline{w}$)		$1 - 2hspq - sq^2$	

Because q would be extremely small in the case of an allele that affects heterozygotes adversely, the average fitness of the population equals (approximately) $1 - 2hsq$.

The frequency of a after selection is

$$q - \tfrac{1}{2}(2hspq) - sq^2$$

or (neglecting all terms involving q^2), approximately

$$q - hsq.$$

The change in gene frequency in this case is

$$\Delta q = -hsq.$$

The loss of a rare mutant gene which adversely affects the fitness of its heterozygous carrier is a function of its frequency, not of its frequency squared. Alleles that express themselves to the detriment of their heterozygous carriers have no place to hide; completely recessive alleles that express themselves only in homozygous individuals are sequestered in heterozygous ones, and (at low frequencies) are exposed to adverse selection primarily as a consequence of the mating of heterozygotes.

Selectively superior heterozygotes

Until now, the algebra of selection has been limited to recessive and incompletely recessive deleterious mutations. These are mutations with which most beginning geneticists have had some experience. The grossly distorted Mendelian ratios yielded by overcrowded classroom cultures of *Drosophila* are a case in point. In F_2 cultures of monohybrid crosses, the student expects to obtain 3 to 1 ratios for autosomal recessives; to the dismay of both student and teaching assistant, crowded, moldly cultures often yield ratios of wildtype to mutant of 5, 6, or even higher to 1. Classroom Drosophila mutants are for the most part monstrosities which, when competing with wildtype flies for food, come off exceedingly poorly. At times, experimental crosses reveal, too, that so-called recessive mutations are not entirely recessive. The dominant mutations that serve as genetic markers in various modified *ClB* procedures are generally lethal when homozygous. That these same mutations are detrimental to the survival of their heterozygous carriers is revealed by the proportions of wildtype flies in the control series of cultures: in these cultures, wildtype flies commonly have frequencies of 34 percent–36 percent instead of the expected $33\frac{1}{3}$ percent (see table 3.11 and figure 3.7); the excess of wildtype flies reflects a deficiency of the genetically marked heterozygotes.

Nevertheless, recalling that *h* in the last section was said to lie between 0 and 1.00, we shall now consider the possibility that *h* is negative. In this case $1 - (- h)s$ becomes $1 + hs$: the heterozygous individuals lie *beyond* the range enclosed by the two homozygous classes. Quantitative geneticists refer to this situation as *overdominance*. However, these same persons remove from this term contributions made by gene-gene interactions (epistasis) and gene-environment interactions. Overdominance remains, then, as a relationship between alleles $(a_1 a_1 < a_1 a_2 > a_2 a_2)$ that is supposedly independent of genotypic background or environmental conditions. Evolutionary biologists tend to be less precise; they acknowledge that genes interact with other genes and with the environment (tables 9.7 and 9.8). Indeed, no allele of any locus is *sufficient* to cause any phenotypic trait whatsoever; at best, certain alleles are *necessary* for the development of a specified aspect of the phenotype. (This statement includes the gene control of amino-acid sequences in protein molecules; both transcription and translation require the cooperation of the products of many genes. Only as an intellectual exercise can one claim that the sequence of codons of a gene is sufficient to specify the sequence of amino acids in a polypeptide chain.) The wildtype allele at the *white* locus in *D. melanogaster* is *necessary* for the development of normal red eyes; neither that allele or any other is *sufficient* to make red eyes, however, because the entire genome is needed for the development of an individual of whom eyes are but one part. Because of the sensitivity of alleles to both gene-gene and gene-environment interactions in natural populations, evolutionary geneticists do not normally use the term overdominance in its rigorous sense; instead they use more poorly defined terms such as "balanced polymorphisms," "heterozygous advantage," "superior heterozygotes," or "heterotic mutations." These terms simply imply that, within the context of a particular discussion, superior fitness resides not with either homozygote, but, rather, with the heterozygous individuals.

The algebra, following the previous examples can be organized as follows (note that $1 + hs$ has been replaced by 1, following a convention which sets the maximum relative fitness of any genotype at 1.00):

Genotype	AA	Aa	aa
Darwinian fitness (w)	$1 - s$	1	$1 - t$
Original frequency	p^2	$2pq$	q^2
Final frequency	$p^2 - sp^2$	$2pq$	$q^2 - tq^2$
Average fitness ($\overline{w}$)		$1 - sp^2 - tq^2 = \overline{w}$	

Frequency of A after selection $(p - sp^2)/\overline{w}$
Frequency of a after selection $(q - tq^2)/\overline{w}$

If the frequencies of A and a are p and q before selection operates and p_1 and q_1 afterwards, then an equilibrium is established if $p_1/p = q_1/q$; the stipulated relationship can be true only if $p = p_1 = \hat{p}$.

We can now use the original and final frequencies given above to determine if there is an equilibrium frequency and, if so, what that frequency is:

$$\frac{p - sp^2}{p\overline{w}} = \frac{q - tq^2}{q\overline{w}}$$

$$1 - sp = 1 - tq$$

$$sp = tq$$

$$\hat{p} = t/(s + t)$$

$$\hat{q} = s/(s + t).$$

The equilibrium established under the selective superiority of heterozygous individuals is a stable equilibrium; it can be approached from either side and tends to be restored if the population is perturbed. The resistance of the equilibrium to change depends upon the values of s and t. A balanced-lethal strain of flies such as the *CyL/Pm* stock used in tests of the second chromosome of *D. melanogaster* cannot be shifted from its equilibrium value of 50 percent *CyL* and 50 percent *Pm*; here, $s = t = 1.00$ and the *CyL/CyL* and *Pm/Pm* homozygotes are unable to survive. On the other hand, if homozygotes have fitnesses of 0.99 ($s = t = .01$), the population would return to the equilibrium frequencies (50 : 50) only slowly if perturbed slightly.

Frequency-dependent selection

Wright (see Wright and Dobzhansky 1946:147) once suggested the following as a hypothetical situation: "[A] species occupies a heterogeneous environment. Each genotype may be favored by selection when rare and unable to occupy fully the ecological niches to which it is best adapted, but selected against when so abundant that it must in part occupy ecological niches to which it is less well adapted than other genotypes." The effect of this suggestion is to insert gene frequencies, p and q, into selection coefficients. If selection coefficients depend upon gene frequencies, selection is said to be frequency dependent. (The change in frequency, Δq, of a recessive mutation with a constant selection coefficient, s, equals $-sq^2$; the change in frequency, consequently, depends upon gene frequency. This, however, does not mean that

s is frequency dependent because, as we have just said, s is a constant that is independent of p and q.)

The algebra for one (simple) case illustrating frequency-dependent selection is presented below:

Genotype	AA	Aa	aa
Darwinian fitness (w)	$1.5 - p$	1	$1.5 - q$
Frequency	p^2	$2pq$	q^2

Average fitness ($\overline{w}$) $\qquad 1 + p^2(0.5 - p) + q^2(0.5 - q)$
Frequency of A after selection $p(1 + 0.5p - p^2)/\overline{w}$
Frequency of a after selection $q(1 + 0.5q - q^2)/\overline{w}$
Letting $p_1/p = q_1/q = \hat{p}/\hat{q}$, $\hat{p} = 0.5$.

The result arrived at above could have been obtained intuitively by noting that, when $p = q = 0.5$, the fitnesses of all three genotypes become 1.00. Also notice that at equilibrium the average fitness ($\overline{w}$) under this scheme equals 1.00. Finally, however, notice that the notation used here violates the convention that we respected in our earlier examples: In all previous examples, the maximum fitness was arbitrarily assigned the value 1.00. In the present example, the fitness of AA individuals approaches 1.5 as p approaches zero; conversely, that of aa individuals approaches 1.5 as q approaches zero.

Subjecting the Algebra to Experimental Test

The algebraic aspects of selection that were described sequentially in the preceding section could have been (and, indeed, largely were) developed without recourse to any experimental verification. The calculations follow from Mendelian inheritance, from the Hardy-Weinberg equilibrium, and from assumptions concerning the relative (Darwinian) fitnesses of individuals of different genotypes. The manipulation of abstract symbols requires skill at mathematics, not necessarily a knack at biological experimentation. On the other hand, the calculations pinpoint events that are subject to experimental verification or quantities (such as h) that can only be estimated empirically. This section deals with some of the experiments or analyses that have been carried out in an effort to confirm or supplement the theoretical calculations on selection in populations.

Figure 10.4 illustrates the fate of a recessive lethal gene in each of two laboratory populations. We have seen that the frequency of a recessive lethal in successive generations decreases according to the following sequence: q, $q/(1 + q)$, $q/(1 + 2q)$, ... or (if $q = 1/n$) according to the sequence $1/n$, $1/(n + 1)$, $1/(n + 2)$, $1/(n + 3)$, ... The second sequence suggests using the reciprocals of these frequencies in graphing the fate of a lethal where $X =$

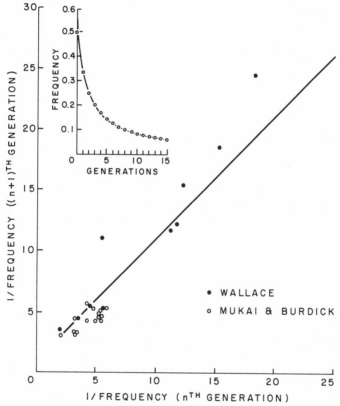

Figure 10.4 The elimination of recessive lethals from populations of *D. melanogaster*. The expected frequencies of a lethal in two successive generations can be written $1/n$ and $1(n + 1)$; therefore, the reciprocals equal n and $n + 1$. The straight line represents the expected changes in frequency as the lethal is eliminated from the population; the points are the outcomes of two experiments: ● Wallace 1963b; ○ Mukai and Burdick 1959. The small inset illustrates the expected change in frequency with generation, the curve which generates the straight line of the larger diagram.

$1/q_t = n$, and $Y = 1/q_{t+1} = n + 1$. An elimination of a lethal proceeding as expected, or nearly so, would result in points lying on or near a straight line with slope 1.00. The solid circles in the figure do follow the theoretical line rather well; they represent the elimination of a second-chromosome lethal from a laboratory population of *D. melanogaster* (Wallace 1963b). The cluster of open circles represents similar data from an experiment (Mukai and Burdick 1959). in which the lethal chromosome failed to leave the population; consequently, the points tend to lie in a closely packed cluster.

Sperlich and Karlik (1968) followed the fate of the *CyL* chromosome in a number of different populations of *D. melanogaster.* The mutation *Curly* is a recessive lethal with a dominant effect on wing shape; it is also associated with inversions that prevent recombination between the *CyL* and its homologous wildtype chromosome. Consequently, in their experiments (figure 10.5), Sperlich and Karlik followed the fate of an entire chromosome, not of a (more or less) single locus.

The data presented in figure 10.5 summarize the data obtained from 15 populations. In some, the *CyL* chromosome was introduced into populations at low (16.7 percent) frequency together with wildtype chromosomes all of which were identical

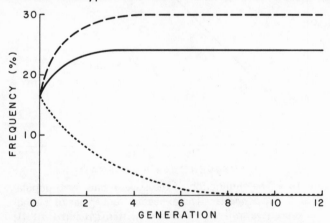

Figure 10.5 The influence of competing wildtype chromosomes on the fate of *CurlyLobe* in experimental populations of *D. melanogaster.* ———: single chromosomes or several from highly uniform laboratory strains; – – – – –: single chromosomes from a newly captured wildtype strain; ·········: several chromosomes from newly captured widltype strains. (After Sperlich and Karlik 1968.)

(isolated as one chromosome by the modified *CIB* technique, or obtained from highly inbred laboratory stocks); in these populations, the frequency of *CyL* increased and stabilized at about 22 percent–23 percent. In other populations, the wildtype chromosomes were single chromosomes freshly isolated from wild populations of *D. melanogaster*; in these cages the frequency of *CyL* rose to 30 percent and remained there. In still other populations, several wildtype chromosomes from natural populations were introduced into the same population, here the *CyL* chromosome was rapidly eliminated. It appears that, with several wildtype chromosomes to draw on, wildtype combinations are assembled through recombination that make *CyL*-bearing heterozygotes inferior in respect to Darwinian fitness; when the wildtype combinations are limited (because only one chromosome is available, or because all chromosomes come from a uniform source), the *CyL*-bearing heterozygotes (homozygous *CyL/CyL* individuals die) are superior. In these latter cases, judging from the equilibrium frequencies of *CyL* chromosomes shown in figure 10.5, the adaptive value $(1 - s)$ of wild-type flies lies between 0.6, and 0.7 (that of *CyL/+* flies = 1.00).

A number of experiments have been carried out in a search for evidence that lethals are not completely recessive (that they *are* completely recessive cannot be proved); such experiments are designed so that large numbers of lethal-free individuals and others heterozygous for "recessive" lethals are compared.

One of the first experiments of this sort was carried out by Stern and his collaborators (Stern et al. 1952) using sex-linked lethals. The crosses were designed so that two types of heterozygous females—$w^a B/+$ and $w^a B$/lethal—developed in competition, each surviving female was then progeny-tested (by examining her sons) to determine whether she carried a lethal or non-lethal wildtype chromosome (w^a = white-apricot eye color, B = bar eye shape). Evidence for partial dominance (that is, incomplete recessivity) was sought in deviations of the actual proportions of the two types of females from the 1 : 1 proportions expected.

The data obtained by Stern et al. were sufficiently detailed both in numbers of lethals tested, numbers of replicate cultures tested for each, and numbers of females tested per culture that a detailed analysis of their results can be carried out. Table 10.6 and figure 10.6 present the results of this analysis: the average frequency of lethal-bearing females was 0.487 instead of the expected 0.500; this deviation suggests that the average viability of

Table 10.6 Estimating the between-lethal variation in the viability effects of sex-linked "recessive" lethals on their heterozygous carriers in *D. melanogaster*. (Data from Stern et al. 1952.)

Total variance	0.00134263
Sampling variance	0.00053813
Experimental variance	0.00022250
Between-lethal variance (residue)	0.00058200
Between-lethal standard deviation	0.0241

lethal heterozygotes may be only 0.487/0.513 or 0.95 that of flies not carrying lethal chromosomes. However, the variation from lethal-to-lethal that remains even after binomial and environmental variances are removed is large enough to suggest that some lethal heterozygotes hatched in proportions exceeding those of their nonlethal competitors. This proportion, shown in figure 10.6 is about 29 percent. Selection within a free-breeding population would quickly distort the distribution pattern in respect to dominance for those lethals remaining in the population. The first to leave would, of course, be those that harmed their heterozygous carriers the most.

The results of a second large test (Hiraizumi and Crow 1960) are summarized in figure 10.7. In this case, two sets of cultures were tested; those yielding lethal-carrying (second chromosome of *D. melanogaster* in this case) flies, and a comparable number of vials yielding non-lethal-carrying flies. Both classes in their respective cultures were expected to occur in frequencies equaling that of a standard, genetically-marked class of flies. As the curves in figure 10.7 reveal, the proportion of wildtype flies carrying lethal chromosomes consistently lagged behind that of lethal-free heterozygotes. The difference between the two sets of cultures is about 1 percent, suggesting that the dominance of "recessive" lethals is about 0.02.

Tables 10.7 and 10.8 present still more data bearing on the question of the partial dominance of lethals. The lethals (lethals

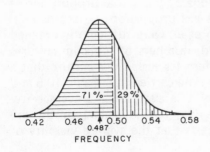

FREQUENCY

Figure 10.6 Frequency of w^aB/le females among the daughters of $+/le$ females and w^aB males (le = lethal). The curve represents variation remaining after binomial and experimental variances are removed; consequently, it represents variation between the frequencies of w^aB/le heterozygotes carrying different lethals. (Data from Stern et al. 1952.)

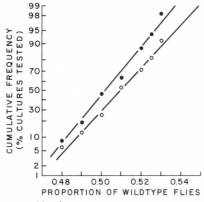

Figure 10.7 Cumulative frequency distributions of cultures containing wildtype flies in different proportions where the expected (Mendelian) proportion is 50 percent. Solid circles: wildtype flies were *cn bw/le;* open circles: wildtype flies were *cn bw/+.* (After Hiraizumi and Crow 1960.)

and semilethals, otherwise referred to as "drastic," or *D*, chromosomes) in this case were not a random sample of newly arisen lethals but, instead, were obtained from laboratory populations of *D. melanogaster.* The numbers in the two tables reflect the numbers of flies of the various classes of flies relative to those of an independent standard class (*CyL/Pm*) which was arbitrarily set at 1.000. Overall, there is no suggestion in these data that chromosomes with lethal or near-lethal effects on viability when homozygous exerted any deleterious effect on the viability of their heterozygous carriers.

Earlier comments (page 228) stressed the importance placed on the role gene-interactions might play in determining the role of

Table 10.7 Comparison of the relative viabilities of *CurlyLobe* and *Plum* flies that carry quasi-normal (*N*) or lethal or semilethal (*D*) wildtype chromosomes. The standard for measuring the relative viabilities of these flies were *CyL/Pm* flies, which were assigned viability 1.000. (After Wallace 1962.)

Population	CyL/N	n	CyL/D	n
5	1.062	122	1.063	483
6	1.091	70	1.091	523
7	1.065	395	1.064	317
17	1.068	361	1.068	367
18	1.060	497	1.057	315
19	1.067	204	1.081	521
Average	1.065	1,649	1.073	2,526
	Pm/N	n	Pm/D	n
5	1.157	124	1.114	481
6	1.180	71	1.165	522
7	1.126	397	1.141	315
17	1.134	360	1.140	368
18	1.134	497	1.126	315
19	1.123	201	1.161	524
Average	1.134	1,650	1.143	2,525

Table 10.8 Comparison of the relative viabilities of wildtype flies (*D. melanogaster*) heterozygous for two different quasi-normal chromosomes (*N/N*), one quasi-normal and one lethal or semilethal (*N/D*), or two different lethal or semilethals (*D/D*). As in the previous table, *CyL/Pm* flies served as the standard (1.000) for measuring relative viabilities. (After Wallace 1962.)

Population	N/N	n	N/D	n	D/D	n
5	1.028	28	1.109	190	1.050	363
6	1.101	8	1.125	125	1.135	444
7	1.047	219	1.048	354	1.058	132
17	1.075	162	1.060	397	1.076	163
18	1.078	301	1.086	392	1.052	116
19	1.078	66	1.097	273	1.133	377
Average	1.067	784	1.079	1,731	1.097	1,595

selection in governing gene frequencies within populations. The purpose of presenting the data in figure 10.8 is to emphasize that obvious interactions need not be the important ones governing the nature of selective forces. The data in this figure concern the fate of the mutant alleles *cinnebar* (*cn*) and *scarlet* (*st*) in populations of *D. melanogaster*. The pioneering experiments of Beadle

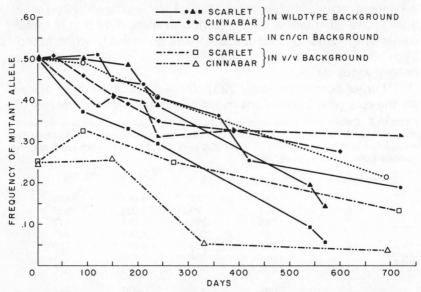

Figure 10.8 Patterns of elimination of the mutant alleles *scarlet* (*st*) and *cinnebar* (*cn*) from wildtype populations of *D. melanogaster* (solid lines) and from mutant populations (*cn/cn* and *v/v* in the case of *scarlet*; *v/v* in the case of *cinnebar*) (dashed lines) which might be expected to render the *st* and *cn* mutant alleles harmless. Contrary to expectation, no striking differences are apparent in these patterns of elimination. (After Frydenberg and Sick 1960, 1962.)

and Ephrussi (1936) in which eye anlagen were transplanted from larvae of one genotype to those of another showed that these mutant loci control the formation of normal eye pigment. The pathway proposed by Beadle and Ephrussi was as follows:

$$A \longrightarrow B \longrightarrow C \longrightarrow D \longrightarrow \text{normal pigment.}$$
$$\qquad v^+ \qquad cn^+ \qquad st^+$$

The implication of this pathway (a pathway that may need revision) is that in the absence of v^+, the presence or absence of cn^+ or st^+ is a matter of indifference; the gene products of these loci should have no substrate on which to work. Similarly, in the absence of cn^+, the presence or absence of st^+ should be of little concern. The data summarized in figure 10.8 show, contrary to obvious expectations, that *scarlet* and *cinnebar* are eliminated from *vermillion* populations and *scarlet* from *cinnebar* ones as rapidly as they are from wildtype populations. The fitnesses of these mutants are not improved relative to their wildtype alleles by homozygosis for a mutant allele farther "upstream" in the proposed biochemical pathway.

Except for a comment on the distorted Mendelian ratios students sometimes encounter in overcrowded cultures, the account of Darwinian fitnesses and selection coefficients presented so far has implied that these quantities are constants. It would be a grave error to leave the impression that selective forces are invariate. Figures 10.9 and 10.10 present data suggesting that the magnitude of selection coefficients is dependent upon the number of flies (*D. melanogaster* in both instances) in the individual culture bottles. In the case of the tumor-bearing larvae studied by Herman Lewis (1954; see figure 10.9), the relative viability of these larvae may have exceeded that of normal larvae at low densities but was reduced to only 30 percent of normal viability at high ones. Moree and King (1961) found a similar dependence of the relative viability of *ebony* flies on the degree of larval crowding (figure 10.10).

Tests for frequency-dependent selection take a number of forms, the most common of which is represented in terms of "input" and "output" ratios of two genotypes. Suppose that any ratio of *A* and *B* parental females (males are excluded so that crosses between *A* and *B* are avoided) that produce offspring under competition in the same culture yields progeny in precisely

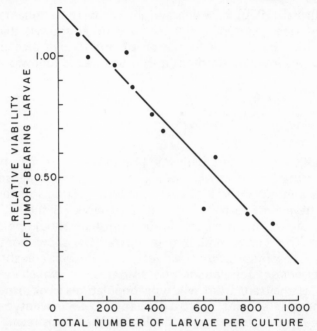

Figure 10.9 A demonstration of the dependence of the relative viability of tumor-bearing larvae (*D. melanogaster*) on the total number of larvae per culture. Total culture size = number of normal larvae $\times \frac{4}{3}$; expected number of tumor-bearing larvae = number of normal larvae $\times \frac{1}{3}$; relative viability of tumor-bearing larvae = number observed/ number expected. (After Herman Lewis 1954.)

the original, parental ratio; a graph of the logarithm of A/B among parents against the logarithm of A/B among progeny would be a straight line of slope 1.00 (45°) passing through points, $Y = X$. Suppose, next, that among the progeny, A always appear in greater proportions than among parents so that (A progeny/B progeny) ÷ (A parents/B parents) = K, a constant. In this case, the logarithm of A/B among offspring minus the logarithm of A/B among parents equals log K or k. Rearranged, log (A/B) output equals log (A/B) input plus a constant k. This relationship when plotted as log (A/B) input versus log (A/B) output would be a straight line of slope 1.00, but displaced upward (by an amount, k) from the previous one. Similarly, if A were always in a deficit among offspring, the line would have slope 1.00, but displaced downward.

Still another possibility is that the slope of a log (A/B) input versus log (A/B) output regression line not equal 1.00 but (let's say) something less than 1—that is, more nearly horizontal. The

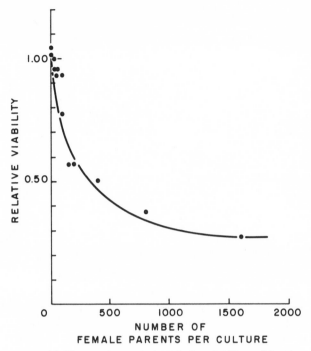

Figure 10.10 The dependence of the viability of homozygous *ebony* (*e/e*) flies on the numbers of heterozygous (*+/e*) parental flies per culture. The calculation of these data takes into account an apparently density-independent viability of *+/e* progeny flies of 1.027 relative to 1.000 for *+/+* flies. (Recalculated from data of Moree and King 1961.)

implication of such a relationship is that when the input ratio of *A/B* is small (*A* is rare), the output ratio is larger than expected. At the other extreme, when the input ratio of *A/B* is large (*B* is rare), the output ratio is smaller than expected (*B* increases its frequency). Now, if *A* becomes more common when it is rare, and *B* becomes more common when *it* is rare, there must be an intermediate frequency at which input and output ratios are equal. This is the situation illustrated in figure 10.11 in respect to the mating success (one of the many components of fitness) of male flies (*D. pseudoobscura*). The figure illustrates what has become known as the "rare male" effect: if one of two types of males is rarer than the other, the rare type frequently is disproportionately successful in mating with available females. Selection of this sort, always favoring the rare genotype in a population, would tend to maintain alleles at intermediate frequencies.

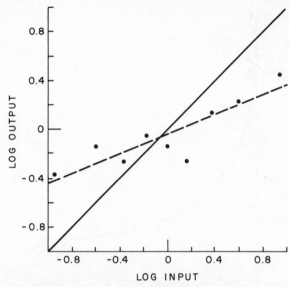

Figure 10.11 The use of log input ratios (of proportions of *CH* to *AR* males) and log output ratios (of matings by the two types of males) to illustrate frequency-dependent mating success in *D. pseudoobscura*. The solid line passes through points at which input and output ratios are equal; the dashed line is the one that best fits the data. Rare males of either sort perform a disproportionate share of all matings. Mating success, like viability, is a component of Darwinian fitness. (After Ayala and Campbell 1974.) Reproduced with permission; from *Annual Review of Ecology and Systematics*, vol. 5; © 1974 Annual Reviews, Inc.

Darwinian Selection in vitro

Evolutionary theory is frequently criticized because it is non-predictive; this charge is leveled most often by persons who, for a variety of conflicting personal reasons, cannot (or will not) admit that present-day forms of life have descended from common ancestors of the past. More recently natural selection has come under criticism, not only by those who interpret molecular changes as being largely haphazard (see page 312) but also by those who, being alarmed at a reactionary social philosophy—"biological determinism"— respond by attempting to weaken one of the claimed underpinnings of that philosophy, evolution through natural selection.

The following pages are not intended as a defense of circular arguments; such arguments, if they are to be rendered useful, must be uncircularized. The material to be presented is intended to illustrate the various outcomes of natural selection in a simple system (the replication of a chemical molecule under its own direction in a lifeless "growth" medium) and to argue that these outcomes were indeed predictable.

The lack of prediction in matters of evolution stems largely from a general unawareness on the part of those who might otherwise have acted as seers: Darwin, in compiling the evidence presented in *On the Origin of Species*, turned to geology, paleontology, embryology, anatomy, geography, natural history (ecology) and—in lieu of genetics—to plant and animal breeding in order to support his arguments. He left no branch of science within which others independently might have set about predicting and testing his ideas; Darwin, in effect, did it all himself. Similarly, before students of evolution became aware of the elegant techniques that biochemistry had placed in the hands of molecular biologists, the latter were well along in their studies of the evolution of proteins; again, a priori predictions were not made. The same is somewhat true of the in vitro studies of natural selection among self-replicating molecules except that Spiegelman (1967) quickly realized the nature of the events he was observing, and then carried out his research with Darwinian evolution in mind.

Among the RNA phages that attack *E. coli* is one known as Qβ (cue-beta) which consists of a single strand of RNA more than 3,600 bases long. When the Qβ RNA first enters a susceptible bacterial cell, the RNA is read by the bacterial machinery as if it were a strand of mRNA; the protein that is synthesized then joins several proteins that are normal constituents of bacterial cells, making a complex polymer which is a polymerase for Qβ RNA. The polymerase attaches to the Qβ RNA and synthesizes a complementary strand of RNA. As more and more polymerase molecules are formed, Qβ strands are copied as complementary strands, and complementary strands are copied in turn to give molecules identical to the original Qβ strands. There are, of course, complex systems of control (all quite well understood by now) leading to the eventual use of another region of the Qβ RNA in synthesizing the coat protein of mature phage particles. These events occur with incredible speed; thousands of phage-mediated molecules are made in the first 15 minutes following the injection of the RNA strand into the bacterium.

Spiegelman (1967) succeeded in devising a culture medium, containing bacterial extracts, supplemented with the nitrogenous bases needed for RNA synthesis, and provided with an energy source, in which he found that $Q\beta$ RNA was copied and recopied. The increase in RNA was initially studied both chemically and biologically: the first measures the increase in quantity of RNA, the second measures the increase in the number of "infective" moelecules—molecules of RNA that are able to infect bacterial cells. Upon serial transfer of small amounts of replicating material from exhausted vials to fresh ones, Spiegelman found that the two measures did not coincide: infective particles did not increase in parallel with the amount of newly synthesized RNA. In fact, the RNA rather quickly became noninfective but, nevertheless, continued to increase in speed of replication (figure 10.12). The explanation lay in the nature of the selective forces to which the RNA molecules were exposed in the culture vials. Those portions of the RNA molecules that coded for polymerase and coat protein molecules served no purpose in the in vitro experiments; they merely prolonged the time required for the synthesis of a new RNA molecule. RNA molecules which lost these now useless regions replicated more rapidly (had shorter generation times) and displaced the large, more slowly reproducing molecules.

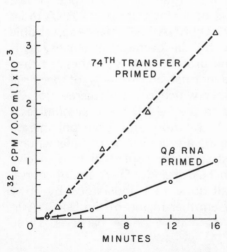

Figure 10.12 A comparison of the kinetics of replication of $Q\beta$ phage particles (single-stranded RNA) and the rapidly replicating RNA molecules that displaced them during 73 serial transfers; the entire process took less than 13 hours. Note that the initial lag period has been greatly reduced. The rate of RNA synthesis is 2.6 times greater for the selected primer than for the $Q\beta$ phage; because the newly selected particles are only $\frac{1}{6}$th the size of the phage particles, the selected RNA particles are being produced 15 times as fast as phage particles were. (After Spiegelman 1967.) Reprinted with permission: from *American Scientist*.

Eventually, about 94 percent of the original molecule was discarded (Figure 10.13), leaving little more of the Qβ particle than a highly efficient attachment site for the polymerase molecules of the growth medium.

The replicating molecules of Spiegelman's studies undergo mutation and, under adverse conditions, some of these prove to have a selective advantage over the prevalent "wildtype" molecules. Among the adverse conditions that have been tested is starvation for one or the other of the four nitrogenous bases. When the amount of any base in the medium is severely limited, growth slows down (i.e., replication time increases) momentarily, but then speeds up once more—not gradually and uniformly, but in steps corresponding to the occurrence of mutational changes. The nature of three mutations that enable the replicating RNA particle to replicate relatively rapidly in the presence of a chemical inhibitor (ethidium) is shown in Figure 10.13. Figure 10.14 compares the replication rates of wildtype and mutant strands under various concentrations of ethidium. The mutant replicates more rapidly in the presence of ethidium; it should because, after all, it did displace the wildtype. The mutant does not replicate as fast as wildtype in the absence of ethidium; again, as expected, because the mutant would otherwise have been established earlier as the wildtype molecule.

Subsequently, Spiegelman and his coworkers have found (Sumper and Luce 1975) that the "growth" medium need not be primed with RNA in order that replicating molecules arise; the polymerase molecule can by trial and error piece together RNA molecules capable of replication. Once more observations fit expectations. If the growth medium is rich in nitrogeneous bases, the replicating RNA molecules that are put together independently (in separate growth tubes), although different in detail, show structural similarities. If the growth medium contains only low concentrations of nitrogenous bases, the replicating molecules of different tests differ considerably. Apparently the structure of polymerase calls for a somewhat limited conformation in the RNA molecule (note in figure 10.13 that the RNA molecule has a complex physical structure) if replication is to proceed rapidly. When raw materials are plentiful, a fast replicating molecule is likely to arise quickly in each tube, thus eliminating previously assembled but more slowly replicating molecules. Under starvation conditions, putting together a replicating molecule is time-consuming; each one that is assembled is likely to be the only replicating molecule of its tube; inefficient as a lone replicat-

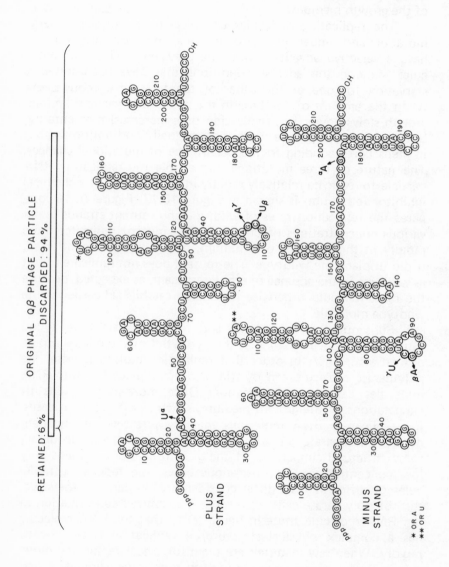

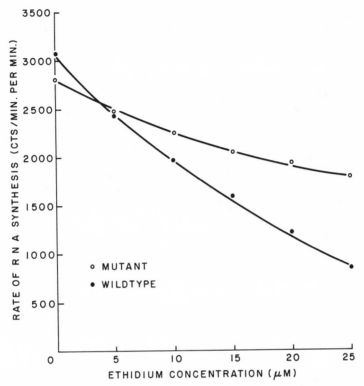

Figure 10.14 The rates of RNA synthesis for "wildtype" and mutant molecules in the presence of various concentrations of a replication-inhibitor, ethidium. The mutant molecule arose, and eliminated its "wildtype" predecessor, in reaction tubes containing ethidium. Note that the mutant molecule does not possess an advantage in the absence of the inhibitor. The composition of both wildtype and mutant molecules is shown in figure 10.13. (After Kramer et al. 1974.)

Figure 10.13 One of the rapidly self-replicating molecules of RNA obtained by Spiegelman and his colleagues from reaction tubes initially inoculated with $Q\beta$, a single-stranded RNA phage. During the course of selection for rapid replication, 94 percent of the original phage particle has been discarded, together with all phagelike properties such as the ability to infect and reproduce in bacterial cells. The three indicated base substitutions form the ethidium-resistant mutant RNA molecule of figure 10.14. (After Kramer et al. 1974.)

ing molecule may be, it uses up the raw materials during replication, and in so doing precludes the origin of a more efficient competitor molecule. These later observations were not possible in the early studies because the $Q\beta$ primer RNA did replicate and in doing so it prevented the origin of de novo competitors. In the early experiments, rapidly replicating molecules arose only as fast replicating fragments of the $Q\beta$ RNA strand, itself.

The experiments carried out by Spiegelman and his colleagues do not concern the origin of life. They should not, however, be lightly dismissed on this account. What these workers have done—and it has been an extremely important task—is to follow the workings of natural selection in a stark physical model and to show that every consequence observed has its counterpart in the theoretical or logical construct of natural selection. There have been surprises of neither omission nor commission.

BOXED ESSAY

In assigning fitness to individuals of genotypes a_1a_1, a_1a_2, and a_2a_2, it is understood that these are based on averages. For example, if either b or c is a recessive lethal, a_1a_2bb or a_1a_1cc individuals die. One assumes that the distribution of b's, c's and similar alleles at still other loci are random in respect to the a-alleles, and that the three genotypes at the latter locus are affected equally by events occurring at others.

During their lifetimes, individuals pass through sequences of differing external and internal environments. As they move about, a_1a_1, a_1a_2, and a_2a_2 individuals encounter patches within their environment within which their relative fitnesses take on different values. Or, as time slips by (days, weeks, months, or even years), the relative fitnesses of these individuals may change; the causative agents may be temperature, humidity, photo period, drought, or any one of many biologically important factors. Even in passing through successive stages of embryonic development, the probability of successfully completing any one stage may vary from genotype to genotype depending upon the stage.

Recognition that fitnesses may fluctuate, even reverse, sequentially has given rise to the concept of "marginal overdominance" (Wallace 1959a and 1968a:213). Imagine n stages in half of which the fitnesses of $a_1 a_1$, $a_1 a_2$, and $a_2 a_2$ individuals are $1 : 1 : 1 - s_i$, and in the other half $1 - s_j : 1 : 1$. Multiplying through all stages, marginal fitnesses $[(1 - s_j)^{n/2} : 1 : (1 - s_i)^{n/2}]$ are obtained. Heterozygous individuals, because of the dominance of the favored allele in each instance, emerge with an overall fitness exceeding that of either homozygote.

Dominance need not be complete in order that heterozygotes have an overall (marginal) superiority. If the individual s's are small, $1 - S$ is very nearly equal to $1 - n\bar{s}/2$. Let the fitness of heterozygotes in each of the n stages be $1 - X_i$; that is the heterozygotes are always somewhat less favored than the superior homozygote. Multiplied through all n stages, the marginal fitness of heterozygotes would equal $(1 - X_i)^n$ or, approximately, $1 - n\bar{X}$. Consequently, heterozygotes will exhibit marginal overdominance if $n\bar{X}$ is smaller than $n\bar{s}/2$, or if $2\bar{X}$ is smaller than $\bar{s}$. Thus, in order to gain an overall superiority in fitness, heterozygotes need only resemble the favored homozygote more closely than the disfavored one; by no means must the dominance of the favored allele be complete.

FOR YOUR EXTRA ATTENTION

Kidwell and Kidwell (1979) followed the fate of a fourth-chromosome lethal in a number of populations of *D. melanogaster*; some of their experimental data are listed below (N = number of individuals examined each generation; H = the proportion of lethal heterozygotes):

Generation	N	H
0	—	1.00
1	794	0.69
2	946	0.57
3	684	0.47
4	614	0.46
5	1,312	0.45

Generation	N	H
6	2,212	0.41
7	2,227	0.42
8	1,181	0.40
9	2,066	0.40
10	1,765	0.37
11	2,439	0.39
12	3,033	0.40

Transform these data into a form that can be plotted in the manner illustrated in figure 10.4. If you see an excuse for doing so, compute the (approximate) relative fitnesses of +/+ and +/le flies (that of lethal homozygotes, of course, equals 0).

These data suggest that the number of flies counted increased by nearly 190 flies per generation. Does it appear that the decline of lethal homozygotes explains this increase in the number of flies examined in each generation?

2. Serious doubts exist in the minds of many whether or not natural selection acts on the allozyme variation in natural populations; no doubts exist, however, on the *existence* of such variation. Hickey (1977) maintained 16 initially similar populations of *D. melanogaster*—8 on starch, 8 on glucose —and followed the frequency of the *amy*[4,6] allele (the initial frequency of this allele in all populations was 0.50). After 20 generations, the frequency of *amy*[4,6] in populations maintained on starch was 47.8 percent ± 5.3 percent, in populations maintained on glucose the frequency was 21.2 percent ± 4.1 percent.

Consider some of the criticisms that might be addressed to Hickey's study, and some additional studies that might answer these criticisms.

3. Wright (1931) says that if selection is against both homozygotes, the equilibrium frequency, $\hat{q}$, equals $(1 - h)/(1 - 2h)$. In this notation, the fitnesses of *aa*, *Aa*, and *AA* individuals are given as $1 - s$, $1 - hs$, and 1. Show that Wright's expression is equivalent to $\hat{q} = s/(s + t)$ given on page 229.

4. Consider the following statements from Place and Powers (1979) relative to the concept of marginal overdominance developed in this chapter: " ... LDH-B^{b}B^b would be a more efficient catalyst at low temperatures than either LDH-B^{a}B^b or LDH-B^{a}B^a. At higher temperatures, LDH-B^{a}B^a should be the most efficient. The heterozygous phenotype, LDH-B^{a}B^b, would be intermediate. Viewed in this light, in an annually fluctuating thermal environment such as that experienced by *F[undulus]heteroalitus*, the heterozygous phenotype would have an advantage with a net heterozygote superiority."

5. Bijlsma (1978) studied the survival of larval *D. melanogaster* in food containing 0.10 percent sodium octanoate. Two allozyme loci—G6PD and 6PGD—were subjected to analysis. (In the table, the percentage survival is shown for each genotype.)

		G6PD		
		FF	FS	SS
	FF	27.6	32.1	19.1
6PGD	FS	26.5	33.8[a]	28.6
	SS	24.7	29.0	30.6

[a]Both FF/SS and FS/SF double heterozygotes were tested and proved to be similar in survival.

A thorough analysis of these data is not simple if one wishes to compute equilibrium frequencies for the various alleles; nevertheless, unweighted marginal averages reveal a great deal.

6. The difficulty in obtaining convincing evidence for the selective superiority of single-locus heterozygotes has been discussed by Young (1979a, 1979b) in two papers reporting on enzyme polymorphisms in the cyclic parthenogenetic *Daphnia magna*. Sexual reproduction in these organisms leads to Hardy-Weinberg expectations; an excess of heterozygotes at many allozyme loci develops during subsequent parthenogenetic reproduction, a manner of reproduction that exaggerates small differences in the adaptive values of differing genotypes (refer to table 4.5).

What explanations might be advanced for the seeming selective superiority of heterozygous individuals?

7. From tests of 1,140 wild-caught females, Snyder and Ayala (1979) obtained 20 strains of *D. pseudoobscura* homozygous for each of two alleles (say, a_1 and a_2) at the phosphoglucomutase (*Pgm*-1) locus. First instar larvae were placed, 60 per vial, in cultures where they competed for survival; five replicate vials were set up for each of eight frequencies ranging from 100 percent a_1a_1 to 100 percent a_2a_2. Data on survival obtained from the "mixed" vials were as follows:

			Frequency			
	.10	.25	.40	.60	.75	.90
a_1a_1	.634	.612	.561	.574	.606	.543
a_2a_2	.640	.594	.572	.562	.536	.528

Notice that the probability of survival for each homozygote is highest when it is rare, and lowest when it is common.

Using elementary algebra, calculate at what frequencies of a_1a_1 and a_2a_2 the percent of surviving larvae of the two genotypes are equal. (One genotype, a_1a_2 is missing in this example; nevertheless, frequency-dependent selection is one that tends to establish stable equilibria.)

8. Fujio et al. (1979) studied the survival of individual oysters (*Crassostrea gigas*) that were homozygous and heterozygous for the *A* and *B* alleles of the catalase locus. They knew the number of starting individuals

and the relative starting proportions of the three genotypes; they also knew the number and genotypes of the survivors:

	AA	AB	BB	Total
Starting numbers	137	353	229	719
Surviving numbers	48	346	113	507
Survival (percent)	35%	98%	49%	71%

Provided that such differential survival determines the fate of the *A* and *B* alleles within oyster populations, what is that fate? Compare your conclusions with item 3 following chapter 5.

9. Occasionally, haploid populations are used in order to illustrate some aspect of population genetics; the complexity of diploidy can be avoided in this manner.

Andrews and Hegeman (1976) studied the selective disadvantage suffered by bacteria (*E. coli*) which make an unnecessary protein (a fragment of β-galactosidase in a noninducing medium); the disadvantage was determined in a continuously growing culture.

The ratio (*R*) of mutant to parental individuals at the outset ($t = 0$) was 1.70; after 21 generations ($t = 21$) it had declined to 0.98. Solve for the disadvantage, s, by noting that

$$R_t = R_0 e^{-st}$$

10. A good deal of interest in the empirical measure of mutation rates and selection intensities resides in a desire to estimate the harm that might accrue to human populations through the exposure of individuals to mutagenic radiations and chemicals.

Newcombe (1979) has suggested that "risk" and "mutation" should be disentangled. He argues that a multigenerational study will be needed to measure the level of harm. While admitting that such a study would not measure mutation accurately because the prenatal loss of mutant zygotes might go unnoticed, Newcombe is unconcerned. The point, he argues, is to measure *tangible* harm of kinds society would consider important.

The point Newcombe has made will impinge on the "genetic load" which will be discussed in chapter 12. It is not too early, however, to consider whether, as Newcombe says, a stillbirth or loss of an early embryo is less important than the birth of a malformed child (see Dobzhansky 1979:192).

Selection II

PREVIEW: The account of selection is continued in this chapter to include its more complex (and more realistic) patterns: directional, stabilizing, and disruptive. The chapter then continues with a discussion of the interplay between selection and mutation, and of nonadaptive selection.

THE MATERIAL COVERED in chapter 10 served to introduce the ideas of selective death and reproduction of individuals in populations; to describe several simple algebraic models dealing with recessive, incompletely recessive, and heterotic mutations as well as those whose fate is subject to frequency dependence; to illustrate each model with simple examples; and (in the closing section) to use self-replicating molecules and the events that befall them in laboratory test tubes as the simplest possible physical model illustrating much that had been discussed earlier.

In this chapter we shall cover more realistic forms of selection, many of them being selection for or against certain phenotypic (rather than genotypic) aspects of individuals. With few exceptions, selection acts on phenotypes: on the relative sizes of individuals, on their health and longevity, on their reproductive vigor. Phenotypes are composites formed by the interplay of many genes. Only because certain genotypes are correlated with certain phenotypes does most selection modify the genetic characteristics of a population.

Directional selection.

A population whose individuals are subjected to selection in respect to a given trait such that those expressing it to the greatest (or least) extent are preferentially chosen as parents is said to undergo directional selection. Selection of this sort may be for the tallest, the shortest, and largest, the smallest, the most rapidly developing, the most resistant, the darkest—in brief the most or the least of any trait imaginable.

Figure 11.1 illustrates the outcomes of four experiments on the flour beetle, *Tribolium castaneum,* in each of which beetles emerging from the heaviest (upward) and lightest (downward) pupae were chosen as parents. Rapid progress was made in each direction within all four experiments. The results were highly repeatable. Within seven generations of selection, pupae of the upward selected lines were consistently twice as large as those of the downward selected ones.

Figures 11.2 and 11.3 illustrate the technique and the results of selecting for geotaxis in *D. pseudoobscura.* In figure 11.2 is shown, in diagrammatic form, the outline of a "Hirsch" maze. Beginning at the origin on the left, a fly walking to the right (where vials, numbered 1-16, with food await it) encounters a series of up-or-down choices. If the fly chooses to move upward at each of the 15 opportunities, it will arrive at vial #1; if it

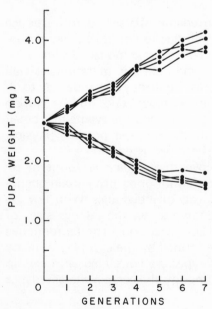

Figure 11.1 The results of a two-way selection experiment in which four replicate lines of the flour beetle, *Tribolium castaneum*, were selected for both increased and decreased pupal weight. Note the consistency of the response among replicates, and the symmetry of the upward and downward responses. (After I. R. Franklin, from Sokoloff 1977: 134.)

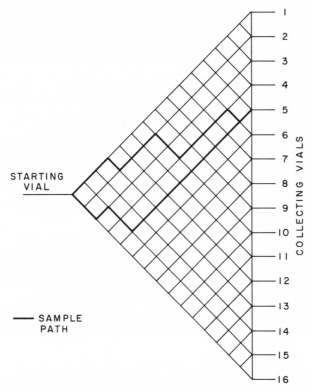

Figure 11.2 A diagram representing the structure of the
"Hirsch" maze used in the study of geotaxis and photo-
taxis in *Drosophila*. The maze is assembled from small
Y-tubes (constructed so that flies cannot retrace their
paths) that lead from the starting vial (the entry point
for flies to be tested) to 16 collecting vials. Flies that
make the same number of "up" or "down" choices (as
have those taking the sample paths in the diagram)
arrive at the same collecting tube. (See Hirsch and
Erlenmeyer-Kimling 1962.)

chooses to move downward at each opportunity, it will arrive at
vial #16. If, as most unselected flies do, it chooses to move upward at some points but downward at others, it will arrive at an
intermediate vial. A given number of downward paths combined
with the remaining upward ones lead to the same vial regardless of
the order in which the choices are made; the two pathways shown
by heavy lines in the figure both contain 4 downward segments
and 11 upward ones, and both lead to vial #5. To select for
positive and negative geotaxis among flies, the maze is placed in

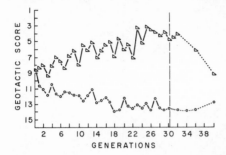

Figure 11.3 The result of directional selection for positive and negative geotaxis in *D. pseudoobscura*. Selection ended at generation 30; subsequent tests reveal the commonly observed tendency for selected populations to regress following the cessation of selection. (After Dobzhansky and Spassky 1969.)

a vertical position, several hundred flies are introduced at the starting vial, and at some time later (the next morning, perhaps) the flies in the uppermost (negative) or lowermost (positive) vials are collected as parents for the next generation. Males and unmated virgin females are, of course, tested separately.

The results of selection for geotaxis are shown in figure 11.3. At the start of the experiment, the flies were practically indifferent as to whether they went up or down at each choice while making their way to the food vials. After 30 generations of selection, flies selected for positive geotaxis emerged in the neighborhood of vial #13; those selected for negative geotaxis emerged in and around vials #4 and #5. The gravitational responses of the selected flies were altered during the course of the experiment. After selection had ceased, the flies gradually converged once again in their response to gravity; within 10 generations, the negatively selected line had returned to the original indifference shown by unselected flies.

What proof exists that the alterations brought about by selection (artificial selection in the examples given here) are genetic changes? The usual proof involves linkage to a known genetic trait. Table 11.1 illustrates an early example. Garden beans

Table 11.1 An early use of *linkage* as proof that factors responsible for a continuously varying quantitative trait are chromosomal and, hence, genetic in the modern (post-Mendelian) sense. The numbers give the average seed weight (in centigrams) of beans in the eyed (colored) and white F_2 segregates of crosses in which the parents differed both in size and in seed-color alleles. (After Sax 1923.)

Parents		Offspring	
Eyed	*White*	*Eyed*	*White*
56	28	39.0	33.8
58	21	26.6	23.4
48	21	31.3*	26.4
48	21	30.2	25.8

*Standard error in this cross = 1.1; otherwise it is 0.6 or less.

with genetically determined seed colors were subjected to selection for large seed size. Once the selection program had succeeded,
crosses were made between large ("eyed") and small ("white")
seeds. As the data in table 11.1 show, large size accompanies the
"eyed" phenotype in the F_2 generation. These F_2 eyed seeds are
not as large as the original heavy parents but, of course, that is
not necessary to prove the point that is being made: despite the recombination and independent assortment of genes in the F_1
generation eyed-ness is still linked to large size.

Linkage studies in *Drosophila* are not limited to linkage of
genetic markers; in these organisms, chromosomes can be manipulated at will by the use of *ClB*-like techniques. In figure 3.9 we
saw proof that genes conferring DDT resistance in *D. melanogaster*
were located on all major chromosomes of that species; the
resistant strains of this species had been obtained by directional
selection. The studies summarized by Hirsch and Erlenmeyer-
Kimling (1961, 1962) showed that genes modifying geotaxis in
D. melanogaster are also located on all major chromosomes of that
fly.

The need for genetic variation as a prerequisite for progress
under a selection program is illustrated in figure 11.4. The data
shown in this figure are the average numbers of bushels of oats
harvested per acre in the United States from 1900 through 1965.
Early in the program seeking to improve the yield of oats in this
country, the best-yielding strains in various parts of the world
were sought out, tested in this country, and then pressed into
agricultural service. Following that early period, some twenty
years were spent during which purelines were established and selection was practiced by picking the best-yielding plants of each
pureline. In retrospect, this period, during which the yield of oats
did not increase, was a period during which selection was practiced
on material containing no genetic variation. Starting in the late
1930s, selection programs were shifted to lines that were obtained by the hybridization of purelines. Because different pure-
lines were not identical, selection in this case was once more
effective. Such genetic variation was maintained by crossing
selected lines of different orgins, and by continuing the selection
of their progeny.

Figure 11.3 illustrates two points of interest concerning selection experiments: first, the selected lines, both upward and
downward lines, plateaued somewhere about generation 24 and
made no substantial progress thereafter; second, after selection

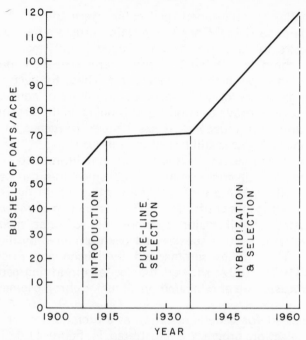

Figure 11.4 An illustration of the ineffectiveness of selection in the absence of heritable variation. During the early 1900s, the yield of oats (bushels/acre) was improved in the United States through the identification and importation of superior varieties from throughout the world. During the 1920s and early 1930s scarcely any improvement occurred when plant breeders sought to improve pure (i.e., homozygous) lines by artificial selection. Since the mid-1930s a steady improvement has resulted once again from selection programs based on hybridization and the selection of superior individuals found among the segregating progenies of subsequent generations. The rates of improvements in bushels/ acre/year during the three periods are 1.8, 0.1, and 1.5. (After Sprague 1967.) Reproduced with permission: from *Annual Review of Genetics*, vol. 1; © 1967 Annual Reviews, Inc.

ceased, the lines regressed toward the original, unselected phenotypes. Why should progress under selection cease? An obvious answer can be found in figure 11.4: genetic variation in respect to the selected trait has been exhausted; the plateau begins when the selected material is effectively homozygous. This explanation repeats that which was just given to explain the failure to increase the yield of oats per acre when selection was practiced on pure-lines (1915–1935). It does not, of course, easily explain the rapid return of lines to their original state once selection ceases (see figure 11.3).

Although selection is ineffective in changing genetically uniform material, plateaus in material subjected to selection are not always explained by the exhaustion of genetic variation. Figure 11.5 illustrates the relationship between the mean weights of eggs produced by chickens that had been selected to produce eggs of large size, and the proportion of eggs produced by those females which were able to hatch when incubated. The hens which produced the eggs with the highest proportion of hatching lag nearly one-half standard deviation below the mean of the selected line in the average size of the eggs they produced. Artificial selection for egg size in this experiment was opposed by natural selection for egg hatch. This appears to be a common phenomenon: the individual that is selected to be a parent but which dies before mating successfully is more often than not the individual with the extreme phenotype.

Here, then, is a plausible explanation for the commonly observed regression of selected lines toward their original state when selection ceases.

Nearly every selection program, even one which attempts to modify a seemingly unimportant aspect of the phenotype, is accompanied by a decline in the health and reproductive ability

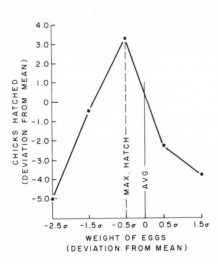

Figure 11.5 A diagram relating the size (weight) of eggs produced by the individual hens of a selected poultry flock and the number of chicks hatched from eggs laid by these hens. Notice that the hens which have the greatest number of chicks produce eggs whose weight is substantially below the average of the flock. A plateau in a selection program can result, then, not only from a loss of genetic variance (see figure 11.4) but also from the opposition of natural selection to the artificial selection imposed by the breeder. (Data from Lerner and Gunns 1952.)

(that is, in the fitness) of the selected individuals. A strain of flies can be maintained for years by choosing each generation three or four pairs as parents for the next generation. Let these three or four pairs be the flies with the highest number of bristles on any one of a half-dozen well-defined segments of the body, or the least bristles on these same segments, and shortly the strain verges on extinction from sterility or poor viability.

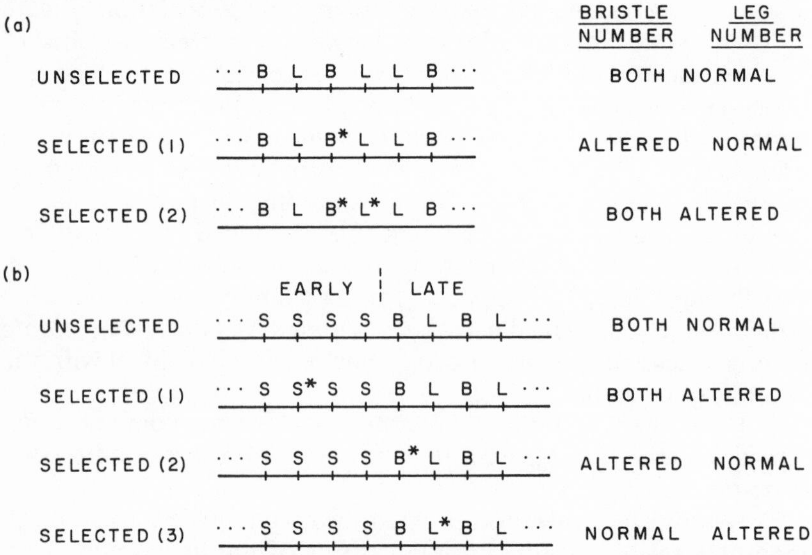

		BRISTLE NUMBER	LEG NUMBER
(a)			
UNSELECTED	··· B L B L L B ···	BOTH	NORMAL
SELECTED (1)	··· B L B* L L B ···	ALTERED	NORMAL
SELECTED (2)	··· B L B* L* L B ···	BOTH	ALTERED
(b)	EARLY ┊ LATE		
UNSELECTED	··· S S S S B L B L ···	BOTH	NORMAL
SELECTED (1)	··· S S* S S B L B L ···	BOTH	ALTERED
SELECTED (2)	··· S S S S B* L B L ···	ALTERED	NORMAL
SELECTED (3)	··· S S S S B L* B L ···	NORMAL	ALTERED

Figure 11.6 Alternative explanations for correlated phenotypic responses to a program of selection (either artificial or natural). The example involves selection for altered numbers of sternopleural bristles in *D. melanogaster*, a program that led inadvertently to missing and malformed legs in one selected line.

(a) One can imagine that the chromosome contains genes that affect both bristles (*B*) and legs (*L*); the starred loci give rise to altered phenotypes. Selection in the first instance affects bristles only; that in the second affects both bristles and legs. Theoretically, the two aberrant phenotypes could be separated by chromosomal recombination.

(b) In addition to the loci (*B* and *L*) described above, the chromosome is now said to carry genes (*S*) that control the early development of cells that are precursors to those affecting both bristles and legs. If one of these is responsible for the correlated response, the two phenotypic abnormalities cannot be separated by recombination.

Many geneticists believe that the genetic control of development involves the physical rearrangement of chromosomal material in somatic cells; a rearrangement of that sort may blur the distinction between the (logically) alternative models outlined in *a* and *b*.

The above facts do not prove that every aspect of the phenotype has been established by natural selection, but they do suggest that developmental processes leading to a given aspect of an organism's phenotype cannot be tampered with casually. In changing the characteristic number of sternopleural bristles of flies from the normal 7-8 bristles to 10 or more, some developmental changes may occur which affect the sternopleural alone but others may occur early enough in development to affect legs and other parts of the fly in addition to the numbers of bristles on the sternopleura. Paraphrasing John Donne, no aspect of the phenotype is an island unto itself; an attempt to alter one can have repercussions throughout the entire individual (figure 11.6).

Stabilizing selection

Most persons are familiar with the history of agriculture and realize that the plants and animals which provide us with food have been subjected to artificial selection for millennia. Most of this selection has been directional: greater total yield in crop plants, more efficient use of plant material by farm animals, and more milk, beef, or pork per animal. Only occasionally is the direction of selection reversed as when the poultry geneticist is told by his administrative supervisor that the eggs no longer fit into the available egg cartons, so please see that they become smaller again.

Natural selection need not be directional; a great many of the failures of wild plants and animals may involve extreme individuals. Haldane (1954) analyzed both the weights of newborn babies and the size of duck eggs in respect to successful survival; in each instance there was an intermediate optimum. The optimal birth weight for human babies according to the data available to Haldane was about 7.1 pounds. Only 1.7 percent of babies of optimal weight died within the first 28 days of their lives whereas 4.1 percent of all babies died in that time; consequently, nearly 60 percent of the babies that died did so selectively in respect to birth weight.

Data presented in figures 11.3 and 11.5 suggest that selection favors intermediate phenotypes and, hence, operates contrary to directional selection. Additional observations on this point are presented in tables 11.2 and 11.3. For reasons that are not at all clear, predation of the stickleback by trout and squawfish does not occur at random in respect to the bony plates on individual stickle-

Table 11.2 Differential predation of the stickleback, *Gasterosteus aculeatus*, according to plate number by trout and squawfish. (After Moodie et al. 1973.)

Plate number	Eaten	Not Eaten	Total
7	34	56	90
5, 6, 8, 9	52	38	90
Totals	86	94	180

NOTE: Chi-square, 1 d.f., = 6.43; p ≈ 0.01.

back fish. Table 11.2 shows that the fish with the intermediate number (7) of plates escaped predation to a greater extent than did those with either smaller or larger numbers.

The data presented in table 11.3 are comparable in many respects to those shown in figure 11.5. Those in the table, however, concern the number of sternopleural bristles on *D. melanogaster* females, and the relation between this number and the estimated fitness of these females. The composite estimate of fitness is highest for females possessing an intermediate number of bristles (about 27); females with numbers of bristles greater or smaller than this optimum have lower estimated fitnesses. The decline in fitness is most rapid on the upper tail of the distribution curve; that is, the decline is more rapid in the direction that selection had been applied unsuccessfully for more than thirty preceding generations.

That selection at times favors intermediate phenotypes does not prove that these phenotypes are under genetic control. They may be, of course, as in the examples drawn from the opposition

Table 11.3 Distribution of sternopleural hairs on females of a laboratory population of *D. melanogaster* which was continually selected for high bristle number but in which there had been no progress in 30–35 generations of such selection. (After Scossiroli 1959.)

Number of Sternopleural Bristles	Percent Females	Mean Number of Eggs per Female	Composite Estimate of Females' Fitness
21	0.3	187	0.82
22	3.3		
23	6.0		
24	12.5	178	0.92
25	16.1		
26	20.3		
27	14.3	235	1.00
28	11.4		
29	10.2		
30	3.6	173	0.80
31	1.8		
32	0.3		

Table 11.4 A differential sensitivity to environmental variation exhibited by flies homozygous or heterozygous for chromosomes derived from various *Drosophila* species. Sensitivity is revealed as a statistically significant Chi-square between the proportions of wildtype individuals in replicated (i.e., genetically identical) *ClB*-type test cultures. (After Dobzhansky and Wallace 1953.)

Species	Chromosome	Homozygotes			Heterozygotes		
		Chi-square	d.f.	p	Chi-square	d.f.	p
D. pseudoobscura	2	144.3	81	<.001	31.9	33	>.30
	3	212.2	86	<.001	33.3	31	>.30
	4	173.9	77	<.001	44.1	30	~.05
D. persimilis	2	225.8	184	~.01-.001	74.9	99	>.30
	3	519.3	329	<.001	67.3	63	>.30
	4	327.4	269	~.01-.001	99.1	66	~.01-.001
D. prosaltans	2	253.1	152	<.001	48.1	47	>.30
	3	290.4	216	<.001	72.4	54	~.05
D. melanogaster	2	449.1	355	<.001	189.6	194	>.30

of natural selection to progress under a regime of artificial directional selection. Haldane (1954), following his analysis of human birth weights, expressed his opinion in these words: "I believe that a large fraction of the selection is of homozygotes for pairs of genes at loci where the heterozygous genotype is fitter than either homozygote. That is to say selection is not mainly counterbalancing the effects of mutation, but those of segregation." The suggestion contained in this opinion is that the extremes largely represent homozygous individuals, and that the optimal intermediate phenotypes tend to represent heterozygous individuals. To this suggestion, we might add the observations summarized in table 11.4: the reactions of homozygotes to environmental differences appear to be considerably larger than those otherwise comparable heterozygous individuals. Here we see that environmental variation, by causing homozygous genotypes to take on extreme phenotypic values, would expose these individuals to the adverse effects of stabilizing selection much more than it would expose the phenotypically more stable heterozygotes (see also Lerner 1954).

Disruptive selection
The final form of phenotypic selection we shall consider in this brief review is that in which individuals of intermediate phenotypes are at a selective disadvantage relative to those of *both* extremes. Disruptive (Thoday 1953) or diversifying (see Dobzhansky 1970:167) selection of this sort poses extremely interesting and

important problems for the population geneticist. Some biologists (see Thoday 1972 for review) see disruptive selection as an artificial situation, devised by experimentalists to place extraordinary strains on selected populations. Others, however, emphasize the graininess of any organism's environment. To the extent that individuals perceive these grains, and to the extent that they might adapt to a life exclusively on one or the other of the different sorts of grains, the population is subjected to discordant selective pressures—in short, to disruptive selection (A discussion of the definition of this term has been given by Mayr 1974a.)

A simple example of disruptive selection is given in figure 11.7. Two strains of corn (*Zea mays*), white flint and yellow sweet, were grown together in the same field. The original strains required almost identical numbers of days for flowering of the tassel (about 72 days) and of the ear (about 75 days). When the strains were grown together, hybrid seeds were produced but these were discarded each generation. Only white-flint and yellow-sweet seeds were chosen each year for planting in mixed stand the next season. The result of this artificial selection for plants representing exclusively intrastrain matings was an alteration in flowering times. After four generations of selection, white-flint plants flowered on the 67th (tassel) and 70th (ear) days, whereas yellow-sweet plants flowered on the 74th–75th (tassel) and 77th (ear) days. The flowering of white flint was advanced about five days, that of yellow sweet was postponed about two. As a result, relatively few hybrid plants were formed even though the two strains continued to be grown together in the same field. An examination of the distributions of flowering times of the original, unselected material suggests that the observed response might have been expected: some white-flint plants were early flowerers, even within the original, unselected strain.

Wasserman and Koepfer (1977) have studied the mating preferences of *D. arizonensis* and *D. mojavensis,* using strains of flies that were captured at different geographic localities (*allopatric*) and at the same locality (*sympatric*). The results of their study are summarized in table 11.5. There is no evidence that males and females of *D. arizonensis* (and only slight evidence that those of *D. mojavensis*) discriminate against members of the other sex on the basis of geographic origin; only one of the first six isolation indices (see page 148) differs significantly from zero.

Mating tests that involve males of one species and females of the other do reveal a strong preference for homogamic matings,

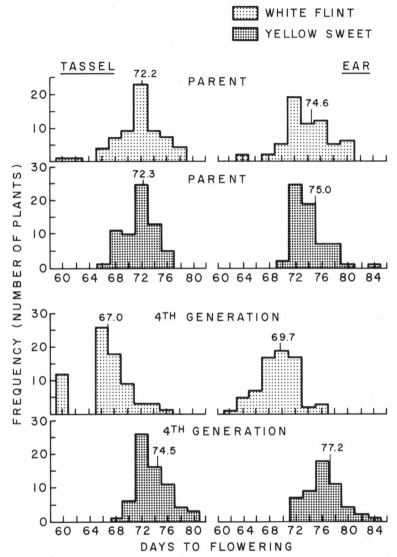

Figure 11.7 The displacement of flowering times (both tassel and ear) of two strains of corn, *Zea mays,* following four generations of artificial selection during which they were grown together but with hybrid progeny being discarded each generation. (After Paterniani 1969.)

Table 11.5 Results of mate-choice tests of sexual isolation between allopatric and sympatric strains of *D. arizonensis* and *D. mojavensis*. *Allopatric*, whether of the same or different species, means that the strains were initiated with gravid females from *different* geographic localities; *sympatric*, whether of the same or different species, means that the strains were initiated with gravid females captured within the *same* geographic locality. (After Wasserman and Koepfer 1977.)

Species	Region	Females of Each Type	Inseminated Females		Isolation Index ± (S.E.)
			Homogamic	Heterogamic	
arizonensis	Allo X Allo	120	90	73	0.104 ± (.078)
	Allo X Sym	480	359	321	0.056 ± (.038)
	Sym X Sym	120	96	88	0.043 ± (.074)
mojavensis	Allo X Allo	120	76	81	−0.032 ± (.080)
	Allo X Sym	480	356	259	0.158 ± (.040)
	Sym X Sym	120	82	63	0.131 ± (.082)
arizonensis	Allo X Allo	480	354	119	0.497 ± (.040)
X	Allo X Sym	960	780	138	0.699 ± (.024)
mojavensis	Sym X Sym	480	363	14	0.926 ± (.019)

matings between males and females of the same species. In addition, however, the data show that this preference is much more pronounced when the tested flies come from the same locality (isolation index = 0.926) than when they come from widely separated localities (isolation index = 0.497). Sympatry in the case of the two Drosophila species corresponds to the two varieties of corn being grown in the same field (figure 11.7). The result has been the same: isolating mechanisms have been reinforced. For the corn plants the reinforcement was a change in flowering time; for the flies it must involve recognition and courtship signals.

Disruptive selection (selection for the tails of a bell-shaped statistical distribution) of one population once started becomes directional selection of opposite sign in the two resulting populations. In the case of the corn varieties illustrated in figure 11.7, for example, the first generation of selection may have resulted in varietal seed (white flint and yellow sweet) that was set throughout the normal flowering times. White flint, however, predominated among the early seed. By the fourth generation the experiment, although carried on by the same protocol as before, was really imposing directional selection on two clearly identified strains: an early-flowering white flint and a late-flowering yellow sweet.

The disruptive selection characterizing the enhancement of the reproductive isolation of two Drosophila species when these live sympatrically is really disruptive at the very outset. Given that there is considerable isolation even between allopatric strains of *D. arizonensis* and *D. mojavensis*, the selection that occurs sympatrically is (for each species considered separately) directional. The same is true for situations such as those illustrated in figures 11.8 and 11.9: the divergence of species and of the ecological niches they occupy. Figure 11.8 shows in diagrammatic form that two species living sympatrically may have nearly identical requirements and may utilize nearly identical resources. Whatever initial differences exist, however, serve to impose selection on each species; the selection may be diverse, even contradictory, within

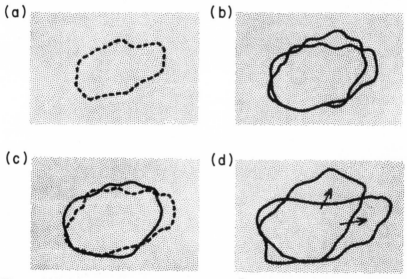

Figure 11.8 The representation (a) of an unoccupied ecological niche as a unique (multidimensional) constellation of many environmental factors in a multitude of possible states (e.g., temperature: $9°$, $9\frac{1}{2}°$, $10°$, ...) that could in theory support a population of organisms. When two similar but not identical niches are occupied by different species (b), the outcome may be the extinction of one species (c), or a divergence of species characteristics that enables the two to occupy less similar niches (d). Reprinted by permission of Prentice-Hall Inc., from Bruce Wallace and Adrian M. Srb, *Adaptation*, 2d ed;© 1964.

each species. The result (neglecting the possible extinction of one or both species) is generally the development of phenotypic differences that enable the species to coexist but with fewer shared requirements. Figure 11.9 gives an example illustrating (by means of fruit-eating pigeons of New Guinea) what was a purely theoretical argument in figure 11.8. The pigeons of figure 11.9 are differentiated by size, by the size of the fruit they eat, and by the size of the branches best able to support them. The sizes of the eight species shown in the figure have not been arrived at by chance as anyone can verify by plotting the logarithms of their weights on semilog graph paper: the different weights form a nearly perfect geometric sequence with each species being 1.48 times as heavy as the next lighter one.

Controversy exists among interested workers as to whether the subjection of a single population to disruptive selection can cause it to split into two noninterbreeding groups without the intervention of an experimenter, as in the earlier case of white-flint and yellow-sweet varieties of corn in which seeds of hybrid origin were identified and manually discarded. This problem will recur later in this book (chapter 23); here we shall present some preliminary data illustrating the nature of the problem.

The data in table 11.6 concern three laboratory populations of *D. melanogaster*: one (NaCl-1) was kept on food containing table salt; a second ($CuSO_4$-1) was kept on food containing copper sulfate; and a third (NaCl + $CuSO_4$-1) was kept on both kinds of food (in separate food cups). Both table salt and copper sulfate are toxic to *Drosophila*. Populations kept on these poisons are automatically subjected to directional selection and respond by developing a resistance to them. After being maintained for two years on the salts (whose level was periodically increased, thus maintaining a continuous selective pressure for the development of resistance), the flies of the two single-poison populations (NaCl-1 and $CuSO_4$-1) had become somewhat resistant to the poison on which each had been raised, while remaining sensitive to the other one. Scarcely any progeny could be reared from $CuSO_4$-1 parents on food containing the three highest levels of NaCl; progeny of NaCl-1 flies were largely killed by the highest concentrations of $CuSO_4$.

Within the population cage that was supplied with food cups containing either NaCl or $CuSO_4$, the flies were subjected to simultaneous selection for resistance to both poisons. The concentrations of NaCl and of $CuSO_4$ in the medium offered to this

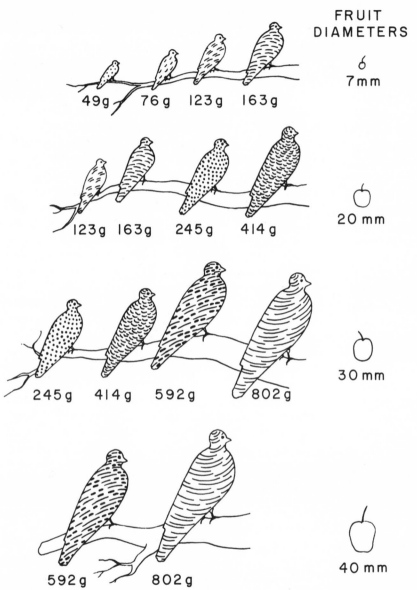

FRUIT
DIAMETERS

7mm

20 mm

30 mm

40 mm

49 g 76 g 123 g 163 g

123 g 163 g 245 g 414 g

245 g 414 g 592 g 802 g

592 g 802 g

Figure 11.9 The segregation of eight species of fruit-eating pigeons (of the genera *Ptilinopus* and *Ducula*) in New Guinea by size. Larger birds can swallow larger fruits, but cannot perch on the smaller branches of trees bearing small fruit. This type of apportionment of available resources, like the sexual isolation between sympatric *Drosophila* species (table 11.5) and the origin of temporal isolation between strains of *Zea mays* (figure 11.7), is regarded as the outcome of natural selection. (After Diamond 1978.) Reprinted with permission: from *American Scientist*.

"two-salt" population were determined by the tolerance achieved by flies in the two cages, NaCl-1 and $CuSO_4$-1; to survive, the flies in the "two-salt" cage had to achieve (as one possibility) resistance to both salts.

Alternatively, the population could split into two: one portion could restrict itself to NaCl-containing food and adapt to it while the other portion could adapt to $CuSO_4$-containing food. A split of that sort, however, requires the elmination of between-group matings; reproductive isolation must arise de novo within a single population. Now, the origin of reproductive isolation presents formidable problems requiring still more selective changes within each postulated subpopulation. A full discussion of these problems will be postponed until chapter 23. The data in table 11.6, although preliminary, place the theoretical problem into a tangible context. Eggs collected from NaCl-containing food (which were then transferred to nontoxic food) gave rise to flies that left relatively more progeny in the presence of NaCl than did similar flies obtained from eggs laid originally on $CuSO_4$-containing food. Conversely, eggs obtained from $CuSO_4$-containing food (which were then transferred to nontoxic food) gave rise to flies that left relatively more progeny on $CuSO_4$-containing food than did comparable flies obtained from NaCl-containing food. The differences are slight, but consistent. They may reflect the effect of differential mortality during the egg-sampling period. Alternatively, they may reflect the effect of a disruptive selection that

Table 11.6 The relative number of surviving progeny (*D. melanogaster*) in cultures containing increasing concentrations, 0-5, of NaCl or $CuSO_4$. The parental flies came from populations that had been maintained on either NaCl- or $CuSO_4$-containing food for two years, or from a population that had from its inception been maintained on both NaCl- *and* $CuSO_4$-containing food (in separate cups). The actual numbers of progeny surviving in the absence of either salt (0) ranged from 1,760 to 2,682.

	NaCl-Containing Medium					
Population	*0*	*1*	*2*	*3*	*4*	*5*
NaCl-1	1.000	1.024	0.935	1.006	1.014	0.773
$CuSO_4$-1	1.000	0.309	0.455	0.054	0.009	0
(Both)–NaCl cup	1.000	0.589	0.396	0.147	0.065	0.040
(Both)–$CuSO_4$ cup	1.000	0.578	0.365	0.099	0.021	0.014

	$CuSO_4$-Containing Medium					
Population	*0*	*1*	*2*	*3*	*4*	*5*
NaCl-1	1.000	0.876	0.427	0.347	0.075	0.011
$CuSO_4$-1	1.000	0.869	0.791	0.383	0.514	0.047
(Both)–NaCl cup	1.000	0.998	0.774	0.733	0.332	0.258
(Both)–$CuSO_4$ cup	1.000	1.043	1.052	0.917	0.674	0.530

is taking the members of one population down two paths even before reproductive isolation arises (in this case, laying eggs on NaCl-containing or on $CuSO_4$-containing food). If the latter is true, the data do not tell us how far down these separate paths the subpopulations may eventually go (see Maynard-Smith 1966).

Selection *versus* Mutation

Populations, as we learned in chapter 3, contain a great deal of genetic variation. More recently, we have learned that individuals differing genetically suffer dissimilar fates during encounters with predators and other hazards of their environment. In this section, we shall pause momentarily in order to see one of the means by which genetic variation is maintained in populations: by the opposition of mutation and selection.

The occurrence of sex-linked lethals in a population of *D. melanogaster* can serve as a simple example of a class of mutant genes arising repeatedly by mutation but, simultaneously, disappearing regularly through the death of their carriers (males, primarily). Figure 11.10 shows the accumulation of sex-linked lethals in an irradiated population of flies. The level of radiation was sufficiently high to make the original frequency of such lethals negligible.

Given that the proportions of males and females in a population are equal, and that the Y chromosome is genetically inert (in respect to concealing sex-linked lethal mutations at any rate), then it follows that in any generation one-third of all X-chromosomes are to be found in males. This ratio applies to lethal-bearing X chromosomes, too, and so one-third of all existing lethal chromosomes are eliminated each generation by the death of their male carriers. Thus, it can be seen that when the frequency of lethals equals three times the rate at which lethals arise by

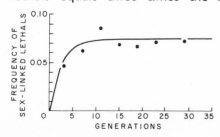

Figure 11.10 The accumulation of sex-linked lethals in an irradiated population of *D. melanogaster*. Because males carry a single X-chromosome, the expected equilibrium equals three times the mutation rate. This population is number 6 of table 11.7 and figure 11.11.

mutation, the rates of origin and of loss are equal:

$$\Delta Q = U - Q/3.$$
$$\text{When} \quad \Delta Q = 0,$$
$$Q = 3U.$$

The curve drawn to fit the points in figure 11.10 is based on the accumulation of lethals when $U = .025$, a value obtained as one-third of the apparent equilibrium frequency.

Table 11.7 and figure 11.11 illustrate the accumulation of lethal and semilethal autosomes (second chromosome) or, conversely, the loss of nonlethal (quasi-normal) autosomes in continuously irradiated populations of *D. melanogaster*. During the early generations the loss of quasi-normal autosomes is exponential as predicted by the equation arrived at previously (page 208)

$$Q = (1 - U)^t = e^{-Ut}.$$

Table 11.7 Frequencies of lethal and semilethal chromosomes in the early generations of chronically irradiated populations of *D. melanogaster*.

Generation	Population 5		Population 6		Population 7	
	n	Frequency	n	Frequency	n	Frequency
0	—	0	—	0	—	0
2	275	0.113	270	0.119	260	0.042
4	261	0.180	263	0.217	265	0.053
6	250	0.292	258	0.333	242	0.083
8	89	0.292	272	0.353	278	0.079
10	269	0.428	255	0.380	291	0.107
14	45	0.533	246	0.524	262	0.141
18	196	0.434	272	0.621	264	0.174
22	203	0.458	179	0.609	170	0.206
24	—	—	20	0.650	20	0.050
25	19	0.474	—	—	—	—
26	—	—	182	0.698	182	0.187
27	180	0.594	—	—	—	—
28	—	—	20	0.650	20	0.350
29	20	0.650	—	—	—	—
30	—	—	79	0.696	78	0.192
31	80	0.588	—	—	—	—
32	—	—	79	0.646	78	0.205
33	80	0.750	—	—	—	—
34	—	—	80	0.563	80	0.163
35	80	0.725	—	—	—	—
36	—	—	80	0.688	80	0.213
37	80	0.688	—	—	—	—
38	—	—	80	0.700	80	0.300
39	80	0.688	—	—	—	—
40	—	—	80	0.725	80	0.213
41	80	0.713	—	—	—	—

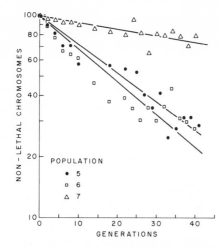

Figure 11.11 The loss of nonlethal (or, conversely, the accumulation of lethal) autosomes in three irradiated populations of *D. melanogaster*. (After Wallace 1956.)

The different values of *U* reflect the different levels of radiation to which the populations were exposed. Eventually, all three irradiated populations arrived at stable, equilibrium frequencies. (The upper-case *Q* and *U* are used here because *chromosomal* frequencies and mutation rates are involved.)

In chapter 10, two relationships were derived for the loss of deleterious genes by selection:

for recessive mutations

$$\Delta q = -sq^2 \, ;$$

for incompletely recessives

$$\Delta q = -hsq.$$

In either case, gene frequencies stabilize when the absolute value of Δq equals *u*:

$$sq^2 = u; q = \sqrt{u/s}, \text{ or}$$
$$hsq = u; q = u/hs.$$

Table 11.8 lists the equilibrium frequencies expected in *large* populations for various mutation rates (*u*), selection coefficients (*s*), and degrees of dominance (*h*), including $h = 0$ in the case of complete recessives.

Empirical studies of Drosophila populations concern almost entirely lethal chromosomes, not lethal genes. The frequencies of

Table 11.8 Equilibrium frequencies attained by recessive and partially dominant mutant alleles in large populations given various values for mutation rate (u), selection coefficient (s) of homozygotes, and degree of dominance (h).

Complete Recessives

		u		
s		10^{-4}	10^{-6}	10^{-8}
1		10^{-2}	10^{-3}	10^{-4}
0.01		10^{-1}	10^{-2}	10^{-3}

Partial Dominants

		u		
s	h	10^{-4}	10^{-6}	10^{-8}
1	0.1	10^{-3}	10^{-5}	10^{-7}
1	0.01	10^{-2}	10^{-4}	10^{-6}
0.1	0.1	10^{-2}	10^{-4}	10^{-6}
0.1	0.01	10^{-1}	10^{-3}	10^{-5}

lethal chromosomes are large enough to study easily; frequencies of individual lethal genes are prohibitively low. Two items that are necessary for understanding the maintenance of lethal chromosomes in populations can be determined with great accuracy: Q, the frequency of lethals in the populations and U, the rate at which lethal chromosomes arise by mutation. (Capital letters are used here, as earlier, to indicate that chromosomes, not individual gene loci, are the object of investigation.) Theoretically, one can also determine empirically the probability (I) that random combinations of lethal chromosomes will themselves prove to be lethal because the lethals they carry are allelic. Actually, the calculation of I proves to be difficult because (as was shown in figure 6.3) the allelism of lethals declines with increasing distance between their points of origin; lethals intercrossed to estimate I must be obtained from flies captured at one spot, at one time. Few workers have been careful in this respect. A second, practical matter also interferes with a precise estimate of I. Because large numbers of heterozygous combinations (lethal-1/lethal-2) must be studied, two lethal chromosomes are said by most workers to be allelic only if *no* heterozygous flies survive. Table 11.9 is based on heterozygous combinations of lethals comparable to those made in studying the allelism of lethals except, because they were made for a different purpose, the flies in each heterozygous culture were actually counted, not just examined and scored as allelic or not. The data listed in table 11.9 show that whereas the standard procedure would have revealed 58 cases of allelism, the

Table 11.9 An analysis of 97 lethal$_1$/lethal$_2$ heterozygous combinations where these combinations themselves are either lethal (A) or semilethal (B). The lethal *chromosomes* of this analysis were classified as "absolute" (no surviving wildtype homozygotes in test cultures) or "near-lethal" (fewer than 10 percent of the expected proportion of homozygous wildtype flies survived). The *combinations* have been classified: (A) three or fewer wildtype flies survive per test cultures and (B) more than three but fewer than one-half the expected number of heterozygous wildtype flies survive.

Lethal Chromosomes Tested	Viability of Lethal$_1$/Lethal$_2$ Combination		Total Tests
	A	B	
Absolute X absolute	58	20	78
Absolute X near-lethal	2	6	8
Near-lethal X near-lethal	3	8	11
Total	63	34	97

elimination of lethals from populations would have been reflected more accurately by the figure 97; the use of expedient test procedures, consequently, result in an estimated value for *I* that is only two-thirds as large as the likely true value.

The diagram shown in figure 11.12 illustrates the relationships between *U*, *Q*, *I*, and *H* in a Drosophila population. The numerical values shown in the figure are from a study on *D. pseudoobscura* by Dobzhansky and Wright (1941). The frequency (*Q*) of lethal third chromosomes in the population under study was 0.153. The mutation rate to lethal chromosomes was found by experiment to be 0.0030. The probability that two lethal chromosomes were allelic was estimated as 0.0311. Consequently, the rate at which lethals were lost from the population by allelism, *IQ²*, was 0.0007. (We might note here that corrections

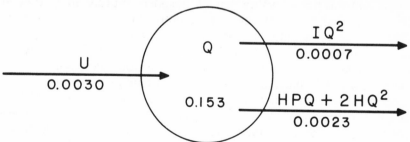

Figure 11.12 The balance attained by the input, accumulation, and outflow of lethals of a population. The experimentally determined value of *I* in this case was 0.0311; the value of *H* required to establish the balance is 0.013. (*U*, mutation to lethal chromosomes; *Q*, frequency of lethal chromosomes in the population; *I*, probability that two lethal chromosomes are allelic; and *H*, the dominance expressed by "recessive" lethals.)

of the sort suggested in figure 6.3 and table 11.9 could easily raise the estimated value of IQ^2 from 0.0007 to 0.0030, a value equal to the observed mutation rate).

Ostensibly, lethals are not leaving the population through homozygosis (IQ^2) as rapidly as they enter it (U); consequently, if the observed value of Q is truly an equilibrium value, some lethals must be eliminated by virtue of their effect on heterozygous carriers. This loss is given as $HPQ + 2HQ^2$ (the first portion is contributed by individuals carrying one lethal chromosome; the second by those carrying two, nonallelic lethal chromosomes).

Obviously, $HPQ + 2HQ^2 = U - IQ^2 = 0.0023$. Upon solving, we find that $H = 0.013$. In other words, if individuals carrying one lethal third chromosome had overall fitnesses (survival and reproduction) only 99 percent as great as that of carriers of two normal chromosomes, the input and outflow of lethals in the analyzed population would be in balance.

Analyses such as those described in figure 11.12 are obviously carried out at the limits of experimental resolution. When operating at the limit of resolution, each worker tends to see what he wants to see; this was the situation, for example, that allowed early cytologists to see a homunculus in the head of each sperm, or an old-time astronomer to identify canals on the surface of Mars. These are also the situations that generate unresolvable controversies. As we have seen, errors in the estimation of I are large enough to allow IQ^2 to equal U, thus causing H to drop to zero. Furthermore, what has been designated H in this discussion is really $H + F$ (combined dominance and homozygosis by inbreeding); the calculated value can be assigned entirely to H only if F is assumed to equal zero, an assumption unlikely to be true. Nor are matters helped considerably by attempts to measure H experimentally (see tables 10.7 and 10.8, and figure 10.7). Dobzhansky and Spassky (1968) with considerable effort showed that estimated values of H are dependent upon the genetic background in which lethal chromosomes are tested (table 11.10); their conclusions were reinforced by Anderson's (1969) independent analysis of their data. No previous worker had made an effort to estimate H in any background other than the artificial one created by crossing flies from the population to be tested with the ClB-like stock which made the manipulation of chromosomes possible.

A final word can be injected into this section on mutation and selection; one emphasizing that the mutations which occur within populations may not be completely random in origin, es-

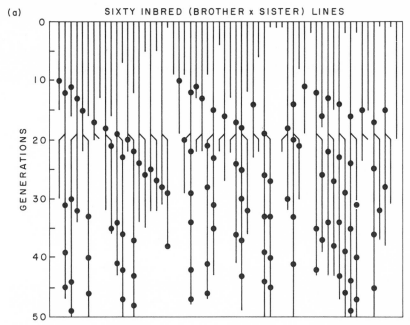

Figure 11.13a A diagram depicting the maintenance of 60 inbred (brother X sister) lines of *D. melanogaster* for 50 generations. During generation 19 when more than half of the lines had been lost through sterility of the inbred flies, each remaining line was subdivided to restore in part the original number of lines. The heavy dots indicate when and in what lines periodic tests for lethals were carried out.

(b)

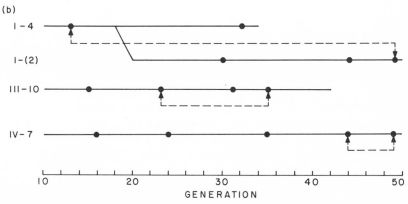

Figure 11.13b Diagrams showing (in terms of both inbred line and generation of inbreeding) the origins of the lethals which proved to be allelic in intraline tests (dashed lines). In two of the three instances (I-4 and III-10), lethals recovered at intervening times within the same line proved *not* to be allelic with those which were. Consequently, intraline allelism depends upon more than mere persistence of a lethal through successive generations of matings. (After Wallace and Madden 1965.)

Table 11.10 The role of background chromosomes in determining whether or not lethal chromosomes exhibit a deleterious effect on the viability of their heterozygous carriers. The numbers are the ratios of experimental (i.e., lethal heterozygotes) to control viabilities. The total number of flies counted in obtaining these ratios exceeded 500,000. (After Dobzhansky and Spassky 1968; see Anderson 1969 too.)

	Background		
Source of Lethal	Arizona	California	Mexico
Arizona	1.001	0.972*	0.983
California	0.991	1.021*	0.956*

*Significantly different from 1.000; $p \leqslant 0.05$.

pecially in small, relatively isolated populations. Figure 11.13a illustrates 60 lines of *D. melanogaster* that were maintained for 50 generations by brother-sister matings, with tests for the presence of second chromosomes lethals being made at times indicated by the numerous dark dots. Among 811 tests for allelism between lethals arising in different lines, 1 proved to be allelic. Among 50 tests for allelism of lethals found at different times in the same line, 3 proved to be allelic. However, as shown in figure 11.13b, in two of these instances, other lethals found at intervening times were not allelic to those that were. Consequently, it seems that within a line, a tendency may exist for the "same" lethal to occur repeatedly. Ives (1949) reported a similar tendency for different strains of *D. melanogaster* to exhibit different patterns of visible mutations.

Nonadaptive Selection

Until now we have spoken of the advantage of this or that allele in terms of the enhanced survival or reproductive ability of its carriers (or, conversely, the reduced survival or reproductive ability of those carrying a contrasting allele). Implicit in this discussion, of course, was the notion that whatever increases the overall fitness of an individual, increases the well-being of the population or the species of which the individual is a member (see Boxed Essay I at end of this chapter). The latter implication may or may not be so; such questions will be discussed in more detail in the second half of this book. Here we want only to emphasize that alleles (or chromosomal regions that behave as alleles) may spread in populations because of *gametic* rather than *zygotic* selection. As Crow (1979) has described it, some alleles circumvent Mendelian inheritance and promote their own spread in populations.

One of the first cases of nonadaptive selection to be described was the sex-ratio (*SR*) chromosome in the *obscura* group of *Drosophila*. Gershenson (1928) found that females of *D. obscura* when mated with certain males produced only female offspring. Further study showed that the basis for these unisexual progeny was sex-linked: males carrying a *SR* (= *sex ratio*) X-chromosome produced only female offspring in contrast with normal males carrying normal (*ST*) X-chromosomes. Females produced *SR* and normal sons according to their genotypes; the *SR* and *ST* chromosomes do not affect the progeny of females.

In laboratory cultures, *SR* males of *D. pseudoobscura* appear to produce as many progeny as *ST* males. The consequences of such aberrant inheritance is illustrated in figure 11.14. The fre-

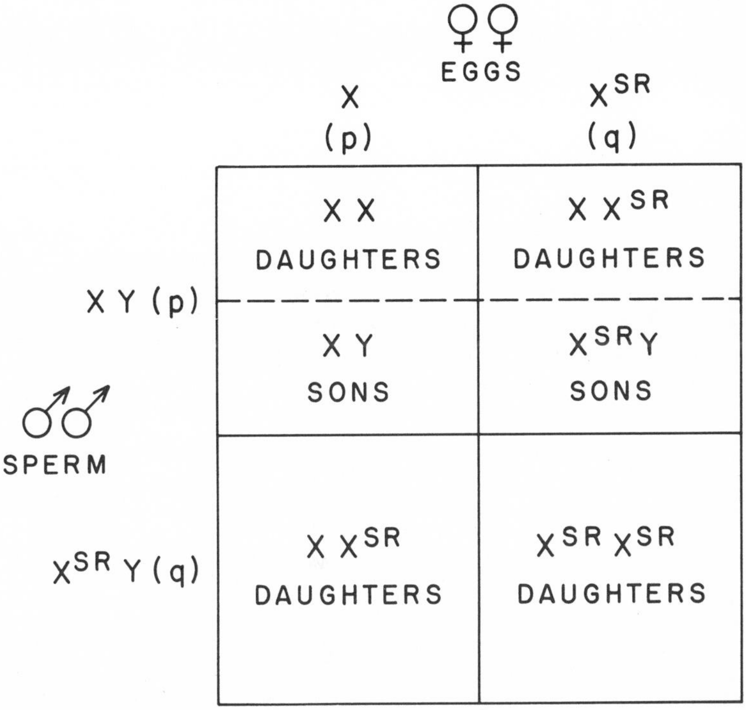

Figure 11.14 The inheritance of *SR* in *D. pseudoobscura*. X = normal X-chromosome; X^{SR} = *sex-ratio* X-chromosome.

quency of *SR* among mothers equals *q*. The frequency among
daughters equals $[q(\frac{3}{4}p + q)/[(\frac{1}{2}p + q)]$. If *q* were very small, the
frequency of *SR* among daughters would be very nearly 1½*q*, an
increase in frequency of 50 percent. No matter how large *q* might
be among mothers, however, the frequency among daughters
will according to the above equation, be larger. Consequently,
Gershenson concluded that *SR* X-chromosomes should increase
in frequency until they drive normal X-chromosomes to extinction, at which time there would be no males and the species
(or population) would become extinct or adopt a parthenogenetic
mode of reproduction. Wallace (1948) found that *SR* males have
quite low reproductive fitnesses as do *SR/SR* females; consequently, the presence of *SR* X-chromosomes in wild populations
seems to depend upon the high fitnesses that characterize *SR/ST*
females. More recently Policansky (1974, 1979) has studied the reproductive fitness of *SR* males in the field and has found that
these males either do not mate with females as often as *ST* males
or that they produce only one-half as many sperm. The question
then as to why *SR* X-chromosomes exist within wild populations
is still unanswered.

Other alleles that promote their own existence through distorted segregation ratios are the *t*-alleles of the mouse, *Mus
musculus*. The *t*-locus in the mouse is one whose mutant alleles
affect the development of the notocord and, eventually, the
tail. Many mutations at this locus are recessive lethals; of these, a
great many give extremely distorted segregation ratios in +/*t*
heterozygotes. Table 11.11 shows that among *t*-mutations newly
arisen in the laboratory, male heterozygotes (+/*t*) produce + and *t*
sperm in proportions that vary widely from the expected 50:50
ratio. Males that are heterozygous for lethal *t*-alleles recovered
from wild populations of mice show extremely distorted segregation ratios, however; it would seem that such lethals are found in
these populations *because* the segregation ratios of heterozygous

Table 11.11 The distribution of segregation ratios (percent *t* sperm produced by
+/*t* males) for various *t*-alleles of the mouse, *Mus musculus*. The alleles have been
grouped as newly arisen (new), lethals obtained from natural populations (wild-lethal),
and male-sterility alleles (wild-sterile). (After Lewontin and Dunn 1960.)

	99–90	89–80	79–70	69–60	59–55	(50)	44–40	39–30	29–20	Total
New	3	1	0	0	2	8	3	2	0	19
Wild-lethal	15	1	0	0	0	0	0	0	0	16
Wild-sterile	1	2	0	0	0	0	0	0	0	3

males favors such alleles. Lewontin and Dunn (1960) showed by computer simulation that extremely small populations of mice could maintain lethal alleles which distorted the segregation ratios of heterozygotes as naturally occurring *t*-alleles do.

A considerable study has grown around a phenomenon known as segregation-distorter (*SD*) in *D. melanogaster* (see Crow 1979 for a summary and references). In brief, *SD* elements favor their own spread over that of homologous alleles; thus, they might be expected to be the normal, wild-type alleles of any population. If they were, however, their distorting effects on segregation would no longer be discernible because all alleles would be equally effective at causing such distortions: 50:50 ratios among heterozygotes would then be the rule. Consequently, the effects of *SD* elements can be observed only by crossing between populations (Hartl 1970; see page 528) or by studying *SD* elements that have an adverse effect on their carriers and which cannot attain a frequency of 100 percent (Crow 1979).

These comments on nonadaptive selection reveal the importance of a "normal" segregation of dissimilar alleles in the gametes of heterozygous individuals to population genetics. The Hardy-Weinberg equilibrium is predicated on the assumption that Mendel was right, and that *Aa* heterozygotes produce *A* and *a* gametes in equal proportions. We have now seen that this "normal" segregation need not be so. However, no fixed alternative arises as a replacement for Mendelian ratios. Furthermore, once an abnormally segregating allele has run its course and has become *the* normal allele, the distortion is no longer discernible. It appears, then, that the possibility that alleles will not exhibit Mendelian ratios among the gametes of heterozygotes must constantly be kept in mind but, because no a priori basis exists for assigning values to non-Mendelian segregational ratios, Hardy-Weinberg expectations are the most valuable expectations, and will remain so in each instance until proven wrong.

Summary

In this and previous chapters we have discussed a number of factors affecting gene frequencies in populations. The effects of

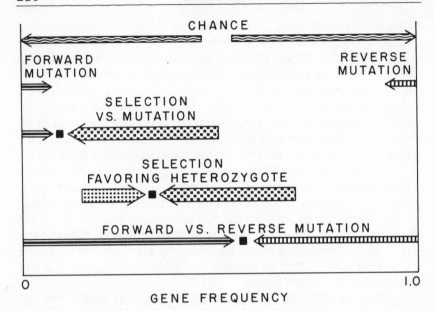

Figure 11.15 A diagram summarizing the roles of mutation, chance, and selection in determining gene frequencies in populations. Stable equilibria at intermediate frequencies are designated (■).

these are summarized in figure 11.15. Chance events are dispersive; left unchallenged they force populations to fixation by the loss of either one allele or another. Mutation, although normally a weak force, keeps 0 percent and 100 percent from being gene frequency "traps"; as long as A can mutate to a (or a to A) the population will have to contend in the long run with both alleles. If left alone (that means in an enormous population in which selection does not affect the carriers of dissimilar alleles) forward and reverse mutation would lead to stable equilibria and intermediate gene frequencies; it seems unlikely that such equilibria normally occur in natural populations. The opposition of mutation to deleterious alleles and selection against such alleles can lead to stable (but low) gene frequencies. Finally, both selection favoring heterozygous individuals and frequency-dependent selection that favors rare genotypes lead to stable equilibria at intermediate gene frequencies.

BOXED ESSAY I

FISHER'S FUNDAMENTAL THEOREM

In his book *The Genetical Theory of Natural Selection*, Fisher describes a beautifully simple relationship between a population's genetic variance in respect to fitness and the increase in the average fitness of that population which occurs in a single generation; he refers to this relationship as the fundamental theorem of natural selection. The theorem can be stated as follows: the increase in fitness in one generation equals the additive genetic variance in respect to fitness. Additive genetic variation assumes that heterozygotes lie exactly halfway between the two corresponding homozygotes, and that the properties of all other gene combinations can be predicted from the properties of genotypes at individual loci. Nonadditive genetic variation is all else, consisting primarily of dominance, overdominance, and epistatic interactions.

Fisher's theorem can be illustrated (as shown in table 11.12) by means of a haploid population in which individuals differ in genotypes (g_i), frequencies (f_i), and fitnesses (w_i). The average fitness $(\overline{w})$ of the original population equals the sum of each fitness times its frequency; this value can be set equal to 1.00. Thus,

$$\sum_{i=1}^{n} f_i w_i = \overline{w} = 1.00.$$

The variance in respect to fitness equals the average squared deviation from the mean (see page 75 for an earlier calculation) or, with some algebraic

Table 11.12 Summary of frequencies (f) and relative fitness (w) of various genotypes (g) in two successive generations of a population from which Fisher's fundamental theorem can be demonstrated.

First generation

Genotypes	g_1	g_2	g_3	$\cdots g_n$
Frequencies	f_1	f_2	f_3	$\cdots f_n$
Fitnesses	w_1	w_2	w_3	$\cdots w_n$

Average fitness $= f_1 w_1 + f_2 w_2 + f_3 w_3 + \cdots + f_n w_n = \overline{w} = 1.00$

Second generation

Genotypes	g_1	g_2	g_3	$\cdots g_n$
Frequencies	$f_1 w_1$	$f_2 w_2$	$f_3 w_3$	$\cdots f_n w_n$
Fitnesses	w_1	w_2	w_3	$\cdots w_n$

Average fitness $= f_1 w_1^2 + f_2 w_2^2 + f_3 w_3^2 + \cdots + f_n w_n^2$

manipulation,

$$\sum_{i=1}^{n} f_i w_i^2 - \overline{w}^2 = \sum_{i=1}^{n} f_i w_i^2 - 1.$$

Selection alters the initial frequency of each genotype; the new frequency equals the product of the original frequency times its corresponding fitness $(f_i w_i)$ because that is the meaning of fitness. The sum of these products, as shown in table 11.12, equals $\overline{w}$ or 1.00.

The new average fitness is calculated as before: it equals the sum of the new frequencies times the corresponding fitnesses, or

$$\sum_{i=1}^{n} (f_i w_i)(w_i) = \sum_{i=1}^{n} f_i w_i^2.$$

The increase in fitness, then, equals

$$\sum_{i=1}^{n} f_i w_i^2 - \overline{w} = \sum_{i=1}^{n} f_i w_i^2 - 1$$

which equals the genetic variance in respect to fitness. The proof is now complete.

An obvious implication of this theorem is that little or no additive genetic variation in respect to fitness will be found in most populations because it will have been exhausted by prior selection. The genetic variation which does exist, then, will be nonadditive; balanced polymorphisms in which heterozygous individuals exhibit the highest fitness provide an example. Furthermore, experience has revealed that artificial selection for improving aspects of the phenotype that are important components of fitness is remarkably ineffective.

A second implication has been pointed out by A. Robertson (1955): if (through artificial selection) one *can* reveal the presence of additive genetic variance within a population in respect to some component of fitness, then either the population is not at genetic equilibrium or the genetic basis of various components of fitness are negatively correlated. The data presented in figure 11.5 and table 11.3 suggest that the second possibility is not impossible.

A third implication is that fluctuations in the environment disturb populations so that there is always a momentary room for improvement, and that this improvement is accomplished by selection for newly revealed additive genetic variance. Laboratory experiments, by being conducted under unnatural conditions, may reveal additive genetic variance which in fact did not exist in the original population. A special case of this suggestion is known as the Red Queen hypothesis (Van Valen 1973): an improvement of the fitness of one species represents the deterioration of the environment for all others; consequently, every species must "run" continuously merely to stay in place—as did the Red Queen in *Alice in Wonderland*.

BOXED ESSAY II

In addition to the genetics of natural populations, the main topic of interest for this book, there exists the quantitative genetics of plant and animal breeders. The methodologies of these agricultural specialists differ; plant breeders subject their results primarily to analyses of variance, animal breeders are more inclined to build and test mathematical models. The procedures of neither group will be discussed here; interested readers are referred to Allard (1960), Lerner and Donald (1966), Mather and Jinks (1971), and Spiess (1977).

Certain concepts are held in common by population and quantitative geneticists. The matter of heterosis and overdominance has already been mentioned on page 227; population geneticists, perhaps because they cannot control the breeding of wild individuals, tend to be less precise in their use of these terms than their colleagues who can carefully control the crosses giving rise to the (statistical, *not* Mendelian) populations they study. Crow (1948) carried out calculations that confirmed those of Haldane (1937) on the effect of variation on fitness; Crow, however, was studying alternative hypotheses of hybrid vigor and, in doing so, had equated yield with fitness.

Because of the continuing debate on the genetics of human intelligence (Jensen 1969; Lewontin 1974b; and Bodmer 1972), readers of this book should acquire some familiarity with an often-used term, *heritability*—what it is, what it is not, and what purpose it legitimately serves.

Heritability is defined as the proportion of the total phenotypic variation that is caused by genetic differences between the members of a population. It is used by plant breeders to predict the advance they can expect from a given level of artificial selection; because advances are made from additive genetic variation, the definition is at times restricted to the proportion of *additive* genetic variance within the total phenotypic variance.

Because heritability reflects a relationship between variances, it is a property of the population—not of the trait (yield or *IQ*). Heritability is a property of the population *in respect to* the trait; furthermore, its numerical value refers to the population at a given moment. Selection alters heritability from one generation to the next. Eventually, if additive genetic variation were to be exhausted, heritability would be reduced to zero. Thus, the heritability of egg weight in an experimental flock of chickens may at the outset be nearly 0.50 but, after a selection program lasting many years, be reduced to zero. This decline alone demonstrates that it is the population, not the weight of eggs, to which heritability refers.

Even though the normal use of heritability is in connection with complex traits that are governed by the interactions of many genes (multifactorial or polygenic inheritance), its calculation can be illustrated by a single-locus model. The example given here illustrates the calculation of total, environmental, and genotypic variances and of heritability. Having calculated heritability (h^2), it is used to predict the outcome of an artificial-selection program. Next, that prediction is verified. The starting gene frequencies are 50 percent *A* and 50 percent *a*; the phenotypic character is the weight of some hypothetical organism. (It is important to remember in verifying the calculations that the frequency of *Aa* individuals in the starting populations is twice that of either homozygote.)

1. The initial population. Because of environmental variation, the phenotypes of each genotype vary as follows

Weight (gms)	AA	Aa	aa
10	5%	—	—
9	10	5%	—
8	20	10	5%
7	30	20	10
6	20	30	20
5	10	20	30
4	5	10	20
3	—	5	10
2	—	—	5

(Confirm that the average weight of *AA* individuals is 7 lbs; that of *Aa* is 6 lbs; and of *aa*, 5 lbs. The grand average equals 6 lbs.)

2. Calculating the total phenotypic variance. (If necessary, refer to Boxed Essay for chapter 3.)

x	f	fx	$(x - \bar{x})^2$	$f(x - \bar{x})^2$
10	(.05)/4	.125	16	.20
9	(.20)/4	.450	9	.45
8	(.45)/4	.900	4	.45

x	f	fx	$(x - \bar{x})^2$	$f(x - \bar{x})^2$
7	(.80)/4	1.400	1	.20
6	(1.00)/4	1.500	0	0
5	(.80)/4	1.000	1	.20
4	(.45)/4	.450	4	.45
3	(.20)/4	.150	9	.45
2	(.05)/4	.025	16	.20
	4.00 /4 = 1.00	6.000 = $\bar{x}$		2.60 = σ^2_{Total}

3. Calculating the environmental variance. (Because all genotypes reacted similarly to environmental variation, these calculations are given for *aa* individuals only.)

x	f	fx	$(x - \bar{x})^2$	$f(x - \bar{x})^2$
8	.05	.40	9	.45
7	.10	.70	4	.40
6	.20	1.20	1	.20
5	.30	1.50	0	0
4	.20	.80	1	.20
3	.10	.30	4	.40
2	.05	.10	9	.45
		5.00 = $\bar{x}$		2.10 = σ^2_E

4. Calculating genetic variance.

x	f	xf	$(x - \bar{x})^2$	$f(x - \bar{x})^2$
7	.25	1.75	1	.25
6	.50	3.00	0	0
5	.25	1.25	1	.25
		6.00 = $\bar{x}$		.50 = σ^2_G

Note that $\sigma^2_{Total} = \sigma^2_E + \sigma^2_G$ (that is, 2.60 = 2.10 + 0.50).

Heritability = $h^2 = \dfrac{\sigma^2_G}{\sigma^2_{Total}} = \dfrac{0.50}{2.60} = 0.192.$

5. The selected population. Individuals weighing 8, 9, and 10 grams are chosen as parents for the next generation. What will be the weight of their offspring?

The average weight of the selected parents can be calculated as follows:

x	f	fx
10	5	50
9	20	180
8	45	360
	70	590

$\bar{x} = \dfrac{\Sigma fx}{\Sigma f} = \dfrac{590}{70} = 8.43$

The expected weight of the progeny of these individuals equals

$$6.00 \quad + \quad [(8.43 - 6.00) \times 0.192] \quad = \quad 6.47 \text{ grams}$$

Grand Average +   Excess weight of selected = Expected average of progeny

parents $\times h^2$

6. The above expectation can be checked by calculating the following three items:

     a. Gene frequencies in selected sample:

	AA	Aa	aa
10	.05	—	—
9	.10	.05	—
8	.20	.10	.05
	.35	.15	.05

0.0875 (= .35 × .25) .0750 (= .15 × .50) .0125 (= .05 × .25)

Total = .1750

$AA = .0875 \div .1750 = .50$
$Aa = .0750 \div .1750 = .43$
$aa = .0125 \div .1750 = .07$
$A = .50 + .215 = .715 = p$
$a = .070 + .215 = .285 = q$

     b. Geneotypes among progeny:

AA	Aa	aa
.511 (= p^2)	.408 (= $2pq$)	.081 (= q^2)

     c. Mean weight of progeny

x	f	fx
7	.511	3.577
6	.408	2.448
5	.081	0.405
	1.000	6.430 grams = $\bar{x}$ of progeny

7. Comparison of prediction and observed average weights:

Predicted: 6.47 grams
Calculated: 6.43 grams.

8. It is left to the reader to show that, with the new gene frequencies, the estimate of h^2 changes in the new generation and that, as a result, predictions for the third generation are also changed.

FOR YOUR EXTRA ATTENTION

1. Wright (1931: 155) summarizes his calculations with this catalogue of symbols: q equals gene frequency, v and u equal mutation rates to and from the gene, m equals the exchange of individuals between the population and its neighbors whose gene frequency equals q_m, s equals the selective advantage of the gene over all other alleles, and N equals the effective population size. He then says that the most probable change in gene frequency equals

$$\Delta q = v(1 - q) - uq - m(q - q_m) + sq(1 - q).$$

He gives the array of probabilities for the next generation as

$$[(1 - q - \Delta q) a + (q + \Delta q) A]^{2N}.$$

Identify each item in these algebraic expressions, and make sure that you understand why each takes the form in which it is given.

2. Latter (1964) established four lines of *D. melanogaster* from a single starting strain; two of these were selected for high numbers of abdominal hairs, two for low numbers. His data can be summarized as follows:

Generation	0	1	2	3	4	5	6	7
High lines	42.4	45.0	46.5	47.2	48.7	50.2	50.4	51.7
Low lines	42.4	41.9	40.7	39.8	39.3	37.8	36.3	35.7

Plot these data, noting whether or not the responses in the two directions are symmetrical. What is the average number of hairs added or removed each generation? Would you expect events to continue as they have if selection were continued over 50 generations?

3. Haldane (1954) has given the following equation for the measurement of the intensity of selection

$$I = \ln \left(\frac{\sigma_1}{\sigma_2}\right) + \frac{(m_1 - m_2)^2}{2(\sigma_1^2 - \sigma_2^2)}$$

where m_1 and m_2 are the means of the original population and of surviving individuals, and σ_1 and σ_2 are the standard deviations of these two groups.

Curtsinger (1976a) found the following in respect to survival and egg length in *D. melanogaster*:

	No.	Mean	Variance	Standard Deviation
All eggs	846	91.6	12.7	3.6
Hatched	783	91.6	11.9	3.4
Unhatched	63	91.0	23.5	4.8

Confirm that I in this study equals 0.057. The proportion of unhatched eggs equals 0.074. Consequently, nearly 80 percent of the mortality of these eggs was selective in respect to their length. Consult a discussion of these data between Roff (1976) and Curtsinger (1976b).

4. If nonsterile soil is treated with an herbicide, the herbicide will remain for a period of 10–80 days (the lag period), and then disappear rapidly (much more rapidly than it would in sterile soil); such soil is called "enriched."

When all traces of the original herbicide have disappeared, additional herbicide added to enriched soil disappears almost at once, with no lag period.

Reconstruct the events that may be responsible for these observations (see Audus 1960).

5. Graham and Istock (1979) have described the role of recombination and selection in the microevolution of the soil bacterium, *Bacillus subtilis*. Sterile soil enriched with histidine and tryptophan was inoculated with two strains of *B. subtilis:*

a. 168-R, resistant to the antibiotics rifampicin (Rfm^r), erythromycin (Ery^r), spectinomycin (Spc^r), and lincomycin (Lin^r).

b. SB-1R, resistant to 3-amino-tyrosine (Amt^r) and 4-Azaleucine (Azl^r), and auxotrophic for histidine (His^-) and tryptophan (Trp^-).

One day after inoculation, soil samples yielded bacteria with 92 identifiably different genotypes (How many were possible?); the most common comprised 9 percent of all bacterial samples.

By the eighth day, samples yielded only 27 different genotypes of which the most common comprised 79 percent of all bacterial samples. This overwhelmingly common genotype was

$$Spc^s Ery^r Rfm^r Lin^r Azl^s Amt^s His^- Trp^-.$$

This genotype was not encountered in samples taken during days 1 and 2.

What is the evidence that recombination took place in the soil? Is there any assurance that the common genotype of day 8 would be the common one on day 30? How does the prevalence of one genotype reduce the probability that a still "better" one will arise by recombination at a subsequent time? Why?

These results have an important bearing on the use of antibiotics in medicine and agriculture, and in understanding the nature of epidemic organisms such as the influenza virus (see Young and Palese 1979). The importance of recombination arises once more in chapter 24 under the evolution of sex.

6. Consider the following in relation to character displacement and the apportionment of environmental resources:

a. Wiens (1977) emphasizes that there is a great overlap in the types and sizes of prey eaten by birds of quite different body or bill sizes. Though the diets may vary through the course of the breeding season, at any moment different species may have high dietary similarity.

Hint: Need selection be all or nothing? Could Wiens's argument be improved by quantitative data?

b. Shapiro (1978) has emphasized the possible absurdity of some selectionist arguments: Balsam fir contains an insect hormone analogue that inhibits the maturation of Pyrrhocorid bugs and, as far as is known, only Pyrrhocorid bugs. There are no such bugs on balsam firs, however. Is the analogue a defensive chemical—one that has become completely effective?

7. Francois Jacob (1977) claims that evolution works like a tinkerer, a tinkerer who does not know exactly what he is going to produce but uses whatever is nearby—string, wood fragments, cardboard, or metal scraps. An engineer, in contrast, has a goal, has specifications, and has tools designed to accomplish that goal.

Jacob manages remarkably well to describe organic evolution, and to place such terms as "optimal" or "maximal" (and terms such as $1 - s$ and $1 - t$) in proper perspective. In the context of organic evolution there is no absolute optimum or maximum, just as no *final* product emerges from a tinkerer's workbench.

The response to the above is the claim that, for example, egg hatch or survival cannot exceed 100 percent, which must be optimal. Can you devise means to circumvent such arguments?

Hint: Females of genotype G_1 lay eggs all of which hatch; those of genotype G_2 lay four times as many, one-third of which hatch. Can survival be treated as if it alone measures fitness?

8. Gallo (1978) reported finding five lethals and semilethals among 203 X-chromosomes tested from a natural population of *D. melanogaster*. This is scarcely 2.5 percent, one-tenth (or less) of the frequency of autosomal lethals one might find in a similar population.

Why is the frequency of sex-linked lethals and semilethals so low?

9. In analyzing the induction of mutations in *D. melanogaster* by *MR* (*male recombination*, a possible transposable element) Green and Shepherd (1979) have made two points that require the attention of population geneticists:

a. *MR* induces lethals at certain preferred gene loci rather than randomly throughout the genome (see figure 11.13b for evidence leading to a similar conclusion).

b. The high frequency of a particular lethal in a population, consequently, need not be evidence for a selective advantage of the heterozygous carriers of this lethal (see table 8.5 for an additional reason for avoiding this conclusion).

Genetic Load Theory

PREVIEW: This chapter concludes part 1. It pulls together material covered in the preceding chapters in explaining the theory of genetic loads. This theory relates the average fitness of a population to selection coefficients and genotypic frequencies where the latter are themselves dependent upon mutation, migration, selection, and chance events. The chapter concludes with an account of the most important corollary of genetic load theory: the neutral mutation hypothesis.

DURING THE DISCUSSIONS of the previous two chapters, we have had occasion to refer to the fitnesses of individual genotypes (W_{AA}, W_{Aa}, and W_{aa} are appropriate symbols) and to the average fitness of populations ($\overline{W} = p^2 W_{AA} + 2pq W_{Aa} + q^2 W_{aa}$). For the most part, the average fitness was expressed in terms of selection coefficients and gene frequencies: for recessive mutations, $\overline{W} = 1 - sq^2$; for incompletely recessive ones, $\overline{W} = 1 - 2hsq$.

In the cases of both recessive and incompletely recessive mutations, stable equilbria result from the opposition of mutation and selection. In the case of recessive mutations, $sq^2 = u$ or $q = \sqrt{u/s}$; in that of incompletely recessive mutations, $hsq = u$ or $q = u/hs$. Haldane (1937), in a paper entitled "The Effect of Variation on Fitness," showed that, because of these relationships, the average fitness of a population can be expressed in terms of mutation rates: for recessive mutations, $\overline{W} = 1 - sq^2 = 1 - u$; for incompletely recessives, $\overline{W} = 1 - 2hsq = 1 - 2u$. The average fitness of a population at equilibrium is reduced by an amount equal to, or twice as great as, mutation rate, depending upon the recessivity of mutant (autosomal) alleles. This reduction holds for each locus, of course; for all loci combined, $\overline{W} = 1 - U$ or $1 - 2U$ where U is the combined mutation rate of all loci.

Haldane's calculations did not have a profound effect on population genetics at the time. Crow (1948) made similar calculations some ten years later while working on a somewhat different problem, but one that led him to the same conclusions. During the course of his investigations whose results were published under the title "Genetics of Natural Populations," Dobzhansky (1947b) had occasion to refer first to Haldane's calculations ten years after their publication.

The terms "genetic load" and "load of mutations" were first used by Muller in his 1949 presidential address before the American Society of Human Genetics (Muller 1950b). This address and the terms used in describing the effect of mutant genes on populations were intended to shock persons, to alert them to the dangers of hereditary diseases and, especially, to the threat of high-energy radiations because of the gene mutations they induce. Utilizing a series of estimates concerning mutation rates and numbers of gene loci, he concluded that 10 percent or more of all germ cells carry a newly arisen mutation, and that 20 percent of all human deaths have an ultimate genetic cause (a fact that is otherwise concealed by the immediate cause of death normally entered on a death certificate).

The picture painted by Muller in his address was a bleak one, indeed. He closed all obvious escape routes. Relax natural selection by medical treatment? No, because that leads to the retention of otherwise harmful mutations. Improve the environment in other ways? No, because continued mutation will simply raise mutant gene frequencies to new and higher levels. As long as mutations occur, human beings are condemned to genetic deaths. As long as radiation-generating technologies proliferate, artificially induced mutations will increase the frequency of genetic diseases in exposed populations.

Of immediate interest here is the application of genetic load calculations to natural populations of lower organisms. The problems we shall encounter even with this restriction are complex enough without taking on the emotion-laden ones of the genetic aspects of public health. The latter must, and will, be debated; this book, however, is not a proper forum for that debate.

The Genetic Load

Muller's treatment of genetic load, despite his calculations, was fundamentally a qualitative treatment. Populations contain high frequencies of mutant genes, most of which harm their homozygous carriers. The evidence for this variation appeared in chapter 3; bases for its existence appeared in chapters 5, 9, 10, and 11. Populations do, in the sense that Muller intended, carry a load, a burden.

In addition to its descriptive meaning, however, genetic load has been defined in a formal, mathematical sense (Crow 1958): the genetic load of a population is the proportional amount by which the average fitness of the population is depressed for genetic reasons below that of the genotype with maximum fitness. Under this definition,

$$\text{Genetic Load} = \frac{W_{max} - \bar{W}}{W_{max}}$$

where W_{max} is the fitness of the optimum genotype and $\bar{W}$ is the average fitness of the population. If, continuing a practice followed in the past chapters, W_{max} is assigned the (arbitrary) value 1.00,

$$\text{Genetic Load} = 1 - \bar{W}.$$

It is important to notice that the genetic load as defined here is no longer a collection of deleterious mutations, of lethals, of sterility genes, or of morphological mutants. It is, rather, a quantity that can be computed according to prescribed mathematical rules. In the sections that follow, our attention will be focused on these questions:

- What types of genetic loads are there?
- How large are they?
- Which individuals contribute to genetic loads?
- What does the load, as defined, mean to the population?

In concluding the chapter we shall take up the most important corollary of genetic load theory: the neutralist hypothesis.

As the chapter progresses, problems and differences of interpretation will arise. Maynard-Smith (1978:24) has suggested that these differences are in part semantic; I disagree. If, as Haldane

(1964) argued in his "Defense of Beanbag Genetics," the value of mathematics in biology is that it demands precision in the use of words, then an absurdity which follows from a mathematical definition cannot be passed over as semantic. On the contrary, the quantity that has been defined must be carefully examined: Is it the quantity that should have been defined? Is it a quantity worth defining? At the moment this digression seems obscure; its meaning will become clearer as the chapter, and the book, progress.

The Mutational Load

A deleterious mutation that is completely recessive will reach an equilibrium frequency in a large popultion such that $q = \sqrt{u/s}$. The frequency of homozygous individuals will equal q^2 or u/s. Each homozygote suffers a loss of fitness (its selection coefficient) of s. Consequently, the total loss of fitness to the population equals $u/s \times s$ or u.

Despite its effect on individuals (s may be 1.00, 0.10, or 0.01), the effect of a recessive mutation on the average fitness of a population equals u, its mutation rate. For many loci combined, the total effect on the average fitness of the population equals U, the sum of all individual mutation rates (or $n\bar{u}$, where n equals the number of loci and $\bar{u}$ equals the average mutation rate per locus).

The equilibrium frequency, q, for a deleterious mutation that is partially dominant ($0 < h \leqslant 1$) equals u/hs where s is its effect on homozygous carriers. The frequency of heterozygotes ($2pq$) for such an allele is very nearly $2u/hs$. Each heterozygote suffers a decrease in fitness of hs, consequently, the average fitness of the population is lowered by an amount equal to $2u/hs \times hs$ or $2u$. The mutational load in this case equals $2u$, twice the mutation rate for the locus. For all loci, the mutational load equals $2U$ (or $2n\bar{u}$).

The increasing use of radiation in industry and medicine and the production of vast quantities of radioactive waste products (including the physical plant itself) by nuclear power plants makes the mutational load an important one for the human population. The consequences of an artificial doubling of mutation rate by radiation exposure is depicted in figure 12.1; doubling the mutation rate doubles the eventual mutational load. Muller (1950b) estimated that as many as one death in five may have an ultimate

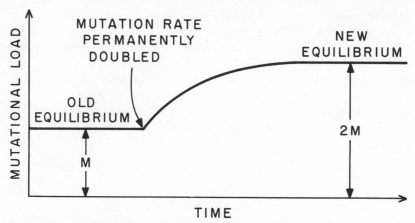

Figure 12.1 The effect on the mutational load of a permanent doubling of the mutation rate. (After Crow 1957.)

(not proximate) genetic basis; under a regime in which mutation rate were permanently doubled, this proportion would climb to 40 percent—nearly one-half of all deaths.

The Segregational Load

The genetic load has been defined as the proportional decrease in the average fitness of a population relative to the optimal genotype. In those instances where heterozygous individuals possess a higher fitness than either homozygote, the average fitness of the population, although higher than that of either homozygote, is less than that of the heterozygotes. Therefore, a balanced polymorphism imposes a genetic load on a population. This load is known as the segregational load (Crow 1958) or the balanced load (Dobzhansky et al., 1960).

The algebra of balanced polymorphisms was presented earlier (page 228). If the fitnesses of *AA* and *aa* individuals (relative to that of 1.00 for *Aa* individuals) are $1 - s$ and $1 - t$, the equilibrium frequencies of *A* and *a* will be $t/(s + t)$ and $s/(s + t)$, respectively. In that case, the average fitness of the population is decreased from its maximum value (1.00) by

$$s\left(\frac{t}{s+t}\right)^2 + t\left(\frac{s}{s+t}\right)^2.$$

The genetic load in this case equals

$$\frac{st}{s+t}.$$

Balanced lethals represent a case in which heterozygotes have superior fitness; s and t in this case both equal 1.00. The segregational load for a balanced lethal system equals 0.50. One half of the population dies because one half of the population is homozygous for either one lethal or the other. Only the heterozygotes (50 percent of all zygotes) live.

The same calculation can be applied to the grouse locust, *Apotettix eurycephalus*, where Nabours and Kingsley (1934) found the homozygous normal (+/+) individuals were too scarce by some 7 percent in crosses where they and lethal heterozygotes (+/*le*) were expected to occur in Mendelian ratios. In this case, $s = .07$ and $t = 1.00$; consequently, the segregational load in an equilibrium population would be (.07) (1.00/1.07) or about 6 percent–7 percent.

These calculations pose serious problems for populations. How many balanced lethal systems, for example, can a population sustain? This question could be rephrased: What is a population's load space? A Drosophila species with three autosomes and a balanced lethal system on each would lose $1 - (\frac{1}{2})^3$ or $\frac{7}{8}$ of all its zygotes to these three pairs of lethals alone; only one egg in eight would hatch. Such flies could be maintained in the laboratory if properly cared for because each female can lay several hundred eggs. On the other hand, three independent, balanced-lethal systems would pose an extremely serious threat to human populations if only one fertilized egg in eight could successfully complete development; surely a balanced lethal system on each of seven to ten chromosomes would make human reproduction impossible.

The problem described above confronts any species. A female *Drosophila* can, as we said above, lay several hundred eggs. The data of Lewontin and Hubby (1966) and others suggest that populations of these flies maintain segregating alleles at several thousand gene loci. If these loci are looked upon as independent (which they are not because most Drosophila species have only four, five, or six linkage groups), the sizes of s and t must be very small if the population is not to be overwhelmed by the segregational load these genes would generate. If, for example, s and t both equaled 1 percent, the segregational load per locus would be (.01) (.01)/[(.01) + (.01)] = 0.005. The surviving fraction of all zygotes produced each generation would be $(0.995)^{1000}$ (assuming that 1,000 loci had such segregating alleles) or, approximately, 0.007. Only 7 zygotes in each 1,000 would survive to face the other hazards of a fly's life. It seems unlikely that Drosophila females produce eggs in numbers sufficient to cope with a load this large. Eventually, we must find a way out of the dilemma posed by Lewontin and Hubby's observations. It is proper at this time to point out that the calculation we have just performed (i.e., 0.995^{1000}) falls under the multiplicative model described in chapter 1 where survival was said to represent the successful negotiation of a series of pitfalls: $(1 - .005)(1 - .005)(1 - .005)$. . . . Whether survival can reasonably be regarded as a series of 1,000 independent events is a question we must eventually answer.

Other Genetic Loads

Mutational and segregational loads are only two of many possible genetic loads. Certain maternal and fetal genotypes are incompatible. *Rh*-positive babies (of necessity *Rhrh*) born of *Rh*-negative mothers (*rhrh*) run the risk of *erythroblastosis foetalis*, the destruction of the fetus's red blood cells because of the leakage of anti-*Rh* antibodies from the mother through the placenta into the baby's bloodstream. Such genetically based incompatibilities give rise to *incompatibility* loads—loads that would not exist if genotypes did not interact with one another in this manner.

At any moment, the members of a species are reasonably well adapted to their way of life, but not necessarily perfectly so. The physical environment is not constant, nor do competing species remain unchanged. Consequently, alleles that were selected earlier under different regimes are continuously replaced by alleles more appropriate for present conditions. The need to replace ill-adapted alleles by other, better-adapted ones leads to a *substitution load.* This load would not exist if the environment were constant, and if the species had responded by creating the genotype best able to cope with perhaps numerous, but unchanging, challenges.

Because of the definition of genetic load, a load must exist whenever phenotypes vary. If the dissimilar phenotypes reflect dissimilar genotypes, the load can be identified by an appropriate taxonomy of genetic loads. If the different phenotypes are not caused by the genotypes of their carriers (the various phenotypes associated with age differences, for example), the load is not genetic but, following the same mathematical logic, can be expressed as a phenotypic load.

The Relative Sizes of Mutational and Segregational Loads

A preliminary estimate of the relative sizes of mutational and segregational loads can be made on the basis of known mutation rates, on the one hand, and polymorphic systems on the other. The known spontaneous mutation rate for sex-linked lethals (X chromosome) in *D. melanogaster* is about 0.003. The second chromosome which is about twice as large has a mutation rate to lethals that is also about twice as large, 0.005. The third chromosome has a lethal mutation rate comparable to that of the second (see table 9.3). If we accept 0.003 as the mutation rate for lethals on the X chromosome and on each of the similar-sized autosomal arms, the total lethal mutation rate amounts to 0.015. Following Muller's estimates (figure 9.7), we can guess that mutations with smaller detrimental effects are four times as frequent as lethals (see, however, Mukai's [1964] estimate of 0.14 described on page 210); the total mutation rate to lethals and detrimentals, consequently, equals 0.075. Now, if these mutations exhibit an appreciable dominance, the load equals $2U$, or 0.150. The mutational load, then, could be as large as 15 percent, although these calculations are subject to many possible errors. Mukai's estimate of the mutation rate for second chromosome detrimental mutations, 0.14, would raise the mutational load to $(0.003 \times 5) + (0.14 \times 2.5)$ or 0.365. Doubling this value for dominance leads to 0.73; that is, three quarters of all deaths of flies would have a genetic basis.

In his studies on the fitnesses of third-chromosome inversion homozygotes and heterozygotes in laboratory populations of *D. pseudoobscura*, Dobzhansky discovered a number of instances in which these fitnesses were (approximately):

AA	AB	BB
0.70	1.00	0.30

In this case, $s = 0.30$ and $t = 0.70$. The genetic load equals $st/(s + t)$ or 0.21. This one system imposes a load on the population as large as Muller's estimate of the total mutational load.

The B/A ratio

An alternative method for estimating the relative contributions of mutation and segregation to the total genetic load was divised by Morton et al. (1956). This method relies upon a comparison of the genetic load (B) that is estimated for a population which has been made completely homozygous and the considerably smaller load (A) which is exhibited by a population under random mating.

If an allele is maintained in a population by recurrent mutation, and if it is somewhat detrimental in heterozygous condition, its equilibrium frequency is u/hs. The genetic load imposed by such an allele on a randomly mating population is $2u$ (= A).

If the population were made completely homozygous with no change in gene frequency, there would be u/hs homozygous mutant individuals whose selection coefficient would be s. The amount by which the fitness of the homozygous population would be lowered (= B) equals $u/hs \times s$, or u/h.

The B/A ratio, the ratio of the loads expressed in homozygous and randomly mating populations, equals $(u/h) \div 2u$, or $1/2h$. Because h is generally small (some estimates have suggested .02 as a reasonable value), $1/2h$ can be quite large—say, 25, or 50, or even more.

The B/A ratio in the case of a locus at which two alleles are maintained by the superior fitness of heterozygotes is calculated in the same manner.

The random load (A) has been calculated earlier (page 295) as $st/(s + t)$.

The frequencies of the two alleles in the case of a balanced polymorphism, have been given as $t/(s + t)$ and $s/(s + t)$. These, then, would be the relative frequencies of the two types of homozygotes in a homozygous population. The homozygous load (= B) in this case would be $ts/(s + t) + st/(s + t)$ or $2st/(s + t)$.

In the case of a locus whose alleles are maintained by the superiority of heterozygotes the B/A ratio equals

$$2st/(s + t) \div st/(s + t) = 2.00.$$

The B/A ratio in such cases equals the number of alleles: had there been k alleles segregating in the example rather than 2, the B/A ratio would have been k.

The calculations given here suggest that a high B/A ratio should be interpreted as evidence that the genetic load is mutational in origin, and that a low B/A ratio should be regarded as evidence that the genetic load is segregational. Unfortunately, a segregational load at one locus (or one associated with an inversion polymorphism) is able to cause a low B/A ratio which conceals the higher ratio that would reflect the mutational load of many other loci. This defect (and others to be discussed shortly) impairs the usefulness of the B/A ratio as an investigative technique: it allows for only one interpretation.

The Genetic Load Concept: A Critique

In his 1950 paper, Muller used the phrase, "our load of muta-tions," in a descriptive or qualitative sense despite the algebraic calculations he performed in estimating its importance for human health. He had in mind, for one thing, the multitude of genetic disorders—individually rare but collectively all too common—that afflict the carriers of mutant genes: the wastage of young chil-dren by Tay-Sachs disease, the life of pain endured by sickle-cell homozygotes, the mental impairment suffered by phenyl-ketonurics, . . .: the list can be extended almost indefinitely. He also had in mind a second aspect of genetic disease: the cause of death given on a death certificate more often conceals than reveals. Garrett Hardin (1971) has pointed out that no one ever dies of overpopulation. Few persons officially die of starvation. Similarly, when a person is weakened by a chronic genetic disorder, the official cause of death becomes influenza, cancer, emphysema, or any one of many other debilitating illnesses. The genetic flaw ultimately responsible is concealed by the proximate, fatal ailment.

In addition to its qualitative usage, genetic load has been given the precise definition discussed in this chapter. Calculations made according to this definition lead to a quantity that *is* the genetic load, whether we wish to call it that or not. Dobzhansky (1964, 1970:193) substituted the term "genetic burden" for a quantity that he suggested be calculated in a different manner, thus admitting that calculations performed according to the rules that we have followed above ($W_{max} = 1.00$; $\overline{W} = 1$ − genetic load) lead to what has been *defined* as genetic load.

The following critique will center not on whether a genetic load exists (because it *does* exist in the two senses just mentioned) but, rather, on th meaning of the genetic load as defined, and on some of the counterintuitive aspects of this formalized genetic load.

The Genetic Load Is a Static Concept

The concept of genetic load is one that cannot embrace change and, consequently, gives the impression that lasting changes do not occur in populations. This feature is illustrated in figures 12.2 and 12.3. In figure 12.2 an equilibrium mutational load is shown on the left. A sudden, permanent improvement occurs in the environment, alleviating much of all genetic ailments. The effect is only transitory. As long as *any* deleterious aspect remains to plague the carriers of these deleterious mutations, the load eventually regains its initial size; the growth of the genetic load reflects the now-increased frequency of each mutant allele in the population.

Figure 12.3 illustrates the effect a favorable mutation has on genetic load. The occurrence of such a mutation imposes a tremendous load on the population because W_{max} (= 1.00) is by tradition assigned to the individual whose genotype has maximum fitness; all other individuals are inferior by definition. As the new, favorable mutation increases in frequency, the genetic load gradually returns to its former value.

In each of these instances, circumstances were altered (by improving the environment, by introducing an advantageous gene mutation); genetic load calculations best describe equilibrium conditions.

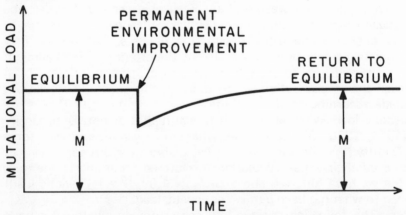

Figure 12.2 The effect of a permanent improvement in the environment upon mutational load. The diagram suggests that the effect of all deleterious recessive mutations has been ameliorated by about one-third; each mutant allele, then, has a new and higher equilibrium frequency which eventually it will attain. (After Crow 1957.)

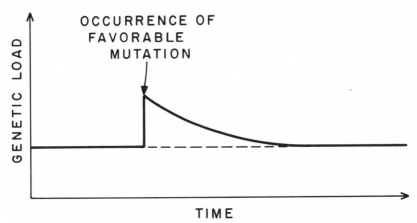

Figure 12.3 The effect of a new, favorable mutation upon mutational load. Because the new mutation reveals that fitnesses twice as great as those which existed in the pre-mutational population are possible, virtually the entire population must now be regarded as grossly unfit. The genetic load decreases as the new mutation increases in frequency within the population. (After Crow 1957.)

Nongenetic Deaths Can Be Labeled Genetic under Genetic Load Calculations

The point made here is that genetic load calculations constitute an exercise in bookkeeping. As is frequently the case among accountants, in balancing ledgers the reality of the entries is secondary to their proper manipulation.

Consider a population that contains only the allele for normal hemoglobin, Hb^A, and which lives in a heavily infested malarial region. One-quarter of all newborn babies die of malaria before the age of four. Imagine now that a mutant allele, Hb^S, is introduced into this population. We shall assume that $Hb^A Hb^S$ individuals are resistant to malaria but that $Hb^S Hb^S$ individuals suffer from a lethal anemia.

Relative fitnesses can be assigned to the three genotypes as follows: $Hb^A Hb^A$, 0.75; $Hb^A Hb^S$, 1.00; and $Hb^S Hb^S$, 0. Because $s = 0.25$ and $t = 1.00$, equilibrium frequencies can be calculated as 0.80 (Hb^A) and 0.20 (Hb^S); the genetic load of the population at equilibrium equals 0.20.

Of what does the genetic load consist? Sixty-four percent (0.80^2) of the population are $Hb^A Hb^A$ homozygotes, of which one-quarter die of malaria; this equals 0.16. Four percent (0.20^2)

of the population die of anemia. The sum (0.16 + 0.04) equals 0.20, the genetic load. However, 0.16 or 75 percent of the genetic load consists of deaths by malaria—the same malaria that caused environmental deaths in the earlier, load-free population.

The largest part of the load can be traced to individuals of the most nearly normal genotype. In the preceding example, three-quarters of the genetic load was ascribed to persons ($Hb^A Hb^A$) possessing normal hemoglobin. That these persons bore the brunt of the load is no (algebraic) accident; although they are the more normal of the homozygotes, they are also by far the most frequent.

If the more nearly normal homozygous individuals (AA) have a fitness $1 - s$ and the more deleterious ones (aa) a fitness of $1 - t$, then the equilibrium gene frequencies are $t/(s + t)$ for A and $s/(s + t)$ for a. The total segregational load in that case equals $st^2/(s + t)^2 + s^2 t/(s + t)^2$. The ratio of the portion of the load attributed to AA individuals to that attributed to aa ones equals $st^2/s^2 t$ or t/s. Consequently, in the example involving the grouse locust where $s = .07$ and $t = 1.00$ (the mutant was lethal when homozygous), the nearly normal homozygotes were responsible for about 13/14ths of the total load; the lethal zygotes contributed only the remaining 1/14th. If s had been smaller, say 0.01, then the deaths of the lethal zygotes would have contributed only 1/100th to the total load; the nearly normal individuals would have contributed 99 percent. To some, this "counterintuitive" distribution of loads has biological significance (Crow 1958, Kimura and Ohta 1971: 55); to others (Li 1963, Wallace, 1970b) it reflects a serious inadequacy of genetic load calculations and of the assumptions on which they are based.

If h is a variable, the B/A ratio leads to absurd results. The B/A ratio was devised as a means for deciding whether an observed genetic load is of mutational or segregational origin. The B/A ratio equals $1/(2h)$ for populations in which deleterious alleles are maintained by recurrent mutation. In calculating this ratio, two assumptions are made: (1) that h is known, and (2) that h is a constant.

Figure 12.4 illustrates some of the difficulties accompanying the empirical measurement of h. If h is assumed to be independent of s, it can, at least in theory, be measured; if, on the other hand, it is assumed that h is a function of s, then empirical measurements based on the relative viabilities (the component of fitness most easily studied) of different types of individuals do not yield interpretable results.

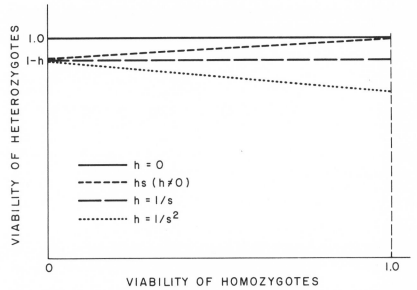

Figure 12.4 A diagram illustrating the futility of estimating *h* by means of a comparison of homozygous and heterozygous viabilities. If *h* were unrelated to *s*, the slope of the regression would (provided that proper precautions were taken to avoid various subtle errors) represent an estimate of *h* (including *h* = 0 when the slope of the regression is 0). If, however, *h* is related to *s* as some persons now claim, the slope of the regression would be 0 if *h* were proportional to $1/s$. If *h* were proportional to $1/s^2$, the slope of the regression would be negative. The experimental procedures do not provide a fixed standard for comparison (such as the Y-intercept) but only a regression reflecting changes of heterozygous viabilities on *s*: consequently, a negative regression would suggest that lethal chromosomes exhibit "hybrid vigor."

So much for *h* being a function of *s*. Suppose, additionally, that *h* is not a constant but, rather, a variable with a certain mean value, $\bar{h}$, and some distribution about this mean (see figure 12.5). The high values of the *B/A* ratio which supposedly rule out balanced polymorphism as a possible cause for observed genetic variation are caused by small values of *h* (as *h* becomes small, $1/2h$ becomes large). If *h* is a variable, however, the smaller $\bar{h}$ becomes, the more often the mutant allele participates in combinations of genes in which the heterozygous carriers have superior fitness. Thus, the firmer one's belief that the observed load is mutational, the more likely it is that a portion is segregational. Figure 12.6 illustrates the abrupt change in the calculated genetic load that occurs as *h* passes from positive to negative values.

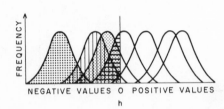

FREQUENCY

NEGATIVE VALUES 0 POSITIVE VALUES

h

Figure 12.5 The possible relation between the recessivity of an allele in respect to fitness and the proportion of background genotypes in which it *enhances* fitness. The diagram emphasizes that h need not be a constant but, rather, that it may have a certain mean value $(\bar{h})$ and variance (σ_h^2). Because of this, the discontinuity of the B/A ratio at $h = 0$ cannot be used in deciding on which side of 0 h actually falls. As the average value of h approaches zero, the mutant heterozygote takes on a selective superiority $(h < 0)$ in an ever increasing proportion of background genotypes.

The Cost of Natural Selection

The replacement of one allele by another, unless it happens entirely by chance, requires that the carriers of the two alleles differ in their reproductive success, either in the probability of surviving until maturity, or in fertility, fecundity, and mating success. Haldane (1957, 1960) has analyzed the process of gene substitution in terms of the "cost" to the population, the cost of natural selection.

Using a halpoid population for illustrative purposes, let us assume that A individuals are extremely common and that a individuals are rare. Perhaps the frequency of a equals u/s, the equilibrium between mutation and selection in a haploid population. Now, however, the environment changes so that a has the maximum fitness while that of A drops to $1 - k$. Let the frequency of A equal q. What are the consequences for the population? The account that follows is taken from Haldane (1957).

1. The fraction of selective deaths in any generation, n, equals kq_n.

2. The frequency of A in any generation, $n + 1$, equals

$$q_{n+1} = \frac{q_n - kq_n}{1 - kq_n} .$$

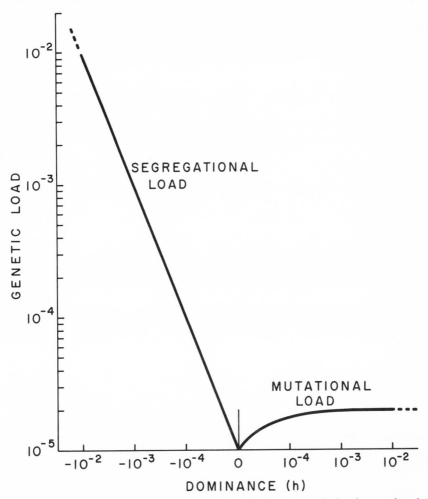

Figure 12.6 The contrast in the sizes of calculated genetic loads associated with a lethal gene exhibiting various degrees of partial dominance (h), or corresponding degrees of superiority in heterozygous individuals ($h < 0$). Whereas on the right the load estimates range between u and $2u$ ($u = 10^{-5}$), on the left the load estimates are (for low values of h) nearly equal to h.

3. The change in gene frequency, $\Delta q_n = q_{n+1} - q_n$ equals

$$- \frac{kq_n(1 - q_n)}{1 - kq_n}.$$

4. The total fraction of selective deaths $= D = k \sum\limits_{n=0}^{\infty} q_n$

5. When k is small,

$$\frac{dq}{dt} = -kq(1-q).$$

6. By integration,

$$D = k \int_0^\infty q\, dt, \text{ or}$$

$$D = k \int_0^{q_0} -q\, \frac{dt}{dq}\, dq,$$

but

$$\frac{dt}{dq} = -\frac{1}{kq(1-q)},$$

therefore

$$D = \int_0^{q_0} \frac{dq}{1-q} = -\ln(1-q_0) = -\ln p_0.$$

The cost of natural selection, which equals the total fraction of selective deaths occurring during the displacement of one allele by another, is independent of the intensity of selection (as long as k is small) but does depend upon the initial frequency of the rare advantageous gene: the rarer the allele, the greater the fraction of selective deaths. Thus, if p_0 equals 0.001, D equals 6.91 $(=-\ln 0.001)$; if $p_0 = 0.0001$, $D = 9.21$; and if $p_0 = 0.00001$, $D = 11.51$. These values of D mean that 7, 9, or 11 to 12 times as many individuals die during the replacement of one allele by another in a haploid population as there are in the population at any moment. The proportion of selectional deaths among diploid individuals is somewhat larger than that for haploids. Haldane (1957), by assuming that no more than 10 percent of all mortality occurring in a population could be expended on gene substitution, concluded that one allele could displace another one only once in 300 generations, that gene substitution (and, hence, evolution) is a slow process.

The cost of natural selection which Haldane computed and

the exceptionally slow rate of gene substitution have been questioned by some (Brues 1964; Maynard-Smith 1968a) and supported by others (Kimura and Crow 1969; Kimura and Ohta 1971: 74). Brues, for example, countered by pointing out that the cost of not evolving, extending as it does over all future time, greatly exceeds the cost of evolving. Maynard-Smith suggests that, if allele substitutions at different loci are not occurring independently of one another, many genes can undergo coordinated substitution.

Kimura, Crow, and Ohta emphasize the importance of maintaining a constant population size throughout the period during which the allele substitution occurs. This insistence is necessitated by Feller's (1967) argument which led him to conclude that D cannot exceed the initial number of A individuals in the evolving population. Kimura and Ohta (1971: 74) suggest "that Feller overlooked an important biological fact that in the process of gene substitution in evolution, the relative proportions of genes change enormously, while the total population number remains relatively constant due to population regulating mechanisms." The latter, of course, include food, space, and competitors. One consequence of Kimura and Ohta's insistence on a constant population size is that the gene substitution under discussion is absolutely inconsequential for the population's existence because its size is said to remain constant for all values of p from 0 percent to 100 percent.

The flaw in Haldane's calculation (and I do believe that the calculation is flawed) can best be illustrated by a numerical example, but one in which the total population size is restored each generation. The example can be presented in tabular form as follows (fitness of $A = 1 - s = 0.99$; population size = 7,000):

Generation	Number of Individuals		Operation	Selective Deaths	"Revivals"
	A	a			
0	6,999	1	Start	—	—
	-69.99	-0	Selection acts	69.9900	—
	6,929.01	1	Sum: 6,930.01	—	—
1	6,998.9899	1.0101	Restore to 7,000	—	69.9799
	-69.9899	-0	Selection acts	69.9899	—
	6,929.00	1.0101	Sum: 6,930.0101	—	—
2	6,998.9798	1.0202	Restore to 7,000	—	69.9798

The two generations of selection tabulated here are sufficient to pinpoint the selective deaths which when summed over these two (and all subsequent) generations equals Haldane's D; in the present case, the sum would be 7,000 D. Not taken into account by Haldane's calculations are the revivals that are needed to maintain a constant population size. The revivals virtually cancel the deaths in each generation; in short, the fraction of selective deaths does not equal kq_n (see item 1, page 304) nor does the sum $\left(k \sum_{n=0}^{\infty} q_n \right)$ equal the total selective deaths in any meaningful sense.

Calculations of the sort outlined above have been continued by computer for 2,000 generations under two regimes (fitness of A equals 0.99 and 0.98); the results are summarized in table 12.1. Consider the population in which the fitness of A equals 0.99: in 400 generations the number of a individuals has grown only to 55 so that that of A individuals equals 6,945. Thus, for 400 generations the number of selective deaths has ranged from 69.99 to 69.45 each generation or about 28,000 in all—a number four times as large as the total population size. Every generation, however, these "deaths" were undone so that, as Feller (1967) claimed, only 55 A individuals actually died and were replaced by a.

The quantity that Haldane calculated as $k \sum_{n=0}^{\infty} q_n$ can be identified in an exponentially growing culture (in which, incidentally, there are *no* selective deaths): it equals s, the rate of increase, times t, the time in generations required to attain a given size. Recalling that for small Θ's, $\tan 2\Theta \approx 2 \tan \Theta$, it can be seen in figure 12.7 that $s \times t = 2s \times 1/2t = st$. The data presented in table 12.1 have also been plotted (figure 12.8); here one can see that the conclusions based on the exponentially growing cultures shown in figure 12.7 are scarcely affected by the need for a to replace A as shown in figure 12.8. Because $p_0 = 1/7000$ or 0.000143, $- \ln p_0 = 8.85$. We have seen above that one-half of these selective "deaths" have occurred by generation 400 for the one population, by generation 200 in the other.

Genes in populations are carried by individuals; the number of individuals carrying each of two alleles can be counted and expressed as a proportion of the total. The replacement of one allele by another represents the elimination of a certain number of one kind of individuals and, if the total population size is to remain unchanged, their replacement by individuals of the other kind. The total number of individuals eliminated, as Feller (1967) suggested, is equal to the initial number; figure 12.9 shows that this is so despite possible irregularities in the course of selective elimination.

Table 12.1 The increase from 1 to 7,000 in the number of *a* individuals within a haploid population of 7,000 individuals under two regimes of selection: fitness of *A* individuals equals (1) 0.99 or (2) 0.98.

Generation No.	Fitness of $A = 0.99$	Fitness of $A = 0.98$
0	1.0000	1.0000
50	1.6527	2.7453
100	2.7313	7.5333
150	4.5134	20.6475
200	7.4569	56.4070
250	12.3168	152.7431
300	20.3348	404.0356
350	33.5473	1,007.8983
400	55.2766	2,211.6612
450	90.8967	3,914.0200
500	148.9780	5,438.4693
550	242.8675	6,337.3485
600	392.5382	6,743.2271
650	625.9019	6,904.2581
700	977.4764	6,964.8278
750	1,480.6596	6,987.1503
800	2,150.3828	6,995.3151
850	2,960.5438	6,998.2932
900	3,834.5896	6,999.3783
950	4,668.4542	6,999.7736
1,000	5,375.7018	6,999.9175
1,050	5,918.1305	6,999.9700
1,100	6,302.9067	6,999.9891
1,150	6,560.9855	6,999.9960
1,200	6,727.6467	6,999.9985
1,250	6,832.6527	6,999.9995
1,300	6,897.7887	6,999.9998
1,350	6,937.8028	6,999.9999
1,400	6,962.2378	7,000.0000
1,450	6,977.1049	7,000.0000
1,500	6,986.1304	7,000.0000
1,550	6,991.6022	7,000.0000
1,600	6,994.9169	7,000.0000
1,650	6,996.9238	7,000.0000
1,700	6,998.1386	7,000.0000
1,750	6,998.8737	7,000.0000
1,800	6,999.3185	7,000.0000
1,850	6,999.5877	7,000.0000
1,900	6,999.7505	7,000.0000
1,950	6,999.8491	7,000.0000
2,000	6,999.9087	7,000.0000

The replacement of one allele by another can be considered (see, for example, Nei 1975: 61) not in terms of selective deaths but in terms of the excess progeny the selectively favored rare individuals must produce if the population is to retain a constant size. Consider, for example, one of the populations that served as the basis of table 12.1: the total population size equals 7,000 of

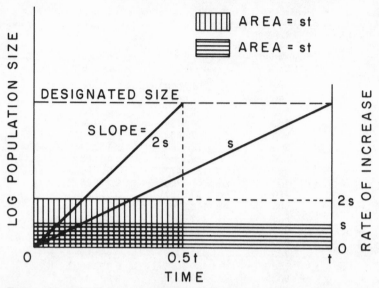

Figure 12.7 A geometric diagram revealing that Haldane's cost of evolution corresponds to the product of the rate of increase of a population and the time required for that population to attain a specified size. This product exists whether the expanding population displaces a preexisting one (thus resulting in deaths) or merely grows in a formerly sterile environment (such as bacteria growing within a laboratory culture tube).

which 6,999 individuals are of the now-selectively-inferior type, *A*, and 1 is the superior type, *a*. Imagine that these are bacteria with a 20-minute generation and that the intensity of selection against *A* is so great that all are killed in a 10-minute interval. Because *a* cannot create 7000 descendants within 20 minutes, the total population size must decrease to one individual and then grow exponentially as shown in figure 12.7. Nei points out that, if population size is not to decrease, the intensity of selection must be limited by the reproductive limit of the favored type. Stated differently, if the less-favored type can successfully occupy the excess spaces not yet required by the expanding number of the more-favored type, the calculated advantage of the favored type is limited by its rate of increase.

The insistence by many on maintaining a constant population size during the replacement of one allele by another is not a view based on biological necessity. Figure 12.10 depicts the fate of wheat rust populations in a diagrammatic way, illustrating the

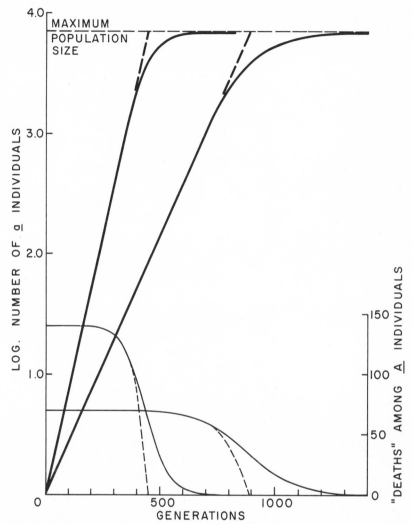

Figure 12.8 Curves comparing the exponential growth of populations of haploid organisms whose growth rates are 1.01 and 1.02 per generation, and the increase in the frequency of a mutant *a* in populations of haploid organisms in which the earlier, common type (*A*) has new and reduced fitnesses of 0.99 and 0.98. Notice that exponential growth (heavy dashed lines) is virtually identical to the increase in frequency of the rare advantageous mutant throughout the early generations. Notice, too, that the curves at the right require about twice as many generations to reach a given population size as do those on the left. The lower, light dashed lines indicate the "deaths" under exponential growth.

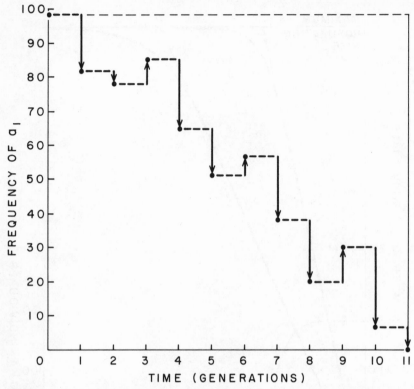

Figure 12.9 A diagram illustrating the irregular replacement of one allele (a_1) by another in a haplid population. Regardless of such irregularities, provided that the size of the populations (N) remains constant, the number of a_1 individuals lost by selective deaths equals 1.00 (the sum of the vertical arrows) times N. This product equals the original number of a_1 individuals.

tremendous decline the rust population suffers each time (generally at five-year intervals) a new rust-resistant variety of wheat is introduced, and the correspondingly tremendous explosion that occurs as once-rare, now-virulent strains of rust successfully attack and grow on each new wheat variety. A constancy of population size is nowhere assured in natural populations.

The Neutralist Theory

In 1968, Kimura proposed a neutral-mutation, random-drift hypothesis to account for the unexpectedly large amount of

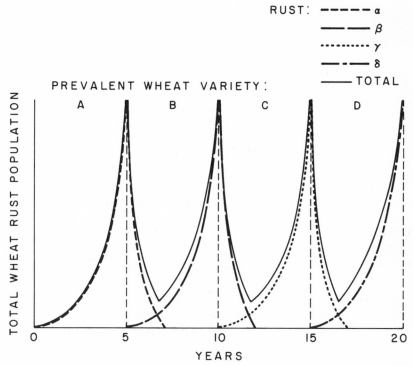

Figure 12.10 A schematic representation of the wheat rust population of the Great Plains states as wheat growers introduce new, rust-resistant wheat varieties every five years. The population of wheat rust does not remain constant as one pathogenic strain (say, β) displaces the prevalent, earlier one (α).

allozyme variation observed within populations of organisms and of differences in the amino acid composition of related proteins found in different species of organisms. This hypothesis is best-known today as the "neutralist" theory or, in Lewontin's (1974a) terminology, the neoclassical theory of evolution.

Kimura acknowledges two influences that led him to his neutralist views. One was described on page 295 where it was shown (Kimura and Crow 1964) that a ubiquitous advantage of heterozygotes results in a segregational load much too large for most organisms, including *Drosophila*, to sustain. The second influence was Haldane's (1957, 1960) calculated cost of natural selection: one allele could replace another only once in 300 generations.

The calculations that Kimura performs which he combines with these general influences are reasonable ones:

In a population of N individuals within which there are $2N$ alleles at any one gene locus, a unique allele has a frequency of $1/2N$. Given that this allele is selectively identical to all others (that is, that it is neutral in respect to selective forces), the probability that it will eventually be *the* allele in the population equals $1/2N$. The reasoning is as follows: First, the inbreeding coefficient, F, will eventually equal 1.00 (see page 163); at that time all alleles in a population will have been descended from a single earlier one. Second, the unique allele has the same chance as any other of becoming the remaining allele in the population at that time.

Another aspect of Kimura's calculations involves mutation rate. Given that the mutation rate of, say, a_1 to a_2 equals u, the number of such mutations in a population sized N equals $2Nu$. If, however, after a long time in which mutations and eliminations are randomized, the rate of fixation per locus equals $1/2N$, the total rate of fixation equals $2Nu \times 1/2N$ or u. The rate of fixation of mutant alleles, then, equals their rate of origin, and is independent of population size.

Kimura's next calculation involves rates of mutation. The average rate of amino acid substitution in hemoglobins, cytochromes, and an enzymatic dehydrogenase has been found to be approximately one amino acid substitution in about 28×10^6 years for every 100 amino acids.

Assuming that the haploid mammalian genome consists of 4×10^9 nucleotide pairs, that 300 nucleotide pairs correspond to 100 amino acids, and that 20 percent of all base-pair substitutions fail to alter the amino acid (are synonymous mutations), the average time for one base-pair replacement can be estimated (see Kimura 1968) as

$$\frac{28 \times 10^6 \times 3 \times 10^2 \times 5}{4 \times 10^9 \times 6} = 1.75 \text{ years.}$$

On the average, it appears that one nucleotide pair is substituted by another once every two years. This contrasts sharply with Haldane's estimate that one substitution could occur every 300 generations. The only alternative Kimura can accept is that the newly substituted alleles are selectively neutral.

A second calculation with a profound effect on Kimura's hypothesis concerns the number of alleles that can be maintained in a population by mutation alone. We have already seen (page 163) that

$$1 - F_t = \left(1 - \frac{1}{2N}\right)^t, \text{ or}$$

$$1 - F_{t+1} = \left(1 - \frac{1}{2N}\right)(1 - F_t)$$

$$1 - F_{t+1} = 1 - 1/2N - F_t \left(1 - \frac{1}{2N}\right)$$

$$F_{t+1} = 1/2N + F_t \left(1 - \frac{1}{2N}\right).$$

For F in this instance to represent not only *identity by descent* but also genetic *homozygosity*, the probability of *not* having mutated must be inserted into the calculations:

$$F_{t+1} = \left[1/2N + F_t \left(1 - \frac{1}{2N}\right)\right](1 - u)^2.$$

The solution to this equation for the equilibrium condition in which $F_{t+1} = F_t = F$, can proceed as follows:

$$F \approx \left[1/2N + F \left(1 - \frac{1}{2N}\right)\right](1 - 2u),$$

or

$$F \approx \frac{1}{1 + 4Nu}.$$

In 1964 when Kimura and Crow first published these calculations, reasonable estimates of population size seemed to lie between 10^3 and 10^4 and those of mutation rates between 10^{-5} and 10^{-6}. Thus, F (which in this case equals true homozygosis) was thought to lie between $1/(1 + 4 \cdot 10^3 \cdot 10^{-6})$ and $1/(1 + 4 \cdot 10^4 \cdot 10^{-5})$, or between $1/1.004$ (= 0.996) and $1/1.4$ (= 0.714). The effective number of alleles, then, would equal $1/F$ or 1.004 and 1.400. Populations, it seemed, would tend to be overwhelmingly homozygous for one (the "normal" or "wildtype") allele at each locus.

The allozyme variation discovered by Harris (1966) and Lewontin and Hubby (1966) altered these conclusions. The presence of many alleles at each of many loci was now an undisputed fact. Consequently, population size was increased in order that $1/F$ might equal 5 or more. For example, if N equals 10^6 and $u = 10^{-5}$, then $4Nu$ equals 40. Because $1/F$ equals $1 + 4Nu$, the number of alleles equals 41, a number considerably larger than that which must be explained. The shift in population size from what once seemed reasonable (10^3-10^4 individuals) to what is now needed to explain the empirical observations (10^6 or more) is accounted for by migration: "In our view, the constant distribution of the same set of alleles only indicates sufficient migration between local populations *to make the entire species or subspecies practically panmictic*" (italics added, Kimura and Ohta 1971: 153). Whereas in 1964, 10,000 was advanced as a generous estimate of population size for *Drosophila* (because of the small bottlenecks these populations encounter during winters and other inclement seasons), the whole species is now said to be a single panmictic unit because the observed genetic variation seemingly demands it.

Lewontin (1974a) presents an evenhandled account of the neoclassical (or neutralist) hypothesis, pointing out repeatedly the difficulties that confront any attempt to prove or disprove the neutralist theory experimentally. It is important to note that the genetic variation under discussion is *existing* variation. Virtually everyone agrees that genetic variation which lowers the fitness of its carriers is eliminated from populations. Formerly, the question concerning genetic variation was whether it was at all extensive; the calculations of Kimura and Crow (1964) suggested that it was not. The work of Lewontin and Hubby (1966) and Harris (1966) showed, on the contrary, that genetic variation is extremely extensive. The neutralist theory arose, then, as a means for admitting that natural variation exists while, simultaneously, operating with the constraints imposed by genetic load calculations and those bearing on the cost of natural selection.

In analyzing their position, Lewontin (1974a) identified two tasks confronting the neutralists: one, to show that the non-neutralist position is untenable and, second, to show that the neutral gene theory can adequately account for the observed genetic variation at the molecular level. I have adopted the complementary position: First, I suggest that the neutralist theory as set forth by Kimura was unnecessary (not necessarily *wrong*, but *unnecessary*). Second, I claim that in showing that neutralism can adequately account for the observed variation, the term "neutral" has been extended beyond the effect of alleles on the fitness of their carriers: small populations are said to cause alleles *to behave* as if they are neutral despite their having large selective effects (despite the semisterility of translocation heterozygotes, these chromosomal aberrations can become fixed in populations of ten individuals or fewer); selection which fluctuates erratically in direction is said to cause an allele *to behave* in the long run as if it were neutral; and linkage relationships which change with time can also cause the selective forces acting on an allele to fluctuate and, hence, *to mimic* neutrality.

The basis for the second set of objections is straight forward: a counterpart was described earlier (page 68) where it was claimed that the frequency of subvital chromosomes should depend upon the frequency of chromosomes exerting a deleterious effect on the viability of their carriers, *not* upon the number of flies an investigator is willing (or able) to count in attempting to detect them. Similarly, the possible neutrality of an allele should depend upon the effect of that allele on the fitness of its carriers—not upon population size, fluctuations in selection coefficients, or linkage relationships.

The matter of whether the neutralist theory is or is not necessary hinges on two points: genetic load theory and the cost of natural selection. The latter has been dealt with in this chapter; the enormous cost of natural selection computed by Haldane resulted from calculations in which kq individuals were killed and then unobtrusively revived each generation, but in which only the proportions killed were summed. The immediate resurrection of these "dead" individuals was ignored.

The next six chapters (chapters 13 to 18) will attempt to show that genetic load theory does not preclude the retention by natural selection of the amounts of allozyme variation observed in natural populations. For the moment it is enough to say (Wallace 1968a: 276): "If the genetic load can be relegated to individuals already destined to die without reproducing, it is of no consequence to a population." This point is, of course, recognized by Kimura who responds (Kimura and Ohta 1971: 155): " . . . it is highly doubtful that natural selection mimics artificial selection in such a way that the total number of heterozygous loci for each individual is counted and the individual is culled if the number of heterozygous loci is less than a certain critical number." It is for the next six chapters to muster the evidence supporting the first of these conflicting views.

FOR YOUR EXTRA ATTENTION

1. Recalling much of the material from earlier chapters as well as references to Haldane's (1957, 1960) "cost of natural selection," edit and improve the summarized overview of population genetics cited below (Hood 1976):

"Population geneticists of the 1920s and 1930s generally agreed on theoretical grounds that the rate of amino acid replacement was very slow—about one substitution every 300 years. These suggested proteins should exhibit very little polymorphism. More recently, Lewontin, Harris and others demonstrated extensive polymorphism in *Drosophila* and man—at perhaps 30 percent or more of the loci studied (H. Harris). Three possible explanations have been offered. First, the calculations of the population geneticists were based on incorrect assumptions. Second, most protein variants are selectively neutral . . . . Third, these polymorphisms have subtle functional differences and, accordingly, have been selected."

2. Richmond (1970) has presented the following argument concerning the possible neutrality of synonymous mutations: "For this synonymous mutation [CUA → CUC] to be truly neutral, the various tRNAs bearing leucine molecules must be present within the cell, in equal supply, and the synthesis of each tRNA species must require approximately equal amounts of energy output by the cell." Consider the merits of this argument, and seek data relevant to it (for example, see Nichols and Yanofsky 1979).

3. On page 296, the proportion of zygotes surviving a segregational load of 0.005 per locus at each of 1,000 loci was said to be $(0.995)^{1000}$, or 0.007. The standard for comparison in this calculation are individuals of the optimal genotype—that is, individuals heterozygous at each of the 1,000 loci. Using the following hints, calculate the proportion of individuals that possess the optimal genotype.

a. What are the equilibrium frequencies of two alleles if $s = t = 0.01$?

b. What is the frequency of heterozygotes at each locus?

c. What is the expected number of loci (among the 1,000 under discussion) at which an individual is expected to be heterozygous?

d. Recalling that the *variance* of a binomial distribution equals *npq* or, in the present case, 1,000 *pq*, calculate the *standard deviation* of the distribution of numbers of heterozygous loci per individual.

e. What is the range of numbers of heterozygous loci within which you might expect 95 percent of the population to fall $(\overline{X} \pm 2\sigma)$? Within which 99 percent of the population $(\overline{X} \pm 3\sigma)$ should fall? By how many standard deviations is 1,000 removed from $\overline{X}$?

f. Calculate the extreme divergence of relative fitnesses among 95 percent of the population as follows: $(0.99)^{4\sigma}$.

Calculations of the sort you have made here can be found in Sved, et al. (1967) and Wallace (1970b: 96–100). Because of the difficulties such calculations raise, Kimura and Ohta (1971: 59) suggest calculating the segregational load not on the basis of the completely heterozygous individual (whose fitness equals 1.00) but on the basis of "the most probable largest normal value"; the discussion on soft selection in chapter 16 will reveal that no single optimal genotype can serve as an adequate standard.

We now begin the second half of *Basic Population Genetics*. The emphasis in the treatment of problems changes considerably. If this were a symphonic work, rather than a textbook, the next six chapters, starting with the struggle to measure the fitness of populations and ending with a reexamination of the concept of population fitness, could easily qualify as its third movement. Continuing the musical metaphor, the last six chapters deal with a number of themes and counterthemes: with forces that tend to integrate the gene pools of Mendelian populations, with counterforces which are divisive and tend to destroy this integration, and finally with the outcome arising from these conflicting forces—the divergent evolution of Mendelian populations.

The Struggle to Measure Population Fitness

PREVIEW: This chapter views the efforts of many persons to measure population fitness, and the procedures which they have used for this purpose. If the procedures seem roundabout, recall that populations of the same species cannot be tested directly, one against the other, because they would merely interbreed and form a new, third population. It may appear that the measures of fitness used by different persons are independent and to some extent ad hoc; to dispel this appearance, frequent cross links between the different measures have been provided.

Introduction

Throughout the twelve chapters of part 1, stress was placed on the possession of hugh stores of genetic variation by natural populations of virtually all organisms. The occurrence, not only within species but also within each local population of a species, of alternative alleles at many gene loci is the rule—not the exception. Questions may remain concerning the "why" of this variation, but questions no longer exist as to the fact of genetic variation itself: genetic variation exists. Furthermore, at least some of this variation affects the viability, sexual activity, and other aspects of the reproductive success of its carriers. That is, among all existing genetic variation, some fraction is variation in respect to fitness.

The relative (or Darwinian) fitnesses of individual members of a population are measured as their adaptive values. By tradition the highest adaptive value is assigned the value 1.00; all other, lesser fitnesses are assigned fitnesses $1 - s$, $1 - s_2$, . . . , where s_1, s_2 , . . . , are selection coefficients. Horrendous as the measurement of these fitness values might be (consisting as they do of a complex joining of viability, sexual activity, fertility, fecundity, longevity, developmental speed, judgment as to where and when eggs are to be deposited, and many other essentially intangible aspects of the organism's biology), Darwinian fitnesses pose no conceptual difficulties. This point is made here because to some skeptical biologists, relative numbers are make-believe numbers. These persons are wrong. If AA and Aa females produce an average of 79 eggs each but aa females produce an average of only 52 eggs, the relative values, 1.000 and 0.658, are as useful in predicting the fate of the two alleles as are the actual numbers of eggs themselves.

The transition from a discussion of the relative fitnesses of individuals within a population (Darwinian fitnesses) to that of the average fitnesses of populations is fraught with difficulties. Ironically, Lewontin (1974a) entitles one chapter, "The Struggle to Measure Variation." That struggle was a struggle to develop suitable experimental techniques; the definition of genetic variation has never been in doubt: two alleles differ if the base sequences within their corresponding segments of DNA are not identical. The segment of DNA involved extends from the transcription control regions through the portions that specify amino acid sequences, including interstitial regions of the structural gene that are not normally transcribed or, if transcribed, are not translated.

The struggle to measure population fitness involves not only techniques but also the concept of population fitness itself. Because an average Darwinian fitness can be assigned to individuals of every genotype within a population, an overall average Darwinian fitness ($\overline{W}$) for the population can be computed; in previous chapters we have shown that $\overline{W}$ equals $1 - U$, $1 - 2U$, and $1 - [st/(s + t)]$ under various circumstances. Because these are the average Darwinian fitnesses of populations, many persons hold that the fitnesses of two or more populations can be compared directly by comparing their $\overline{W}$s. If exposure to mutagenic radiation were to double the mutation rate, then the average fitness of a population would drop from its present value of $1-U$ (or, if mutations are partially dominant as the rule, $1-2U$) to $1-2U$ (or, in the second case, $1-4U$). Because of its mutagenic effect, radiation would lower the fitness of the population.

On the other hand, the fitnesses that are averaged in computing $\overline{W}$ are relative fitnesses. What does it mean to average relative values? Let 90 percent of the females in one population produce 60 eggs each and 10 percent produce 35; the average fitness in respect to egg production equals $[(.90 \times 60) + (.10 \times 35)] \div 60$ or, approximately, 0.96 ($=\overline{W}_1$). In a second population, let one-half of all females produce 100 eggs and the other 50 eggs; the average fitness of this population ($\overline{W}_2$) in respect to egg production equals 0.75. Here we see that $\overline{W}_1$ (0.96) exceeds $\overline{W}_2$ (0.75) although the average number of eggs produced by females of population 1 (57.5) is considerably smaller than the average produced by females of population 2 (75). Averages of relative fitness values in the opinion of many persons are meaningless.

Are average population fitnesses toys that have been invented for the amusement of population geneticists? I think not, despite the claim that populations do not exist, that only individuals have meaning. For one thing, sexual reproduction creates a network of gene descent that ties together the reproducing members of a population. The individual in a sexually reproducing population of higher organisms is meaningless by itself; at least two individuals are required to maintain the population. Furthermore, certain attributes are population attributes, attributes that cannot be ascribed to individuals. Among these are the gene frequencies of the Hardy-Weinberg equation; only three gene frequencies—100 percent, 50 percent, and 0 percent—make sense in reference to diploid individuals. Means and variances are measures that pertain to populations; consequently, if individual values are known (such as Darwinian fitnesses), means and variances can be computed. And, having been computed, they apply to the population, not to any single individual member of that population. Consequently, sexual reproduction forces us to acknowledge that individual characteristics exist only as samples drawn one by one from a population which itself has a characteristic mean and variance.

To know the relative fitnesses of certain genotypes within

populations is to lose one's innocence in respect to the fitness of populations: one has become wiser, but not all knowing. If a study reveals, for example, that the average Darwinian fitness of *ST/AR* individuals in a population of *D. pseudoobscura* exceeds that of both *ST/ST* and *AR/AR* homozygotes, one can no longer feign innocence when asked to predict the outcome of a projected study of the competition between polymorphic and monomorphic populations of *D. pseudoobscura* and a population of a second Drosophila species. Past knowledge dictates that the odds favor the polymorphic population because only in that population are the *ST/AR* individuals generated. A certain humility must accompany such a prediction, however. Remember, the Philistines had no logical alternative when they dispatched Goliath to do battle with David.

The metaphor "to do battle" is overly simplistic in discussing the fitness of populations, but the reference to Goliath does emphasize one important point: extinct populations, by definition, have zero fitness. To have a fitness value, a population must exist, and existence requires persistence. The continued existence of a population, its persistence through time, is a vital component of population fitness (Thoday 1953). It is a component, however, that can be evaluated only after the fact. To a great extent, the various estimates of fitness to be described below were made in an effort to *predict* which of the various populations studied would most likely persist successfully in some unspecified competitive situation. Extinction has the effect of reducing all other estimates of fitness, no matter how excellent in themselves, to naught.

Irradiated Drosophila Populations: Cold Spring Harbor

The relation between the average fitness of a population and mutation rate—that is, the effect of variation of fitness—was first described by Haldane (1937). Several irradiated laboratory populations of *D. melanogaster* were set up at Cold Spring Harbor during 1949 and 1950 in an effort to confirm Haldane's calculations. The newly developed atomic weapons and the projected use of high-energy radiation and radioactive substances in newly developed industrial and medical technologies suggested that human beings were about to enter an extended period of unprecedented radiation exposure, a period during which radiation-induced mutations as well as "spontaneous" ones would lead to genetic disease and disability. Despite the twenty or more years that had elapsed since Muller (1927) had discovered the mutagenic effects of radiation, radiation biology was largely unknown outside genetic laboratores. As late as 1955, the Governor of Colorado could publicly recommend that two University of Colorado professors be arrested for suggesting that radiation from the Nevada atomic-bomb tests might be dangerous. The laboratory populations of *D. melanogaster* were expected to provide information about radiation-exposure and the average fitness of irradiated populations that might be useful in deciding between alternative courses of action in human affairs. In retrospect, *average* fitnesses provide poor guides in human behavior where equal worth is assigned to each individual (see Dobzhansky 1970:192 for an extended discussion of this problem).

The populations with which we shall be concerned are five in number; they are identified as populations 1, 3, 5, 6, and 7. Each was started with a mixture of flies carrying sixteen different lethal-free second chromosomes isolated initially from a culture of Oregon-R flies that had been maintained for a decade or more by the mass transfer of parents every two or three weeks. The times at which the populations were started, together with information concerning their irradiation histories, are listed in table 13.1. During the interval between the start of populations 1 and 3 and that of 5, 6, 7, the sixteen chromosomal strains were maintained separately; lethal-free chromosomes were reextracted from each strain to set up the last three populations.

Table 13.1 Short histories of five experimental populations of *D. melanogaster* studied at Cold Spring Harbor. (r, the *roentgen*, is a measure of radiation dose; gamma rays are similar to X rays but are emitted by many radioactive substances, including radium.) (After Wallace 1952.)

Population	Approx. Size (no. of adults)	Starting Date	First Sample Number	Irradiation
1	10,000	7/25/1949	1	Original ♂♂ 7,000-r X-rays Original ♀♀ 1,000-r X-rays
3	10,000	7/25/1949	1	None
5	1,000	4/1/1950	20	2,000 r/gen., chronic, gamma rays
6	10,000	4/15/1950	21	2,000 r/gen., chronic, gamma rays
7	10,000	4/15/1950	21	300 r/gen., chronic, gamma rays

The calculations of Haldane (1937) apply only to equilibrium populations. The experimental populations studied at Cold Spring Harbor were not necessarily at equilibrium. Nevertheless, the following reasoning applies even to nonequilibrium populations: The greater the frequency of deleterious mutations (whether they are equilibrium frequencies or not), the greater the frequencies of their homozygous carriers, and, consequently, the greater the anticipated reduction in average fitness of the population. The argument applies to incompletely recessive mutations as well as to those that are completely recessive; the depressing effect of a partially dominant (that is, incompletely recessive) deleterious allele on the average fitness of a population should be very nearly proportional to its frequency. At equilibrium, the effect of a partially dominant mutant allele at any one locus equals $2(1-u/hs)$ $(u/hs)(hs)$, or roughly $2u$.

How shall the fitness of a population be measured? It is impossible (or impossibly laborious) to measure fitness directly, so it is necessary to measure instead an important component of fitness and to assume that, as a rule, other components of fitness are correlated with the one actually measured. With *Drosophila* it is possible (see figure 3.6) to manipulate chromosomes so that one can estimate the preadult viabilities of flies carrying random combinations of different chromosomes from a given population. The estimates are based on the relative frequencies of genetically marked and wildtype flies obtained from crosses such as $CyL/+_1$ $\times CyL/+_2$. The survival of $+_1/+_2$ flies through egg, larval, and pupal stages relative to that of the $CyL/+$ flies can be calculated from the deviation of genotypic ratios from the $2CyL/+:1+/+$ expected on the basis of Mendelian segregation. A large number of cultures of this sort yields information on a large number of chromosomal combinations within which different ones are formed with frequencies proportional to those with which they arise in the population. Thus, the average survival obtained from a study of a large number of chromosomal combinations can be taken as a measure (in a relative, not an absolute, sense) of the average fitness of flies in the population itself.

The first indication that the frequency of deleterious genes in a population does *not* bear a direct relationship to the relative fitness of that population came from a comparison of populations 1 and 3. Population 1 had been subjected at the outset to a heavy exposure of X-radiation; about 95 percent of first generation zygotes in this population were killed by dominant lethals. What

Table 13.2 The observed frequencies of lethals (L) and semilethals (SL) in populations 1 and 3.

Sample	Population 1		Population 3	
	n	% L + SL	n	% L + SL
1	131	0.214	133	0.008
2	156	0.173	52	0
3	173	0.156	183	0.022
4-5	178	0.118	212	0.014
6-7	202	0.158	263	0.057
8-9	248	0.177	285	0.035
10-11	226	0.199	283	0.057
12-13	289	0.218	386	0.062
14-15	367	0.207	377	0.061
16-17	389	0.272	408	0.081
18-19	402	0.284	409	0.064
20-22	246	0.207	289	0.090
24-26	275	0.273	434	0.120
28	63	0.254	—	—
30	284	0.257	278	0.144
34	252	0.302	254	0.197
38	278	0.284	253	0.206
42	180	0.244	179	0.162
44	20	0.300	20	0.300
46	180	0.244	177	0.186
48	80	0.275	80	0.275
50	79	0.266	78	0.141
52	79	0.215	80	0.263

happened after this exposure can be seen in tables 13.2 and 13.3. Table 13.2 lists the frequencies of lethal and semilethal chromosomes observed in the irradiated (#1) and control (#3) populations over the first two years of their existence (a sample interval in these studies was two weeks; time throughout the course of the experiments was measured in these two-week intervals). Lethals gradually accumulated in population 3 even though there were none present initially (some of these same data are listed in table 9.4); lethals in relatively high frequencies were present in population 1 from the beginning because of the X-radiation.

Not all deleterious mutations are lethal or semilethal in their effects on homozygous individuals; many lower the viability of their carriers by smaller amounts. Amounts, in fact, small enough that it would take a tremendous amount of work to identify every deleterious mutation of this sort. Although they cannot be identified individually in the study of populations 1 and 3, the presence of these deleterious mutations as a group can be readily demonstrated. In table 13.3 are listed the average frequencies of wildtype flies in cultures containing quasi-normal homozygous wildtype

Table 13.3 The average frequencies in F_3 test cultures of quasi-normal wildtype homozygotes in populations 1 and 3.

Sample	Population 1	Population 3	#3 − #1 (Sign Only)
1	0.3066	0.3208	+
2	0.3194	0.3165	−
3	0.3111	0.3146	+
4–5	0.3039	0.3149	+
6–7	0.3079	0.3214	+
8–9	0.2963	0.3164	+
10–11	0.2985	0.3140	+
12–13	0.3081	0.3179	+
14–15	0.2999	0.3084	+
16–17	0.2990	0.3087	+
18–19	0.2926	0.3225	+
20–22	0.3033	0.3277	+
24–26	0.3063	0.3204	+
28	0.3104	−	No test
30	0.3083	0.3277	+
34	0.2972	0.3019	+
38	0.2960	0.3100	+
42	0.3107	0.3159	+
44	0.2906	0.3158	+
46	0.3135	0.3317	+
48	0.2987	0.3137	+
50	0.3078	0.3127	+
52	0.3138	0.3139	+

flies—wildtype homozygotes whose frequencies exceed those of lethals or semilethals. The lower the frequency of such wildtype flies, the higher the frequency of deleterious mutations among the tested chromosomes. Of the twenty-two comparisons listed in table 13.3, twenty-one show that the average frequency of quasi-normal homozygotes of population 3 (the control) is higher than that of similar flies of population 1. There are, consequently, higher proportions of subvital mutations, as well as lethals and semilethals, in the irradiated population than in its control. These findings agree with expectations: radiation induces mutations and these, in general, are deleterious.

The effect of deleterious mutations on Mendelian populations is manifested in individuals that arise through random mating, not individuals made homozygous by specially designed mating procedures. In table 13.4 are listed the average frequencies of wildtype flies in "heterozygous" cultures in which these flies are heterozygous for two different wildtype chromosomes picked at random from a population. Knowing that the chromosomes of population 1 carried the higher frequencies of lethals, semilethals, and other deleterious chromosomes, it was expected that random

Table 13.4 The average frequencies in F_3 test cultures of wildtype flies heterozygous for two different wildtype chromosomes obtained either from population 1 or 3. (*All* refers to all heterozygous combinations including those in which the wildtype flies were rare or missing; *normal* refers to heterozygous combinations excluding lethals and semilethals ones.)

| | *All* | | *Normal* | | *All* | *Normal* |
Sample	#1	#3	#1	#3	3-1	3-1
28	0.3533	—	0.3533	—		
32	0.3419	0.3312	0.3438	0.3324	–	–
36	0.3431	0.3356	0.3439	0.3356	–	–
40	0.3690	0.3482	0.3690	0.3482	–	–
42	0.3514	0.3484	0.3514	0.3484	–	–
44	0.3547	0.3407	0.3547	0.3426	–	–
46	0.3529	0.3652	0.3529	0.3652	+	+
48	0.3505	0.3487	0.3505	0.3487	–	–
50	0.3471	0.3546	0.3515	0.3546	+	+
52	0.3489	0.3492	0.3489	0.3492	+	+

combinations of these same chromosomes would exhibit the greater effects of these deleterious mutations. This is not the case, however. As the data listed in table 13.4 show, the frequency of wildtype flies in test cultures of population 1 was frequently higher than that in the control (population 3) cultures. Table 13.5 shows that at two widely separated times (1951 and 1954), the average frequency of wildtype flies in the test cultures of population 1 exceeded that of the control cultures by 3 percent to 4 percent. The differences are highly significant ($p < .001$). Table 13.6 presents a more detailed analysis of the experimental data that had been gathered through 1954; the bottom portion of the table shows that the average frequencies of wildtype flies in the heterozygous test cultures of populations 1 and 3 did not change relative to one another throughout the course of the study.

Despite the greater numbers of mutant alleles in population 1 (alleles that were in the population as a consequence of its exposure to X radiation), the chromosomes found within the popu-

Table 13.5 The average frequencies and the relative (adaptive) values of flies heterozygous for different wildtype chromosomes from five experimental populations of *D. melanogaster*. (After Wallace 1956.)

| | Percent Wildtype | | Adaptive Value | | Number of Tests | |
Population	1951	1954	1951	1954	1951	1954
1	35.02	34.80	1.04	1.03	994	3832
3	33.75	33.75	1.00	1.00	722	3762
5	31.07	31.97	0.92	0.95	390	3176
6	31.95	33.17	0.95	0.98	707	3318
7	32.83	33.26	0.97	0.99	730	3391

Table 13.6 A detailed analysis of the frequencies in F_3 test cultures of wildtype flies heterozygous for two different chromosomes from each of five experimental populations of *D. melanogaster*. n, number of samples studied; $\bar{y}$, average frequency of wildtype heterozygotes in all samples; b, the slope of the regression of y on generation; S_b, error of the slope; p, level of significance ($\beta = 0$); d, difference between control population (#3) and other populations; and S_d, error of the difference.

Population		n	Number of Combinations	$\bar{y}$	b	S_b	p
All	1	37	3,832	0.34799	−0.000154	0.000060	<0.02
	3	37	3,762	0.33748	−0.000298	0.000073	<0.001
	5	39	3,176	0.31972	0.000001	0.000078	~ 1
	6	37	3,318	0.33171	−0.000018	0.000058	>0.5
	7	37	3,391	0.33261	−0.000116	0.000047	<0.03
Normal	1	37	3,825	0.34842	−0.000159	0.000060	<0.02
	3	37	3,755	0.33798	−0.000283	0.000070	<0.001
	5	39	3,090	0.32900	0.000144	0.000060	<0.02
	6	37	3,287	0.33483	0.000012	0.000019	>0.50
	7	37	3,378	0.33392	−0.000077	0.000046	0.09

		d	S_d	p
All	1–3	0.000144	0.000095	0.10–0.15
	3–5	0.000299	0.000107	0.005
	3–6	0.000280	0.000094	0.003
	3–7	0.000182	0.000087	0.04
Normal	1–3	0.000124	0.000092	0.15–0.20
	3–5	0.000427	0.000092	0.0001
	3–6	0.000295	0.000092	0.001
	3–7	0.000206	0.000084	<0.02

lation are those which in random combinations confer high fitness on their carriers. The chromosomes remaining are not a random sample of those present in the newly irradiated populations, of course; this can be seen, for example, in the rapid early drop in the frequencies of lethals in population 1. The irradiated population either retained or assembled through recombination a collection of chromosomes that conferred high fitness in random combinations. The fitness of homozygotes such as those which can be created by the *CyL*-technique is of little importance in a large population of randomly mating individuals because such homozygotes arise rarely, if ever.

Populations exposed continuously to gamma radiation underwent changes in average fitness which resembled that observed in the case of population 1. The accumulation of lethals and semilethals in these populations has already been illustrated in figure 11.11; this figure shows the initial rate of *decrease* of nonlethal chromosomes. The eventual equilibrium frequencies of lethals and semilethals in the three populations were approximately (5) 80 percent, (6) 85 percent, and (7) 35 percent. Lethal chromosomes in these populations had considerably higher frequencies than those in populations 1 and 3. The average frequency of quasi-normal wildtype homozygotes offers a basis for estimating relative frequencies of chromosomes with only slight detrimental effects on the viability of their carriers. The average frequency of quasi-normal wildtype flies in the homozygous test cultures of populations 5 and 6 combined was 0.2986 (samples 20 through 71), while that of population 3 during the same time interval was 0.3161. Presumably, then, populations 5 and 6 had higher frequencies of subvital mutations. The mean frequency of quasi-normal homozygotes in population 7 during the interval from samples 21 through 71 was 0.3154; this value is not significantly lower than that of the control population.

Despite the obvious increase in the frequencies of lethals and detrimental genes in the continuously irradiated populations, the average viability of individuals carrying random combinations of these same chromosomes increased, rather than decreased, with time. This change can be seen in table 13.5 where comparable data from early and late in the history of these populations are compared. The same change is revealed even more clearly in table 13.6 where regressions have been computed for the viabilities on sample interval over some two hundred weeks, or nearly four years. Only relative changes have meaning because there are no absolute values to serve as reference points. Consequently, in the lower portion of table 13.6 every regression slope has been compared to that observed for the control population (#3).

In the case of the continuously irradiated populations—5, 6, and 7—the mean frequencies of flies heterozygous for random combinations of wildtype chromosomes were lower than those of the control population. Rather startling, though, are the slopes of the regressions: relative to the control, each is positive. The lowest estimates of viability for individuals carrying chromosomal combinations representative of these populations occurred in the early samples when the cumulative effects of radiation-induced mutations were small. As the cumulative effects of continuous radiation grew, the estimates of fitness also grew. This is contrary to what seemed to be the most likely expectation based on the theoretically expected behavior of deleterious mutant genes as described in algebraic terms in chapters 10 and 12.

Faced with the conflict between expectations and observations, an alternative scheme upon which to base expectations was erected. According to this scheme (Wallace and King 1952; Wal-

lace and Madden 1953; Wallace 1956) selection for a new mutation in a population proceeds on the basis of the fitness of its
heterozygous carriers. The decision concerning the retention of
the mutation in or its rejection from a population is made when
the mutation is rare and is found almost entirely in heterozygous
condition. Only later, when the retained mutation becomes more
frequent, does the fitness of homozygous individuals become important. Even then, the viability of the new homozygotes affects
only the equilibrium frequency for the new mutation, not the
question of its being in the population. As a result of shifting the
initial responsibility for retention of mutant alleles onto heterozygous carriers, the lack of correlation between the fitness of
homozygotes and of heterozygotes becomes understandable. A
more quantitative treatment of this type of selection has been developed by Bodmer and Parsons (1960) and Parsons and Bodmer
(1961). Underlying the argument is the implied assumption that
the superiority of heterozygotes over the two corresponding homozygotes is not a vanishingly rare phenomenon.

   In pursuing this account of the Cold Spring Harbor populations, we must recall once more that the entire endeavor was intended to reveal average fitnesses of the experimental populations.
The measure of fitness described here was the preadult (egg
through pupa) survival and the speed of development of flies
heterozygous for random combinations of chromosomes extracted
from the experimental populations. The fitness estimates arrived
at in this manner could be related to one another (with population
3 being assigned a fitness of 1.000) only because the wildtype flies
of each tested culture were pitted against a common competitor:
$CyL/+$ or CyL/Pm flies. An assumption is made that high preadult
viability (given that all else is equal, of course) is correlated with
the otherwise (unmeasureable) ability of a population to maintain
itself successfully through time.

   Because the ability of a population to maintain itself is not
measureable, estimates of fitness are limited to quantities which
to the investigator would appear to be important components of
this ability. Table 13.7, for example, lists the relative numbers of
offspring per fertile pair produced by numerous such pairs obtained by sampling the five Cold Spring Harbor populations. Flies
obtained directly from egg samples taken from the population
cages (no inbreeding) produce progeny in proportions that roughly
parallel the earlier estimates of fitness (also listed in table 13.7
for convenience); population 7, however, exceeds both popula-

Table 13.7   A comparison of two measures of fitness to which the Cold Spring Harbor populations of *D. melanogaster* were subjected. (After Wallace 1956, 1959b.)

Population	1954 F_3 Tests (see table 13.5)	Measure Based on the Number of Progeny per Fertile Pair	
		Generation 1 (no inbreeding)	Following ten generations of inbreeding
1	1.03	1.02	0.85
3	1.00	1.00	1.00
5	0.95	0.84	0.79
6	0.98	0.82	0.69
7	0.99	1.07	0.92

tions 1 and 3 in respect to the number of progeny its females could produce. Following ten generations of intensive inbreeding, the numbers of progeny produced by flies taken from the X-rayed populations decreased markedly. Pairs from populations 1 and 7, both of which had radiation histories, now produced fewer progeny than did those from 3, the control population. Again, the contrast between the early and late (post-inbreeding) results suggests that the characteristics of flies in the irradiated populations depend upon heterozygous combinations of alleles, combinations that are largely lost under a regime of intensive inbreeding. (Still other procedures for estimating the relative fitnesses of these populations are given in Wallace 1952.)

The algebric calculations presented in chapters 10 and 12 by their very nature assume an all-prevailing environment within which all individuals and all populations must develop and carry out their existance. Obviously, this assumption is false; the environment, even the moment-by-moment environment of an individual, is patchy. Admitting that the assumption is in error, however, the question remains: Is the error important? Table 13.8 and figure 13.1 present data suggesting how seriously wrong the assumption may be: the average numbers of flies hatching in test cultures throughout the duration of the Cold Spring Harbor experiments more than doubled in the course of some four years. Without speculating on either the genetic basis of this change or its consequence relative to the genetics of the populations involved, an obvious fact can nevertheless be admitted: life in a laboratory population cage leads to progressive, adaptive changes in the cage populations, one manifestation of which is the enormous increase in the number of flies supported by a limited amount of nutrient medium.

Table 13.8 The average total number of flies per F_3 test culture calculated for intervals of 10 samples (20 weeks) for 5 experimental populations. The test cultures included in these data are those yielding wildtype flies heterozygous for two different second chromosomes.

Sample Period	Population				
	1	3	5	6	7
30s	164.8	147.1	165.4	154.2	157.2
40s	237.1	221.1	276.3	273.5	287.2
50s	203.4	215.1	205.9	204.6	204.6
60s	323.7	314.7	301.6	306.9	295.4
70s	397.8	374.5	374.4	379.8	347.5
80s	340.8	397.9	296.6	362.8	336.8
90s	373.2	408.3	388.5	430.1	392.0
100s	420.6	399.5	353.7	376.4	357.9
110s	463.1	477.4	404.4	418.4	394.8
120s	404.6	455.2	374.7	383.2	364.8

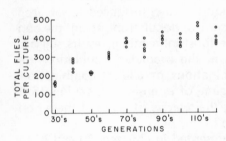

Figure 13.1 The average number of flies per F_3 test culture at various times during the analysis of five experimental populations of *D. melanogaster*. The doubling of the mean culture size over a period of 50 generations presumably reflects genetic changes occurring under the intensely crowded conditions of ex-perimental Drosophila populations.

Irradiated *Drosophila* Populations: Bonnier's Populations

Irradiated populations of *D. melanogaster* were studied during the 1950s by Bonnier and his colleagues at Stockholm (Bonnier et al. 1958). These populations were not maintained in precisely the same manner as those studied at Cold Spring Harbor. Furthermore, the information upon which a judgement as to the fitness, or "well-being," of each population was based differed from the measures used at Cold Spring Harbor.

The populations studied by Bonnier were maintained as shown in figure 13.2. In one series (P), adult flies were irradiated (1,500 r to 2,000 r) and were then shaken into a plastic box. Eggs were collected from these flies on petri dishes of culture medium. Following the egg collection, 5,000–6,000 young larvae were trans-

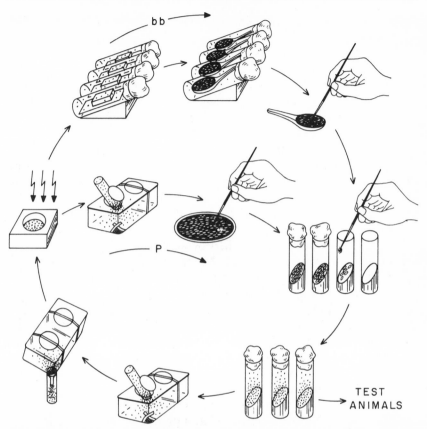

Figure 13.2 Technical procedures by which Bonnier maintained his irradiated populations of *D. melanogaster*. Starting with the irradiation of flies for the *P*-populations, the X-rayed flies were placed in a box and allowed to oviposit on media in petri dishes. Eggs were transferred to vials (25 per vial in series *P25;* 200 in *P200*). As the adults emerged, they were placed in a common cage (to encourage matings between flies from different vials) until they were collected for X-radiation once more. The *bb*-populations differed only in the collection of eggs. The irradiated parents were separated into vials containing five pairs each; only 25 eggs were collected from any five females. These were placed in culture vials: 25 per vial in series *bb-25*; 200 in *bb-200*. The purpose of the *bb* series was to eliminate the possible disproportionate contribution of one female to the population's pool of genes. (After Bonnier et al. 1958.)

ferred to small culture tubes—25 per tube in the *P25* series, 200 per tube in *P200*.

As the adults emerged in the culture tubes, they were shaken into a plastic cage for storage (and to allow more or less random mating between flies of different tubes). Three weeks after the irradiation of the original adult flies, the new generation was shaken from the storage cage, irradiated, and the entire process was repeated.

A variation on the above scheme was introduced in an effort to mimic more closely human reproductive patterns. Under the system described immediately above, all 5,000 to 6,000 transferred larvae might have been the progeny of very few females since no control was exercised over the total number of progeny contributed by any one female. The number of children per couple in man is rather small, however. To imitate man's lower fecundity, Bonnier subdivided the irradiated flies into lots of five pairs each; only twenty-five larvae were collected from any one lot of parents. Consequently, one female could contribute at the very most only 25 larvae to the 5,000 to 6,000 picked for the following generation. The populations maintained in this special manner were designated (*bb*); as in the earlier series, the larvae were placed 25 (*bb25*) or 200 (*bb200*) to a culture tube to develop.

One of the measures of the well-being or fitness of the irradiated populations was the actual number of larvae collected following the irradiation of the parental adults. Quite often fewer than the desired 6,000 were actually obtained. The charts presented in figure 13.3 show that in some populations—especially in *bb25*— the number of larvae collected was much below the goal. Interestingly, however, the populations improved with time. The effect of radiation was greatest in the early generations; in subsequent generations, despite the cumulative radiation dose and the accumulation of deleterious mutations, the populations tended to recover. This pattern resembles that revealed by the data in tables 13.5 and 13.6. Bonnier and his colleagues suggest, as we did above, that mutations beneficial to their heterozygous carriers arise and spread rapidly in these experimental populations.

The results obtained by the Swedish workers resemble those obtained at Cold Spring Harbor in still another respect: When chromosomes were tested to determine the frequencies of lethals, the average number of flies hatching per culture was related to the larval density of the tested population. This is shown in table 13.9,

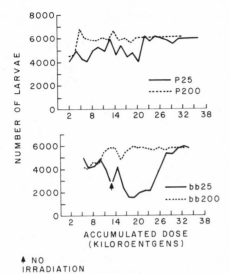

Figure 13.3 The total number of larvae collected per generation (an estimate of population fitness) in Bonnier's experimental populations. (After Bonnier et al. 1958.)

where the average numbers of *CyL*/+ flies per test culture are listed. Both *P200* and *bb200* yielded considerably more *CyL*/+ flies per culture than did *P25* or *bb25*. Furthermore, populations in which no limit was placed on the contribution of individual females to the following generation (series *P*) had higher numbers of *CyL*/+ flies in subsequent test cultures than did those of the *"bb"* series; the difference between *P25* and *bb25* if highly significant but that between *P200* and *bb200* is not. These observations parallel those described in discussing figure 13.1 and table 13.8.

Relative to the struggle to measure fitness, the interesting features of Bonnier's studies lie in (1) a measure of fitness that is dependent upon the availability and number of young larvae, and (2) the similarity in the sequences of events transpiring in both the Cold Spring Harbor and Stockholm populations despite the differences in experimental techniques.

Table 13.9 The average number of *Cyl*/+ flies in F_3 test cultures containing chromosomes from experimental populations of *D. melanogaster* maintained at different population densities. (After Bonnier et al., 1958.)

Number of Larvae in Each Maintenance Vial	Offspring per Parental Female	Average Number of CyL/+ in Test Cultures
25	(P) Unlimited	85.75
25	(bb) Limited	69.88
200	(P) Unlimited	110.35
200	(bb) Limited	103.93

Irradiated *Drosophila* Populations:
Sankaranarayanan's Populations

A number of irradiated populations of *D. melanogaster* were studied at Columbia University by Sankaranarayanan (1964, 1965, 1966). Some of these populations received 0 r (control), 2,000 r, 4,000 r, and 6,000 r each generation (only males were irradiated); the parental males of one population (VII-OR) were exposed to 7,000 r with no subsequent exposure of the flies of this population to irradiation. Unlike the Cold Spring Harbor populations, the flies in those studied by Sankaranarayanan were not allowed to oviposit freely on food cups; in this respect they resembled Bonnier's populations.

The measures of fitness used by Sankaranarayanan were two: the proportion of eggs that hatched, and the overall egg to adult viability under uncrowded conditions. Within 10 generations, the egg to adult viability in population VII-OR equaled that in the two corresponding control populations; the proportion of eggs hatching among those laid by VII-OR females was similar to, but somewhat less than, that of the control populations. The populations in which males were exposed to 2,000 r, 4,000 r, and 6,000 r each generation had drastically reduced egg hatch and egg to adult viability; subpopulations of each quickly approached the controls when further exposure to X-radiation ceased.

In these experiments still another measure of population fitness is employed—still another attempt to measure a quantity that virtually defies definition. The argument, of course, would go as follows: provided that all else is equal, the fitness of a population should be correlated with percentage egg hatch, or with egg-to-adult viability. All else, however, is not always equal. The ratio of egg-to-adult viability to percentage of egg hatchability in Sankaranarayanan's two control populations, for example, differ significantly (average difference of ratios = 0.0163, $\sigma_{\bar{d}}$ = 0.0068, d.f. = 14, p = .01 – .05); thus, post-hatching events in these two populations, both unirradiated controls, cannot be the same. Unfortunately, a similar uncertainty hovers over every effort to measure population fitness experimentally.

Population Size and Fitness

Living organisms characteristically transform portions of their environment into more organisms of their own sort. Green plants utilize nonliving portions of the environment in creating more green plants; *Drosophila*, for the most part, transform consumed yeast into flies; and large carnivores live by consuming herbivores. Carson (1957, 1958a) proposed that either biomass or population size might be used as a measure of population fitness; a number of persons have since followed his suggestions.

Studies by Francisco Ayala offer excellent examples of the uses of population size in assigning fitness values to various experimental populations. Table 13.10 lists the sizes of various laboratory populations of *D. serrata*. In this table, the important comparisons are those contrasting populations of hybrid origin with those that were started with flies of single geographic localities. At either temperature, the mean size of each hybrid population exceeds the average of the two corresponding single-locality populations by 20 percent to 40 percent. These results agree with expectations based on one's knowledge of hybrid vigor. Starting with the lethality of many chromosomes when tested by modified *ClB* procedures, a panorama can be created extending from lethal and semilethal homozygotes on one side through the relatively poor viability of virtually all homozygotes relative to corresponding random heterozygotes, and then to Ayala's demonstration that hybrid populations exceed in number those of non-hybrid origin.

Once the hybrid populations are accepted as those possessing the highest fitnesses, the logic which led to that conclusion can be applied to the strains of the three geographic origins. Populations containing flies descended from those captured at Cooktown are considerably larger than those containing flies of the other two geographic origins. Consequently, if population size is used as an

Table 13.10 The mean size of laboratory populations of *D. serrata* maintained at 25° and 19°. Populations were established using flies from single localities or interpopulation hybrids. (After Ayala 1968.)

Temperature	Population	Population Size
25°	Sydney (S)	1,782 ± 76
	Cooktown (C)	2,221 ± 80
	Popondetta (P)	1,828 ± 90
	S × C Hybrid	2,360 ± 74
	S × P Hybrid	2,541 ± 117
	C × P Hybrid	2,419 ± 76
19°	Sydney	1,803 ± 87
	Cooktown	2,017 ± 84
	Popondetta	1,580 ± 52
	S × C Hybrid	2,418 ± 171
	S × P Hybrid	2,448 ± 86
	C × P Hybrid	2,227 ± 172

Table 13.11 The average number of individuals produced per week in laboratory populations, irradiated and control, of *D. serrata* and *D. birchii*. (After Ayala 1966.)

	Individuals Produced Per Week
D. serrata (25°C)	
Control	595 ± 21
Experimental 1	941 ± 30
Experimental 2	1,133 ± 27
D. serrata (19°C)	
Control	236 ± 28
Experimental 1	499 ± 19
Experimental 2	519 ± 18
D. birchii (25°C)	
Control	533 ± 22
Experimental 1	933 ± 32
Experimental 2	833 ± 31
D. birchii (19°C)	
Control	473 ± 28
Experimental 1	495 ± 16
Experimental 2	557 ± 29

empirical estimate of population fitness, Cooktown flies must be assigned a higher fitness than those from Sydney or Popondetta.

The notion that population size and related measures are indicators of population fitness can now be extended to irradiated populations. Table 13.11 lists the average number of individuals produced per week in irradiated and control (nonirradiated) populations of the two sibling species, *D. serrata* and *D. birchii*. In each comparison, the irradiated populations exceed their control in this estimate of fitness; in only one instance (*D. birchii*, Experimental 1 at 19°) is the comparison not statistically significant. These results suggest that radiation-induced genetic variation can be utilized by populations in adapting to laboratory cage (or serial transfer culture bottles) conditions—that is, to improving their fitnesses. These results, consequently, confirm and extend those obtained by studying the irradiated population of *D. melanogaster*.

Dobzhansky and his colleagues (Wright and Dobzhansky 1946, Dobzhansky 1948b) have carried out numerous experiments on the Darwinian fitnesses of inversion homozygotes and heterozygotes in experimental populations of *D. pseudoobscura*. As a rule, individuals heterozygous for different gene arrangements have superior fitnesses; dimorphic populations generally establish stable polymorphisms.

Do polymorphic populations of *D. pseudoobscura* have higher population fitnesses than monomorphic ones? Table 13.12 presents data obtained by Dobzhansky and Pavlovsky (1961) on this point. Populations polymorphic for the two gene arrange-

Table 13.12 The number of adult flies and their total weight in six (two sets of three each) experimental populations of *D. pseudoobscura*. Both numbers of flies and biomass have been used as measures of population fitness. (After Dobzhansky and Pavlovsky 1961.)

	Sex	AR + CH	AR	CH
Numbers of Flies				
Set 1	Females	217.5	155.4	216.8
	Males	139.9	106.0	132.0
	Total	357	261	349
Set 2	Females	184.3	139.1	169.2
	Males	115.3	74.3	101.8
	Total	300	213	271
Total biomass (mgs)				
Set 1	Females	265.2	192.9	247.3
	Males	124.3	92.3	110.9
	Total	390	285	358
Set 2	Females	221.9	168.3	204.1
	Males	103.4	62.8	89.3
	Total	325	231	293

ments *AR* (Arrowhead) and *CH* (Chiricahua) tend to be larger than corresponding monomorphic ones in terms of both numbers of individuals and the total weight of these individuals (biomass). Following the arguments of Carson and Ayala which claim that large population size (numbers of biomass) is an indication of high population fitness, the polymorphic populations appear to have higher fitnesses than monomorphic ones. This conclusion agrees with that which is based on relative Darwinian fitnesses: a polymorphic population in which heterozygotes have superior fitness tends to maximize its average Darwinian fitness.

The Capacity to Increase as a Measure of Fitness

In 1964, Dobzhansky and his colleagues introduced what they referred to as "a quite different measure of the population fitness"— the capacity of the population to increase in numbers of individuals under a given set of environmental conditions. This capacity is measured by r_m in the equation

$$a \sum_{0}^{t} e^{-r_m(X+F)} l_x m_x = 1$$

Table 13.13 Estimates of the capacities of six experimental populations of *D. pseudoobscura* to increase at 25°C. The capacity to increase (r_m) calculated in these studies was used as a measure of population fitness. (After Dobzhansky et al. 1964.)

Genotype	Source	1961	1962
AR + CH	Piñon	0.217	0.222
AR + PP	Mather	0.209	0.219
AR	Mather	0.198	0.215
AR	Piñon	0.202	0.208
CH	Piñon	0.200	0.185
PP	Mather	0.154	0.187

where a = the proportion of eggs surviving to adults,

       F = developmental time (days) from egg to adult,

       l_x = the proportion of adults alive on day x,

       m_x = the number of eggs laid by a female on the xth day of life, and

       t = the observed life-span.

This measure of population fitness was applied to monomorphic and polymorphic populations of *D. pseudoobscura*, populations for which standard calculations yield highest average Darwinian fitnesses for polymorphic ones. The measures of the capacity to increase (r_m) are listed in table 13.13. Ostensibly, the calculated values of r_m for the polymorphic populations exceed those of the corresponding monomorphic ones in each instance. While it is true that the estimated fitness of the *AR-PP* population of Mather, California origin is only slightly higher than that of the monomorphic *AR* population, this slight superiority was observed in tests that were carried out in 1961 and again in 1962.

The data listed in table 13.13 suggest for five of the six cases that the capacity to increase in numbers (r_m) increased with time; only the calculated value for the *CH* (Piñon) monomorphic population decreased from 1961 to 1962. This change is consistent with the observation that laboratory populations of *Drosophila* tend to increase in size with time, a tendency which Ayala (1966) measured and referred to as the "evolution of fitness."

Interspecific Competition as a Measure of Population Fitness

If the ability to persist through time is an important component of population fitness, then the ability of the population to avoid ex-

Table 13.14 The logarithm of relative fitness (log W) and relative fitness (W) of three strains of *D. serrata* relative to each of three other *Drosophila* species. (Ayala 1969.)

Strain of D. serrata	Competitor					
	pseudoobscura		melanogaster		nebulosa	
	log W	W	log W	W	log W	W
Popondetta	0.301	1.35	−0.375	0.69	−0.270	0.76
Cooktown	0.643	1.90	−0.272	0.76	−0.074	0.93
Sydney	0.265	1.30	−0.349	0.71	−0.173	0.84

tinction at the hands of a competitor can serve as still another measure of population fitness. Table 13.14 illustrates the use of interspecific competition in estimating the relative fitnesses of three strains of *D. serrata*; these strains were forced to coexist with each of three other *Drosophila* species.

The fitness (W) in each case is computed as the antilog of the change in the (log) ratios of the competing species (see Ayala 1969). According to the estimates of fitness given in table 13.14, the strain of *D. serrata* from Cooktown (Australia) is clearly superior to the other two; the latter are nearly equal. The same conclusion was reached earlier (table 13.10) on the basis of population size in single species populations. At both 19° and 25°, the populations of Cooktown flies were clearly larger than those of the other two strains; no clearcut superiority existed in reference to the Sydney and Popondetta strains.

Table 13.15 presents data similar to that of table 13.14, but in reference once more to the fitnesses (W_1) of monomorphic and polymorphic populations of *D. pseudoobscura*. The results agree with those obtained in respect to population size and biomass (table 13.12) and capacity to increase in numbers (table 13.13): the fitness of the polymorphic populations, as estimated by interspecific competition, is higher than that of either monomorphic population.

Blaylock and Shugart (1972) used interspecific competition as a means for studying the relative fitnesses of irradiated popu-

Table 13.15 The relative fitness (and its logarithm) of *D. serrata* (s) with respect to monomorphic (AR or CH) and polymorphic ($AR + CH$) populations of *D. pseudoobscura* (p). For clarity, the fitness of *D. pseudoobscura* with respect to *D. serrata* has been calculated as $W_p = 1/W_s$. (After Ayala 1969.)

Population of D. pseudoobscura	ln W_s	W_s	W_p
AR	0.5530	1.739	0.575
CH	0.6685	1.952	0.512
AR + CH	0.4827	1.620	0.617

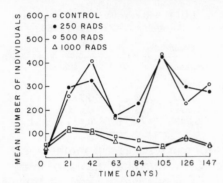

Figure 13.4 The mean numbers of *D. simulans* in competition with *D. melanogaster* in laboratory populations. *D. simulans* males were exposed to 0 (control), 250, 500, or 1000 rad of gamma radiation at the start of the experiment and at subsequent 3-week intervals. (After Blaylock and Shugart 1972.)

lations of *D. simulans*; *D. melanogaster* served as the interspecific competitor in this instance. As the results summarized in figure 13.4 show, the populations in which male *D. simulans* were exposed to 250 r and 500 r competed consistently better with *D. melanogaster* than did either the nonirradiated (control) flies or those that were exposed to 1,000 r. Although the means for measuring fitness differs from those of earlier studies, the results are consistent with those obtained for irradiated populations of *D. melanogaster* (tables 13.5, 13.6, and 13.7) and of *D. serrata* (table 13.11).

Summarizing Comments

In contrasting Darwinian fitness and population fitness (adaptedness), Dobzhansky (1968) claimed that the first can be known only in relative terms, whereas the latter can be assigned absolute values. I am not nearly that sanguine. On the other hand, I am not prepared to claim that adaptedness takes on only two values: that existing populations have a fitness of 1.00, while nonexistent ones have a fitness of zero. Biologists know that island species are under a constant threat from related species that may be accidentally introduced from continental areas; it is illegal, for example, to import *Drosophila* into Hawaii where about one thousand exotic species of this genus have arisen in isolation from the rest of the world.

The measures of population fitness discussed in this chapter are measures of those properties of organisms which are necessary for the maintenance of populations. Most of these properties are equally important, however, in determining both Darwinian and population fitness (adaptedness): these properties include, for example, viability, fertility, fecundity, and speed of development. Interspecific competition can be viewed as a complex aggregate of all these traits.

The sad fact remains, however, that attempts to measure population fitness experimentally cannot be all inclusive; consequently, the supposition "all else being equals" lurks unspoken in the background. At this time we shall not discuss "all else" except to say that all else can be exceedingly complex.

Population flushes (periods of marked increase in population numbers) are often followed immediately by crashes during which the oscillating population verges on extinction (Carson 1968). Now, extinction would reduce population fitness to zero; consequently, the assignment of high population fitness to large population size in this case makes little sense. Similarly, the assignment of high fitness to populations with high capacities to increase in numbers must likewise be questioned. One by one, the items that were measured in an attempt to assign fitness values to populations can be shown to be irrelevant under certain circumstances. This claim includes the ability to exterminate neighboring species.

With what are we left? One, with correlations between measurements made by different means. Second, with the consistent finding, based on several experimental procedures, that the adaptedness of polymorphic populations exceeds that of monomorphic ones. (This consistency emphasizes the futility of basing ecological predictions on calculated genetic—especially segregational—loads.) But, third, with the necessity to explain how demonstrably deleterious, radiation-induced mutations can contribute to high population fitness.

BOXED ESSAY

A number of studies involving irradiated mammalian populations have been carried out; each of these in its own way has asked, "How is the fitness of a population affected by newly induced mutations?"

Because of the complexity of radiation exposures, the mating procedures by which the populations were maintained, and the interpretation of the experimental observations, no attempt is made here to review and summarize the results of these experiments. The reader is referred instead to Blair and Kennerly (1959), to a review article by E. L. Green (1968), and to a symposium entitled "The Effects of Radiation on the Hereditary Fitness of Mammalians," which was published in *Genetics* in 1964.

Blair studied the effect of radiation on natural populations of the woodmouse (*Peromyscus leucopus*). The other investigators (among those participating in the symposium or reviewed by Green), their experimental animals, and the sites at which their studies took place are tabulated below:

Investigator	Mammal	Location
Gowen	mouse	Ames, Iowa
Carter	mouse	Harwell, England
Sugahara	mouse	Misima, Japan
Green	mouse	Bar Harbor, Maine
Lünning	mouse	Stockholm, Sweden
Touchberry	mouse	Berkeley, Calif.
Spalding	mouse	Los Alamos, N.Mex.
Brown	rat	College Station, Texas
Newcombe	rat	Chalk River, Canada
Chapman	rat	Madison, Wisc.

FOR YOUR EXTRA ATTENTION

1. Soulé (1979) has drawn attention to still another basis for estimating fitness: individual asymmetry. A number of persons have regarded developmental asymmetry (of normally symmetrical organisms, of course; fiddler crabs and other normally asymmetric organisms are excluded from consideration) as a sign of low Darwinian fitness (see Mather 1953, and Thoday 1955; see, too, table 11.4). What are your predictions concerning the level of allozyme variation in a population (a measure of heterozygosity) and (1) *inter*-individual morpological variation and (2) *intra*-individual variation (asymmetry)? Check with Soulé's results and earlier studies.

Chapter

14

Hybrid Vigor

PREVIEW: Studies of the Cold Spring Harbor irradiated populations suggested that new mutations are retained in, or rejected from, populations on the basis of the fitnesses of their heterozygous carriers; only later do their effects on homozygotes become important and determine their ultimate fate: balanced polymorphisms at intermediate gene frequencies or (presumably rarely) the replacement of the original ("wild-type") alleles. The experiments described in this chapter are intended to answer the question (which the populations could not answer): How frequent are mutations which appear to enhance the fitnesses of heterozygous carriers?

THE DESCRIPTIONS GIVEN in chapter 13 of events taking place in irradiated populations led to the suggestion that some muta-tions are retained in populations and increase in frequency because their heterozygous carriers possess a selective advantage over that of other individuals. The result was an ostensible increase in the average fitnesses of some irradiated populations. This improve-ment did not lead to the production of superflies; it was, rather, an improvement over the average fitnesses exhibited by the same populations during the early generations of radiation exposure.

Heterotic mutations—mutations that are favored in hetero-zygous individuals, at least within the populations in which they arise—are apparently frequent enough to confound the obvious predictions about population fitness. In predicting the fitness of population 1, for example, it was of no use to know that it contained more deleterious mutations than did population 3; the fitness of random heterozygotes of population 1 was higher than that of population 3. Similarly, it was not enough to know that the irradiation of the Swedish populations had decreased their "fitness" (measured in this case as the ability to produce larvae for the next generation) in early generations; upon continued exposure to radiation the populations improved in this respect.

The studies of irradiated populations were started with the belief that high frequencies of deleterious mutations must mean low average fitness. As the frequencies of lethals and other deleterious mutations went up in a population, the average fitness of that population was expected to go down. This belief was not borne out. Now we have a new problem. We have had to postulate that a certain proportion of mutations are heterotic in the sense that they improve the fitness of their heterozygous carriers but not that of individuals homozygous for them. Having made such a postulate, it is necessary to determine the frequency with which such mutations occur. If it is possible, we would like to estimate "frequency" in terms of both loci and alleles per locus.

The material presented in this chapter constitute a "historical" approach to the two questions: What is the extent of genetic variation in populations? At what proportion of all gene loci are the individual members of populations heterozygous? The experiments to be described have been rendered obsolete in reference to these questions by electrophoretic and other molecular techniques that (in theory, at least) allow us to detect genetic variation locus by locus. The newer techniques have by no means rendered the older experiments obsolete in reference to Darwinian fitnesses; the "neutralist–selectionist" controversy of the past decade or more attests to the inability of molecular techniques to resolve matters pertaining to the survival and reproduction of individuals. Consequently, the experiments described in this chapter provide a necessary background against which the newer observations must be viewed.

The Viability Effects of Random Mutations

Many experiments have been carried out in which radiation-induced mutations have been assessed for their effects on viability.

The original experiments of Muller (1927, 1930) can serve as examples. Muller developed special techniques (including the *ClB* technique) for detecting and measuring the frequency of newly induced mutations. Moreover, the mutations Muller chose to study were, for the most part, those that killed their homozygous and hemizygous carriers. Muller's experiments and the multitude of those that were carried out by others revealed that radiation-induced mutations are deleterious—especially in homozygous condition.

Still other experiments have been carried out in an effort to measure the extent to which deleterious mutations (lethals, primarily) affect their heterozygous carriers. Such studies were discussed earlier (page 233). In these early studies, the genotypes of the control individuals were of no concern to the investigator as long as they matched those of the experimental organisms. In the case of Drosophila experiments, the background genotypes contained chromosomes from a variety of wildtype and genetically marked strains.

To determine the effects of new mutations in heterozygous condition, and to obtain data useful in understanding the genetics of populations, we shall develop two models around otherwise homozygous individuals: new mutations will be studied within an otherwise homozygous genetic background.

Model 1: homozygote superiority.

The first model to be proposed is one that is consistent with the material presented in chapters 10 and 12. According to this model, there exists at each locus a normal allele as well as a number of relatively rare mutant alleles. The frequencies of these rare mutants can be given in terms of mutation rates and selection coefficients: complete recessives have frequencies (at equilibrium) equal to $\sqrt{u/s}$, while those whose deleterious effects are partially dominant have (lower) frequencies equal to u/hs.

Under this model, the ideal genotype—one achieved by few if any individuals of a population (see page 68)—would be that shown in figure 14.1A: every locus would be occupied by the unique, normal allele.

Within a population of cross-fertilizing individuals, mutations accumulate; consequently, individuals possessing the ideal genotype most likely do not exist. Instead, individuals carrying mutant genes at various loci make up the bulk of any population. These mutations are generally carried in heterozygous condition so the

A   IDEAL GENOTYPE

A	B	C	D	E	F	G	H	⋯
A	B	C	D	E	F	G	H	⋯

B   "ORDINARY" GENOTYPE

A	B	C	D	e	F	G	H	⋯
a	B	C	D	E	F	g	H	⋯

C   ARTIFICIAL HOMOZYGOTE

A	B	C	D	e	F	G	H	⋯
A	B	C	D	e	F	G	H	⋯

D   ARTIFICIAL HOMOZYGOTE HETEROZYGOUS FOR NEWLY INDUCED MUTATION

A	B	C	D	e	F	G	H	⋯
A	B	C	D	e	F	G	h*	⋯

Figure 14.1 Population model based on the notion of a single normal, wild-type allele at each locus. All mutant alleles (lower case letters) are deleterious when homozygous and, depending upon the degree to which they are dominant, when heterozygous as well. (After Wallace 1959a.)

phenotype of the ordinary individual (figure 14.1B) is very much like that expected for the "ideal"—somewhat worse, perhaps, because of the partial dominance of some mutant alleles and because of occasional homozygosis for others.

By means of standard CIB-type techniques, a chromosome can be taken from a population, and used in the synthesis of homozygous individuals (figure 14.1C). Such homozygotes have been shown repeatedly to have poorer egg-to-adult viabilities than comparable individuals heterozygous for two different chromosomes (see figure 3.7). By no means does the poor viability of these homozygotes imply that all loci are occupied by deleterious alleles. On the contrary, a few deleterious alleles in homozygous condition are the limiting factors in determining the viability of homozygous individuals; a recessive lethal allele at a single locus is sufficient, when homozygous, to kill an otherwise genetically superb individual.

At this point we should consider in some detail the frequency of mutant loci among all loci in the artificially synthesized homozygote. Lethals do not concern us here because we shall deal experimentally only with viable homozygotes. If we consider mutations for which s equals 0.1 and u equals 10^{-5}, then $\sqrt{u/s}$ equals 1 percent; 1 percent of all loci are expected to be occupied by alleles of this sort while alleles at the remaining 99 percent would be more nearly normal. If these genes ($s = 0.1$) were partially dominant (say, $h = 0.05$), then their expected frequency, u/hs, equals 20×10^{-4}; in this case, only 2 loci of 1,000 would be occupied by such mutations. These examples could be extended but the point is already clear: Homozygotes under this familiar model carry *very* few mutant alleles; the bulk of all loci are occupied by "normal" alleles.

The rarity of mutant alleles is important in the discussion that now deals with the individuals represented in figure 14.1D. The crosses by which this type of individual can be arrived at are described elsewhere (Wallace 1958, 1959a). For our present discussion it is enough to say that by a judicious use of *ClB*-type mating procedures, one can irradiate the wildtype chromosome in one line and then proceed to synthesize individuals identical to those of figure 14.1C except that one (not both) wildtype chromosomes has been irradiated. Presumably the radiation induces mutations at one locus or another in a fraction of the exposed chromosomes. Presumably, too, these mutations affect normal alleles since, under this model, the normal alleles occupy 99 percent or more of all loci. Finally, because the model is specific on this point, each mutant allele *must* be a deleterious form of its corresponding unmutated, normal allele.

Can we predict the effect of the newly induced mutations on these otherwise homozygous individuals? Recall that the control (figure 14.1C) and experimental (figure 14.1D) individuals are homozygous for precisely the same (supposedly rare) mutant alleles but, in addition, the experimental ones carry additional, newly induced ones. There is no alternative under this model: the newly induced mutations should have a depressing effect on viability.

Model 2: heterozygote superiority

As an alternative to model 1, we can devise a scheme such as that shown in figure 14.2. Remember that our task is to estimate the frequency with which alleles show an advantage in heterozygous condition. For purposes of this alternative model, we shall assume that large numbers of alleles exist at every locus such that the viability of homozygotes for the various alleles is somewhat lower than that of heterozygotes carrying two different alleles. We are now in a position to set up four types of individuals in figure 14.2 which are analogous to the four discussed for the preceding model.

According to model 2, the "ideal" genotype (figure 14.2A)

A IDEAL GENOTYPE	$\dfrac{A_1}{A_7}$	$\dfrac{B_9}{B_2}$	$\dfrac{C_2}{C_8}$	$\dfrac{D_7}{D_1}$	$\dfrac{E_4}{E_9}$	$\dfrac{F_5}{F_3}\ \cdots$

A IDEAL GENOTYPE

$$\frac{A_1\quad B_9\quad C_2\quad D_7\quad E_4\quad F_5\ \cdots}{A_7\quad B_2\quad C_8\quad D_1\quad E_9\quad F_3\ \cdots}$$

B "ORDINARY" GENOTYPE

$$\frac{A_1\quad B_6\quad C_5\quad D_2\quad E_5\quad F_8\ \cdots}{A_1\quad B_7\quad C_8\quad D_4\quad E_5\quad F_3\ \cdots}$$

C ARTIFICIAL HOMOZYGOTE

$$\frac{A_1\quad B_6\quad C_5\quad D_2\quad E_5\quad F_8\ \cdots}{A_1\quad B_6\quad C_5\quad D_2\quad E_5\quad F_8\ \cdots}$$

D ARTIFICIAL HOMOZYGOTE HETEROZYGOUS FOR NEWLY INDUCED MUTATION

$$\frac{A_1\quad B_6\quad C_5\quad D_2\quad E_5\quad F_8\ \cdots}{A_1\quad B_6\quad C_5\quad D_*\quad E_5\quad F_8\ \cdots}$$

Figure 14.2 Population model based on the superiority of certain heterozygous combinations of alleles at each locus. Notice that the differentiation of this model from that shown in the previous figure lies in the comparison of *C* with *D*; new mutations in heterozygous condition are expected to be deleterious in the earlier model (figure 14.1), but may not be so in the present one. (After Wallace 1959a.)

would be any one of a great number of genotypes in which two different alleles are found at every gene locus. This would be the "ideal" since the model states that individuals homozygous for any one allele are inferior in fitness to heterozygotes.

Within a cross-fertilizing population, the most frequently encountered genotype (figure 14.2B) would of necessity be homozygous at a number of gene loci. This would be true because (1) the number of alleles theoretically possible at each locus is presumably limited and (2) Mendelian inheritance provides no mechanism for maintaining large numbers of alleles in populations (recall the loss of lethals described in table 8.5).

Through the use of modified *CIB*-techniques, individuals can be created that are homozygous for a particular chromosome (see figure 14.2C). These homozygous individuals would be expected to have lower fitnesses (or egg-to-adult viabilities) than those of individuals possessing dissimilar chromosomes (figure 14.2B). Notice that this expectation does not differ from the corresponding one described under the first model.

The consequences of irradiating a single chromosome of an otherwise homozygous pair (figure 14.2B) *could* be quite different from those postulated under the earlier model, however. Under model 2, all loci are homozygous in the absence of an induced change; consequently, every induced mutation leads to heterozygosity at the locus involved. Under model 2, superior fitness is to be found in heterozygosity.

What are the expected effects of radiation-induced mutations on viability under the present model? Unfortunately, because of alternative possibilities the model cannot make a firm prediction. The overall, average effect of the new mutations would depend upon (1) the proportion of all possible radiation-induced mutational changes that enter into advantageous heterozygous combinations with the particular allele carried by the sampled chromosome, (2) the proportion of all loci at which alleles behave in the manner described by the model, and (3) the relative magnitudes of the increased and decreased viabilities of heterozygotes carrying the newly mutated allele.

To discriminate between the two models experimentally, we proceed as we do when we perform a Chi-square or any other statistical test: We temporarily accept the model that leads to a definite prediction, we compare the data with the expected results, and we reject the model if the data do not fit. In the present case, model 1 is the only one that makes a prediction; it says that the induced mutations will reduce the viability of their carriers. We accept this model and then proceed to test it experimentally.

The experimental data

The viabilities of flies are determined experimentally by noting deviations from Mendelian proportions that arise under somewhat overcrowded culture conditions. One genotypic class is taken as a standard against which to compare the others.

Unfortunately, we are interested here in comparing two types of wildtype flies: those homozygous for a given chromosome and those identical in all respects but heterozygous for some newly induced mutations as well. These two types of flies cannot be allowed to develop in the same culture because they could not be separated for counting; the flies that are to be compared must be grown in separate cultures and compared through a common standard.

The experimental procedure used to test the effect in heterozygous condition of radiation-induced mutations on their otherwise homozygous carriers culminates in two series of cultures that yield flies as shown in figure 14.3. In both the X-ray and control series, *CyL/Pm*, *CyL/+*, *Pm/+*, and *+/+* flies develop. The *CyL/Pm* flies serve as the control (viability is said to equal 1.00) so that the

CONTROL $\dfrac{CyL}{Pm}$: $\dfrac{CyL}{+}$: $\dfrac{Pm}{+}$: $\dfrac{+}{+}$

VIABILITY   1   1−r   1−s   1−t

EXPERIMENTAL $\dfrac{CyL}{Pm}$: $\dfrac{CyL}{(+)}$: $\dfrac{Pm}{+}$: $\dfrac{+}{(+)}$

VIABILITY   1   1−r'   1−s'   1−t'

Figure 14.3 The four genotypes appearing in the final (F_3) cultures that test the effect of newly induced mutations on their (otherwise homozygous) heterozygous carriers. The irradiated wildtype chromosomes are enclosed in parentheses.

viabilities of the other classes can be measured in relative terms. The only differences between the two series are (1) the wildtype flies of the X-ray series carry one irradiated chromosome and (2) one other genotypic class (*CurlyLobe* or *Plum*, depending upon the nature of the cross) also carries an irradiated chromosome. The relative viabilities of the corresponding genotypic classes can be determined by comparing their viabilities which have both been determined relative to that of *CyL/Pm* (1.00).

The results of the first large series of tests of the sort described above are given in tables 14.1, 14.2, and 14.3. All in all, seven tests were made, each of which involved from 600 to 800 control and a similar number of experimental cultures. More than three and one-quarter million flies were counted in these tests.

Table 14.1 lists the relative viabilities of those genotypes (*Plum* in most tests, *CurlyLobe* in others) that carried nonirradiated wildtype chromosomes even in the experimental series of cultures; a glance at figure 14.3 shows that there are such genotypes. The data in table 14.1 reveal that the viabilities of these flies in the two series do not differ. The viability listed in the "control" column is higher than that in the "X ray" column in three of the seven experi-

Table 14.1 The viability effects of identical, nonirradiated wildtype chromosomes on genetically marked heterozygous carriers in the control and experimental test cultures illustrated in figure 14.3. The standard viability, 1.000, is that exhibited by *CyL/Pm* flies in these test cultures; see figure 14.3. (After Wallace 1959a.)

	Pm/+		*Pm/+*	
A	1.146	(766)	1.137	(764)
B	1.139	(676)	1.140	(672)
C	1.137	(636)	1.145	(637)
D	1.143	(596)	1.136	(598)
G	1.215	(839)	1.222	(837)
	CyL/+		*CyL/+*	
E	1.127	(639)	1.125	(637)
F	1.189	(499)	1.201	(496)
Average:				
Unweighted	1.157	(4,651)	1.158	(4,641)
Weighted	1.158	(4,651)	1.159	(4,641)

Table 14.2 The viability effects of newly induced mutations in heterozygous condition in flies (*D. melanogaster*) otherwise homozygous for their second chromosomes. The irradiated wildtype chromosome is marked ('). (After Wallace 1959a.)

	$+_1/+_1$		$+_1/+'_1$	
A	1.008	(766)	1.033	(764)
B	1.000	(676)	1.007	(672)
C	0.989	(636)	1.015	(637)
D	0.979	(596)	0.989	(598)
E	0.983	(639)	0.990	(637)
F	0.992	(499)	1.002	(496)
		3,812		3,804

Average diff.:	1.5%
Error:	0.5%
Probability:	0.002

ments. The averages of all seven experiments, whether weighted or unweighted, differ only in the third decimal; the probability of seeing the observed differences in viability by chance alone, given that the two series do not differ, is greater than 0.50.

The comparisons of the viabilities of wildtype flies in the control and experimental (X-ray) series of cultures are listed in table 14.2. These results are quite different from those of the preceding table. In all six of the large experiments involving homozygous control flies and their counterparts in the X-ray series, the observed egg-to-adult viabilities of the wildtype flies of the X-ray series are higher than those of the corresponding control. Overall, the probability that the observed difference (0.015) is the result of chance alone is no greater than 0.002. These results cast serious doubt upon model 1, the model based upon the selective superiority of individuals homozygous for *the* normal

Table 14.3 The viability effects of newly induced mutations on their genetically marked heterozygous carriers. The irradiated wildtype chromosome is marked ('). (After Wallace 1959a.)

	CyL/+		CyL/+'	
A	1.094	(766)	1.115	(764)
B	1.093	(676)	1.108	(672)
C	1.105	(636)	1.110	(637)
D	1.100	(596)	1.108	(698)
G	1.185	(839)	1.182	(837)
	Pm/+		*Pm/+'*	
E	1.127	(639)	1.125	(637)
F	1.189	(499)	1.201	(496)

Average:				
Unweighted	1.129	(4,651)	1.136	(4,641)
Weighted	1.127	(4,651)	1.135	(4,641)

allele at each locus, because the model is incapable of accounting for an increased viability of flies carrying an irradiated chromosome. Model 2 was unable to predict that an increase would be seen because it admitted of other alternative possibilities; however, an increase in viability is compatible with this model.

Not all the data obtained in these tests agree with expectations based on model 2; the discrepant data are given in table 14.3. Why are these data unexpected? We said that the results (table 14.2) of the viability analyses in the case of wildtype flies did not agree with model 1 but were compatible with model 2. Fine; so we can assume for the moment that model 2 as described in figure 14.2 is correct. We must then admit, however, the individuals carrying two different chromosomes (figure 14.2B) are probably heterozygous at many loci. In that case, however, the $CyL/+$ and $Pm/+$ flies of our test cultures must also be heterozygous for genes at many loci. And, if this is so, not every radiation-induced mutation should lead to increased heterozygosity in these flies since many of these mutations will fall at loci already occupied by dissimilar alleles. However, the results shown in table 14.3 reveal that the *CurlyLobe* or *Plum* flies bearing the irradiated wildtype chromosome have higher viabilities than their controls in five of the seven experiments. The overall probability for the observed difference is only 0.05; this level of significance would not be sufficient to claim with assurance that the two series differ, but it is too low to be brushed aside without comment.

Additional experimental tests: favorable results

The experimental results described in the paragraphs immediately above (together with the studies on irradiated populations that preceded them) provided the first suggestion that the selective superiority of genetically heterozygous individuals may not be limited to a relatively few instances of striking polymorphisms (see Ford 1965) but, instead, may extend to a multitude—perhaps a majority—of gene loci. Thus, the fundamental concept—that of *the* normal, wildtype allele—that underlies the calculations of Haldane (1937), Crow (1948, 1952), and Muller (1950b) proves to be but one of two (or more) equally plausible ones. Because of the importance (and the unexpected direction) of the experimental results, a rather large number of similar studies have since been made. These will be reviewed at this point, starting with those whose results support those presented in tables 14.1, 14.2, and 14.3.

Table 14.4   The effect on viability of radiation-induced mutations in heterozygous condition on flies (*D. melanogaster*) homozygous (*AA* versus *AA'*) or heterozygous (*AB* versus *AB'*) for background genotypes. The irradiated chromosomes are marked ('). (After Mukai and Yoshikawa 1964.)

Experiment 1: 2nd and 3rd chromosomes, 150 r

Genotype	Number of Chromosomes	Number of Flies	Viability
AA	330	51,500	0.984
AA'	330	51,700	1.002
AB	430	71,300	1.283
AB'	430	80,100	1.252

Experiment 2: 2nd chromosome, 500 r

AA	292	224,000	1.015
AA'	291	230,000	1.041
AB	324	394,000	1.147
AB'	312	371,000	1.142

Mukai and Yoshikawa (1964) carried out tests in which the effects of newly induced mutations were studies in a homozygous background (as in the experiments described above) and in a heterozygous (wildtype chromosomes of two different origins) background. In one test, both the second and third chromosomes were manipulated simultaneously; in the other, the study was limited to the second chromosome alone. The results (table 14.4) agree with those expected on the basis of the results described earlier: newly induced mutations in heterozygous condition but carried in otherwise homozygous individuals, tend to increase viability; those carried by individuals already heterozygous at many loci tend to lower viability.

In a second, larger series of tests, Mukai, et al. (1966) once more tested the effect of newly induced (X-ray) mutations on the viability of their heterozygous carriers. The results shown in table 14.5 reveal that, in one instance, the newly induced heterozygosity increased the viability of its carriers significantly; this increase occurred only in the case of flies that were otherwise homozygous for a chromosome obtained from an inbred strain of flies (*D. melanogaster*). Corresponding increases were not noted when the control flies were heterozygous for two different chromosomes nor, interestingly, when they were *homozygous* for a "multiple-strain" chromosome obtained by a systematic tournament-style mating scheme that allowed the recombination of 64 wildtype chromosomes of separate origins.

More recently, Maruyama and Crow (1974) have published the results of an extensive experiment testing the effect of new mutations on the viability of their heterozygous carriers, but in a background that changed systematically in the degree of homo-

Table 14.5 The effect on viability of radiation (500 r)-induced mutations in heterozygous condition on flies (*D. melanogaster*) homozygous or variously heterozygous for background genotypes. A_1, A_2, and B are three isogenic strains maintained by brother-sister matings; M is a multiple strain synthesized from 64 wildtype strains. (After Mukai et al. 1966.)

Genotype	Number of Chromosomes	Total Number of Flies	Average Viability
A_1A_1	287	220,023	1.014
A_1A_1'	286	225,866	1.044*
MM	315	533,849	1.010
MM'	306	515,224	1.011
A_1A_2	308	282,957	1.063
$A_1'A_2$	306	269,280	1.068
A_1B	317	384,055	1.156
$A_1'B$	313	369,450	1.155

zygosis. The experimental procedure will be described below because, by eliminating the need for an independent series of control cultures, it represents a much more efficient test than those described so far.

The mating scheme used by Maruyama and Crow is shown in figure 14.4. The males that are subjected to irradiation are heterozygous for the second-chromosome recessive mutations *brown* (*bw*) and *cinnebar* (*cn*); these phenotypically wildtype males carry the mutants in the repulsion pattern.

The irradiated males are mated with two kinds of females: those homozygous for *bw* but heterozygous for *cn*, and those homozygous for *cn* but heterozygous for *bw*. From each type of mating, phenotypically wildtype males (+*bw*/*cn*+) are collected, and these males are mated individually with virgin females homozygous for both *cn* and *bw*. As shown in figure 14.4, the irradiated chromosomes in the two matings differ: in mating 1, the +*bw* chromosome has been exposed to 1,000 r; in mating 2, the *cn*+ chromosome has been irradiated.

Let us suppose that the proportions of *cinnebar* and *brown* flies in generation 2 are A and $1 - A$ in cultures obtained by mating 1, and B and $1 - B$ in those obtained by mating 2. These are empirical proportions obtained by classifying and counting progeny. Let us also suppose that the expected proportions of *cinnebar* and *brown* flies in the absence of irradiation are p and q where $p + q = 1$. The effect of radiation is said to change p in mating 1 to $p(1 + t)$, and q in mating 2 to $q(1 + t)$; in brief, the effect of radiation is unrelated to the effects of the *brown* and *cinnebar* mutations themselves. Expressing the empirical propor-

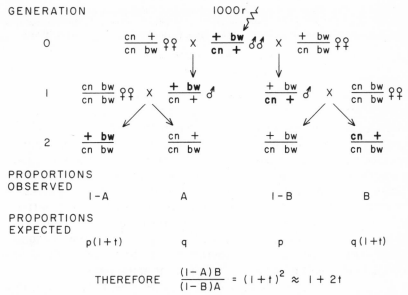

Figure 14.4 A mating scheme leading to progeny (generation 2) whose proportions reveal the effect of irradiated chromosomes on their heterozygous carriers. Explanation of the algebraic manipulations is given in the text. (After Maruyama and Crow 1974.) Reproduced with permission: from *Mutation Research* (1974), 27:241–48; © 1974 Elsevier/North Holland Biomedical Press.

tions in terms of their theoretical expectations, we find that

$$\frac{B(1 - A)}{A(1 - B)} = \frac{p(1 + t)\, q(1 + t)}{pq} = (1 + t)^2 \approx 1 + 2t.$$

Thus, the observed proportions of *cinnebar* and *brown* flies in the two types of cultures, when combined, provide information on the effect of radiation-induced mutations on their heterozygous carriers.

For reasons too complex to be described here, the early tests conducted by Maruyama and Crow were tests of random mutations on their heterozygous carriers when the mutations were induced in an otherwise heterozygous background. As the experiment proceeded, the background was made more homozygous (co-isogenic) by means of a coordinated system of inbreeding.

The results obtained by Maruyama and Crow are presented in table 14.6. In the original heterozygous background, the average effect of newly induced mutations was to *decrease* the egg-to-adult

Table 14.6 The effect of newly induced mutations (1000 r) on the viability of their heterozygous carriers when these carriers are largely heterozygous, intermediate, or homozygous in respect to background genotype. (After Maruyama and Crow 1974.)

Background	Number of Flies	Number of Cultures	Effect (t)
Heterozygous	540,913	4,824	−0.017
Intermediate	460,809	5,452	−0.000
Homozygous	456,956	4,440	0.016

viability of their carriers by 1.7 percent. In the homozygous background of the co-isogenic lines, the average effect of newly induced mutations was to *increase* the survival of their (heterozygous) carriers by 1.6 percent. During an intermediate period before co-isogenicity was complete, a test was performed whose results revealed that newly induced mutations, on the average, left the viabilities of their carriers unchanged. Speaking of their results and those of earlier workers, these authors say ". . . it is hard to ignore three independent sets of experiments, in three different laboratories, each significant and pointing in the same direction."

Organisms other than *D. melanogaster* have been used in testing the effect of newly (X-ray) induced mutations on their heterozygous carriers. Crenshaw (1965) studied the number of progeny produced per week by females from a highly inbred strain of *Tribolium confusum* (the four beetle) whose fathers had been exposed either to 500 r of X-rays or to none (controls). The results of these experiments are summarized in table 14.7; females carrying an irradiated genome produced substantially more progeny on the average than did the control females.

Still another study of a similar sort has been carried out on inbred Japanese quail (Shinjo et al. 1973). The fitness index in this study is a composite of egg production, egg hatch, and the survival of the newborn chicks. The effects of irradiation (150 r of gamma

Table 14.7 The mean weekly productivity of live offspring per culture by female progeny of irradiated (500 r) males (experimental) and of unirradiated males (control) in *Tribolium confusum*; 50 replicates were tested in each set. (After Crenshaw 1965.)

	Mean Productivity (± S.E.)	
1st week	2d week	3d week
	Experimental Cultures	
31.1 ± 1.7	55.7 ± 2.1	48.5 ± 2.2
	Control Cultures	
33.2 ± 1.5	45.7 ± 2.1	39.4 ± 1.9
	Overall:	

Irradiated: 135.3 progeny/culture
Control: 118.3 progeny/culture

Table 14.8 The effect of gamma radiation on the fitness of inbred Japanese quail. The fitness index includes egg production, proportion of eggs hatching, and survival to four weeks (After Shinjo et al. 1973.)

Test	Mating System	Dose	Inbreeding Coefficient Parents	Offspring	Fitness Index
	Random	0	0	0	0.307
a.	Full sib	0	0.424	0.549	0.062
	Full sib	150 r	0.424	0.549	0.139*
	Random	0	0	0	0.564
b.	Random	150 r	0	0	0.575
	Full sib	0	0.424	0.549	0.156
	Full sib	150 r	0.424	0.549	0.208
	Random	0	0	0	0.516
c.	Random	150 r	0	0	0.429*
	Full sib	0	0.549	0.643	0.150
	Full sib	150 r	0.549	0.643	0.179
	Full sib *I*	0	0.549	0.643	0.244
d.	Full sib *I*	150 r	0.549	0.643	0.190
	Full sib *G*	0	0.549	0.643	0.012
	Full sib *G*	150 r	0.549	0.643	0.209*

NOTE: a. First irradiation following 6 generations of inbreeding; b. First postirradiation generation; c. Second postirradiation generation; d. Details on two families (*I* and *G*) of section c.
   *Significant difference.

rays) are revealed by comparing the fitness indices of irradiated and control inbred and outcrossed material. Table 14.8a reveals that six generations of inbreeding had lowered the fitness index to roughly 20 percent of its control value (0.062 *versus* 0.307), but that the irradiated lines were substantially better (0.139) in this respect. Two generations following irradiation the fitness index of the irradiated outbred lines was significantly lower than that of the corresponding controls (0.429 *versus* 0.516); during neither the first nor second postirradiation generations did the inbred lines exhibit a similar decrease. The data presented in table 14.8d show that different inbred lines reacted differently to the radiation exposure (and, presumably, to the radiation-induced mutations); the increase of average fitness displayed by the highly inbred material represents, for the most part, an increase in the fitness index of especially poor inbred lines (0.012 versus 0.209). Inbred lines characterized by high fitness indices were not improved as the result of radiation.

Additional experimental tests: inconclusive results
A glance at the data presented in tables 14.1 through 14.6 reveals that attempts to distinguish between the two models described

earlier using *D. melanogaster* as the experimental material requires prodigious amounts of work. Three-and-one-quarter million flies hatching in more than 9,000 cultures were counted in order to obtain the data presented in the first three tables of this chapter. Mukai and his colleagues (tables 14.4 and 14.5) examined over four million flies. Maruyama and Crow (see table 14.6) counted one and one-half million flies but, because of the efficiency of the experimental design, their data are the equivalent of the earlier experiments combined.

Clearly, the experimental techniques are scarcely adequate to reveal the small differences that might distinguish between the two models. When working at the limit of his or her ability to resolve small differences, the experimenter has recourse to two alternative procedures: (1) to manipulate experimental techniques so that the sought-for differences will be larger or (2) to amass data sufficient to reveal even the smallest differences. Each procedure has its dangers. In attempting to enlarge the expected differences (for instance, by increasing the level of radiation in the present case), assumptions that were implicit in the original models might be violated. We shall return to this point later. In amassing data, negative results tend to be ignored with the result that positive ones are overemphasized. Homunculus, the small person once depicted in human sperm heads, presumably represented a "positive" observation; failure to see this tiny creature was a negative one.

Not all attempts to measure the average effect of newly induced mutations on their heterozygous carriers have yielded statistically significant results. A number of these experiments will be described here more to assemble a body of information than to criticize or to explain away. One can question, for example, whether the X-chromosome which is hemizygous in male *Drosophila* should follow the logic described in figure 14.1 and 14.2, and which deals with autosomal loci. One can also question experimental results for which experimental errors are enormous; a crude instrument is admittedly incapable of resolving minute differences.

Tables 14.9 and 14.10 present the results of two studies designed to measure the average effect of newly induced sex-linked mutations on male (hemizygous) and female (heterozygous) carriers. Three radiation doses were used in both studies: 500 r, 1,000 r, and 2,000 r. The results presented in table 14.9 involve females heterozygous for two different X-chromosomes: the wildtype chromosome that was irradiated and a balancer chromosome

Table 14.9 The average effect of irradiated X-chromosomes on the viability of hemizygous (+) males and heterozygous (+/*B*) females. Total flies counted: 709,713. (Wallace, exps. 53–83; unpublished.)

Dose	Type of Culture	X-Ray			Control		
		Number	+ ♂♂	+/B ♀♀	Number	+ ♂♂	+/B ♀♀
500 r	All	432	57.2%	59.1%	458	58.2%	58.3%
"	Nonlethal	424	57.8	59.2	457	58.2	58.3
1,000 r	All	450	56.3	59.0	443	57.9	58.8
"	Nonlethal	434	57.3	59.0	443	57.9	58.8
2,000 r	All	450	55.6	58.6	445	57.8	59.4
"	Nonlethal	429	56.7	58.7	442	58.0	59.3
Average	All	1,332	56.4	59.0	1,346	58.0	58.8
Average	Nonlethal	1,287	57.2	59.0	1,342	58.0	58.8

(Muller-5 or *Basc*) labeled with the dominant mutation, *Bar* eye. One effect of the X-radiation is apparent in the data on lethal mutations that can be extracted from table 14.9. Among the control chromosomes, there were 4 lethals in 1,346 tests (0.30 percent); this is a commonly observed mutation rate for this chromosome. Among the irradiated chromosomes the mutation rates were 1.85 percent (8/432) for 500 r, 3.56 percent (16/450) for 1,000 r, 4.67 percent (21/450) for 2,000 r. These observations are consistent with those commonly observed by radiation geneticists.

An examination of the frequencies of males in the control and irradiated cultures listed in table 14.9 reveals that even in nonlethal cultures, the frequency of wildtype males (among all males) is smaller in the irradiated than in the control series. X-radiation induces deleterious mutations whose effects fall short of lethality; the occurrence of such "viability-mutations" was also discussed in reference to table 9.6. Turning our attention to the females,

Table 14.10 The average effect of irradiated X-chromosomes on the viability of their hemizygous (+) males and heterozygous (+/+') females. Note that these wildtype females are homozygous for a particular X-chromosome of which only *one* has been exposed to radiation. Total flies counted: 719,437. (Wallace, exps. 84–111; unpublished.)

Dose	Type of Culture	X-Ray			Control		
		Number	+ ♂♂	+/+ ♀♀	Number	+ ♂♂	+/+ ♀♀
500 r	All	429	56.6%	50.2%	427	57.6%	50.0%
"	nonlethal	424	57.0	50.2	426	57.6	50.0
1,000 r	All	430	56.9	49.8	428	57.6	49.8
"	nonlethal	421	57.4	49.8	428	57.6	49.8
2,000 r	All	428	56.1	49.6	428	57.6	49.8
"	nonlethal	410	57.2	49.6	428	57.6	49.8
Average	All	1,287	56.5	49.8	1,283	57.6	49.9
Average	nonlethal	1,255	57.2	49.9	1,282	57.6	49.9

we see that the frequency of $+/B$ females (among all females) was ostensibly greater in the 500 r cultures than in their controls, about equal in the 1,000 r cultures and controls, and ostensibly smaller in the 2,000 r cultures than in their controls. In each instance, however, the observed difference is too small to be reliable. The differences between the frequencies of females with and without an irradiated wildtype X-chromosome, for example, are no greater than the differences between the frequencies of $+/B$ females in the control (0 r) cultures of the different series (58.3 percent, 58.8 percent, and 59.3 percent).

Table 14.10 presents additional data on the effects of irradiated X-chromosomes on their heterozygous (females) carriers. The mutation rate to lethals in the controls in this series of experiments was 0.08 percent (1/1283); among the X-rayed chromosomes the rates were 1.17 percent (5/429), 2.09 percent (9/430), and 4.21 percent (18/428) for 500 r, 1,000 r, and 2,000 r respectively. Not only do the lethal mutations agree with expectations based on past observations of radiation geneticists, but the frequencies of wild-type males among all males in the nonlethal irradiated cultures is consistently smaller than their corresponding controls. Radiation, as commonly believed, induces mutations which are debilitating to their hemizygous (and homozygous) carriers. The data in table 14.10, however, fail to reveal any consistent effect of irradiated chromosomes on their heterozygous (female) carriers. The gradual switchover from a 0.2 percent excess at 500 r to a 0.2 percent deficiency at 2,000 r is much too small to be statistically significant. The best that can be said of the data in tables 14.9 and 14.10 is that they do not reveal a marked debilitating effect of irradiated X-chromosomes on their heterozygous (female) carriers.

Experiments involving the X-chromosome, such as those described immediately above can, as a result of nondisjunction of the sex chromosomes, give misleading results. The results, for example, could be quite misleading if irradiation of males led to high frequencies of exceptional eggs among F_1 females. This problem is confronted in figure 14.5 where the crosses that gave rise to the cultures in table 14.9 (using *bar*-eyed-B/Y males) and table 14.10 (using wildtype–$+Y$ males) are shown in diagrammatic form.

The proportion of exceptional (XX and 0) eggs is said to be X; that of normal eggs, consequently, is $1-X$. Let the relative viability of wild-type males be $1-t$ and that of their *bar*-eyed broth-

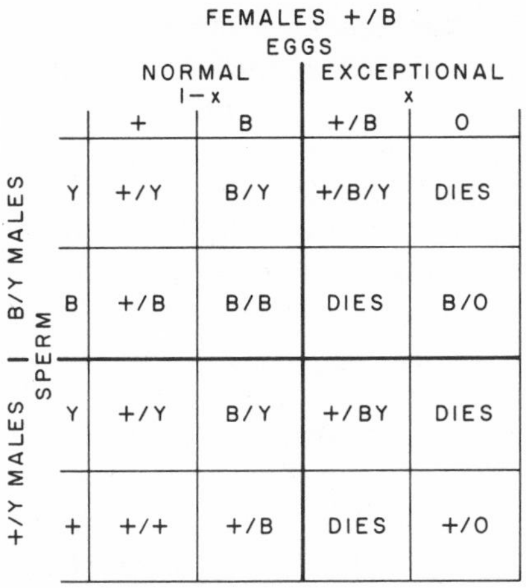

FEMALES +/B
EGGS

		NORMAL $1-x$		EXCEPTIONAL x	
		+	B	+/B	0
B/Y MALES	Y	+/Y	B/Y	+/B/Y	DIES
	B	+/B	B/B	DIES	B/0
+/Y MALES	Y	+/Y	B/Y	+/BY	DIES
	+	+/+	+/B	DIES	+/0

(SPERM)

Figure 14.5 A diagram illustrating the consequences of nondisjunction on the ratios of various types of offspring in the crosses of wildtype $(+/Y)$ and *Bar* (B/Y) males to $+/B$ females. The relative proportions of wildtype males among all males in these two crosses allow an estimation of X, the frequency of nondisjunction (see text for computational details).

ers be 1.00. Ignoring the product, tX, the expected proportion of wildtype males in the first (top) series of crosses equals $(1-t-X)/(2-t-X)$. The corresponding proportion in the second (bottom) series equals $(1-t)/(2-t-X)$. Consequently, the ratio of the proportion of wild-type males in the first series to that in the second equals $(1-t-X)/(1-t)$ or, very nearly, $1-X$. The observed ratio is somewhat greater than one (58.0/57.6); consequently, nondisjunction cannot be an important factor affecting the outcome of the experiments involving the X-chromosome.

The results of still another series of inconclusive tests are presented in table 14.11. These experiments dealt with the second chromosome of *D. melanogaster.* The (ambitious) experimental design included six types of homozygous genotypes (two chromosomes were obtained from each of three populations—one experimental [#3] and two "natural" [Louisiana and California]),

Table 14.11 The relative viabilities of three categories of wildtype flies carrying one radiation-exposed chromosome (except for 0 r, the control). Homozygous: these flies were homozygous for any one of six chromosomes obtained from three different populations (Experimental population #3; New Orleans, La.; and Riverside, Calif.). Intrapopulation: these flies were heterozygous for two chromosomes obtained originally from the same population. Interpopulation: these flies were heterozygous for any of the previously mentioned chromosomes and one obtained from Capetown, Africa. Total flies counted exceeded 2,350,000. (After Wallace 1963c.)

	Radiation Exposure			
Category	0 r	250 r	750 r	2,250 r
Homozygous	1.175	1.158	1.184	1.182
Intrapopulation	1.376	1.400	1.396	1.378
Interpopulation	1.488	1.472	1.460	1.479

three genotypes heterozgyous for chromosomes obtained from the same population, and six genotypes heterozygous for chromosomes obtained from different populations. Furthermore, including the controls, four radiation levels were analyzed: 0 r, 250 r, 750 r, and 2,250 r. In all, nearly two and one-half million flies were counted in order to obtain the data listed in table 14.11. Overall, the effects of radiation exposure in these experiments were miniscule. Not shown in the table are the *Curly-Lobe* heterozygotes of all cultures whose combined egg-to-adult survival was significantly *decreased* by the presence of an irradiated wildtype chromosome; this result disposes of the questionable increase in the viability of these flies noted in table 14.3. On the other hand, except for mere suggestions, irradiated chromosomes in these tests neither raised the viability of their carriers that were otherwise homozygous (although 1.184 and 1.182 ostensibly exceed 1.175 and 1.158) nor lowered the viability of carriers that were otherwise heterozygous (except for the *Curly-Lobe* heterozygotes mentioned above). Again, the data fail to show an average deleterious effect of irradiated chromosomes on their heterozygous carriers. The nonsignificant trends, however, are (1) an ostensible increase among homozygotes, (2) no effect on intrapopulation heterozygotes, and (3) a depressing effect on interpopulation heterozygotes; this suggested trend parallels the results of Maruyama and Crow (see table 14.6).

To make this account complete, mention must also be made of an experiment in which the second and the third chromosomes of *D. melanogaster* were rendered homozygous simultaneously (table 14.12). Following exposure of both second and third chromosomes to 500 r, wildtype flies heterozygous for these irradiated chromosomes (and for any newly induced mutation) had

Table 14.12   The average effect of X-radiation (500 r and 150 r) on the viability of wildtype flies (*D. melanogaster*) carrying one irradiated second and one irradiated third chromosome. The *ClB*-type balancer in this case consists of *CyL* (second chromosome) and *Ubx* (third chromosome) which have undergone a reciprocal translocation. Only a small proportion of the F_3 generation produced by mating (*CyL-Ubx*)/(+ +) flies survive to be counted under this test.

500 r:	Cultures Tested	Number of Flies Counted	Frequency of Wildtype
Control	1,442	249,323	28.06 percent
X-ray	1,442	249,272	28.21 percent
150 r:			
Control	756	127,131	25.37 percent
X-ray	756	127,247	25.20 percent

a frequency of 28.21 percent wildtype flies in the corresponding control cultures had a frequency of 28.06 percent. Following an exposure to 150 r, wildtype flies in the experimental series had a frequency of 25.20 percent, whereas the control wildtype flies had a frequency of 25.37 percent. No evidence is to be found in these tests for either an increase or decrease in the viability of (otherwise homozygous) carriers of irradiated second *and* third chromosomes.

An alternative to the accumulation of massive amounts of data is, as we saw above, an attempt to exaggerate the difference between the average viabilities of carriers of irradiated chromosomes and their corresponding controls. This approach was made explicit by Muller and Falk (1961): "In order further to increase the significance of results involving any given total number of flies, it was decided to raise the effective dose as much as practicable, . . . it was found that flies of some vigorous types, of both sexes, could still give a few offspring when crossed to nonirradiated flies even after receiving as much as 24,000 r . . . " In effect, if the mutations induced by 500 r increase the egg-to-adult survival of their heterozygous carriers by $1\frac{1}{2}$ percent (see table 14.2) in an otherwise homozygous background, then the mutations induced by 24,000 r should increase that survival by $1\frac{1}{2}$ percent × 48 or 72 percent.

The first experiments carried out by Falk (1961) are summarized in table 14.13. Sophisticated statistical analyses are not presented in this table; only the overall effects are listed. The estimated viabilities of two of three groups of experiments in the first set of experiments show that those of the irradiated material are lower than those of their controls; the other group exhibits a difference that is nearly as large but in the opposite direction. The

Table 14.13 The average effect of irradiated (24,000 r) X-chromosomes on the viability of their female carriers. (After Falk 1961.)

Control			Irradiated		
Number of Lines	Females Counted	Viability	Number of Lines	Females Counted	Viability
1st set:					
61	2,044	0.495	62	2,257	0.489
61	2,140	0.508	62	2,413	0.511
61	2,110	0.501	62	2,252	0.495
2d set:					
58	14,476	0.526	71	17,577	0.525
58	16,592	0.517	71	20,916	0.514
51	8,846	0.498	67	10,612	0.496

second set of experiments are consistent: all three groups yield smaller viabilities for the irradiated series––0.001, –0.003, and –0.002. The confidence interval calculated in the original publication of these results was sufficiently large to include the entire 2.5 percent–0.5 percent range that is consistent with the data listed in table 14.2.

In subsequent experiments, Falk and his colleagues studied a variety of radiation levels, used the X- as well as the second chromosome, and employed a number of investigative techniques. Table 14.14 lists the results of experiments in which X-chromosomes were exposed to 3,000 r. In Series 1, the viabilities of the flies carrying irradiated chromosomes are consistently lower than that of their controls; in Series 2, however, the results are markedly different. Unfortunately, when the results of large studies differ one from the other, the experimental error must (by definition) be large.

Table 14.15 summarizes the results of experiments by Falk and Ben-Zeev (1966) in which second chromosomes of *D. melanogaster* were exposed to 2,000 r. The analysis of the results in this case was analogous to that used later by Maruyama and Crow (1974): the product of relative proportions of two kinds of flies with irradiated chromosomes divided by the product of the same two kinds of flies with nonirradiated chromosomes yields a measure of the average effect of the irradiated chromosomes on the viability of their heterozygous carriers. The results presented in table 14.15 are inconclusive; the observed decrease in viability is only slightly larger than the standard error of this decrease; a two-fold difference would be required if the observations were to be statistically significant.

Table 14.14 The average effect of irradiated (3,000 r) X-chromosomes on the viability of their female carriers. The letters *a–f* refer to subdivisions of each of the series of experiments. (After Falk et al. 1965.)

	Control		Irradiated	
Experiment	*Number of Chromosomes*	*Viability*	*Number of Chromosomes*	*Viability*
Series 1:				
a + b	120	0.562	91	0.542
d	88	0.556	110	0.543
e + f	181	0.533	97	0.526
Series 2:				
a + b	159	0.495	133	0.502
c + d	185	0.496	234	0.505
e + f	506	0.514	250	0.497

Table 14.15 The average effect of irradiated (2,000 r) second chromosomes (*D. melanogaster*) on the viability of their heterozygous carriers. In these tests chromosomes marked with the recessive mutant *black* were irradiated in one set of cultures, those marked with *purple* in another. Viability is estimated (as in figure 14.4) by the ratio of cross products. (After Falk and Ben-Zeev 1966.)

Experiment and Type of Culture	"Black"		"Purple"		
	Number of Chromosomes	Number of Flies	Number of Chromosomes	Number of Flies	Viability[a]
A. Bottles	518	103,222	576	94,602	0.996
B. Vials	314	44,561	348	42,440	0.988
C. Vials	283	19,930	289	20,545	1.005
A'. Vials	130	91,009	124	87,278	1.007
B'. Vials	128	48,133	124	47,448	0.988

[a]In all, 58 estimates of the viability of flies carrying an irradiated chromosome are available. Their unweighted average equals 0.9936; their standard error equals 0.0043. Therefore, the observed decrease in viability (0.0064) is not significant. Furthermore, the flies carrying the irradiated chromosomes were not otherwise homozygous as the test of models 1 and 2 requires.

Table 14.16 summarizes the results of still more experiments performed by Falk (1967a). In this instance, it would appear that irradiated chromosomes consistently increase the viabilities of their carriers; furthermore, the increases in the case of both 6,000 r and 12,000 r appear to be significant. Here, however, it is necessary to take note of an experimental procedure routinely employed by Falk and his colleagues: "While there were often dead flies in the vials, most of them could be reliably classified and were

Table 14.16 The average effect of irradiated (3,000 r, 6,000 r, and 12,000 r) second chromosomes on the viability of their carriers. As in table 14.14, each of two genetically marked chromosomes (*brown* and *purple*) were exposed to radiation in one set of cultures but not in another; viability is estimated by a ratio of cross products as in figure 14.4. (After Falk 1967a.)

Generation	Dose	Culture	"Brown"		"Purple"		Viability
			Number of Chromosomes	Number of Flies	Number of Chromosomes	Number of Flies	
F_2	3,000 r	bottles	64	27,444	63	27,510	0.998
F_4	3,000 r	vials	47	5,898	55	7,468	1.025
F_2	6,000 r	bottles	61	25,220	62	23,498	1.033
F_4	6,000 r	vials	56	34,725	50	30,268	1.012
F_2	12,000 r	bottles	188	83,958	186	78,303	1.015
F_3	12,000 r	vials	185	26,502	188	23,522	1.020
F_4	12,000 r	vials	176	34,536	173	31,521	1.007

Pooled Estimates of Viability

3,000 r	6,000 r	12,000 r
1.002 ± 0.013	$1.020 \pm 0.006*$	$1.014 \pm 0.003*$

*Statistically significant ($p < 0.01$).

thus counted" (Falk et al. 1965). A possible consequence of this procedure is illustrated in figure 14.6. In the figure, a genetically marked class of flies serving as a standard for comparison is shown (above) with rapidly developing wildtype flies and (below) with slowly developing ones. The wedge-shaped segment in each diagram represents the probabilities that flies will die and dissappear in culture bottles (early-hatching flies run the greatest risk); the disappearance results from the consumption of dead adults by develop-

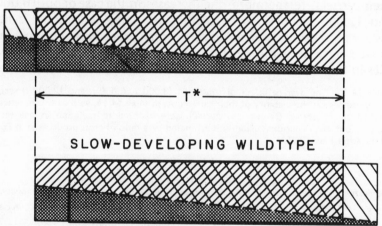

FAST-DEVELOPING WILDTYPE

T*

SLOW-DEVELOPING WILDTYPE

* T = TIME DURING WHICH
STANDARD FLIES EMERGE
FROM PUPAE

Figure 14.6 Diagrams showing how the practice of counting dead flies in determining the relative proportions of two genotypes in test cultures can cause slowly developing flies to appear in a seemingly greater proportion among identifiable flies than they were among the flies that actually hatched in the test cultures.

ing larvae or by decomposition. The diagrams presented in figure 14.6 clearly show that, if dead flies are classified and counted, slow development raises the proportion of wildtype flies in the final tabulation. In support of this contention is Falk's (1967b) later observation that viability and fitness are negatively correlated (a correlation also stressed by Crow and Kimura 1979). High "viability," when determined by counting classifiable dead flies, is a spurious viability that is correlated with slow development; slow development, of course, lowers fitness in a Mendelian population.

The most interesting of the many experiments carried out by Falk (1967b) utilized population cages in estimating the fitnesses of flies carrying irradiated chromosomes. Once more, as in the calculations presented in figure 14.4 and immediately above, the estimates of fitness are arrived at by the ratio of two products—one the frequencies of two classes of flies with irradiated chromosomes, the other the product of the same classes but without irradiated chromosome.

The results listed in table 14.17 clearly show that irradiated chromosomes lower the fitnesses of their carriers in freely breeding populations. From the cumulative W (a measure of the fitness of the carriers of irradiated chromosomes) one can compute that these chromosomes lower the overall fitnesses of their carriers by $2\frac{1}{2}$ percent; lethal chromosomes lowered the fitnesses of their

Table 14.17 The effect of irradiated (12,000 r) second chromosomes on the overall fitness of their heterozygous carriers as determined in long-term population studies. Both *P* and *A* cages contained *Cypr/bw* and *Cypr/pr* flies of both sexes; *pr* chromosomes in *P* had been irradiated previously, *bw* in *A*. The *bw* and *pr* chromosomes carried the same recessive male and female sterility genes (*fes* and *ms*); consequently, only *Cypr/bw* and *Cypr/pr* flies could breed in either type of population cage. At 17-day intervals ("cycles"), the proportion of *Cypr/bw* and *Cypr/pr* flies were calculated in each set of cages, and the overall fitness (*W*) of the carriers of the irradiated chromosomes was calculated as in figure 14.4. (After Falk 1967b.)

	Number of Cages		Number of Flies Counted		Cumulative
Cycle	P	A	P	A	W
1	105	106	9,989	10,078	0.937
2	102	106	22,230	23,305	0.980
3	101	101	22,598	23,305	0.900
4	100	99	22,738	22,569	0.912
5	99	97	20,678	20,857	0.891
6	97	97	19,805	20,107	0.858
7	97	95	18,325	18,327	0.849
8	96	92	19,877	19,114	0.827
9	95	87	20,248	18,705	0.792
10	95	83	20,311	17,447	0.743
11	94	83	20,453	17,756	0.749
12	92	81	17,182	17,205	0.733

carriers by some 7 percent. It is necessary to point out, however, that this reduction is of carriers that are otherwise heterozygous and, hence, does not bear on the models described at the onset of this chapter. Furthermore, one should recall that the radiation exposure in the case of the experiments described in table 14.17 was 12,000 r.

In concluding this section on inconclusive data, some not-so-ambivalent data (Hiraizumi 1965) are reported in table 14.18. Hiraizumi studied the fecundity of females, developmental rates, and the relative viability of larvae. In addition to his unirradiated control flies, Hiraizumi studied those which were heterozygous for second chromosomes that had been exposed to 500 r of X rays, and still others that were *homozygous* for irradiated chromosomes. The latter flies, those homozygous for irradiated chromosomes, were inferior in all three respects: fecundity, developmental rate, and viability. However, as the data in table 14.18 show, flies carrying a single irradiated chromosome developed more rapidly than did their nonirradiated counterparts.

Hiraizumi's results are interesting because speed of development is included in the usual estimate of egg-to-adult viability. A culture, for example, that yielded $CyL/Pm : CyL/+ : Pm/+ : +/+$ flies in the numbers $75 : 75 : 75 : 76$ instead of an expected $75 : 75 : 75 : 75$ would nearly account for the $1\frac{1}{2}$ percent increase in wildtype flies stressed in table 14.2. This point emphasizes the resolution being demanded of the experimental techniques: the increase of $1\frac{1}{2}$ percent in the observed "viability" of flies heterozgyous for an irradiated chromosome (see tables 14.2 and 14.6) corresponds roughly to the presence of *a single additional* fly at the time the cultures are counted. The faster developmental rate observed by Hiraizumi (table 14.17) is nearly enough to provide that additional fly (0.050 days equals 1.2 hours or, approximately, 0.6 percent of the total developmental time of *D. melanogaster*). We shall return to developmental rate in speculations to be presented in chapter 15.

Table 14.18 The effect of irradiated (500 r) chromosomes (+') on the developmental rate of their carriers; developmental rate is measured as the mean number of days from egg laying to eclosion. Except for genetic changes induced by the radiation exposure, + and +' chromosomes are identical. (After Hiraizumi 1965.)

Set	+/+–+/+'	+/+–+'/+'
X-rayed	0.050 ± 0.017*	−0.031 ± 0.020
Control	0.009 ± 0.025	0.019 ± 0.040

*Statistically significant ($p < 0.01$).

Reconciling conflicting data

Data presented in the preceding three sections have supported, contradicted, or have been ambiguous (even irrelevant) concerning the effect on individual survival of random mutations in heterozgyous condition but in an otherwise largely homozygous background. The last of these stipulations is important—"in an otherwise largely homozygous background"—because there is no real ambiguity otherwise: radiation-induced mutations tend to be detrimental.*

In an attempt to resolve the conflicting results, a model can be described which will in many respects anticipate matters to be discussed in the following chapter. The model (see figure 14.7) recognizes that many of the data favorable to mode 2 (page 353) were collected following low-level radiation exposures (150 r, 500 r, and 1,000 r), while many of the contradictory results were obtained following large exposures (2,000 r, 3,000 r, 6,000 r, 12,000 r, and even 24,000 r). The latter would include the 2,500 r, 5,000 r and 10,000 r exposures used by Simmons (1976) in experiments not reviewed above.

According to the model in figure 14.7, the gene consists of two regions: 1 and 2. The probability of mutating these portions by irradiation equals p_1 and p_2 or, for a particular dose, D, of radiation, $k_1 D$ and $k_2 D$. The average *enhancing* effect of a mutation in region 1 (that is, its effect on heterozgyous carriers) equals h provided that region 2 remains unchanged. The average effect of a mutation in region 2 on its heterozgyous carrier, whether region 1 is unchanged or not, is a *reduction* in viability of s. The overall effect of radiation exposure, E, is approximately

$$hk_1 D(1 - k_2 D) - sk_2 D \quad \text{or}$$

$$D(hk_1 - sk_2) - hk_1 k_2 D^2 .$$

At low exposures, provided that hk_1 exceeds sk_2, E is positive. At higher exposures, because of the term involving D^2 (the square of the radiation dose), E becomes negative as shown in the parabolic curve in figure 14.7. Recalling that estimates of the parabolic

---

*Lest any confusion on this issue remain, it should be pointed out that high levels of continuous radiation (1,000 r/day, for example) will exterminate experimental Drosophila populations. Furthermore, mutations induced by X or any other ionizing radiation are harmful to normal individuals however low the radiation level to which these individuals are exposed may be. Only in the special case where radiation induces genetic change where little or none existed before does it appear to enhance at least some components of fitness. The same change tested in another homozygote, however, may not have exhibited the same effect.

(a)

	REGION I	REGION 2
PROBABILITY OF HIT :	$p_1 = k_1 D$	$p_2 = k_2 D$

(b)

REGION I	REGION 2	PROBABILITY	AVG. EFFECT
NO HIT	NO HIT	$(1-p_1)(1-p_2)$	0
HIT	NO HIT	$p_1(1-p_2)$	$+h$
NO HIT	HIT	$(1-p_1)p_2$	$-s$
HIT	HIT	$p_1 p_2$	$-s$

$$\text{GRAND AVERAGE} = E = D(hk_1 - sk_2) - hk_1 k_2 D^2$$

(c)

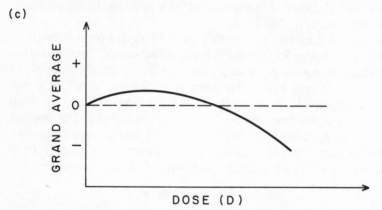

Figure 14.7 Diagrams illustrating a model under which chromosomes that have been exposed to low levels of radiation might lead to an enhancement of the viability of their heterozygous carriers, while those exposed to higher levels of radiation might lead to an adverse effect.

curve are accompanied by statistical errors, the model accounts rather well for the discrepancies in the various experimental data, especially discrepancies that are associated with different levels of radiation exposure.

The model outlined in figure 14.7 will recur in the speculations that are presented in the following chapter. It may be appropriate here to point out that the notion of a bipartite gene locus, although suggested as a possibility by Wallace (1963a), had no compelling recommendation until a plausible molecular model for gene control had been developed (Britten and Davidson 1969).

Using Inversions to Study Heterosis

Inversions provide a convenient means for analyzing the frequency and location of genes whose alleles are involved in the development of hybrid vigor. Dobzhansky and Rhoades (1938) suggested that inversions might be used for this purpose in maize; very little has been done with their suggestion (however, see Sprague 1941 and Chao 1959). Vann (1966) has made a systematic study of the hybrid vigor that accompanies inversion heterozygosity in *D. melanogaster*. Although his techniques do not lead directly to an estimation of the proportion of loci at which heterotic alleles occur (indeed, they cannot distinguish between the two alternative causes of hybrid vigor—dominance and overdominance), the experimental results are certainly pertinent to any serious discussion of the causes and magnitude of heterosis.

Vann's experiment was built around a number of disparate facts: (1) X-radiation can be used to induce chromosomal aberrations whose positions in the genome are determined by more or less randomly occurring breakage points. (2) Crossing-over is suppressed in inversion heterozgyotes within the limits of the inversion. (3) Inversions with average deleterious effects on fitness tend to be eliminated from freely breeding populations; inversions whose heterozygotes have superior fitness (heterosis in respect to fitness or "euheterosis," Dobzhansky 1952) will be retained in these populations. (4) The genetic similarity of a number of strains of flies can be governed somewhat by a careful choice of geographic origins combined with various inbreeding regimes.

The experimental material of Vann's study consisted initially of two stocks of *D. melanogaster* descended from different females captured at Riverside, California; two from Raleigh, North Carolina; and one from Bogotá, Colombia. Highly inbred lines were obtained from each of these stocks. Two inbred lines were obtained from each of the California stocks (*RC*-3 and *RC*-9); furthermore, after ten generations of brother-sister matings each of these was subdivided so that eight inbred lines (four pairs of sublines) were available for study (figure 14.8).

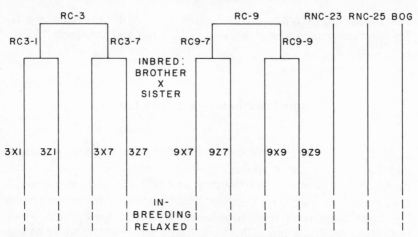

Figure 14.8 A diagram representing the derivation of the inbred lines of *D. melanogaster* that were used to study the relation between levels of hybridity and heterosis in respect to fitness. (After Vann 1966.)

Inversions were obtained by irradiating males (4,000 r to 5,000 r of X-rays) of the inbred California lines and mating them with females of the *same* inbred line. From the progeny of this cross, many *single-pair* matings were set up; the male and female of each pair carried single irradiated genomes. To find out whether either of these genomes contained a radiation-induced aberration, a dozen larval offspring were sacrificed so that their salivary chromosomes could be examined cytologically. If no aberration was found, the culture was discarded; if one was observed, the culture was saved. The adults that hatched were used in a series of mating most easily described by examples based on figure 14.8.

1. *Same culture.* This cross was made between males and females taken from the aberration-containing culture itself. These parental flies were very similar genetically for they were brothers and sisters produced by a single-pair mating of highly inbred material.

2. *Same subline.* If the aberration-containing culture was obtained by irradiating males of *RC3X1*, for example, this cross would involve flies (males or females) from that culture and others from the stock culture of *RC3X1*. These stock cultures, following fifty generations of inbreeding (single pairs of brothers X sisters), were maintained by the use of two nonvirgin females and an additional male. Consequently, the genetic difference between the parents of "same subline" crosses was somewhat greater than that of the "same culture" crosses.

3. *Other subline.* Using the example given above, this cross would consist of matings of flies from the aberration-containing culture (*RC3X*1) and those of the other subline, *RC3Z*1. The parental flies of this cross differed only in respect to alleles that were still segregating after ten generations of brother-sister matings that preceded the isolation of the two sublines.

4. *Intrastrain.* This cross would consist of matings between *RC3X*1 and *RC3X*7 or *RC3Z*7 (two crosses). The parental flies in this cross would differ only in respect to alleles that were segregating in the *RC3* wild-type strain of flies.

5. *Interstrain.* This type of mating would involve, for example, *RC3X*1 and *RC9X*7, *RC9Z*7, *RC9X*9, or *RC9Z*9 (four crosses). Genetic differences between the parents in these crosses would reflect the genetic variability of flies captured at Riverside, California.

6. *Unrelated.* This mating, to continue using *RC3X*1 as the aberration-containing line, would involve that line and flies descended from females captured at Raleigh, North Carolina or Bogotá, Colombia. The geographic origins of these strains are remote; the genetic dissimilarities in this case reflect the genetic variation that distinguishes widely separated populations of *D. melanogaster*.

Cultures that were set up according to the scheme outlined above were allowed to produce many offspring; flies from each aberration-containing culture were used in every one of the different crosses. The entire progeny from each culture (several hundred in many instances) were transferred to a fresh culture as parents for the following generation. This mass transfer of progeny flies from old to new cultures was continued for nine or ten generations. Within such overcrowded cultures, larval mortality is extremely high. To the extent that hybridity for genes contained within inverted gene segments enhanced survival, the structural heterozygotes would possess a selective advantage. Recombination outside the limits of each inversion would tend to destroy the hybridity of noninverted chromosomal segments.

Following the nine or ten generations during which each population was maintained by the mass transfer of all adult progeny each generation, thirty or more larvae of each culture were examined to determine the fate of the aberration it contained. The results are summarized in table 14.19. Translocations and pericentric inversions are listed in the table although they do not really concern us. Both of these types of aberrations adversely affect the re-

Table 14.19 The average frequencies (%) of individuals heterozygous for various types of radiation-induced chromosomal aberrations retained in experimental populations of *D. melanogaster*. The degree of genetic similarity (or dissimilarity) between the structurally altered and normal chromosomes depends upon the nature of the original cross (details in text). (After Vann 1966.)

Original Cross	Inversions		Translocations
	Paracentric	Pericentric	
Same culture	1.0	0	0
Same subline	5.3	0	0
Other subline	7.4	0	0
Intrastrain	14.1	0.2	0
Interstrain	22.9	3.7	1.7
Unrelated	28.8	5.6	1.6

productive ability of heterozygous individuals; both types are extremely rare in natural populations. Only a few remained in the experimental cultures. Paracentric inversions, on the contrary, were not necessarily rare after the series of transfers. Furthermore, a marked relationship exists between the final frequencies of these inversions and the genetic dissimilarity of the original parental flies: the more dissimilar the parents, the higher the frequency of the inversion heterozygotes in the experimental populations.

The frequencies of structural heterozygotes listed for paracentric inversions can be converted to "gene" frequencies; through a series of calculations (see Vann 1966) these can be converted to estimates of H, the amount by which the fitnesses of inversion heterozygotes were raised or lowered in the experimental populations. The estimated selection coefficients, H, are listed in table 14.20 (following the tradition of population genetics, negative values represent an enhancement of fitness). Apparently the radiation-induced breakage points harm their heterozygous carriers considerably (an observation made by Paget 1954, as well); consequently, the calculated values of H can be adjusted to reflect the true contribution of heterozygosity to the fitness of inversion

Table 14.20 The estimated average effect (H) of radiation-induced paracentric inversions on the relative overall fitnesses of their heterozygous carriers in populations of *D. melanogaster*. As in the previous table, the degree of genetic similarity (or dissimilarity) of the inverted and non-inverted chromosomes depended upon the nature of the original cross. (After Vann 1966.)

Original Cross	Calculated	Adjusted
Same culture	0.35	0
Same subline	0.15	−0.20
Other subline	0.08	−0.27
Intrastrain	−0.07	−0.42
Interstrain	−0.18	−0.53
Unrelated	−0.25	−0.60

heterozygotes. This contribution increases steadily from "same subline" crosses to "unrelated" ones.

The inference that the differences in the calculated values of *H* are related to differences in genetic dissimilarities of the parental flies of the various matings is strikingly borne out by the observed relationship between *H* and the lengths of the individual radiation-induced inversions.

Figure 14.9 shows the regressions of estimates of *H* for the individual inversions on the cube-root of their lengths (an empirical transformation that results in linear rather than curvilinear regressions). These regressions have been calculated separately for "intrastrain," "interstrain," and "unrelated" crosses. All three regressions intersect the Y-axis (zero inversion length) at virtually identical points, points that are also nearly identical to 0.35, the value of *H* for "same culture" populations. The significance of the convergence of the three regressions on the common point, 0.35, is clear: there can be no heterozygosity within inverted chromosomal segments of zero length. Without heterozygosity, such short inversions should correspond to the highly inbred (same culture) material which also lacked any heterozygosity.

Although his experiment was not designed to differentiate between the two models described at the outset of this chapter, an attempt to explain Vann's observations by means of model 1

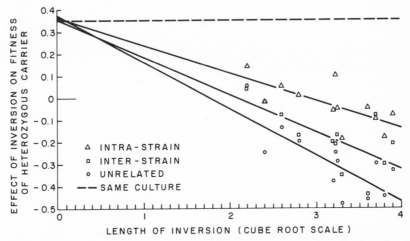

Figure 14.9 The conversion to one point of curves that relate the lengths of inversions, the genetic similarity or dissimilarity of inversion contents, and fitness heterosis. The convergence of these lines toward a single point lying on the Y-axis suggests that each represents a particular variation of a common theme. (Data from Vann 1966.)

seems to lead to absurd results. The lengths of the inversions in Vann's study were measured in units representing 1/120 of a chromosome arm. For inversions 10 units long, the rate of increase of H in the case of "unrelated" populations was approximately 2 percent per additional unit. Under the traditional mutation versus selection model (i.e., model 1), a block of genes containing about 400 loci is needed to explain a contribution of 2 percent to hybrid vigor. Thus, to retain model 1, we must assume that 1/120th of a chromosome arm contains 400 loci; under this assumption, *D. melanogaster* possesses nearly 100,000 loci per chromosome. Thus, Vann's observations are not convincingly explained by the traditional model.

Concluding Remarks

The experiments described in this chapter were undertaken for the most part before electrophoretic techniques had been applied to a study of the genetic variability of natural populations. The results of the experiments—at least of those whose results were unambiguously positive—suggested that many loci were occupied by dissimilar alleles, and that perhaps one-half or more of all gene loci are polymorphic within any one population (Wallace 1959a).

When these data were summarized previously (Wallace 1968a), they were presented somewhat apologetically: the arguments on which the interpretation of the experimental results were based were exceedingly complicated compared with the direct observation of gene variation made possible by gel electrophoresis. The newer techniques, although they have revealed an appropriate amount of gene variation, have not resolved the question: What forces maintain the observed variation? Consequently, the old (and new) data have been presented here with considerably less apology than before. The observations suggest not only that considerable genetic variation exists but also that selection is an important factor in its maintenance. This suggestion raises two new questions: How can random mutations have beneficial consequences? How can populations sustain seemingly intolerable genetic loads? These questions are taken up sequentially in the succeeding chapters.

FOR YOUR EXTRA ATTENTION

1. Schaal and Levin (1976) found in a study of the herb, *Liatris cylindracea*, that the average amount of heterozygosity at 27 allozyme loci increased with (1) the plant's age, (2) its size, (3) its reproductive potential, and (4) its ability to flower in its second year. They conclude that the case for heterozygote advantage is compelling, and the individual heterozygosity is being selected for.

Singh and Zouros (1978), in a study of oysters (*Crassosterea virginica*), found that the level of heterozygosity of five loci increased with size among individuals one year old; unlike Schaal and Levin, Singh and Zouros consider the possibility that decreasing heterozygosity may reflect increasing homozygosity through inbreeding. The superiority (in size) of the heterozygotes among oysters, in that case, could reflect merely an absence of inbreeding depression. Are these views really divergent?

Wallace (1975b) has discussed the relationship between the levels of heterozygosity at two portions of the same gene locus.

2. Schultz (see Bulger and Schultz 1979 and included references) has studied extensively unisexual fishes in the genus *Poeciliopsis*. These unisexual fish are always females, and always arise from the mating of *P. monacha* females with males of any one of several other related species. During meiosis only *monacha* chromosomes are included in the egg; the sperm of the male results once more in all-female, hybrid offspring.

Triploid unisexual females, also of hybrid origin, "use" sperm only to initiate cleavage of an already triploid egg: mothers-daughters-sisters in this case are all genetically identical.

What would be your guess as to the level of allozyme heterozygosity of these hybrid fish relative to their "parental" species? If you were told that the desert streams in which these fish are found undergo extreme temperature fluctuations on both daily and seasonal cycles, where (and *how*) would you seek an explanation for the origin and maintenance of such a bizarre method of reproduction? [This explanation, to be complete, would draw on information of the sort described in table 20.7.]

15

Gene Action in Higher Organisms: Speculation

PREVIEW: A number of studies carried out by different investigators in different laboratories and using different techniques have suggested that the *average* effect of random mutations in heterozygous condition but in otherwise homozygous genetic backgrounds need not be detrimental to the viability or fitness of their carriers. The present chapter attempts to account for these unexpected results using currently accepted notions concerning the control of gene action in higher organisms.

> There is no question that genes determine the nature of developmental reactions and the direction in which they will lead—but what can be said about *how* genes do these things?
>
> A. H. Sturtevant and G. W. Beadle
> *An Introduction to Genetics*, 1939

AS IN 1939, the question of how genes act is one of the major unsolved problems in biology. From the earliest days, *genes* and *characters* (genotype and phenotype) were recognized as separate entities, but the means by which the one produced the other was a mystery. The assertion that the genotype plus the environment in which it operates produces a given phenotype fails to identify the means by which genes exert their control over developmental processes. Neither the observations of Garrod (1909) nor the experiments of Beadle and Ephrussi (1936, 1937), and Beadle and Tatum (1941) which collectively led to the one gene–one enzyme hypothesis included mechanisms accounting for the proper operation of genes at the proper times and in the proper places.

The first utilization of formal genetic procedures in the study of the normal control of genes by one another was described by McClintock in a series of publications dating from the mid-1940s (see McClintock 1951). In her writings, the words "control" and "inactivation" occur, replacing the earlier concepts of "mutable" or "unstable" genes. The changed outlook is important: genes that may be *able* to act may not do so because their action is blocked by genes at other loci. In a series of summarized conclusions, McClintock (1965) made the following points explicitly: The action of a structural gene may come under the control of an element that is foreign to the gene locus; the control arises by the insertion of the foreign element at the locus; and the inserted element may respond to signals from still another genetic regulator.

A hint as to the manner in which a gene is controlled can be found in Lederberg's (1951) observation that wildtype *E. coli* cannot metabolize neolactose (altrose-galactoside) but that certain mutants can: neolactose is not an inducer of β-galactosidase (see figure 9.2); Lederberg's mutants were *constituitive*—that is, they did not require induction. Ten years later (Jacob and Monod 1961), the essential features of the control of lactose utilization by *E. coli* had been revealed.

Population geneticists, as a rule, paid little or no attention to the experimental results obtained by those microbial geneticists and others who were concerned with the control of gene action. In part, this indifference reflected the price one pays for specialization; no one can remain abreast of all scientific advances. In part, however, indifference reflected a faith in the power of the symbolism of population genetics: alleles that *behave* as if they are different are assigned different symbols. Thus, a gene, G, that can fall under the control of dissimilar controls—C_1, C_2, C_3—can be represented by the symbols A_1, A_2, and A_3. In discussing this and related matters, Wallace (1963a) pointed out that as evolution led to more and more complex organisms, the problems associated with the integrated control of gene action must have increased in complexity at a vastly higher rate—exponentially so. In that early

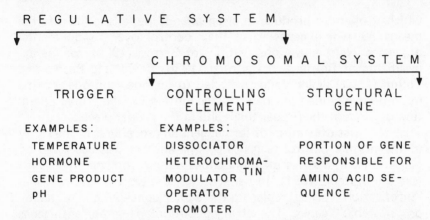

REGULATIVE SYSTEM

CHROMOSOMAL SYSTEM

TRIGGER	CONTROLLING ELEMENT	STRUCTURAL GENE
EXAMPLES:	EXAMPLES:	
TEMPERATURE	DISSOCIATOR	PORTION OF GENE
HORMONE	HETEROCHROMA-	RESPONSIBLE FOR
GENE PRODUCT	MODULATOR TIN	AMINO ACID SE-
pH	OPERATOR	QUENCE
	PROMOTER	

Figure 15.1 A diagram that emphasizes the dual nature of gene control systems. One portion is entirely chromosomal; it follows the formal rules of inheritance (e.g., Mendelian inheritance in diploid organisms). The second portion may extend from an external trigger, through genes whose products serve only to activate genes at other loci, to the structural gene by way of its adjacent transcription controls. (After Wallace 1963a.) Reproduced by permission of the Genetics Society of Canada.

discussion, the dual nature of regulated gene-controlled processes was described (figure 15.1). The Mendelian gene is shown as a small segment of DNA that follows the rules of chromosomal transmission; this unit consists of the structural gene and the nearby controlling elements that permit transcription to occur. The controlling element is the terminus of a regulative system that is only partly genetic, whose genetic components are not necessarily linked to the structural gene locus, and which is partly nongenetic because of initiating triggers that may lie in the individual's external environment. The complexity of gene control, and the difficulty in untangling its mechanisms, experimentally, resides in the essential continuity that in this instance joins the environment with Mendelizing genes.

Examples of Gene Control

Intuitively, persons conversant with elementary genetics are familiar with the notion that gene action is under some sort of control. Hemoglobin is manufactured in red blood cells and, normally, nowhere else. Nevertheless, the simplest hypothesis of gene replication suggests that the gene loci specifying the structure of the globin molecules are present in every cell of the body.

An example illustrating that alleles at two unlinked loci can be experimentally brought under the control of a third, also unlinked locus and that, when this is done, the now-regulated alelles act synchronously is illustrated in figure 15.2. The mutant allele at one locus, *waxy*, is unable to synthesize amylose in endosperm tissue of corn unless it "reverts" to the dominant form; this reversion can be controlled by an unlinked regulator element. Similarly, the mutant allele at still a third locus (*anthocyaninless*) is unable to lead to anthocyanin production in the aleurone layer of the corn kernel; this allele is under the control of the same regulator element. Figure 15.2 shows the pattern of reversion for the two loci in a single corn kernel: the darkly stained wedge of endosperm cells containing amylose (and, therefore, the activated *waxy* allele) leads to a patch of aleurone cells on the surface of the kernel within which the activated allele at the *anthocyaninless* locus now leads to pigment formation. The control of the *ability*

Figure 15.2 A demonstration showing that genes with different functions and located at different loci, but whose actions have fallen under the same control system, however, behave in a coordinated manner. The solid wedge of color in the interior (endosperm) of the corn kernel reveals a cluster of cells which are descended from a single ancestral one in which the control of an amylose-forming gene (*waxy*) has been "activated"; no amylose is present in the bulk of the endosperm. The cells of this kernel carry an allele (a^{m-3}) at the

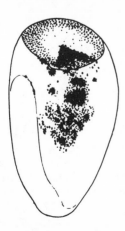

A locus (the allele, A, is needed for the production of a dark pigment) which normally does not produce color, but does so sporadically in small patches of aleurone cells in the presence of an "active" control. Notice that the area of aleurone that is spotted with color is superimposed almost perfectly over the outer face of the sector of starch-forming endosperm cells. The gene for pigment formation has responded to the same control as that which resulted in amylose synthesis. (After McClintock 1965.)

of the two alleles to act traces to a single event within the developing kernel; the action of each allele, however, takes place in the appropriate tissue.

A second example (figure 15.3) illustrates the induction of

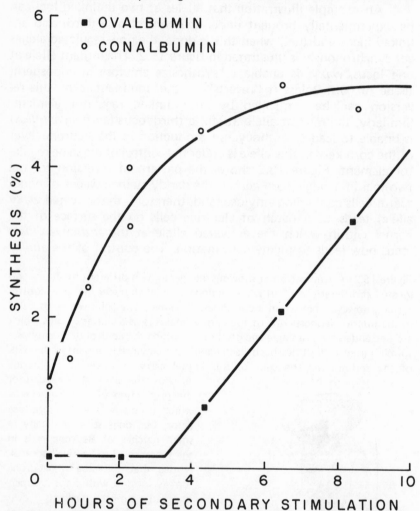

Figure 15.3 The induction of ovalbumin and conalbumin synthesis in the chick oviduct by estrogen. Note that conalbumin synthesis starts almost immediately following secondary stimulation but that ovalbumin synthesis is delayed approximately three hours. (After Palmiter et al. 1976.) Reproduced with permission: from R. D. Palmiter et al. *Cell* (1976), 8:557–72; © M.I.T. Press.

gene products (the proteins conalbumin and ovalbumin) in oviduct cells of White Leghorn or White Rock chicks following an injection of estrogen. The production of both proteins requires the estrogen trigger, but the appearance of ovalbumin is delayed almost three hours while that of conalbumin takes place immediately following the hormone injection.

Gel electrophoresis has permitted not only the detection of allelic differences at many gene loci of most organisms studied (see chapter 4) but also the demonstration that various gene loci are active only during certain stages of development. In discussing gene variation within natural populations, different times of action for alleles at various loci make it necessary for the investigator to specify precisely which *esterase* (or other enzyme) he is studying and what life stages he analyzed. In terms of individual development, however, the appearance and disappearance of enzymes reveal sequential patterns of gene control. Two examples are shown in figures 15.4 and 15.5. Figure 15.4 reveals the patterns

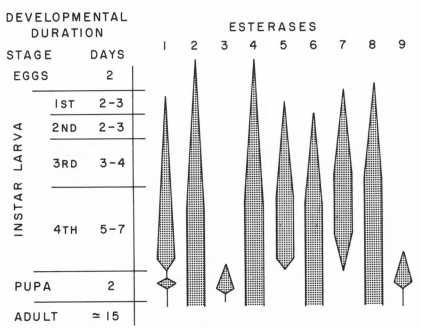

Figure 15.4 The patterns of esterase (*Est*) activities in the mosquito, *Anopheles albimanus*, during different stages of development. (After Vedbrat and Whitt 1975.) Reproduced with permission: from S. S. Vedbrat and G. S. Whitt, (1975), *Isozymes*, vol. 3; © 1975 Academic Press.

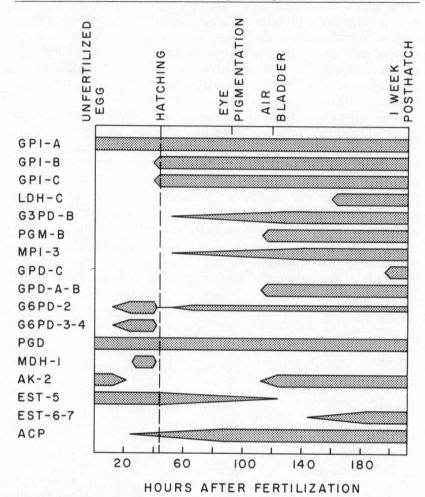

Figure 15.5 Patterns of appearance and disappearance of the activities of 17 enzymes during the first 200 hours of the development of the green sunfish, *Lepomis cyanellus*. Notice several instances in which the activities of several enzymes appear simultaneously, as well as instances in which the cessation of activity of one enzyme accompanies the appearance of that of still other enzymes. In one instance, an enzyme that disappears at hatching reappears considerably later during larvae development. (After Champion et al. 1975.) Reproduced with permission: from M. J. Champion et al. (1975), *Isozymes*, vol. 3; © 1975 Academic Press.

of esterases that are present at various times (egg to adult) during the development of the mosquito, *Anopheles albimanus*. Two of these enzymes are present from the egg through all life stages to adulthood; five others appear only during larval life. With the formation of the pupa, three esterases cease production abruptly; one of these is subsequently reactivated but two other seemingly new loci now become activated. Four loci remain active in the adult.

Figure 15.5 presents the patterns of activity of seventeen enzymes during the development of the green sunfish (*Lepomis cyanellus*) from the unfertilized egg through one week of post-hatching life. Examination of this figure reveals several stages at which genes cease functioning (that is, they no longer produce a functional enzyme) or start functioning. At times these events seem synchronized (at the time of hatching, for instance, or at 120 hours). Complex as this analysis appears to be, hundreds or even thousands of unrecorded events were occurring concomitantly; the picture presented in figure 15.5 represents a very small portion of the events taking place during the development of this fish.

The Britten-Davidson Model

The most inclusive model for the control of gene action in higher organisms has been constructed by Britten and Davidson (1969); subsequently (Davidson and Britten 1979), they have proposed certain modifications of their original model, although leaving much of it intact. For the purposes of this chapter, it is not necessary that Britten and Davidson's model be correct in all details (indeed, it probably cannot be in the sense that many genes will probably exhibit idiosyncrasies; in bacteria, at least, the controls of no two gene loci have proven to be identical); rather, it is important to understand the nature of the problem, and the sorts of genetic machinery that might solve its various facets.

Underlying the Britten-Davidson model is a frequently encountered need for the coordinated action of genes at several loci. Table 15.1 lists a number of functionally linked enzyme systems found in mammalian liver cells; the average number of

Table 15.1 A number of functionally linked enzyme systems in the mammalian liver. The number of enzymes listed for each system represents the minimum number of genes whose activities must be called for in a coordinated manner. (After Britten and Davidson 1969.)

System	Number of Enzymes
Glycogen synthesis	5
Galactose synthesis	6
Phosphogluconate oxidation	11
Glycolysis	12
Citric acid cycle	17
Lecithin synthesis	8
Fatty acid breakdown	5
Lanosterol synthesis	10
Phenylalanine oxidation	8
Methionine to cysteine	10
Methionine to aspartic acid	10
Urea formation	10
Coenzyme A synthesis	6
Heme synthesis	9
Pyrimidine synthesis (to uridine monophosphate)	6
Purine synthesis (to adenosine monophosphate)	14
16 systems	147
Average (approximate)	9

enzymes needed for each system in this list is about nine. The mammalian liver is the organ that is responsible for the breakdown and detoxification of noxious chemicals; to be successful, detoxification must proceed in a highly coordinated manner. As many workers using carbontetrachloride have learned to their dismay, this substance inhibits the second enzyme in the detoxification of ethyl alcohol: the first step in the detoxification of ethyl alcohol leads to the formation of acetaldehyde, a corrosive substance which, if it is not immediately broken down, destroys the cells in which it is formed.

Table 15.2 lists a number of alterations occurring in uterine cells as the result of stimulation by estrogen. Two effects of estrogen on chick oviducts were described earlier in figure 15.2; in table 15.2 we have a catalogue of some 15 effects, each of which is sufficiently complex to require the coordinated action of many genes, and whose nearly simultaneous occurrence must call for still more complex levels of genetic coordination.

Data shown in figure 15.6 illustrate still more evidence for the existence of genetic regulation. The figure shows the patterns of activity of 25 enzymes in cells of eight organs or tissues. Not

Table 15.2 Some effects of estrogen on uterine cells. See figure 15.2 for two effects of estrogen on the chick oviduct. (After Britten and Davidson 1969.)

- Increase in total cell protein
- Increase in transport of amino acids into cell
- Increase in protein synthesis activity per unit amount of polyribosomes
- Increased synthesis of new ribosomes
- Alteration of amounts of nuclear protein to nucleus
- Increased amount of polyribosomes per cell
- Increase in nucleolar mass and number
- Increase in activity of two RNA polymerases
- Increase in synthesis of contractile proteins
- Imbibition of water
- Increased synthesis of many phospholipids
- Increased de novo synthesis of purines (dependent on new enzyme synthesis)
- Alteration in membrane excitability
- Alteration in glucose metabolism
- Increase in synthesis of various mucopolysaccharides

one of the 25 enzymes shows high activity in all tissues studied. Even though a great deal of information is lacking, it is clear that the patterns of enzyme activity differ among cells of different sorts. Such diverse patterns suggest, in turn, that genetic control mechanisms exist by means of which diverse types of cells can call for the production of enzymes by alleles at precisely specified gene loci.

Models for the coordinated control of gene action at many loci are presented in figure 15.7 and 15.8 (Britten and Davidson, 1969). Both models shown in figure 15.7 extend from sensor genes which "read" the environment by way of hormones, temperature-dependent substances, and other signals, through integrator gene products, to the synthesis of structural gene products; any number of these products can be called for simultaneously. Figure 15.8 merely shows that the simple systems illustrated in figure 15.7 can themselves be integrated into coordinated systems of higher orders. These more complex systems are those that would be required for the proper development of diverse tissues, organs, and organ systems in higher plants and animals.

This is not the place to present the Britten–Davidson model in detail; many of their speculations will eventually be discarded as untrue. Here, it suffices to point out the redundancies which either of the models presented in figure 15.7 must possess: either structural genes must be preceded by batteries of dissimilar pro-moters or transcription initiation sites (referred to subsequently as "sensors"; see figure 15.1) or the sensor genes, themselves, must be followed by batteries of DNA segments, each of which

ENZYME	LIVER	KIDNEY	SPLEEN	HEART	SKELETAL MUSCLE	SMALL INTESTINE	PANCREAS	BRAIN
3-HYDROXYBUTYRATE DEHYDROGENASE	■	O					O	
MALATE DEHYDROGENASE	O		O	■	O	O	O	O
PYRROLINE-2-CARBOXYLATE REDUCTASE	O	■			O	O	O	■
CATALASE	■	■	O	O		O		
PEROXIDASE			■			■	O	
HOMOGENTISATE OXIDASE	■	■					O	
CATECHOL METHYLTRANSFERASE	■	O					O	
DIMETHYTHETIN-HOMOCYSTEINE METHYLTRANSFERASE	■	O		O		O	O	
RIBONUCLEASE			O			O	■	O
CARBOXYLESTERASE	O			O		O	■	O
PHOSPHOLIPASE	■	O	■			■	O	
ACETYLCHOLINESTERASE			O	O		O	O	■
CHOLINESTERASE	O		O	■		■	O	O
ALKALINE PHOSPHATASE		■		O		■		
ACID PHOSPHATASE	O	O	■			O	O	O
GLUCOSE-6-PHOSPHATASE	■	O				O	O	
β-MANNOSIDASE	■	■	O			O	■	
β-ACETYLAMINO DEOXYGLUCOSIDASE	O	■	O			O		
β-GLUCURONIDASE	■	O	■	O		O		O
ARGINASE	■	O						
GUANINE DEAMINASE	■	■	■	O		O	■	
ADOLASE					■	O	O	
CITRATE SYNTHASE			O	■		O	O	O
ACONITATE HYDRATASE	O	■	O	■	O	O	O	
GLUTAMINE SYNTHETASE	■			O		O		■

Figure 15.6 The distribution of various enzymes in tissues of the rat. Shaded areas, high enzyme activity; blank areas, very low enzyme activities; open circles, data absent or low-level activity. (After Britten and Davidson 1969.) Reproduced with permission: from *Science* (1969), 165: 349-57; © 1969 by American Association for the Advancement of Science.

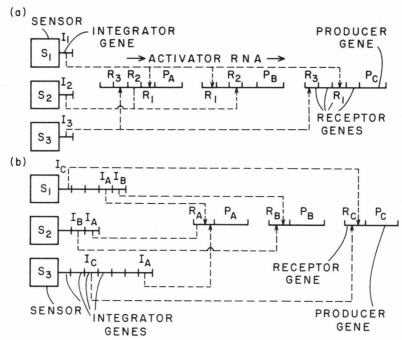

Figure 15.7 Two models for the control of gene action in higher organisms. Sensor genes, *S*, respond to the triggers in figure 15.1; producer genes, *P*, correspond to the structural gene of the earlier figure; receptor genes, *R*, correspond to the adjacent controlling element of that figure. The integrator gene, *I*, produces a signal that may call for gene activity at a multitude of gene loci. Note that each model calls for redundant elements: either each producer gene is preceded by an appropriate number of receptors or the sensor gene activates an appropriate number of integrators. (After Britten and Davidson 1969.) Reproduced with permission: from *Science* (1969), 165:349–57; © 1969 American Association for the Advancement of Science.

makes a unique integrator molecule. The redundancy which the model demands is the feature that forms the basis of the speculations that follow.

The Wallace-Kass Model

If random mutations have an average beneficial effect on viability of otherwise homozygous individuals as a number of experiments

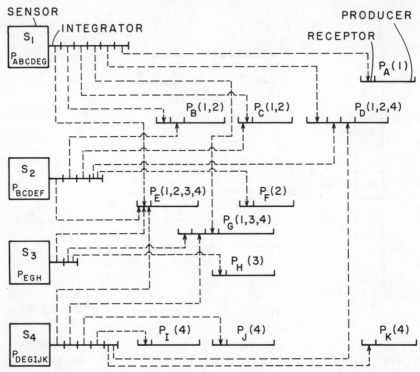

Figure 15.8 An extension of the models presented in figure 15.7 illustrating the means by which overlapping batteries of producer genes can be activated in appropriate combinations by sensor genes that control specific combinations of integrator genes. (After Britten and Davidson 1969.) Reproduced with permission, from *Science* (1969), 165:349–57; © 1969 American Association for the Advancement of Science.

have suggested, then a plausible model must be built to account for these experimental observations. The lack of such a model over a period of nearly twenty years has exacerbated the uncertainties and ambiguities of the experimental data that were collected during that time. The Britten-Davidson model has at last provided a rationale by which the average (slight) beneficial effect of radiation-induced mutations on their heterozygous carriers can be explained; the necessary modification of the Britten-Davidson model was described by Wallace and Kass (1974).

The redundancy in the transcription initiation sites (sensors) preceding each structural gene can be represented as in figure 15.9. Following a suggestion made by Crick (1971), the sensors that receive specific molecular signals (*A*, *B*, *C*, *D*, *E*, and *F*) from inte-

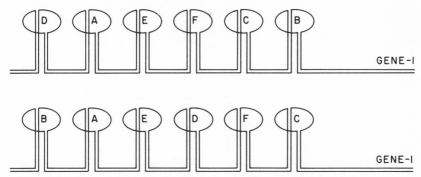

Figure 15.9 The alignment of sensors (receptor genes, in Britten and Davidson's terminology) adjacent to each of the two homologous structural genes of a diploid organism. The reasons for listing these sensors in dissimilar sequences are explained in the text. The physical representation follows a suggestion of Crick (1971): each loop represents the opening of the otherwise double-stranded DNA into single-stranded sensor regions. (After Wallace and Kass 1974.)

grator genes are shown as pedestals of double-stranded DNA, the tops of which are thrown open as single strands for ease of recognition. Restricting our attention to the upper strand, we see that a signal (say, *delta*) that attaches to sensor *D* would initiate transcription at the leftmost portion of the diagram. Transcription would then proceed along the DNA strand through the structural gene segment of DNA at the right. The mRNA would be processed by being cleaved from the large RNA molecule, while the remainder would be degraded and recycled within the cell nucleus. In contrast, signal *beta* would initiate transcription at sensor *B* immediately adjacent to the structural gene; nevertheless, the small excess RNA would be cleaved from the mRNA and recycled within the nucleus.

Suppose that the control region containing the transcription-initiation sites for a given locus is of considerable length; long enough, for example, that the time required for transcription to proceed from the leftmost sensor to the gene is not negligible. In that case, a diploid organism whose two alleles at a given gene locus possessed dissimilar sequences of alleles would, on the average, respond more rapidly to a call for gene action than would one whose two alleles carried identical sequences of sensors. The reason for the more rapid response of the control-region heterozygote can be seen in figure 15.9. Although sensors *D* (top) and *B* (bottom) occupy the most remote positions in the control re-

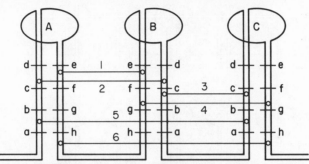

Figure 15.10 A diagram representing a number of possible recombinations within a length of DNA whose structure consists of overlapping (tandemly duplicated) reverse repeats; within each of the repeated segments is a sensor that is capable of responding to a specific signal and of initiating the transcription of the adjacent structural gene (not shown here). The outcomes of the six recombinations are listed in the text. (After Wallace and Kass 1974.)

gions of the two alleles, the cell does not need to wait long for gene-1 to respond to the signals *delta* and *beta:* the top allele responds immediately to *beta*, and the bottom one rather quickly to *delta* (see Wallace and Kass 1974 and Wallace 1976 for more detailed analyses of the consequences of a dissimilarity in the sequences of sensors at each gene locus).

A population geneticist, hearing of the possible advantage of dissimilarity in the sequences of sensors within the control regions of two homologous alleles (that is, the advantage of heterozygosity for control regions), would quickly ask, How can such dissimilarities be maintained? If given sequences were fixed for all time, chance events would soon lead to the retention of only one such sequence in any population. If, on the contrary, the answer to the question lies in the *mobility* of sensors within the control regions, then a specific physical structure of these regions is called for: overlapping reverse repeats such as those shown in figure 15.10 (*abcdefgh · h'g'f'e'd'c'b'a' · a b c d e f g h ·* etc.). The sensors *A, B,* and *C* are shown as DNA segments embedded within this sequence of reverse repeats. Six intrachromosomal recombinational events are illustrated in the figure. By tracing each of these events (being careful to move consistently along the small letters) one finds that each has its own consequence:

Recombination Event	The Order of Retained Sensors			Lost Sensors
1	A	B	C	none
2	B	A	C	none
3	A	B	C	none
4	A	C	B	none
5		C		A and B
6		A		B and C

Two (numbers 2 and 4) of the crossover events have resulted in an interchange of neighboring sensors: A with B in one, B with C in the other. Obviously, if neighboring sensors can exchange position, the entire sequence of sensors within a control region is fluid, provided that the control region is built of repetitive sequences of reverse repeats. Recombination events 5 and 6 reveal the importance of specifying *reverse* repeats (palindromes): These two recombinations occur within *tandemly* oriented sequences (number 5, for example, occurs between—a · b—and—a · b—); consequently, they excise circular fragments of DNA which are lost, leaving the chromosome deficient for certain transcription-initiation sites.

As was the case with the Britten–Davidson model itself, this is not the place to mount a defense of Wallace and Kass's suggested modification; this defense has been presented elsewhere (Wallace and Kass 1974; Wallace 1975b; Wallace 1976). Nevertheless, two items that would not normally be of concern to population geneticists might be mentioned here: First, the excess RNA that is processed from mRNA—that is, the nRNA that is normally recycled within the nucleus—is, to a large extent, double-stranded RNA (dsRNA) (see, for example, Robertson et al. 1968; Jelinek and Darnell 1972; Dunn and Studier 1973). This type of RNA can only arise through the transcription of palindromic (reverse repeat) DNA. Consequently, the structure that has been postulated in order that the sensors be able to move about within control regions is also the structure that leads to the observed configuration of unprocessed nRNA.

A second consequence is illustrated in figure 15.11. In this diagram, the consequences of an interchange of two sensors is shown in detail. Presumably, before the exchange of positions, each site guided the transcriptase (RNA polymerase) molecule toward the structural gene. Assume, as in figure 15.11, that this signal is asymmetrical and, hence, unidirectional. After the two sensors exchange places, the signals for transcribing point in the

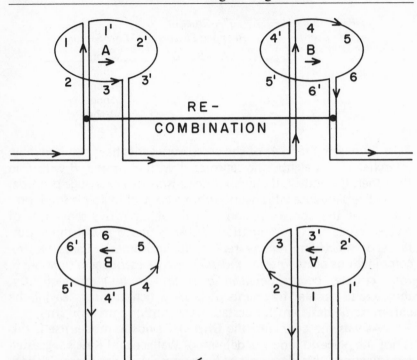

Figure 15.11 A diagram illustrating the need to postulate palindromic (i.e., reverse repeat) sensors if these regions are to exchange positions within the gene control regions as the Wallace-Kass model suggests. Transcription normally follows the DNA strand which is marked by arrows pointing to the right. Asymmetric sensors (1-2-3 and 4-5-6) would serve to initiate transcription on that strand following receipt of appropriate signals. Following the indicated recombination, however, both 1-2-3 and 4-5-6 are on the opposite strand, pointing in the opposite direction, and each would initiate transcription away from the gene. To avoid this problem, the sensor regions must be written 1-2-3-3'-2'-1' and 4-5-6-6'-5'-4'-; in this case, rotation does not alter the orientation of the sensor region because it is structurally symmetrical.

wrong direction: transcription would occur on the *wrong* DNA strand and *away* from the structural gene. Because of this problem, it was necessary to suggest that transcription-initiation sites are themselves structurally symmetrical palindromes embedded within even larger tandem duplications of reverse repeats. This suggestion was made coincidentally with Gilbert and Maxam's (1973) demonstration that promoter regions are in fact palindromic in structure.

On the Beneficial Effect of Random Mutations

The Wallace-Kass model has been introduced in this chapter in order to explain the (sometimes disputed) observation that randomly induced mutations may enhance the viability of their heterozygous carriers. How does the model explain this seemingly inexplicable effect?

The answer to the above question is presented in figure 15.12. At the top of the figure is shown a gene locus which, in the control cultures, is homozygous for not only the structural gene but also for the sequence of sensors within the gene control regions. The modified *CIB* procedures lead to homozygosis not

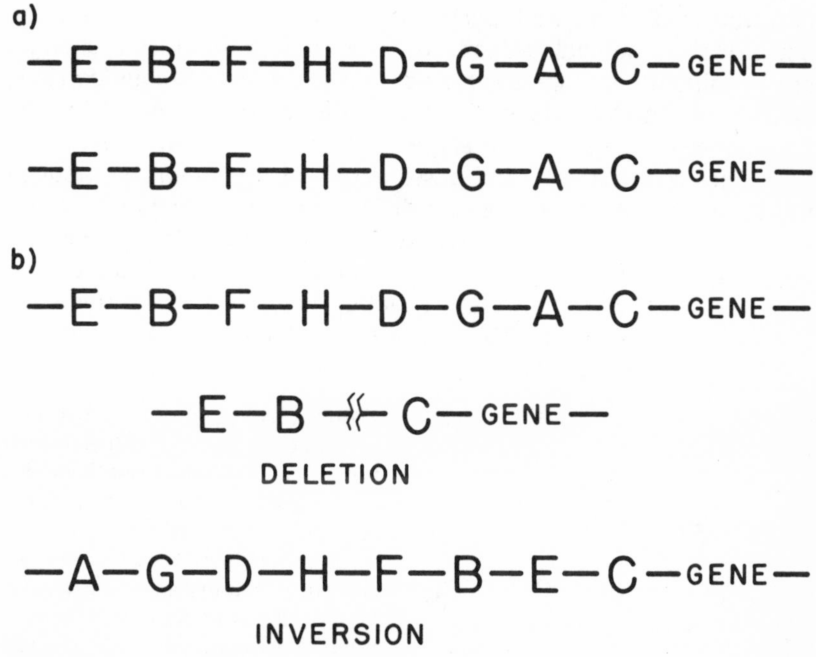

a)

$$-E-B-F-H-D-G-A-C-\text{GENE}-$$

$$-E-B-F-H-D-G-A-C-\text{GENE}-$$

b)

$$-E-B-F-H-D-G-A-C-\text{GENE}-$$

$$-E-B-\text{\textbardbl}-C-\text{GENE}-$$

DELETION

$$-A-G-D-H-F-B-E-C-\text{GENE}-$$

INVERSION

Figure 15.12 Explaining the beneficial effect of a newly induced mutation present in heterozygous condition in an otherwise homozygous individual. The diagram assumes that the functioning of this gene (in a homozygous background that is not illustrated) is improved by the transposition of sensors *E* and *B* closer to the structural gene; this can be accomplished by either a deletion of part of the control region or by a small inversion. Note that a mutagen that causes base-pair alterations but which does not cause breakage and rearrangement within the DNA molecule would not cause "beneficial" mutations of this sort.

only for structural genes but also for all their associated control regions; consequently, the order of the transcription initiation sites for the two alleles are identical. Homozygosis implies genetic identity.

The lower portion (b) of figure 15.12 represents a locus at which the exposure of one (not both) chromosome to irradiation has resulted in an aberration—either a small deletion or an inversion. The aberration may have one of two results. If the transfer of sensors *E* and *B* to a position nearer the structural gene represents an improvement over the homozygosis of the control flies, an enhancement of the viability of heterozygotes results. On the other hand, if the radiation-induced rearrangement does not improve the original homozygosis, the unaltered gene of the nonirradiated allele still performs its functions. In a sense, the occasional beneficial effects of the radiation-induced changes are dominant and are expressed in their heterozygous carriers; neutral or possibly deleterious alterations are concealed by the dominance of the unaltered allele on the other chromosome.

The account of the possible effect of radiation-induced mutations in heterozygous condition on their otherwise homozygous carriers has an immediate bearing on a theoretical account of gene action described originally by Fisher (1930, ch. 1), and later modified by Wallace (1968a: 448). The two models are illustrated in figure 15.13; Fisher's original model is above, the modified version below.

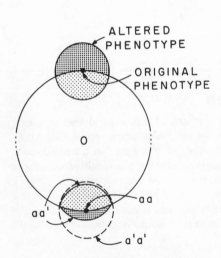

ALTERED PHENOTYPE

ORIGINAL PHENOTYPE

O

a a

a a'

a'a'

Figure 15.13 The effect on fitness of mutations that cause small phenotypic modifications. Fisher's model (upper portion of the diagram) predicts that with small modification, a mutation has about a 50:50 chance of causing an improvement in an otherwise imperfect phenotype. The modified model (lower portion) suggests that, with dominance in respect to the advantageous aspects of the phenotype, mutations in heterozygous condition may have an even greater than 50:50 chance of improving fitness. This is the situation, for example, depicted in respect to the small deletion shown in figure 15.12.

Fisher claimed that mutations with small phenotypic effects may possess a 50:50 chance of being beneficial—the smaller the effect, the closer the odds approach 50:50. His reasoning is illustrated in the figure. For a given set of circumstances, there exists a best or optimal phenotype (0). Any ordinary individual is somewhat imperfect; its phenotype can be represented as lying at some point on a circle somewhat removed from 0. Fisher now imagines that a small change in this imperfect phenotype occurs, displacing the phenotype slightly but in an unknown direction. The mutant phenotype, consequently, will lie somewhere on the small circle. Moreover, if this circle is very small, nearly half of its circumference lies closer to 0 than did the original, unaltered phenotype.

The modified model includes the effects of mutations on heterozygous individuals, and assumes, as we did immediately above, that the beneficial actions of mutated genes tend to be expressed while the harmful ones to be hidden by the alternative, unchanged allele. In this case, as the lower portion of figure 15.13 shows, the inner (beneficial) portion of the spindle-shaped figure can be considerably (approximately, $\pi/2$) longer than the outer (detrimental) one. This could be so even if each new mutation, when homozygous, were unmistakably deleterious.

Some Consequences of the Wallace-Kass Model

The model of gene control described in the proceeding section has consequences many of which parallel experimental observations made by population geneticists. Not all of these are to be paraded for inspection here because the model is speculative and, for those who are interested, many consequences have been discussed in detail elsewhere. For example, an advantage residing with heterozygosity of control regions confers an apparent advantage for heterozygosity of their associated structural genes, even though the products made by these alleles may be selectively equivalent (Wallace 1975b, 1976); this effect is the equivalent of the hitchhiking effect described by Ohta and Kimura (1971) in which selectively neutral alleles establish balanced polymorphisms because of nearby loci at which heterozygous individuals are favored by selection. Ohta and Kimura's calculations dealt with closely linked gene loci; the present model deals with more-or-less adjacent regions of single loci.

The Wallace-Kass model is especially pertinent to a phenomenon known as "coadaptation"; a discussion of this aspect of the model will be postponed until chapter 20. At the present time, we shall consider the model in relation to matters discussed in chapter 3: the relative egg-to-adult viabilities and developmental rates revealed by the modified *C/B* techniques.

The basis of the following discussion (and of figure 15.14) is a hypothetical organism containing seven structural genes (1-7) that are incorporated in various combinations into five regulatory systems (*A-E*). The genetic organization of this organism can be described by either of the systems listed below (compare with figure 15.9):

(a) Regulatory Systems	Included Genes
A	1,3,4,5,7
B	2,3,4,5,7
C	1,2,3,6
D	1,2,4,6,7
E	2,3,5,6,7

(b) Structural Gene	Associated Transcription— Initiation Sites (Sensors)
1	A,C,D
2	B,C,D,E
3	A,B,C,E
4	A,B,D
5	A,B,E
6	C,D,E
7	A,B,D,E

According to the gene control model that we have been discussing, the sensors at a locus can be arranged in any order; those of gene 5, for example, can be ordered in six ways, *ABE*, *AEB*, *BAE*, *BEA*, *EAB*, and *EBA*. Because at each of four loci (1, 4, 5, and 6) the control regions have six possible orders, and at the remaining three loci there are 24 possible orders of 4 sensors, the total number of combinations of different sensors orders equals $6^4 \times 24^3$ or nearly 18,000,000. That an obviously simplistic genetic model can generate such an enormous number of combinations emphasizes the complexity of normal developmental systems.

The five system-seven gene model was used to measure "developmental" time. Development is said to be complete when all five systems, from *A* through *E*, have activated their associated structural genes. No system can start functioning, however, until all genes of the previous one (starting with *A*) have been activated. Time in these calculations is measured in units corresponding to remoteness of the senors from the structural genes (assumed to lie at the left of the control region).

To illustrate the calculations outlined above, consider the following two randomly generated sequences of sensors:

Gene	I	II
1	A D C	A C D
2	E C D B	C D B E
3	A E C B	C B A E
4	D A B	D A B
5	B E A	B A E
6	C E D	E C D
7	E B D A	B A D E

Consider first homozygous individuals. Under example I, system *A* would require four time units to run its course (because *A* occupies the fourth position in gene-7), system *B* would require four time units (because of gene-3), *C* would require three, *D* would require three, and *E* would require two. The total time required would be 16 units. An individual homozygous for genes as organized in the second example (II) would require 15 time units for the proper functioning of all five systems. In contrast, an individual heterozygous for the two arrays of genes illustrated in the example, would require only 12 time units for the five systems to run their courses. The saving involves several of the five systems, including *A* (where gene-7 is no longer the last to act) and *B* (where gene-3 is no longer the last).

Figure 15.14 illustrates the distributions of developmental times of 1,000 homozygous individuals and 999 heterozygous ones according to arrays of sensors generated in a random fashion by computer (the 1,000 represents a random sample of the 18,000,000 possible arrays!). Clearly, heterozygotes as a class "develop" more rapidly under the terms of the model than do homozygotes. The patterns resemble the findings of Dobzhansky and Spassky (1963) that were mentioned in chapter 3. They also are reminiscent of Hiraizumi's (1965) finding (table 14.17) that the presence of an irradiated chromosome shortened the developmental time of otherwise homozygous individuals. That finding, as was pointed out earlier, is nearly sufficient to account for the commonly observed 1 percent to $1\frac{1}{2}$ percent increase in egg-to-adult viability.

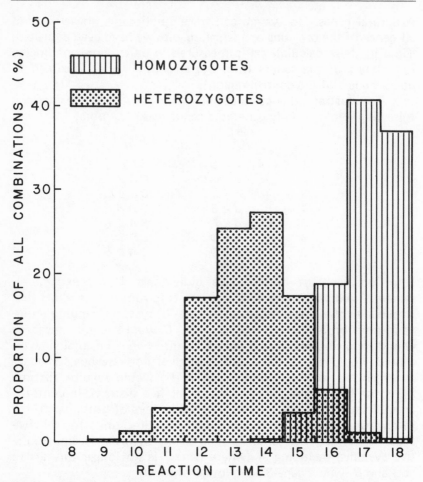

Figure 15.14 The time (in arbitrary units) required for 5 regulatory systems (*A* through *E* as described in the text) to complete their sequential functions if (1) the sensors in the control regions of homologous genes are identical (homozygotes) or (2) that the sensors in homologous control regions occupy positions that differ between the two by chance alone (heterozygotes). Reproduced with permission: from B. Wallace (1976) in *Population Genetics and Ecology*; © 1976 Academic Press.

Concluding Remarks

The title of this chapter emphasized the speculative nature of the models that were to be discussed. Speculations, in the hands of the unwary, have a tendency to assume the appearance of proven facts; let the unwary be warned once again of the material discussed here. On the other hand, speculation in the eyes of some dedicated experimentalists is useless—even sinful. That attitude, if carried to its extreme, is also wrong. It leads some workers to verify Mendel's laws, still again, a century or more after Mendel's original publications; their colleagues meanwhile are investigating the relation between base sequences in DNA, gene function, and gene regulation. Speculation is needed if experimentation is to move forward onto new paths and into new areas.

Experimental data presented in chapters 13 and 14 suggested that populations can utilize radiation-induced genetic variation in increasing their average fitnesses whether the latter be measured in terms of egg-to-adult survival, population numbers and biomass, interspecific competition ability, or innate capacity for increase. A number of large subsequent studies have suggested that mutations induced by low levels of radiation can increase the survival or developmental rate of their heterozygous carriers, provided that the now-heterozygous locus would otherwise have been homozygous. The discussion in chapter 15 has been predicated on the assumption that such results have a rational explanation. The result has been the creation of a model which calls for a structure of DNA resembling many of the findings of molecular geneticists, and which in various manipulations generates consequences resembling many experimental observations. Moreover, the model generates verifiable predictions for future study. Little more can be asked of any speculation.

FOR YOUR EXTRA ATTENTION

1. Following a chapter in which speculations on gene control were rampant, it is only fair to quote the following sentences from Haldane (1964) for your consideration:

"The beanbag geneticist need not know how a particular gene determines resistance of wheat to a particular type of rust, or hydrocephalus in mice, or how it blocks the growth of certain pollen tubes in tobacco, still less why various genotypes are fitter, in a particular environment, than others. If he is a good geneticist he may try to find out, but in doing so he will become a physiological geneticist."

HINT: An understanding as to how a gene works (in the eyes of a population geneticist) depends upon one's interpretation of a symbol. Scute mutations in D. melanogaster *are listed as separate alleles because of the independence of their origins and because of their characteristic patterns of missing bristles (of which there are 20). Alternatively, one can say that the gene at the* scute *locus can "cause" the presence of absence of any bristle out of 20. How many alleles at this locus are possible? In the absence of additional information, one must say* 2^{20} *or 1,048,576. Only if the population geneticist is willing—and able—to incorporate such enormous numbers of potential alleles in his models can he say that* how *a gene acts is a matter for physiological geneticists.*

2. Populations that are subjected to artificial selection in respect to some feature of their phenotype often undergo a serious loss of fitness; this was illustrated in figure 11.5 and table 11.3. Lerner (1954: 10) reports that a regime of selection for long shank (= drumstick) length in chickens lasting 14 years and adding less than 1.5 cm to the length of the shank reduced the index of fitness to 20 percent to 25 percent of control birds. Latter (1962) has carried out a similar study, with similar results, in *D. melanogaster*.

Consider the various means by which the phenotype and fitness of individuals may be related. This requires a consideration of matters raised in chapters 8 and 11 as well as more speculative thoughts based on material presented in this chapter (see Wallace 1976).

3. Following the text of chapter 10, references (Young 1979a, 1979b; Snyder and Ayala 1979) were cited in which the authors admitted that observed patterns of selection need not have been caused by the allozyme variants under observation but, possibly, by closely linked "invisible" loci.

Several workers have now shown that the time, place, and rate of action of structural genes are often controlled by nearby "loci": Doane (1977), Abraham and Doane (1978) and Powell and Lichtenfels (1979) for the amylase (*amy*) locus in *D. melanogaster* and *D. pseudoobscura;* Johnson and Judd (1979) for the *cut* locus in *D. melanogaster;* McCarron et al. (1979) for the *rosy* locus in *D. melanogaster;* and Berger et al. (1979) for *Mus musculus*.

Given a model of nearby control regions such as those described by Britten and Davidson (1969), how might one interpret the ever present importance of nearby (but invisible) loci in matters of fitness?

4. Consider the following observations of Symonds and Coelho (1978) in the light of material discussed in this chapter: One-half of the cells in a lysogenic culture of phage *Mu* yield no viable phage after induction. The *G* segment of *Mu*, a region of some 3,000 base pairs, seems to have some role in this matter. The *G* region can be found in either orientation; its inversion is mediated by inverted repeats of about 50 base pairs at its two ends. Phage particles derived from the induction of lysogenic cultures contain equal numbers of particles with the *G* segment in either orientation. In contrast, virtually all progeny phage derived from infection have their *G* segments in *one* of the two possible orientations.

Hard and Soft Selection

PREVIEW: Studies which suggested that random mutations might have an average beneficial effect (when heterozygous) on otherwise homozygous individuals posed two problems; (1) formulating a credible explanation couched in genetic terms, a task undertaken in the previous chapter, and (2) avoiding the restrictions imposed by genetic load calculations on the amounts of genetic variation that populations can maintain under natural selection. In the present chapter, the implicit and explicit assumptions of genetic load theory are examined. The following conclusion is reached: If the survival of individuals of one genotype depends (as it almost always does) on the numbers and frequencies of nearby individuals of competing genotypes, genetic load calculations are of little use in predicting how much selectively non-neutral genetic variation a population can successfully maintain.

ONE PROBLEM IS solved only to bring a second one into view. In the preceding chapter a model of gene action was proposed which (except that it leads to specific, testable predictions) recalls the "advantage of heterozygosity *per se*" of an older age (East 1936; see Lerner 1958: 95–103). The model explains only how the advantage of heterozygotes *might* arise: it provides a mechanism for heterosis that is rooted in DNA and in the control of gene action. The model, even if approximately true, does *not* explain how populations can support an ubiquitous superiority of heterozygotes. Genetic load calculations (see, for example, Kimura and Crow 1964) suggest that multiple balanced polymorphisms, each of which is maintained by means of the selective superiority (superior fitness) of heterozygous individuals, lead to excessively low probabilities of survival—probabilities so low that offspring cannot be produced in numbers sufficient to replace their parents. A population that continually decreases in size eventually becomes extinct.

In essence, the purpose of this chapter is to present an acceptable alternative to genetic load calculations. This requires a brief review of the assumptions on which genetic load calculations are based and a search for a substitute model. The task calls for a discussion of *hard* and *soft* selection.

Segregational Load (Review)

The calculation of the average fitness of populations when heterozygotes are superior in fitness to their corresponding homozygotes has been presented in both chapter 10 (calculation of $\overline{W}$) and chapter 12 (calculation of genetic load, $1 - \overline{W}$). For the two-allele one-locus system in which the fitnesses of $a_1 a_1$, $a_1 a_2$, and $a_2 a_2$ are $1 - s$, 1.00, and $1 - t$, $\overline{W} = 1 - [st/(s + t)]$ and the segregational load equals $st/(s + t)$. If $t = s$, the segregational load becomes $s^2/2s$, or $s/2$, where 2 corresponds to the number of alleles. Consequently, if 10 alleles existed at a given locus with all heterozygotes having equal fitness (1.00), and with the ten homozygotes having equal fitnesses $(1 - s)$, as well, the segregational load would equal $s/10$: the reduction in the fitness of homozygotes multiplied by the fraction of the population that is homozygous.

When extended to encompass multiple loci, genetic load calculations proceed according to the multiplicative model (chapter 1): the proportions of survivors remaining for each locus are multiplied. Thus, if the average fitness at each of three loci is $\overline{W}_1$, $\overline{W}_2$, and $\overline{W}_3$, then the overall fitness would equal $\overline{W}_1 \times \overline{W}_2 \times \overline{W}_3$. Given that conditions at all loci are similar and that at each one every homozygote has a fitness of $1 - s$ relative to heterozygous individuals, then for l loci and n alleles at each, the overall average fitness equals $(1 - s/n)^l$.

Table 16.1 illustrates the seeming impossibility of maintaining numerous (independent) balanced polymorphisms in popula-

Table 16.1 The approximate average number of eggs that a female of a biparental species must produce in order to maintain balanced polymorphic systems involving a given number of independent gene loci and alleles which when homozygous have specified deleterious effects. Upper figures give the number of eggs required if only 2 alleles exist; lower ones, if 20 alleles interact heterotically. These calculations have been made according to the multiplicative model. (After Wallace 1963d.)

	Number of Loci			
Disadvantage of Homozygotes	1	10	100	1000
1	4	2000	$>10^{30}$	$>10^{300}$
	3	3	340	$>10^{22}$
.5	3	36	$>10^{12}$	$>10^{124}$
	3	3	25	$>10^{10}$
.1	3	3	340	$>10^{22}$
	3	3	4	300
.01	3	3	4	300
	3	3	3	4

tions. In the table it is assumed that each female of a biparental species must produce two viable zygotes if the population is to avoid extinction. The table shows the number of zygotes each female would have to produce before selection operated in order that two viable zygotes would remain later. For example, $\left(1 - \dfrac{0.5}{20}\right)^{100} \times 25$ is approximately 2.00; or, $\left(1 - \dfrac{0.1}{2}\right)^{100} \times 340$ is also about 2.00. The numbers listed in table 16.1 demonstrate the tremendous cost of maintaining balanced polymorphisms under the multiplicative model; figures such as 10^{10} and 10^{22} are absurd, and even the need of an organism to produce some 300 eggs in order that two will survive *intrinsic* destruction would pose an enormous problem. Remember, the calculations on which table 16.1 is based do not take *extrinsic* factors such as accidental or random deaths into account.

The numbers presented in table 16.1 are obtained by routine calculations based on genetic load theory; prior to 1966 they were of academic interest only. Most persons believed that populations were exceedingly homogeneous in their genetic makeup; few persons were motivated to examine the calculations closely in order to question their conclusions. In 1966, the enormous levels of electrophoretic variation in populations were revealed for the first time (Harris 1966, Lewontin and Hubby 1966). Lewontin and Hubby identified the paradox that the observations raise: if the observed variation were maintained by selection, the genetic load should be impossibly large; if not, chance loss of alleles should reduce the variation to levels much lower than those observed.

The Unit-Space Model

The multiplicative model assumes that the contribution of each locus to the overall mortality in the population occurs independently of those of all other loci. The model applies exceedingly well to mortality that occurs, for example, between oviposition (in the case of an insect such as *Drosophila*) and the hatching of the egg. Or, where the death of the zygote is virtually assured— even by lethal mutations that act during early larval stages. Thus, a balanced lethal system in *D. melanogaster* such as *CyL/Pm* (second chromosome), *Ubx/Sb* (third chromosome), or ey^D/ci^D (fourth chromosome) results in the death of about one-half of all zygotes. Any combination of two of these systems permits only about one-quarter of all zygotes to live. All three systems combined would kill seven-eighths of all zygotes (actually more because of epistatic interactions), leaving one-eighth to survive.

The multiplicative model fails in those cases where the fate of an individual depends upon its relation to others which are nearby and which interact with it by contact, by shading (in the case of plants), or by chemical substances secreted into the environment. Birds and mammals establish and maintain territories by a variety of vocal signals; these determine in part the fate of listeners without any need for physical contact.

Figures 16.1 and 16.2 illustrate the unit-space model (chapter 1)—a model in which the fates of individuals are not assumed to be events occurring independently of one another. The material presented in the two figures concerns the fate of silver maple seeds (and seedlings) which fell into a collection of 107 small (5 cm X 5 cm) plastic flowerpots. These pots were arranged in four columns of 27 rows each (one pot of the 108 was missing) on a bench beneath a silver maple tree. A count of the seeds that had fallen in the pots on June 8, 1978 is shown in figure 16.1a. The essential randomness of the distribution of seeds is suggested by its mean (1.32) and variance (1.31); the mean and variance are equal, a property of the Poisson distribution (see page 22).

About one month later (July 12), seedlings were distributed among the small pots as shown in figure 16.1b. Once more, the distribution seems to be a random one ($\overline{X} = 0.24$; $\sigma^2 = 0.28$), suggesting that the successful germination of one seed in a small pot does not inhibit that of others in the same pot. The probability of successful germination among these seeds was 0.24/1.32, or

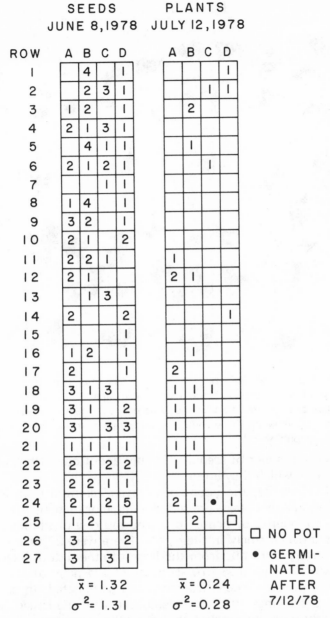

Figure 16.1 Schematic representation of the chance distribution of silver maple seeds (*Acer saccharinum*) among 107
small pots, and of the distribution of small seedlings 35
days later. The similarity of the mean and variance of each
distribution suggests that both are Poisson distributions.
(Wallace, unpublished.)

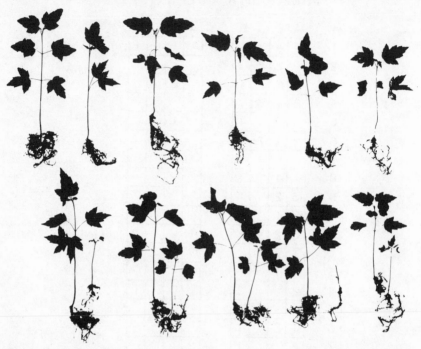

Figure 16.2 The relative sizes of the maple seedlings found in the small pots 10 days after the second count recorded in figure 16.1. Notice that of the five pairs of seedlings found in the same pot, one member of each is as large as the singlets while the second is usually much smaller (even moribund). (Wallace unpublished.)

0.18. The number of seedlings is only about one-fifth that of the seeds from which they arose.

On July 22, the young plants were removed from the pots, roots and all, and were pressed. A sample of these plants is shown in the silhouettes of figure 16.2. The top row of silhouettes is of plants growing singly in their pots. There were 22 such plants of which every third one from the largest to the sixteenth largest are represented. The lower row consists of the pairs of plants from the five pots in which two seeds had germinated. In each case, the two seedlings differ considerably in size: the larger compares favorably with the singlets shown in the upper row, the smaller in most cases is moribund. The advantage possessed by the larger member of each pair may have resided in being the first to germinate (although the growth rates of seedlings seem to differ; plant C24 germinated after July 12 but by July 22 was among the larger singlets even though its cotyledons were still attached). The moribund members of the five pairs did not become moribund through a sequence of independent events (as called for by the multiplicative model). Each one suffered because its larger neighbor cast more and more shade and, in doing so, received more and more sunlight itself. The situation in the case of the doublets was an unstable one in which the large became larger and the small, smaller; the result would have been the eventual survival of a single individual in each small pot. Later, if the plants had not been removed, shading would have occurred among plants of different pots. Events as they occurred in this small experiment parallel those discussed in figure 1.5 and table 1.1.

Maple seedlings illustrate the nature of unit biological spaces—spaces in which only one individual can survive and develop normally—exceptionally well. As do most other higher plants, they tend to remain in one place, thus pinpointing the unit territories. Figure 16.3 illustrates a similarly obvious territoriality in Drosophila cultures. Larvae tend to pupate in preferred zones within these cultures. The figure shows the frequency of pupae encountered at different distances from the surface of the medium for two strains of *D. melanogaster* raised in 25 mm × 100 mm shell vials. Relatively few larvae (under the prevailing laboratory conditions) pupated near the cotton stopper. Most larvae preferred sites near the surface of the food; larvae of one strain pupated closer to the food than the other.

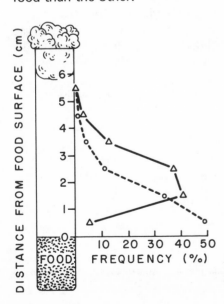

Figure 16.3 The distributions of pupae (*D. melanogaster*) at various heights above the food surface on the sides of culture vials. Although pupae tend to be found in favored zones, the location of the zone varies somewhat with the strain of flies (solid and broken lines). (Wallace, unpublished.)

Table 16.2 The proportion of drowned pupae in vial cultures of *D. melanogaster* from different geographic localities. (Wallace, unpublished.)

Strain	n	Drowned	% Drowned
Pyrenees	509	115	22.6
Madeira Wine Association	747	193	25.8
Madeira, Reid's Hotel	574	99	17.2
Madeira, Public Market	261	32	12.3
bw^D/+ (Riverside, Calif.)	923	137	14.8
bw^D/+ plus +/+ contaminant	121	14	11.6
Riverside, Calif.	883	274	31.0
Total	4,018	864	21.5

Pupation behavior in Drosophila cultures containing somewhat liquified medium has rather serious consequences regarding survival. Third-instar larvae when seeking pupation sites on the glass wall of the vials often dislodge pupae that were formed earlier; these are then pushed into and beneath the surface of the medium. Flies in these dislodged pupae invariably drown. Table 16.2 shows that the proportion of pupae that are drowned in various cultures depends upon the strain of flies (*D. melanogaster*).

Because *D. melanogaster* larvae tend to pupate on the walls of shell vials in a preferred band, one can imagine the existence of pupation territories: a given area on the vial wall should be occupied by a rather constant number of pupae for a specified strain of flies. Table 16.3 shows that this expectation is correct. The pupae in a vial, even those that have been drowned by submersion in the medium, can be removed from the vial and counted. Within the same vials, those pupae from which adult flies have emerged can also be identified and counted. One may then calculate the variation observed between replicate vials of the same strain. In

Table 16.3 A summary of Chi-square tests for homogeneity of surviving pupae in replicated vial cultures of *D. melanogaster* of various geographic origins. (Wallace, unpublished.)

Surviving Pupae			Total Pupae	
Chi-Square	Degrees of Freedom	Strain[a]	Chi-Square	Degrees of Freedom
0.16	2	Pyrenees	0.36	2
0.60	4	Wine Assn.	9.86	4
0.35	1	Public Market	0.46	1
2.83	4	bw^D/+	8.41	4
1.95	3	California	8.23	3
5.89	14		27.32	14

[a]Omitted from the above analysis: Madeira, Reid's Hotel; surviving pupae chi-square (4 degrees of freedom) 49.43, total pupae chi-square (4 degrees of freedom) 47.90.

table 16.3, the Chi-squares and degrees of freedom for five strains of flies have been combined. In the case of the total pupae, the sum of all Chi-squares equals 27.32 and that of degrees of freedom equals 14. The probability of observing the differences in the total number of pupae among replicate vials (*within* strains, recall) is about 0.01; the total numbers of pupae in replicate vials differ significantly.

The case for *surviving* pupae (those which had yielded adults or were about to do so) is strikingly different. Here the sum of Chi-squares is only 5.89; with 14 degrees of freedom. The probability of obtaining a Chi-square this large or larger by chance is greater than 95 percent. Conversely, the probability of obtaining a Chi-square this small or smaller is less than 0.05: the Chi-square is *too* small, and significantly so. Thus, the two sets of data listed in table 16.3 confirm that the number of favored pupation sites or territories per culture vial is limited, and that they are actively sought out and occupied by third-instar larvae. Replicate vials that contain different total numbers of pupae were forced by some agency to yield numbers of surviving pupae that were too similar to be explained by chance. The responsible agency is the space available for pupation, given the preference for the restricted area which the larvae exhibit.

Mobile organisms such as Drosophila larvae also inhabit "territories." In the case of these actively crawling and burrowing larvae, access to a certain volume of medium within a given period of time constitutes the equivalent of a territory. If an individual larva is constrained by others so that it does not have access to that volume, or if its movements are curtailed so that the rate at which it acquires nutrients is reduced, that individual may fail to survive. The extent to which it is constrained or its movements are curtailed depends upon both the age (size) and genotypes of its neighbors (see figure 16.4).

The essential difference between the model on which genetic load calculations have been based (the multiplicative model) and the one we are now discussing (the unit-space model) is to be found in the conclusion of the preceding paragraph: the multiplicative model assumes that decisive events relative to an individual's survival and reproduction occur as independent events; the single-unit model recognizes that the fate of an individual depends upon the phenotypes and genotypes of its immediate neighbors. These two contrasting types of selection have been given the names "hard" and "soft" (Wallace 1968c). As is the case with all

Figure 16.4 Evidence that the *CyL*/+ flies (*D. melanogaster*) of the routine viability tests (see figure 3.5) which would otherwise die manage to emerge successfully when an expected wildtype class is missing. The vertical axis shows the ratio of the sizes of "homozygous" and "heterozygous" cultures: solid circles are the ratios calculated using cultures containing lethal homozygotes; open circles, using those containing quasinormal homozygotes. The horizontal axis represents time measured in sample times of 2 weeks each, 130

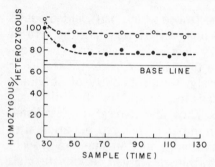

weeks in all. The base line represents the 66.7 percent of "heterozygote" cultures composed of *CyL*/+ flies. If the missing wildtype flies in the lethal homozygotes were not replaced by *CyL*/+ flies, the solid circles should fall on this base line. Replacement of this sort is virtually absent from similar cultures of *D. pseudoobacura*. (Wallace, unpublished.)

convenient, but undefined terms, the words *hard* and *soft* have been used in a variety of senses. Eventually, we shall develop a definition that is suitable for some levels of selection.   First, however, we shall consider two procedures by which a plant breeder can obtain seed while carrying out a program of artificial selection for increased yield. These two procedures, both commonly used, serve as excellent examples of hard and soft selection.

A wheat breeder plants his experimental seed in a randomized series of test plots. When the plants are mature and the heads of grain are to be harvested, the breeder must decide which of the many heads will be saved for the next planting. Because increased yield is the goal of his selection program, the breeder saves the largest heads. Figure 16.5 illustrates the two procedures by which the selected heads might be chosen. Because of variation in soil conditions and accidental differences in the genotypes of the planted seeds, the populations of wheat growing in the experimental plots may vary in average kernel number as positions of the bell-shaped curves in the figure show. The breeder may decide (figure 16.5a) to retain for replanting all heads containing 60 kernels or more. The retained heads in this case would be those falling in the shaded areas of four of the bell-shaped curves. No heads at all would be saved from the remaining four plots. A careless breeder could set his sights too high and, as a consequence, discard all of his material. This procedure imposes hard selection on the population of wheat.

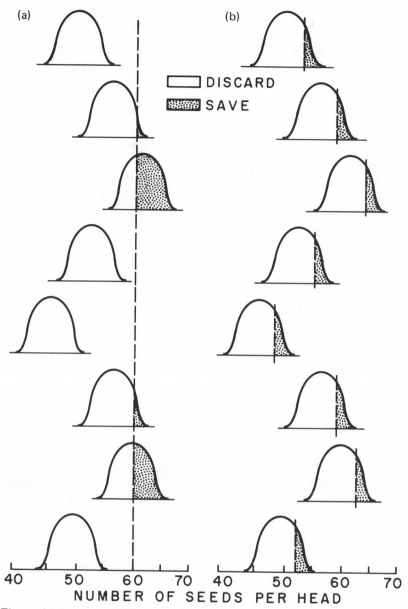

(a)   (b)

DISCARD
SAVE

40   50   60   70   40   50   60   70
NUMBER OF SEEDS PER HEAD

Figure 16.5 Two artificial selection schemes that illustrate the difference between hard and soft selection. In (a) the small-grain breeder saves all heads bearing 60 or more seeds for planting and future selection; some experimental plots are entirely discarded under this scheme. In (b), the breeder first samples a few heads from plants of each experimental plot, determines the statistical distribution of seeds per head for each plot, and then harvests what is estimated to be the best 2 percent to 3 percent of all heads of each plot for planting and further selection. Under this scheme, a few heads are saved from each plot regardless of its average number of seeds per head. Soft selection resembles the second scheme, hard selection the first.

The second procedure available to the breeder is illustrated in figure 16.5b. A small sample of heads is taken from each plot and used to determine the mean kernel number per head and the variation around this mean for each plot. A cutoff point is then calculated for each plot separately; consequently, the best heads of each plot are saved even though some of the plots yielded heads with much smaller kernel numbers than others. The fate of the individual head of wheat under this procedure, like that of the Drosophila larvae discussed earlier, depends upon the phenotypes and genotypes of its neighbors. The second selection procedure, therefore, corresponds to soft selection. One can imagine, of course, intermediate procedures in which fewer heads of grain were taken from the poorer experimental plots than from the better ones; these procedures would also quality as "soft" because the fate of each head still depends upon its kernel number vis-á-vis those of its neighbors. Hard selection, as exemplified by the first of the breeder's procedures, is unique: a decision regarding the cutoff point is made, and acceptance or rejection of each head is made on the basis of the characteristics of that head alone.

Frequency-Dependent and Density-Dependent Selection

Because the fate of an individual that is subjected to soft selection depends upon the phenotypes (and genotypes) of its neighbors, it would appear that soft selection is related to the frequency-dependent and density-dependent selections discussed in chapter 10. In the earlier discussion, frequency-dependent selection was defined as selection in which the selection coefficient is a function of p (or q)—that is, of gene frequency. (Note that selection under which the *change* in gene frequency is a function of gene frequency is not necessarily frequency-dependent selection; Δq may depend upon q even when s is a constant as in $\Delta q = -sq^2$.) An analogous definition of density-dependent selection would insist that the selection coefficient be a function of the number of individuals (N).

The present chapter has frequently touched upon "territories"—the unit biological spaces within which trees, cabbages (see figure 1.6), Drosophila pupae, and Drosophila larvae grow and, if all goes well, survive. The discussion can be expressed algebraically in the following way: Let the frequencies of A and a in a population of N individuals be p and q. The survival and successful reproduction of an individual requires that it occupy a territory of which there are L (for lairs) in all. Individuals of all genotypes can find and occupy territories if vacancies exist; AA and Aa individuals, in addition, can displace aa individuals who must then seek new territories. Three cases are of special interest: (1) when the number of territories exceeds that of individuals in the population ($L > N$), (2) when the number of territories is less than the total number of individuals but exceeds the number of those that are AA and Aa. $[N > L > N (1 - q^2)]$ and (3) when the number of territories is smaller than that of AA and Aa individuals $[L < N (1 - q^2)]$.

Case 1: $L > N$

Genotype	AA	Aa	aa
Frequency	p^2	$2pq$	q^2
Fitness	1	1	1

Comment: all individuals survive because each one has a territory of its own.

Case 2: $N > L > N(1 - q^2)$

Genotype	AA	Aa	aa
Frequency	p^2	$2pq$	q^2
Fitness	1	1	$\dfrac{L - N(1 - q^2)}{Nq^2}$

Comment: All individuals of genotypes AA and Aa have access to territories; only some individuals of genotype aa succeed in occupying those remaining.

Case 3: $L < N(1 - q^2)$

Genotype	AA	Aa	aa
Frequency	p^2	$2pq$	q^2
Fitness (absolute)	$\dfrac{L}{N(1 - q^2)}$	$\dfrac{L}{N(1 - q^2)}$	0
Fitness (relative)	1	1	0

Comment: Not all individuals of genotypes AA and Aa find territories; consequently, not all survive. These two types of individuals do survive, though, with equal probabilities; aa individuals cannot acquire and retain territories and, therefore, do not survive.

Case 2 deserves further comment. The three fitnesses given in the above summary can be regarded as the more usual array of fitnesses $(1:1:1 - s)$ that has been used in earlier calculations (chapter 10). In this case

$$1 - s = \frac{L - N(1 - q^2)}{Nq^2}$$

Rearrangement of the right-hand term leads to $(L - N + Nq^2)/Nq^2$ or $Nq^2/Nq^2 - (N - L)/Nq^2$ or $1 - (N - L)/Nq^2$. Thus, $s = (N - L)/Nq^2$. Because the selection coefficient involves both gene frequency (q) and population size (N), selection under this model is both frequency- and density-dependent according to the definitions given above. The relative fitness of *aa* individuals ranges from 1.00 to 0 through intermediate values that are functions of N and q (and L). This has led (Wallace 1975c) to the definition of soft selection as selection that is both frequency- *and* density-dependent; conversely hard selection has been defined as frequency- and density-*in*dependent selection.

    The terms "hard" and "soft" selection have been applied to cases that are not covered by the definition that has been given here: Christiansen (1975) has used these terms in reference to selection within a population whose subpopulations are linked by migrant individuals; Karlin (1976) has used the two terms for a similar purpose. Dobzhansky (1970; 226) substituted his terms "rigid" and "flexible" for "hard" and "soft" as if these terms are identical. In fact, they are not. Flexible selection, as Dobzhansky uses the term, means either that selective pressures fluctuate or that the frequencies of identifiable genetic elements (alleles, gene arrangements or others) fluctuate in response to (usually unidentified) environmental changes (see Crumpacker and Williams 1974). Fluctuations of this sort may, of course, reflect only that the selection coefficient, s, is dependent upon the environment; they do not prove that the selection coefficient is a function of either population size or gene frequency. Selection coefficients have meaning only in respect to specified environments. Gromko (1977), in a critical review of frequency-dependent selection, overlooked the relations between territories, population size, gene frequencies, and selection coefficients summarized in the calculations given above; consequently, his arguments are not compelling.

Selection and the Limit to Population Size

Individuals are mortal, and populations of mortal beings exist because expiring parents leave offspring in their stead. If, as we saw in Figure 1.4, parents consistently leave too few progeny, the population dwindles until it eventually disappears. If, on the other hand, parents leave offspring in greater numbers than their own, the population grows in size. A population cannot grow without limit, however; certain resources such as nutrients, water, nesting sites, and other necessities are available in finite quantities. As these resources are divided among more and more individuals, the probability that an individual will succeed in leaving an adult offspring becomes less. The result, as we saw in figure 1.4 and as we see again in figure 16.6, is that a population tends to expand until it arrives at a stable size. When this size is exceeded, offspring number less than their parents and the population dwindles; if the population falls below its equilibrium size, progeny exceed their parents in number and the population swells once more. This account pretends that the environmental resources are constant which, of course, they are not; as the quantities of these resources vary, the "stable" population size varies as well: swelling to consume newly available resources, shrinking as resources vanish.

Figure 16.6 illustrates an aspect of populations that does not appear in standard genetic load (hard-selection) calculations. The population illustrated in the figure initially had stabilized in size at N_1 individuals; at that size, each mother, on the average, had successfully reared one adult daughter as her replacement before dying. At some subsequent moment, a new cause of mortality was introduced into the environment; this event is represented by the dashed line. Because mortality has been increased, the multiplica-

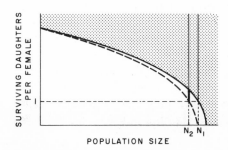

POPULATION SIZE

Figure 16.6 The stabilization of population size through the equalization of daughters to mothers (daughters/ mothers = 1.00). The shaded area represents the loss of zygotes in achieving an earlier stabilization at population size, N_1. The dashed line represents the addition of a novel source of mortality, and the subsequent restabilization of the population at size, N_2. The heavy vertical segment represents zygotes that are eliminated by the new cause of mortality; compensatory reductions have occurred in one or more of the long-standing causes of death. Compare this figure with figure 1.4.

tive model (which ignores the reality of maintaining a population) would suggest that the fecundity of reproducing individuals must be increased in order that there still be one adult daughter surviving for each mother; without this increase, the higher mortality would lead to the population's demise. The diagram in figure 16.6 shows otherwise, however. A second option lies in the lowering of population size (to N_2). The reduction in population size results in a lessening of mortality from causes that were originally present—a lessening which equals the length of the heavy-line segment in the figure. That segment also represents the proportion of deaths ascribable to the new cause of mortality. Changes in numerical size, as well as in fecundity, offer a means by which populations can respond to new, threatening circumstances. (At this point it may be wise to recall, however, that populations can be eradicated; the model presented in figure 16.6 cannot be used as a justification for the pollution of the environment by enormous amounts of toxic or radioactive substances.)

The new cause of mortality that was introduced into figure 16.6 (dashed line) was not identified. It could represent, for example, the introduction of the automobile and the subsequent highway slaughter of innumerable organisms of all sorts—from flying insects and birds to turtles, snakes, and mammals. Alternatively, it could represent a new but long-lasting change in the environment, such as a drought of several decades duration. It could also represent mortality arising from genetic causes: an increase in genetic load through increasing radiation exposure, for example, could be accommodated by a change in population size such as that illustrated in figure 16.6.

Figure 16.7 illustrates that the acquisition of a genetic load is not necessarily accompanied by a decrease in population size. The three curves designated A, B, and C represent the daughter-producing abilities of three homozygous ($a_1 a_1$, $a_2 a_2$, and $a_3 a_3$) strains of some hypothetical species under different conditions of crowding (population size): Strain C cannot maintain itself under any circumstance. Strain B, although it nearly equals A at intermediate levels of crowding, does much more poorly as density increases; consequently, populations of A exceed those of B in final size. Heterozygous females of all three sorts ($a_1 a_2$, $a_1 a_3$, and $a_2 a_3$) produce more offspring than do homozygous ones, and more importantly, continue to do so at levels of crowding under which the homozygotes cease reproducing successfully (zero fitness). A dimorphic population containing any two of the three

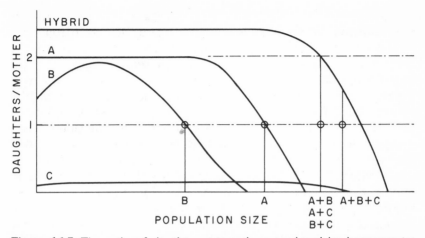

Figure 16.7 The ratio of daughters to mothers produced by homozygotes $(a_1/a_1, a_2/a_2, \text{ or } a_3/a_3)$ of three hypothetical monomorphic populations $(A, B, \text{ and } C)$ and of hybrid $(a_1/a_2, a_1/a_3, \text{ and } a_2/a_3)$ females at corresponding densities. Notice that the monomorphic population a_3/a_3 (C) is incapable of maintaining itself because the number of daughters produced never equals the number of females that produced them. Note also that a dimorphic population involving any pair of alleles would stabilize at a size where both homozygotes would have zero fitness $(s = t = 1.00)$ and, therefore, where each heterozygous female must leave an average of two daughters. The trimorphic population would stabilize where each female left an average of 1.5 daughters.

alleles would grow in size until the average number of daughters produced by each hybrid female equals two. Because these females constitute only one-half of the population $(0.25\, a_1 a_1 : 0.50\, a_1 a_2 : 0.25\, a_2 a_2$, for example), the average number of daughters per mother equals 1.00 at that time $(\frac{1}{2} \cdot 2 = 1)$. The population at that time, although larger than those of the homozygous strains, would have a segregational load of 0.50—a load not possessed by the smaller homozygous populations. A trimorphic population containing all three alleles would, according to the diagram in figure 16.7, be still larger. It would stabilize when one-third of the population consisted of lethal (or sterile) homozygotes and two-thirds heterozygous individuals, that is, when the population size was such that each heterozygous female left an average of one and one-half daughters $(\frac{2}{3} \times \frac{3}{2} = 1)$. In figure 16.7, then, we have seen that the acquisition of a genetic load need not be accompanied by a smaller size (going from monomorphism to dimorphism), although the dimunition of the load in going to trimorphism was accompanied by a still further increase in size.

A diagram such as that shown in figure 16.7 has within it the elements of a three-dimensional figure: fitness (daughters/ mothers), gene frequency, and population size. Except for the difficulty in visualizing the events represented, such a three-dimensional figure could allow fitness (the Z axis) to vary with frequency (the Y axis) and with density or population size (the X axis). No effort to reproduce such a complex example is made here. Instead, figure 16.8 shows the "floor" and one side of such a figure. At the top, the fitnesses (daughters per mother) of *AA*, *Aa*, and *aa* individuals are shown for various population sizes; the use of short, straight-line segments implies that we are interested only in near-equilibrium conditions. Equilibrium sizes for the two monomorphic populations and for the polymorphic one are indicated by triangles on the line representing a daughter-to-mother ratio of 1.00. Gene frequencies in the equilibrium polymorphic population can be estimated from the relative positions of the line segments; because the fitness of *AA* individuals slightly exceeds that of *aa* ones, the frequency of A ($\hat{p}$) exceeds 50 percent.

Projections lead from the equilibrium population sizes of the upper diagram in figure 16.8 to the corresponding gene frequencies in the lower one. Now it becomes clear that a polymorphic population can be out of equilibrium in either or both of two respects—gene frequency and total size. Monomorphic populations, in contrast, can only be too large or too small. The short arrows in the lower diagram show changes in both gene frequency and size that nonequilibrium populations would undergo in a single generation; in the case of polymorphic populations, these arrows seldom point directly at the final equilibrium frequency or size.

The data listed in table 16.4 are presented in order to provide substance to what, for the past several paragraphs, has been a rather abstract discussion. Here are listed (table 16.4a) the adaptive values calculated by Wright and Dobzhansky (1946) for the three genotypes in each of the dimorphic populations (Standard-Arrowhead, Standard-Chiricahua, and Arrowhead-Chiricahua) of *D. pseudoobscura*, and for the six genotypes in the corresponding trimorphic population. Not shown are adaptive values for the three monomorphic ones. Because, by tradition, the highest adaptive value in these studies is set at 1.00 (see the exception (*ST/AR*), however, in the case of the trimorphic population), $\overline{W}$ must always be less than 1.00. The values listed in the lower part of table 16.4 (b) have been adjusted so that $\overline{W}$ equals 1.00. The

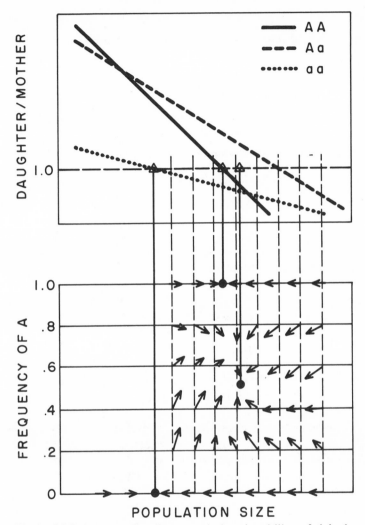

Figure 16.8 A composite diagram relating the ability of *AA*, *Aa*, and *aa* females to produce adult daughters (i.e., the mothers of the next generation), equilibrium size at which the daughter/ mother ratio equals 1.00, and the frequencies of the alleles *A* and *a* within this stabilized population. The short arrows indicate the estimated changes in population size and gene frequencies that would occur for various nonequilibrium conditions.

Table 16.4 The relative adaptive values of various genotypes in experimental populations of *D. pseudoobscura* monomorphic or polymorphic for the *ST*, *AR*, and *CH* gene arrangements from Piñon Flats, Calif. (After Wright and Dobzhansky 1946.)

ST/ST	ST/AR	ST/CH	AR/AR	AR/CH	CH/CH	Average
(a) Original data, polymorphic populations only:						
0.81	1.00	—	0.50	—	—	0.86
0.85	—	1.00	—	—	0.58	0.89
—	—	—	0.86	1.00	0.48	0.89
0.43	1.30	1.00	0.05	0.71	0.21	0.80
(b) Data adjusted so that the mean fitness of each population equals 1.00:						
1.00	—	—	—	—	—	1.00
—	—	—	1.00	—	—	1.00
—	—	—	—	—	1.00	1.00
0.94	1.16	—	0.58	—	—	1.00
0.96	—	1.12	—	—	0.65	1.00
—	—	—	0.97	1.12	0.54	1.00
0.54	1.63	1.25	0.06	0.89	0.26	1.00

individual adaptive values for the various genotypes can now be regarded as the number of adult daughters each of these females leaves on the average. Here we see that monomorphic populations can also be assigned values of 1.00 because such populations can be maintained with ease. The values listed in table 16.4b can be used to demonstrate that each dimorphic population at equilibrium achieves a size larger than its corresponding monomorphic ones, and that the trimorphic population achieves the largest size of all as was illustrated in figure 16.7 (see Wallace 1968c for details).

The discussion of population size and of the attainment of a stable number at which each mother leaves (on the average) one daughter as her replacement emphasizes a point made by Turner and Williamson (1968) and again by Wallace (1970b): genetic load is simply a portion of a larger set of individual mortalities that can be looked upon as the *ecological* load. At some point in the growth of a population of bisexual individuals, all zygotes but two per parental pair (on the average) will be destroyed; at that point the population will have attained its equilibrium (stable) size. No degree of genetic excellence will forestall this destruction of zygotes; therefore, *soft selection is not* to be regarded as *gentle selection*.

Because the excess individuals will be destroyed at any rate, any device that places individuals with the lowest fitnesses among those that are discarded alleviates the problem of genetic load. That device is, of course, the lower fitness itself. Milkman (1967) and, more recently, Wills (1978) have stressed this point. According to Wills, the number of polymorphisms that can be maintained by a population, far from being small as suggested by table 16.1, increases as the *square* of population size. Lewontin and Hubby's (1966) paradox need not be a paradox after all.

In addition to the *necessary* destruction of the excess zygotes (where "excess" means on the average, all except two per pair), soft selection renders the bookkeeping by which the ecological load is apportioned between genetic and nongenetic causes arbitrary. This point was discussed in chapter 12 (page 301) where malaria was shown to create a nongenetic (environmental) load in the absence of the allele for sickle-cell hemoglobin (Hb^s), but a genetic load in its presence.

Population Flushes and Crashes

The above account of the stabilization of population size through limitations imposed by the availability of resources may lead to an erroneous impression that populations are relatively constant in size. For example, Nei (1975: 39) in discussing the growth and regulation of populations states: "In practice, N_t would rarely become larger than K, since K is the carrying capacity by definition." Later he reaffirms that the application of the logistic equation (his page 54) would be restricted to the range of $N_t < K$. (N, in this case, is population size.)

While it is true that populations of many organisms do remain relatively constant over long periods, no population has a truly constant size. During the course of a single summer, populations of short-lived insects swell tremendously in numbers only to crash again at the first frost. "Bad" years are frequently noted for a variety of insect pests; during these years the numbers of individuals may be one or two orders of magnitude larger than those normally observed. Even during a series of "normal" years, however, populations do not remain constant in size. Should K be variable (and there are few constants in complex biological systems), then the size of any population exceeds the *prevailing K* about half the time. The number of newly formed zygotes—especially among fecund organisms such as most insects, many rodents, and most higher plants—almost always exceeds the carrying capacity of the environment. Population flushes, those sporadic periods during which numbers of individuals grow enormously, reflect the sporadic moments when, for some unforeseen circumstance, the carrying capacity (K) of the environment increases enormously and, consequently, a large proportion of all young zygotes survive. When normal environmental conditions prevail once more, usually rather soon, the abnormally large population tends to fall beyond its normal level (it "crashes") before becoming relatively stable once more.

Although Carson (see, especially, Carson 1968 and 1975) has most recently stressed the role of flush-crash sequences in shaping the nature of populations, an early set of important observations was published by Ford and Ford (1930). Figure 16.9 is an attempt to reconstruct in diagrammatic form the narrative provided by the Fords in describing fluctuations in the size of an isolated colony of the Marsh Fritillary (*Melitaea aurinia* or, in modern nomenclature, *Euphydryas aurinia;* see Ford 1945: 268). From old notes, collection records, and pinned specimens, it appears that the Cumberland colony studied by the Fords was quite large in the early 1880s but grew considerably in size during the next twenty years, reaching a maximum about 1900. The colony then crashed abruptly in size, becoming so low in numbers that only occasional individuals were seen during the years before and after 1910. Starting about 1920, the colony grew rapidly in size once more so that by 1930 it had returned to a size approximating that which had been characteristic in the early 1880s. The point of interest at the moment is the tremendous variation in color and morphology that the individuals of the population exhibited during the early 1920s as the population flush began. Phenotypes, it seems, that normally would have been eliminated unnoticed survived to adulthood when the environment became permissive. Significantly, this newly exposed variation vanished *as the population size stabilized* once more; in the terminology used on page 417, the number of individuals once more exceeded the number of territories available, even though the number of territories in the late 1920s and early 1930s exceeded those available during the crash of the early 1900s. These events are consistent with the description of soft selection presented above. The events described by the Fords (at least as I have interpreted them) differ interestingly from those envisaged by Carson (1975); Carson imagines that selection effectively ceases from the start of a population flush

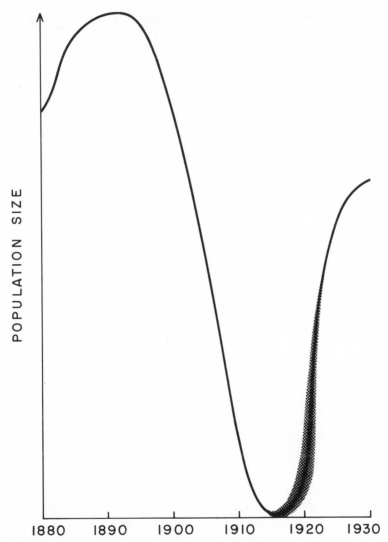

Figure 16.9 A representation of the fluctuations in the size of a colony of the butterfly, *Melitaea aurinia*, from 1880 to 1930 as described by Ford and Ford (1930). For reasons described in the text, the phenotypic variation (represented by the width of the line) which was greatest as the population began to recover in the early 1920s suggests that frequency- and density- (i.e., soft) selection operated in this colony. Later, as the population stabilized at a large size, the individual-to-individual variation vanished.

until the new stabilization that follows the crash. That is, the crash itself as well as the subsequent return to a more or less normal population size takes place in the absence of any appreciable natural selection. This view is quite different from one that I shall present in a subsequent chapter (chapter 18); there I shall argue that during a crash it is precisely the phenotypic variation in fitness that offers the population a means for self-thinning. That is, selection operates during crashes as well as during periods of relative stable population size.

Logistic Growth and Selection

The Lotka-Volterra model of population growth in the case of two competing species leads to the following equations (chapter 1, page 11):

$$\Delta N_1 = r_1 N_1 \frac{(K_1 - N_1 - \alpha N_2)}{K_1}$$

$$\Delta N_2 = r_2 N_2 \frac{(K_2 - N_2 - \beta N_1)}{K_2} .$$

In these equations, α measures the extent to which an individual of species number 2 uses resources that would otherwise be used by individuals of species number 1; β measures the reverse. K_1 and K_2 are the carrying capacities of the environment in respect to the two species; when the number (N) of either equals the corresponding carrying capacity, population growth ceases. The parameter, r, is one less than the average number of progeny that females of either species produce under the prevailing circumstances.

Clarke (1973a, 1973b) has applied a similar model to an analysis of two phenotypes (AA and Aa versus aa) within a single species. According to his model:

$$\Delta N_1 = N_1 \left(\frac{W_1 k_1}{R_1 + W_1 N_1 + \alpha_1 W_2 N_2} - 1 \right)$$

$$\Delta N_2 = N_2 \left(\frac{W_2 k_2}{R_2 + W_2 N_2 + \alpha_2 W_1 N_1} - 1 \right) .$$

In this model, W corresponds to r ($W = r + 1 = R$), of the Lotka-Volterra equations, k corresponds to carrying capacity, α_1 and α_2 correspond to α and β, and N corresponds to population size. The values N_1 and N_2 equal $N(1 - q^2)$ and $q^2 N$ because the model concerns two phenotypes in a population where the frequency of $A = p$ and that of $a = q$.

Figure 16.10 summarizes the consequences of Clarke's analysis (see Wallace 1975c and 1977b). If $\alpha_1 = \alpha_2 = 1$ and $k_1 = k_2$, all section is density- and frequency-independent, or *hard* as this term has now been defined. Notice that in order for selection to be hard, (1) the two phenotypes must be identical in respect to the manner in which they utilize resources, (2) the addition of each individual as N grows must place demands on dwindling resources that are identical to those created by the addition of an individual at an earlier or later time, and (3) the carrying capacity of the environment must be identical in respect to the two phenotypes. In short, for hard selection to operate exclusively, the two phenotypes must be indistinguishable in respect to their ecological behaviors. In truth, this identity is not sufficient because, even in the case of a single phenotype, the effect of adding one individual as N approaches K is unlikely to be identical in terms of resource utilization as the addition of one individual when N is small; α, in other words, probably never does equal 1.00.

If hard selection requires such precise equality of competition coefficients and of carrying capacities, what gives rise to soft (density- and frequency-dependent) selection? All else! Imagine a rectangular surface whose vertical and horizontal axes correspond to α_1 and α_2. On this surface is a line (slope = $45°$) corresponding to $\alpha_1 = \alpha_2$. On this line, in turn, is a point at which $\alpha_1 = \alpha_2 = 1$. At every point other than that one, the selection operating on the population is soft—that is, both frequency- *and* density-dependent. Therefore, under Clarke's model at least, all selection affecting the individual members of a population must be soft selection.

Experimental Data and Selection

Evidence for the operation of selection within natural populations is frequently obtained not in the field but in laboratory

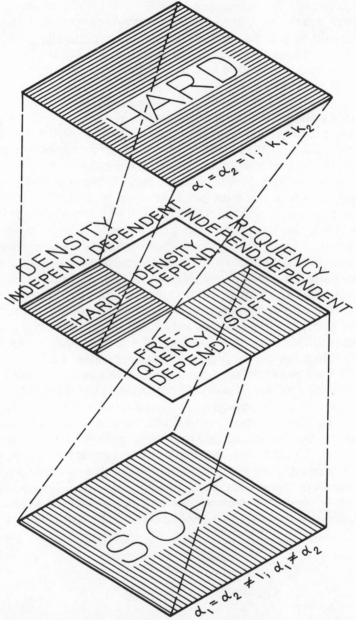

Figure 16.10 A geometric representation of the nature of selection under various relationships of coefficients of competition (α_1 and α_2) and carrying capacities of the environment (k_1 and k_2) for individuals of genotypes AA (= Aa) and aa under a model described by Clarke (1973a, 1973b). (After Wallace 1975c and 1977b.)

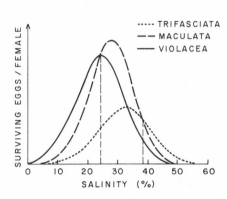

Figure 16.11 The relative number of surviving eggs per female at different salinities for three genotypes of the marine copepod, *Tisbe reticulata* (Battaglia 1965). The dependence of the relative fitnesses of these genotypes upon the environments which they encounter has been called *flexible* selection by Dobzhansky (1970: 225-27). The multidimensional surface formed by the relative fitnesses of individuals of a particular genotype under all possible environments and with all possible background genotypes constitutes the "norm of reaction" of that genotype.

tests under a variety of experimental conditions. Figure 16.11 presents data on egg survival for a marine copepod (*Tisbe reticulata*) under varying salinities. The three curves represent different genotypes: $V^v V^v$ = *violacea*, $V^m V^m$ = *maculata*, and *vv* = *trifasciata*. Each genotype is superior to the other two in some range of salinities—*violacea* at low salt levels, *maculata* at intermediate ones, and *trifasciata* at high levels. Vertical lines erected at various points reveal the relative fitnesses in respect to egg hatch at various salinities. At 24 parts per 1,000, the fitnesses of *violacea* and *maculata* in respect to egg survived are equal, and about three times as great as that of *trifasciata*; at 38 parts per 1,000, *maculata* and *trifasciata* are equal, and nearly twice that of *violacea*. The changes in the relative fitnesses of these three genotypes that occur with alterations in salinity represent *flexible* selection in Dobzhansky's terminology; the data presented here have no bearing in themselves on the matters of hard or soft selection.

Table 16.5 also presents data illustrating flexible selection: eight geographic strains of *D. funebris* exhibit different survivals in laboratory cultures when forced to compete with *D. melanogaster* at different temperatures. The data show that, as a rule, *D. funebris* survives better at low than at high temperatures. In addition, however, it appears that northern European (Russian and Scandinavian) strains excel at low temperatures, and that Spanish and African strains excel (relatively) at high ones. Once more, however, the data do not bear on density-dependent or frequency-dependent selection nor, consequently, on soft selection.

Unlike the previous two tables, table 16.6 presents data

Table 16.5 Survival of different geographic strains of *D. funebris* at different temperatures, expressed as a percentage of the survival of a competing strain of *D. melanogaster*. The dependence of the survival of *D. funebris* on temperature makes this aspect of fitness *flexible* in Dobzhansky's terminology. (After Timofeeff-Ressovsky, from Dobzhansky 1937.)

| | Temperature, °C | | |
Strain	15	22	29
Egypt	68	46	30
Tripoli	64	47	31
Spain	69	48	30
France	80	44	25
Sweden	88	40	21
Moscow	101	43	28
Perm	98	41	26
Tomsk	96	42	28

which, in the absence of additional information, could be interpreted as an example of hard selection. Tantawy and Mallah (1961) measured the percent emergence (egg-to-adult survival) of *D. melanogaster* and *D. simulans* at different temperatures. Areas under the smooth curves fitted to their data have been used in assembling the table: summed over the entire temperature range studied, *D. melanogaster* has the greatest survival; within the more restricted central portion of the temperature range, *D. simulans* exhibits the highest proportion of emergence. Because temperature is independent of either the densities or the

Table 16.6 Measures of the percent emergence of *D. melanogaster* and *D. simulans* over two temperature ranges, calculated as the relative areas under the observed distribution curves. These results imply that the relative fitness of *D. melanogaster* would exceed that of *D. simulans* in a climate that frequently included temperature extremes, but that the reverse would be true under a more moderate and stable temperature regime. (Experimental data from Tantawy and Mallah 1961.)

| | 10–31°C | |
Source	D. melanogaster	D. simulans
Lebanon	1,379	1,161
Luxor, Egypt[a]	1,287	1,078
Alexandria, Egypt	1,213	1,089
Wadi-el-Natroon, Egypt	1,097	1,070
Uganda	1,027	1,053
	18–25°C	
Lebanon	529	520
Luxor, Egypt[a]	488	490
Alexandria, Egypt	459	484
Wadi-el-Natroon, Egypt	413	478
Uganda	396	473

[a] *D. simulans* from Beni-Swef, Egypt.

relative frequencies of these two species in nature, the different emergences tabulated here *could* be evidence for hard selection; in the field, however, the exposure of members of the two species to the prevailing temperature (or to rain, or sand, or any other aspect of the environment affecting survival) may well depend upon total numbers and relative frequencies of the flies. For selection to be density- and frequency-*in*dependent, it is *necessary* that the agent of selection be independent of density and frequency; however, that the agent is independent of density and frequency is not *sufficient* to guarantee hard selection because other events intervene in determining who lives, who dies, and who becomes sterile.

Concluding Remarks

In attempting to define the terms "hard" and "soft" selection more precisely, at least in respect to intrapopulation selection, it was necessary to differentiate between these terms and those ("flexible" and "rigid") used by Dobzhansky (1962) for quite different purposes. The attempt to differentiate between the two pairs of terms should not be interpreted as an attempt to attach greater importance to one pair than the other. All persons should appreciate the implications of flexible selection and of still another concept introduced by Dobzhansky—"the norm of reaction."

The phenotype of an individual is specified by both its genotype and the succession of environments which it encounters during its lifetime. Selection coefficients are functions of phenotypes; consequently, a selection coefficient can be assigned to a genotype only in respect to a given environment or a given sequence of environments. Rigid selection can be viewed, then, as selection in which the selection coefficient remains unchanged over a series of obviously differing environments. *Drosophila* males with white eyes have poor visual acuity and, as a result, have difficulty finding mates; as a consequence, the fitness of the *white* alleles is lower than that of the wildtype allele at the same locus (Reed and Reed 1950). Now, this aspect of fitness is unlikely to be altered by changes in temperature or, let's say, humidity. An investigator who altered only these aspects of the environment might refer to the selection against the *white* allele as rigid selection. On the other hand, if visual acuity is the actual cause of the lowered fitness of white-eyed males, these males may be as efficient as wildtype males in finding mates in the dark. If this is indeed the case, the investigator who studied the effect of light intensity on the fitness of white-eyed males would conclude that selection in this case is flexible.

The norm of reaction is a conceptual view of all possible phenotypes an individual of a given genotype could possess under all possible environments and temporal sequences of environments. It would represent the multidimensional surface generated by thousands of genetically identical individuals, each of whom was reared in a sequence of environments differing from those in which all other individuals were raised.

Lewontin (1974b) has given an excellent account of the importance of the concept of the norm of reaction in interpreting the results obtained by standard analyses of the genotypic and environmental contributions to phenotypes. The standard analyses reveal the relative contributions under existing conditions; in Lewontin's words they are "local" analyses. They are not designed to investigate fundamental causes. The two investigators working independently on the fitness of white-eyed males, for example, would have reached two diametrically opposed conclusions: to one, fitness is genetically determined (rigid); to the other fitness is a function of the environment (flexible). To learn about the "characteristic" fitness of individuals of a given genotype, it is necessary to deliberately survey all possible environments in order to reveal that surface which is the genotype's norm of reaction.

FOR YOUR EXTRA ATTENTION

1. The following observations appear in Yoda et al. (1963): Self-thinning is a density-dependent process; high density increases the variance of growth rate of individual plants, accelerates competition, and results in

high mortality. Furthermore, when the process of self-thinning is prevented, the total stand may collapse before maturity. They go on to suggest that self-thinning is a necessary self-regulatory mechanism for overcrowded populations, one that assures sufficient seed for the next generation although it is detrimental to individual plants. Relate these observations to the concepts of hard and soft selection.

2. Relate the following account (Popescu 1979) of melanism and its maintenance in a species of psocid (winged relatives of book lice) to the account of soft selection (frequency- *and* density-dependent selection) presented in this chapter. According to Popescu, predators may maintain color polymorphisms provided that there is competition among individuals for appropriate resting sites. Each morph, when rare, has ample sites; as its number increases, individuals are forced to occupy sites best suited for some other morph. In addition, Popescu suggests that the number of resting sites may be less numerous than they appear to be because wind, other insects, birds, and other organisms have preempted many sites that otherwise seem to be suitable. (Refer to item 5 following chapter 1 for the possible role of an imperfect searching pattern in accounting for these seemingly suitable but unused sites.)

3. E. D. Ford (1975) has suggested that populations of higher plants grown in pure stand under severe competition always exhibit a high degree of genetic variation, and that variety confers a selective advantage.

Consider both aspects of this suggestion: Compare the first part with figure 1.6. Restate the second part so that the nature of "selective advantage" becomes clearer.

17

Interspecific and Intraspecific Competition

PREVIEW: This chapter represents a brief digression. Matters to be discussed in chapter 18 are valid if patterns of competition between members of the same species differ systematically from those between members of different species. Although this difference is intuitively obvious in extreme cases such as competition between snails *versus* that between snails and wolves, it need not be at all obvious when species of the same genus are involved. In the pages that follow, data bearing on these problems will be presented.

DURING THE DISCUSSION of population flushes and crashes (especially the latter) in the previous chapter, phenotypic variation in fitness was said to provide a means by which extraordinarily large numbers of individuals can be successfully culled to the small numbers that are consonant with the carrying capacity of the local environment. Such variation is needed for thinning not only during the "bust" portion of a boom-and-bust cycle but also for the thinning of young zygotes when these are produced in enormous numbers. Matters such as these are to be discussed at greater length in the following chapter. At the moment, however, a somewhat different matter must be considered: a comparison of intraspecific and interspecific competition.

The remainder of the present chapter will be spent in an attempt to show that competition between individuals of the same species and that between individuals of different species are quite different. If this attempt is successful, then it follows that overcrowding of an area by young .seedlings, by hatchlings, or by adults following a population flush, and the need to drastically reduce the numbers of these individuals are primarily intraspecific concerns.

The term "competition," like so many other useful biological terms, has acquired a variety of meanings. Sammeta and Levins (1970: 473), when reviewing the topic, had this to say: "We would use the term 'competition' despite its drawbacks as it is a simple word and portrays the mechanism involved. Competition implies a similarity in the requirements of competing individuals. As the genetic resemblance between two individuals increases, so too will the similarity in their requirements, which, in turn, leads to more extensive competition between those two individuals. It follows that the most widespread and intensive competition would be intra-specific and intra-varietal." That passage summarizes the intended scope of the present chapter.

Darwin (1859) made much the same point in *The Origin of Species* (Modern Library, page 60): "But the struggle will almost invariably be most severe between the individuals of the same species, for they frequent the same districts, require the same food, and are exposed to the same dangers. In the case of varieties of the same species, the struggle will generally be almost equally severe, and we sometimes see the contest soon decided: for instance, if several varieties of wheat be sown together, and the mixed seed resown, some of the varieties which best suit the soil or climate, or are naturally the most fertile, will beat the others and so yield more seed, and will consequently in a few years supplant the other varieties."

Before proceeding to buttress Darwin's claims with specific examples, a discordant note should be interjected. According to many modern biologists, such as Hamilton (1964a, 1964b), Maynard-Smith (1964, 1976), Wilson (1975), Alexander (1974), and especially Dawkins (1976), alleles can increase in frequency within populations not only by conferring high Darwinian fitness on their individual carriers but also by "compelling" their carriers to perform altruistic acts favoring the survival of siblings and other near-relatives who are likely to possess the identical allele. This view of the genetic basis for altruistic behavior leads to the expectation that close relatives (or, more generally, genetically similar individuals) will exhibit a more pronounced altruistic or cooperative behavior than will unrelated ones (Hamilton et al., 1980). In support of this view, one can cite social wasps, where the queens' helpers are generally nonreproducing sisters, or the scrub jays, (*Aphelocoma coerulescens*) where unmated progeny of one year in the following one protect their parents nest and help rear their younger siblings. Wilson (1975) cites numerous examples such as these. At this time, alturistic behavior is mentioned only to *emphasize* (not to *resolve*) divergent points of view: Sammeta and Levins (and Darwin) emphasize the greater competition between genetically similar individuals; Hamilton, Alexander, and others stress the greater cooperation between genetically similar individuals.

Intraspecific Competition

Darwin, in the passage quoted above, described the rapid alterations that would occur in the proportions of wheat varieties if one were to sow mixed seed from the same stand year after year. The results of precisely that sort of an experiment with barley are given in table 17.1. Here we see that, of six barley varieties tested, four different ones proved to be the superior competitor at one or the other of four localities. After 11 or 12 years during which the seeds planted each year were a sample drawn from the mixed harvest of the year before, three varieties have come to constitute 70 percent to 80 percent of the entire crop. The varieties obviously differ in competitive ability, but what constitutes high competitive ability also depends on the environment of the geographic region in which the tests were performed.

Table 17.1 The proportions (percent of total harvest) of barley varieties from the same initial mixture after repeated (11–12) annual harvestings and sowings at different localities. (From Stebbins 1950.)

Variety	New York	Minnesota	Montana	Oregon
Coast	13	20	20	73
Hannchen	7	75	4	7
White Smyrna	0	1	56	14
Manchuria	78	0	5	0
Getami	2	4	14	0
Meloy	0	0	1	6

The competitive ability of a grain such as barley can be viewed as being related to yield; or to the components of total yield such as kernels per head, heads per plant, or even kernel weight in case this affects germination. Now, an important point made in reference to soft selection concerned the dependence of the fate of individuals of one genotype on the presence of neighbors of other genotypes. The important role such interactions may play is illustrated for barley in figure 17.1.

At the right in each of the diagrams of figure 17.1 is shown the performance of the two pure stands (VV and vv, where V is a dominant allele resulting in two rows of seeds; vv plants have six rows); in the diagrams a and b, the two pure stands are distributed symmetrically above and below their average. The remainder of each diagram illustrates the effect the three genotypes VV, Vv, and vv, have on one another in mixed stand; upon moving from left to right, the heterozygous plants constitute a decreasing proportion of each mixture. For total yield (figure 17.1a), we see that the positions of VV and vv in mixed stand are the reverse of those in pure stands. Heterozygous plants are more constant in their yield than homozygous ones; the yield of VV plants increased only as the proportion of heterozygous plants decreased.

The presence of heterozygotes depressed the homozygotes in respect to heads of grain produced per unit area (figure 17.1b); as heterozygotes decreased in proportion, the effect of VV on vv became more pronounced. The interaction of the two homozygotes makes the outcome in the mixed stand quite different from what might have been expected on the basis of their performance in pure stands.

The remaining diagrams (figure 17.1c and d) illustrate a point that shall resurface in chapter 18: some aspects of reproduction in plants are remarkably constant. Despite the observed variation in yield (and, most likely, corresponding variation in plant size and morphology) within the various mixed stands, the number of kernels per head for the three genotypes is quite uniform throughout. Even more uniform, however, is kernel weight itself. The weights of kernels borne by the three genotypes differ considerably (that of VV is nearly twice that of vv individuals), but for each genotype this character scarcely reflects any of the differences in planting schemes. Even the kernel weights observed in pure stands correspond precisely to those in the mixed stands. The high yields that are characteristic of vv pure stands rests largely on the enormous numbers of kernels borne per head, even

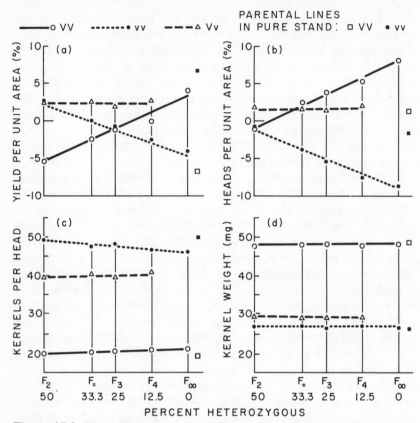

Figure 17.1 The effects on total yield and its components (heads/area, kernels/head, and weight/kernel) that are observed when two nearly identical varieties of barley (*VV*, two rows of seeds; *vv*, six rows of seeds per head) and their (*Vv*) hybrids are planted together in various proportions (F$_2$ = 1:2:1, F$_=$ = 1:1:1, F$_2$ = 3:2:3, F$_4$ = 7:2:7, and F$_\infty$ = 1:0:1). In diagrams *a* and *b*, the data for each genotype are plotted relative to (above or below) the population mean. Notice in *a* that the total yields of *VV* and *vv* plants in mixed (F$_\infty$) and pure stands are reversed. Notice, too, the remarkable constancy of the mean kernel weight (d) under all circumstances. (After Wiebe et al. 1963.) Reproduced by permission of the National Academy of Sciences, Washington, D.C. from *Statistical Genetics and Plant Breeding*.

though each kernel is rather small. The high yield of *VV* individuals in mixed stand (F$_\infty$) results from the ability of these plants to produce many heads of grain at the expense of *vv* plants.

The role of competitors in determining the characteristics of a given rice (*Oryza sativa*) plant has been demonstrated by Sakai (1955). Having determined in preliminary studies that "red" rice is a better competitor than the "upland" variety, Sakai surrounded individual red rice plants by a hexagon of 6 competing plants (all distances between plants were 11.5 cm). The six competing plants were of all combinations from 6 upland, 5 upland: 1 red, . . . , to 1 upland: 5 red, and 6 red plants. The results are illustrated in figure 17.2 for two characters (plant weight and number of seeds). With

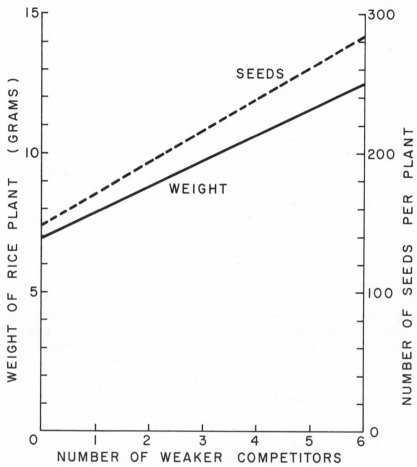

Figure 17.2 A diagram summarizing the effect of lessened competition on the weight and the number of seeds of red rice plants (*Oryza sativa*). The data concern individual plants of the red variety which are surrounded by six competing individuals planted in a hexagonal pattern, 11.5 cm from each other and from the center plant. As red competitors are replaced by upland ones, the size and the number of seeds borne by the center plant increase. (After Sakai 1955.)

increasing number of red plants among the six competitors, the performance of the center plant declines. When all six of the surrounding competitors are red rice plants, the weight of the center plant and the number of seeds it bears is reduced by nearly one-half. The *degree* of reduction is most likely a reflection of the between-plant distances (11.5 cm) that were chosen for the experiment; from what we have seen for maple seedlings (figure 16.2) and barley (figure 17.1), we might expect the reduction to have been greater had the distances between plants been smaller.

Interspecific Competition

The account of interspecific competition is introduced by a sequence of three figures, all presenting data on territories established by nesting birds; the examples range from a seeming indifference of the two species to one another, to a mutual exclusion as precise as might be expected between nesting pairs of one species.

Figure 17.3 illustrates the territories established by nesting Abert and brown towhees (Marshall 1960); the data cover two years at one site, the San Xavier Reservation in California. The important points to notice in this figure are these: (1) The territories of pairs of birds belonging to the same species are tightly packed, with little or no overlap; the borders are generally contiguous but not carelessly so. (2) The territories of the two species seem to be superimposed with no apparent restrictions; between years the patterns within each species changes considerably, but the brown towhee, for example, establishes territories with little or no regard as to those of the other species.

The territories established by redwinged and yellowheaded blackbirds are not as haphazardly related as those of the brown and Abert towhees. Figure 17.4 shows the territories established by redwinged blackbirds around the periphery of a marsh; their territories do not intrude into the marshy area. Three or four weeks later, the yellowheaded blackbirds arrive and establish territories within the unoccupied marsh. The yellowheaded blackbirds, however, intrude upon and confiscate portions of adjacent redwing territories. The eventual outcome is a series of mutually exclusive territories—mutually exclusive both within and between species. The bulk of the area occupied by the yellowheaded redwings would otherwise have been unoccupied.

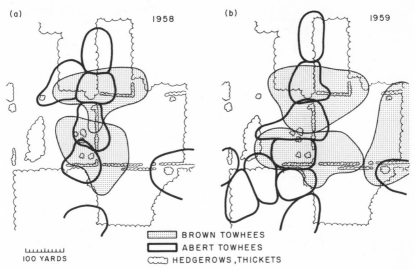

(a)   1958   (b)   1959

⊔⊔⊔⊔⊔⊔⊔⊔
100 YARDS

▨ BROWN TOWHEES
▭ ABERT TOWHEES
◠◠◠ HEDGEROWS,THICKETS

Figure 17.3 This and the following two figures illustrate a gradation in the extent to which the territories established by nesting pairs of one bird species interfere with those of pairs of a second species. In this figure, territories established by Abert and brown towhees overlap haphazardly, even though territories within each species are precise in their exclusion of one another. (After Marshall 1960.)

▨ REDWING

▨ YELLOWHEAD

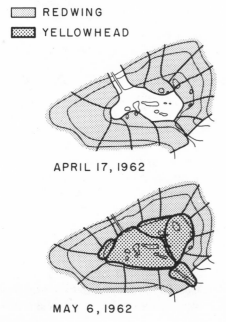

APRIL 17, 1962

MAY 6, 1962

Figure 17.4 Territories held by pairs of redwing and yellow-headed blackbirds (*Agelaius phoeniceus* and *Xanthocephalus xanthocephalus*) at two times during a single season. The yellow-heads arrive late and occupy territory that was left vacant by the earlier arriving redwings. The yellow-heads encroach upon and occupy nearby segments of the original redwing territories, however, thereby creating a cluster of mutually exclusive territories belonging to members of both species. Reprinted with permission: from G. H. Orians and M. F. Willson (1964), *Ecology* 45:736–45; © 1964 Ecological Society of America.

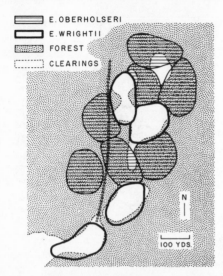

E.OBERHOLSERI
E.WRIGHTII
FOREST
CLEARINGS

N

100 YDS.

Figure 17.5 This diagram illustrates a series of mutually exclusive territories established by nesting pairs of two sibling species of flycatchers (*Empidonax wrightii* and *E. oberholseri*). Note that the former birds occupy clear areas and the latter pine-fir forest (stippled). (After Johnson 1963.)

Figure 17.5 illustrates the territories occupied by sibling species of flycatchers. A map of territories alone would suggest that all territories of both species are essentially equivalent. When superimposed on a map of the plant life of the area, the territories of the two species fall clearly on different environments: one species perfers cleared areas, the other pine-fir forest.

Ornithologists and amateur bird watchers have amassed a tremendous amount of information concerning birds and the territories they establish. Schoolchildren can collect valid data as a laboratory exercise for their biology courses. Near my home, one would have to be tremendously preoccupied not to realize the approximate extent of each cardinal's territory from the virtually incessant calling during the spring. With patience and ingenuity, territories can be demonstrated for other, less obvious organisms. Table 17.2 summarizes the results obtained by N. W. Moore (1964) in a study during which he captured male dragonflies at one pond and released them at another. Males of these insects normally patrol on a set route so that it is quite easy to determine the number patrolling at a pond on a given day. The released males approximately doubled the number of patrolling males but by the following day the initial number was more or less reestablished. Native males had about twice the probability of remaining at the release point than did the ones that were brought in from distant ponds. As the note in table 17.2 says, the importation and release of males of one species had no effect on the numbers of patrolling males of a second. In this sense,

Table 17.2 The effect of adding male dragonflies (*Ceriagrion tenellum*) to those already patrolling a small pond. (After N. W. Moore 1964.)

	Original Number	Added	Combined Total	Number on Following Day[a]*
Total	130	210	340	147[b]
Daily average	4.81	7.78	12.59	5.44[b]

[a]The introduction of a second species had no effect on the numbers of male *Ceriagrion tenellum*.

[b]Even the slight increase on following day is largely explained by two experiments in which the following day was cloudy; the numbers were reduced on the second day.

*The probability that an original male remained = 0.62. The probability that a newly added one remained = 0.31.

it appears that male dragonflies establish independent territories either by ignoring one another or by patrolling subtly different beats.

Still other examples of different "territories" being occupied by different species are illustrated in figure 17.6. In figure 17.6a, the distributions of two genera of planaria are shown relative to the water temperature of the streams in which they are found. In streams where only one or the other of the two genera is found, the distribution ranges overlap broadly, being nearly identical in the cool range ($6°-8°C$) but with *P. gonocephala* extending considerably farther into the warmer water. In streams harboring both genera, the distribution patterns are nearly mutually exclusive: *P. montenegrina* occupies the cooler portions of the streams, and *P. gonocephala* occupies the warmer. A strictly analogous pattern is found in the intestines of rats that are infected with the wormlike parasites, tapeworms and acanthocephalans. The distributions in the intestine when only one type of worm is present are rather similar; under high-density, one-species infections, the distribution areas increase but not markedly. In mixed infections, the two types of parasites take up distinct (though somewhat overlapping) areas of residence in the gut: Acanthocephalans infects the anterior portion of the gut; tapeworms are found more posteriorly.

Intraspecific and Interspecific Competition at Close Quarters

The examples that we have considered in the previous sections have shown that the two types of competition—intraspecific and

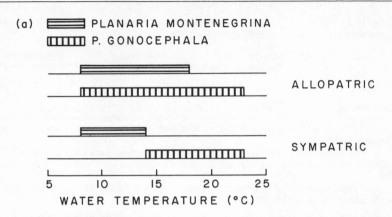

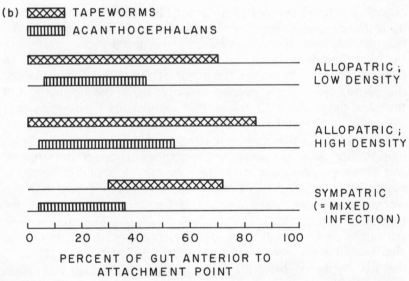

Figure 17.6 The restriction imposed on distribution patterns by the presence of a competing organism. (a) The distributions of two species of Planaria (*P. montenegrina* and *P. gonocephala*) along streams in which each species occurs alone (allopatry), or in the presence of the other (sympatry). Water temperature seems to be important in determining these distributions. (b) The distributions of two internal parasites in the gut of wild rats. The distributions of either genus in single infections (allopatry) are broader than in mixed infections (sympatry), allopatric distributions are also broader under high-density infections than under low-density ones. (After Miller 1967; Colwell and Fuentes 1975.)

interspecific—differ. As Darwin said, members of the same species frequent the same districts and require the same food; therefore, they are thrown into contact and conflict more often than are members of different species. The above discussion of different species has dealt with "districts" of rather large size: clearings within pine-fir forests; marshes and surrounding meadows; temperature gradients along mountain streams and even regions along the rather long gut of the rat. In the present section, the examples will deal with finer subdivisions of the environment.

The reduction in numbers of overcrowded poppy plants serves to demonstrate the greater effectiveness of self-thinning than that of thinning by aliens (Harper 1961). Two species of poppies were used in one of these studies: *Papaver rhoeas*, and *P. lecoqii*. Seeds of these poppies were planted, mixed, at three densities (approximate): 540 seeds/m^2, 2,400 seeds/m^2, and 4,300 seeds/m^2. At the extreme densities, the two types of seeds were mixed in equal proportions; at the intermediate density, two mixtures of seeds were used: 1 to 8 and the inverse, 8 to 1.

Figure 17.7 shows how the two species of poppy fared under competition. On the left, *P. rhoeas* is the alien and *P. lecoqii* the native. The diagram shows the probability that a native seed will produce a mature plant; this probability decreases much more rapidly under self-thinning (8 native seeds to 1 alien one) than under alien-thinning (8 alien seeds to 1 native one). The diagram on the right shows that the inverse is also true: when *P. rhoeas* is the native and *P. lecoqii* is the alien, the probability that a native seed will produce a mature plant is much less under self-thinning (8 natives : 1 alien) than under alien-thinning (1 native : 8 aliens). In each case, the most rapid drop in the probability of the survival of a given species is caused by increasing the density of that same species; competition with one's own kind has more severe consequences than does competition with individuals of another species. These differing patterns of competition imply that the individuals of two different species utilize the local environment (that is, sunlight and soil) in somewhat different ways.

The greater severity of intraspecific than of interspecific competition illustrated by the experiments on poppies has been confirmed by studies on *Drosophila* (figure 17.8). The measure in the case of flies was the number of progeny produced in small (25 mm X 95 mm) shell vials when the number of (nonvirgin) female parents ranged from 1 to 10 (and 15). The relation between number of progeny per female and number of parents is, as

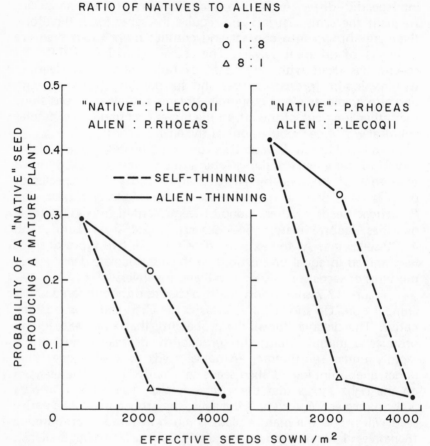

Figure 17.7 The relative degrees to which densely-seeded mixed populations of poppies are thinned by competing members of their own species (self-thinning), or by those of the alien competitor (alien-thinning). At the intermediate density, two proportions of competitors were tested: 8 natives:1 alien and 1 native:8 aliens. In both instances shown in the figure, it is the high proportion of native competitors that most severely reduces the probability of obtaining a mature native plant. Similarly, considering the high probability of a native surviving under heavy alien competition at the intermediate density, it is the large number of native competitors in the most densely sown populations that reduces the probability of maturing successfully to the low value observed under that level of competition. (After Harper 1961.)

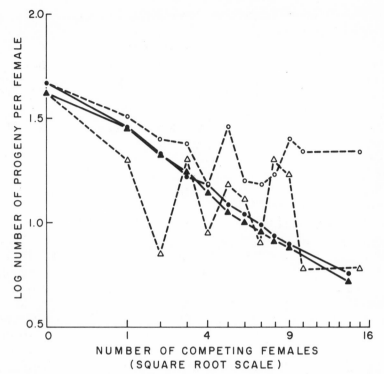

Figure 17.8 Diagram contrasting the consequences of intraspecific and interspecific competition in *Drosophila*. The scales of the ordinate and abcissa have been chosen so that the per capita progeny production of *D. melanogaster* (●) and *D. simulans* (▲) in pure cultures form relatively straight lines with increasing numbers of females. The progeny produced by a single female of either of these species (open symbols) in the presence of increasing numbers of competing *D. pseudoobscura* females does not decrease linearly; at some point the number of progeny produced by these females proves to be relatively unaffected by further increases in the numbers of *D. pseudoobscura* female competitors. (After Wallace 1974.)

one might expect, complex; nevertheless, the data points become reasonably linear when plotted as the *logarithm* of progeny number against the *square root* of number of female parents. The number of females in each culture can also be converted to the number of *competing* females by merely subtracting one: 1 female equals 1 female + 0 competitors, 2 females equals 1 female + 1 competitor, etc. The results obtained for *D. melanogaster* and *D. simulans* in pure cultures (intraspecific competition) are shown as the solid lines in figure 17.8. In contrast to the solid lines are the broken ones which show the number of progeny produced by one *D. melanogaster* or one *D. simulans* female in the presence of increasing numbers of *D. pseudoobscura* females. These latter lines are not linear; after the number of competing females has reached three or four, the number of progeny produced by the *melanogaster* or *simulans* female becomes nearly constant. The additional *pseudoobscura* females are without effect on the number of *melanogaster* or *simulans* progeny produced. (The same effect, to a smaller degree, also holds *within* a species, of course; that, presumably, is the reason why it is necessary to plot the square root of the number of female parents in order to generate even an approximately straight-line relationship between progeny and parents.)

The experiments with *Drosophila* included many more species than the three (*D. melanogaster*, *D. simulans*, and *D. pseudoobscura*) mentioned in discussing figure 17.8; not all of the data collected are suitable for graphic representation, however. Nevertheless, the data can be analyzed (see Boxed Essay) and summarized as in table 17.3. The numbers in the table reflect the decrease in the per capita progeny production (number of progeny per female) effected by the addition of *one more female:* (1) a female of the same species as the progeny being scored; (2) a female of a competing species. The data have been summarized over all experiments

Table 17.3 The contrasting effect on the per capita progeny production of adding (1) one more female of the same species (or strain) or (2) one more female of a different species (or strain). The special case of going from *one* female to *two* is shown as well. Many different species were involved in these studies; consequently, the only numbers to be compared are the corresponding numbers under (1) and (2).

	(1)	*(2)*
a. Experiments with *D. melanogaster*, *D. simulans*, and *D. pseudoobscura* (1 to 10 females per vial)		
Total reduction in progeny	472.3	141.4
Reduction per female per vial	8.75	2.62
The effect of going from 1 female to 2	30.50	4.08
b. Experiments involving *D. funebris*, *D. virilis*, *D. melanogaster*, *D. athabasca*, *D. willistoni*, *D. pseudoobscura*, *D. nebulosa*, and *D. simulans* in many (but not all possible) combinations (1 to 5 females per vial).		
Total reduction in progeny	557.1	277.9
Reduction per female per vial	7.74	3.86
The effect of going from 1 female to 2	14.91	7.10
c. Experiments involving three strains of *D. melanogaster*: apricot-Bar ($w^a B$), sepia (*se*), and wildtype (1 to 5 females per vial).		
Total reduction in progeny	2,063	1,398
Reduction per female per vial	21.49	14.56
The effect of going from 1 female to 2	43.58	19.46

involving the species listed in the table. In brief, the addition of one more female of the *same* species (an increase in intraspecific competition) has a much greater effect in reducing per capita progeny production than does the addition of one more *alien* female. This effect is especially pronounced in the special case of increasing the number of females from 1 to 2. These results entirely support those obtained by Harper (1961), which were illustrated in figure 17.7: overcrowding is primarily an intraspecific affair.

The third portion of the table (table 17.3c) is especially important in relation to the frequent use of competition as a measure of fitness. One can argue (see Jungen and Hartl 1979, Hartl and Jungen 1979) that interspecific competition does not serve adequately as a measure of fitness because two different species affect one another only to the extent that their needs overlap. Intraspecific competition, on the contrary, measures fitness accurately because under this type of competition the needs of the two classes of flies are identical. The data presented in table 17.3c show that the last assumption (individuals of the same species have identical needs) is not necessarily true. Here we see that three strains of flies which have been maintained as separate cultures for many years duplicate the pattern seen earlier in the interspecific studies: the addition of one female of the *same* strain reduces per capita progeny production more than does the addition of one female of a *second* strain. Consequently, although the needs of the members of one species are probably more similar than are those of members of different species, members of a single strain (or variety) have more similar needs than those of different strains. That statement merely repeats the passage from Darwin's *The Origin of Species* that was cited earlier (page 441).

The discussion of intrastrain *versus* interstrain competition can be extended to include a study made by Lewontin (1955) for a somewhat different purpose but which can be used to demonstrate that *D. melanogaster* strains differ in their needs—that is, in the demands they make of the culture medium in which they are grown. Each of the strains of wildtype flies listed in table 17.4 was grown in pure cultures and in cultures where larvae of wildtype and *white* were present in equal numbers. Overall, the probability of survival was approximately 6 percent higher in the mixed than in the average of pure cultures of *white* and wildtype flies; this probably reflects a corresponding difference in the demands made for available resources by the two types of competing larvae.

Table 17.4 The survival of larvae of *D. melanogaster* in overcrowded cultures. Twenty strains of wildtype flies were tested either alone or in competition (1 : 1 proportions) with a mutant (*white*) strain; the *white* strain was also tested alone as a control. The values listed under "pure" are the averages of the wildtype strain and of the *white* strain when tested separately; those under "mixed" are the total survivals when wildtype and *white* larvae developed in the same vial. The difference (*d*) between the "pure" and "mixed" values for each strain has been computed (mixed – pure); the average of these differences (*d̄*) differs significantly from zero. Larval survival in "mixed" cultures consistently exceeds the average of the two corresponding "pure" cultures. (Data from Lewontin 1955.)

Strain	Pure	Mixed	Difference (d)
1	0.392	0.380	−0.012
11	0.530	0.685	0.155
18	0.265	0.425	0.160
22	0.543	0.620	0.077
24	0.619	0.745	0.126
25	0.554	0.552	−0.002
28	0.616	0.623	0.007
33	0.497	0.433	−0.064
34	0.679	0.615	−0.064
41	0.541	0.653	0.112
45	0.681	0.580	−0.101
46	0.589	0.628	0.039
47	0.469	0.783	0.314
48	0.614	0.668	0.054
50	0.588	0.618	0.030
54	0.529	0.523	−0.006
62	0.669	0.678	0.009
67	0.453	0.622	0.169
74	0.592	0.660	0.068
Swedish-B	0.585	0.680	0.095

$\bar{d} = 0.0583$; $s_{\bar{d}} = 0.0218$; $t = 2.67$ (19 d.f.); $p = 0.01–0.02$

Population Models and Competition

The Lotka-Volterra equations (chapter 1) dealing with the coexistence of two (or more, because the equations can easily be extended) species contain equivalency terms, α and β, that denote the extent to which an individual of, say, species number 2 utilizes resources in a manner comparable to one of species number 1. Thus, if species number 1 were to live in isolation, population growth would cease when the number of individuals, N_1, equaled the carrying capacity of the environment, K_1. In the presence of a second species, the number of individuals of species number 1 would cease when $N_1 + \alpha N_2$ equaled the carrying capacity, K_1. The term, α, reflects the extent to which a member of species number 2 resembles one of species number 1. If α equaled zero, species number 1 would grow and stabilize as if species number 2 did not exist. If α equaled 1.00, individuals of the two species would place equal demands upon available resources. The preceding sections have attempted to show that, on the average, the equivalency coefficients do not equal 1.00, that members of one species do not make demands identical to those of another. Indeed, we have seen reason to believe that as the number of individuals of a single species increases, the *intra*specific equivalency term does not remain 1.00; the evidence for this was the need to use the *square root* of the number of female parents in linearizing the per capita production data (figure 17.8).

A standard procedure for studying the competition between two species of organisms is that based on the replacement series, a procedure leading to graphic representations of data known as "de Wit diagrams" (de Wit 1960). The fundamental diagram, one that would be obtained if there were neither intraspecific nor interspecific interactions within or between the tested species, is shown on the left in figure 17.9a. The horizontal axis extends from 100 percent A to 100 percent B with a linear displacement of A by B as one moves from left to right. The vertical axis measures progeny, yield, or some other aspect of production; on the left is the production of pure stands (or cultures) of A, and on the right is that of B. Because (by definition) there are no interactions either within or between species, the total production of each falls off linearly with its decreasing density. The total production, $A + B$, for any density of the two falls on the line connecting the productions of the two pure stands. The last point can be seen if (imagining we are moving from left to right on the diagram) we recall that the slope of A's partial yield is $-A$ (so that the yield becomes zero when percent $A = 0$) and that of B's partial yield $= B$ (so that the total yield when $B = 100$ percent is B). The total at any intermediate point, X, equals $Y = A - AX + BX = A + (B - A) X$. This is a linear function extending from A (when $X = 0$) to B (when $X = 1$).

The basic, noninteractive form of the de Wit diagram is most often an unrealistic one. In figure 17.1 we saw that the total yield of two barley strains were reversed in mixed as opposed to pure stands. The diagram at the right in figure 17.9a illustrates a commonly observed interspecific interaction: a few individuals of species B depresses the production of A considerably, and vice versa (recall figure 17.8). The effect on total production is illustrated by a U-shaped curve falling consistently below the theoretical one discussed above. Figure 17.9b, illustrating the outcomes

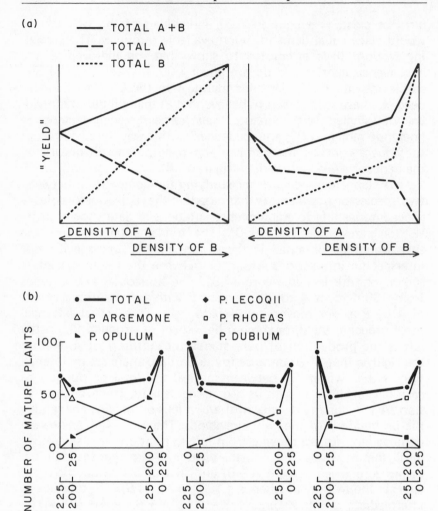

Figure 17.9 The use of replacement series (and their resultant "de Wit diagrams") in the analysis of intervarietal and interspecific competition. (a, left) A theoretical diagram illustrating the total production ("yield") of graded mixture of two species whose individual members interfere with neither one another nor with those of the second species. (a, right) A diagram showing, in contrast to the above, the total production of two species whose members do interfere with each other and with those of the competing species as well. (b) Illustrative examples chosen from experiments on poppies (*Papaver*). (After Harper and McNaughton 1962.)

of interspecific competition in poppies (Harper and McNaughton 1962), shows that results resembling the theoretical expectation are not at all uncommon.

Yet another graphical representation, permitting an analysis of interspecific (or intervarietal) competition was described in chapter 10 (see figure 10.11): log A/B input (parents) versus log A/B output (progeny). If the two species or varieties are precisely equivalent, a plot of the outcome of tests involving several ratios will be a line whose slope equals 1.00, passing through points such that log $Y = $ log X. A constant advantage (or disadvantage) of one species over the other will, as explained in the earlier chapter, generate a line, also of slope 1.00, that is displaced upward (or downward) from the first. In this case, log $Y = $ log $X + K$ or log $Y = $ log $X - K$.

Tables 17.5 and 17.6 present data on the outcomes of competition between mutant strains of *D. melanogaster* (see table 17.3c) that can be analyzed by means of a log input vs. log output diagram (figure 17.10). In the case of each competition, (1) *apricot-Bar* vs. *sepia* and (2) *sepia* vs. wildtype, the outcome of the competition appears to reflect a constant competitive superiority

Table 17.5 The total number of *apricot-Bar* (w^aB) and *sepia* (*se*) progeny produced by various numbers of mutant parental females (nonvirgin) in mixed cultures (totals are for four replicate vials per combination). (Wallace, unpublished.)

Sepia Mothers	*Apricot-Bar Mothers*									
	1		2		3		4		5	
	w^aB	*se*	w^aB	*se*	w^aB	*se*	w^aB	*se*	w^aB	*se*
1	153	256	243	240	252	210	437	210	446	154
2	69	483	175	432	208	368	350	319	416	294
3	51	527	167	470	236	427	305	453	329	397
4	82	682	133	550	194	626	283	407	374	388
5	75	621	138	596	217	557	275	463	321	478

Table 17.6 The total number of *sepia* (*se*) and wildtype (+) progeny produced by various numbers of parental females in mixed cultures (totals are for four replicate vials per combination). (Wallace, unpublished.)

Sepia Mothers	*Wildtype Mothers*									
	1		2		3		4		5	
	+	*se*	+	*se*	+	*se*	+	*se*	+	*se*
1	308	216	463	221	660	152	848	109	1,066	63
2	288	258	404	272	517	190	830	171	1,003	99
3	274	321	382	412	597	255	845	173	961	177
4	317	392	383	321	586	268	829	186	987	148
5	260	417	375	392	511	316	742	301	920	213

of one strain over the other. The slopes of both regression lines are exceedingly close to 1.00; both are displaced vertically. The open triangles and squares surrounding the lower regression line (*sepia* vs. wildtype) represent the outcomes of tests in which the numbers of parental females differ from six. The open triangles lying predominantly (9 of 10 points) above the regression line represent the outcomes of tests involving 2–5 parental females; the open squares lying predominantly (also 9 of 10 points) below the line represent the outcomes of tests involving 7–10 parental females. Because the distributions of the triangles and squares differ significantly, they show that the downward displacement of the regression line depends upon the numbers of parental females placed in the cultures. The outcome of competition between *sepia* and wildtype flies, therefore, is density-dependent.

Interspecific competition leads at times to regression lines (log input vs. log output) whose slopes are less than 1.00 (Ayala 1971, Wallace 1974); the outcome of competition in these cases depends upon the relative frequencies of the competing species—upon frequency-dependent selection. A regression with slope less than 1.00 implies that the two species will establish a stable equilibrium at which the two will coexist indefinitely. A regression

Figure 17.10 The use of the logarithms of input (abcissa) and output (ordinate) ratios in analyzing interstrain competition in *D. melanogaster*. The light solid line represents points at which input and output ratios are identical. The upper solid line and solid circles represent the outcome of an experiment (table 17.5) in which *sepia* and *apricot-Bar* females in ratios 1:5, 2:4, 3:3, 4:2, and 5:1 produced progeny in laboratory cultures. The stepwise dashed line indicates the course by which *sepia* would be expected to displace *apricot-*

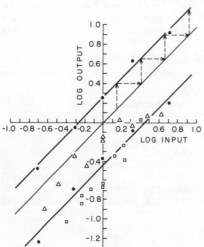

Bar flies if the carriers of the two mutants were reproductively isolated (different species). The lower solid line and solid circles represent the outcome of an experiment (table 17.6) in which *sepia* and wild-type females in ratios 1:5, 2:4, 3:3, 4:2, and 5:1 produced progeny in laboratory cultures. The open triangles and squares clustered about the lower solid line represent the outcomes of cultures in which the numbers of parental females were greater or less than six; the significance of these symbols are explained in the text. (Wallace, unpublished.)

with slope greater than 1.00 suggests that there exists a point of unstable equilibrium; on one side of the equilibrium point, one species will displace the other while on the other side the second will displace the first.

Competition between two species that is frequency-dependent and leads to a stable equilibrium is excellent evidence that the two species utilize available resources in dissimilar manners (see page 267). The diagram in figure 17.11 illustrates how a stable equilibrium between two species, A and B, can be established by the different efficiencies with which they use six resources (R_1-R_6). The relationships are shown as reciprocal: Species A can use resource number 1 with ease; B cannot use it at all; at the other end of the diagram, species B can utilize resource number 6 with ease whereas A cannot use it. Resources number 2 to number 5 are shared but, nevertheless, are not utilized with equal ease by the two species. The diagram is not as simplistic as it appears at first glance. Omitted from it are resources that neither species can utilize and those which both species use with ease. The latter (although they affect the outcomes of one-generation experiments comparable to those of tables 17.5 and 17.6) are not important in determining the outcome of long-term competition; one species

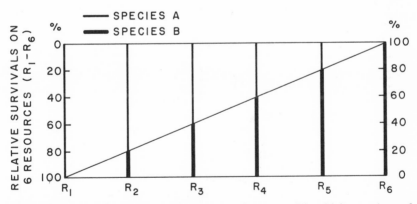

Figure 17.11 A diagram illustrating the relative ease with which members of two species (A and B) utilize each of six resources (R_1-R_6). A given amount of R_1, for example, permits the survival of all of a given number of A zygotes; the same quantity of R_2 permits only 80 percent of the same number to survive; the same amount of R_3 permits only 60 percent survival; and similarly with the remaining resources, R_4, R_5, and R_6. Precisely the reverse order is postulated for the utilization of these resources by members of species B. The relations between these resources, the competitive indexes of the Lotka-Volterra (and similar) equations, and frequency-dependent selection are discussed in the text.

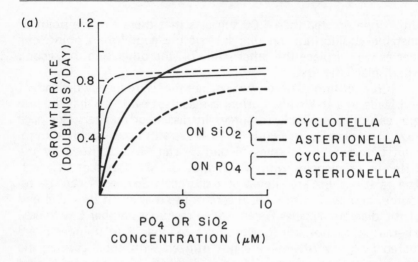

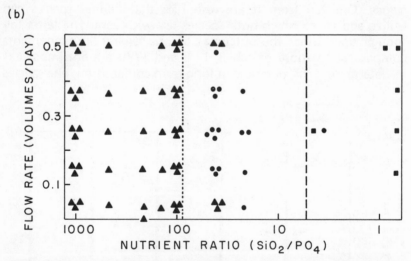

Figure 17.12 The outcome of growing mixed cultures of two species of algae (diatoms) under various $SiO_2:PO_4$ ratios. (a) The growth rate of the two algae plotted against the concentration of either SiO_2 or PO_4 alone. Note the reversal in the limiting effects of these two ions. (b) The outcome of 73 mixed-species experiments. In the left-hand zone, *Asterionella* eliminated *Cyclotella* (▲) as expected in all cases; in the right-hand zone, *Cyclotella* eliminated *Asterionella* (■) as expected in six of seven experiments (the remaining culture retained both species—●; in the center zone, both species were retained (●) as expected in 16 of 21 experiments (in the remaining five, *Asterionella* displaced its competitor—▲). (After Titman 1976; see Tilman 1977, as well.) Reproduced with permission: from *Science* (1976), 192:463–65; © 1976 American Association for the Advancement of Science.

will eventually displace the other no matter with what ease both utilize a given resource. Ultimately, an equilibrium involving two species must be based on differential utilization of two or more resources such as those illustrated in figure 17.11.

The needs of higher organisms are generally so complex that one can name their needed resources only with difficulty. Words, themselves, compound the difficulty because they can be used with a precision greater than that of the organisms' power of discrimination. Thus, in attempting to analyze the competition of two species, an investigator risks overlooking some important resources entirely while spending unnecessary time examining potentially different resources which are not discerned as different by the organism.

Matters are somewhat simplified by the use of organisms whose nutritional requirements are extremely simple. Levin et al. (1977), for example, have studied mixed populations of asexually reproducing *E. coli*; Smouse (1976; Smouse and Kasuda 1977) has done a considerable amount of work on populations of this organism. The results of competition studies using two species of diatoms are shown in figure 17.12 (Titman 1976; see Tilman 1977, as well). The two diatoms (*Asterionella* and *Cyclotella*) utilize silicates and phosphates in different manners as shown in figure 17.12a. Under phosphate-limited growth, *Asterionella* grows more rapidly than *Cyclotella*; under silicate-limited growth, *Cyclotella* grows more rapidly than *Asterionella*. From the data presented in figure 17.12a, predictions can be made concerning the outcomes of competition in mixed cultures of these diatoms in cultures containing various proportions of phosphates and silicates (see figure 17.12b). The diagram is divided into three sections: on the left, *Asterionella* should displace *Cyclotella*; on the right, *Cyclotella* should displace *Asterionella*; in the center, the two diatoms should coexist because they differentially use the two resources, silicate and phosphate. For the most part, the outcomes of 73 tests were as expected (triangles should appear exclusively at the left, circles in the center, and squares on the right); six exceptional results were obtained in all.

Concluding Remarks

The purpose of this chapter has been to illustrate that the demands made of the environment by different species are not

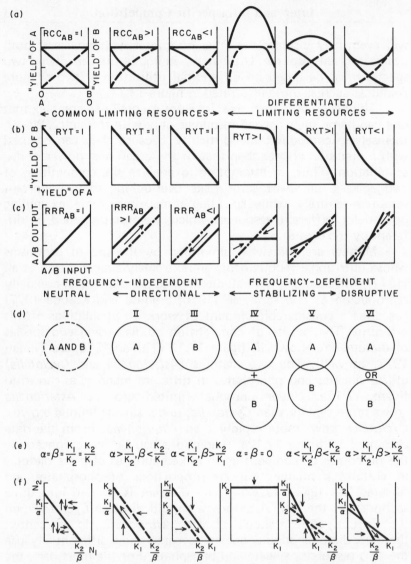

Figure 17.13 Diagrams illustrating the correspondence between various models that are used in analyzing the competitive relationships between species. The diagrams in each vertical column illustrate the same circumstance. The rows represent different models: (a) de Wit diagrams based on replacement series, (b) plots of joint yields, (c) log input *versus* log output graphs, (d) diagrams which represent pictorially what are otherwise verbal descriptions of resources (or niches; see, for example, figure 11.9), and (e) graphic analyses of the Lotka-Volterra equations. RCC: Relative crowding coefficient; RYT: Relative yield total (sum of relative yields of each species); RRR: Relative replacement rate. (After Harper 1977.) Reproduced with permission: from J. L. Harper; *Population Biology of Plants*; © 1977 Academic Press.

identical, and therefore that during periods of overcrowding, members of a species compete primarily with one another. Other nearby species can take "refuge" in resources that favor their survival; individual members of the same species to a great extent are denied such refuges.

Demonstrating that intraspecific and interspecific competition do, in fact, differ has occupied the bulk of the preceding sections. These demonstrations have relied on replacement diagrams, input/output ratios, and selection coefficients. The pictorial models used in illustrating the various examples are not unrelated; they attempt to make the same points but in different ways. Figure 17.13 which has been taken from Harper (1977: 303) presents the various models in a manner such that the diagrams in each vertical column, different as they appear to be, interpret the same competitive situation involving two species. This figure deserves careful study because it represents a Rosetta Stone of ecology.

BOXED ESSAY

The data presented in table 17.3 require an explanation too lengthy for the table's legend, but too extraneous for a discussion within the text. Consequently, this essay serves to explain the origin of the data summarized in table 17.3.

Imagine that gravid females of two species, A and B, have been placed in vials in all possible combinations ranging from 1 to 5 of each. Imagine, too, that having averaged many replicate vials of each combination, the following summary emerges (upper figure: progeny/female of species A; lower figure: progeny/female of species B):

Species B	Species A (# Females)				
(# Females)	1	2	3	4	5
1	43	25	18	15	11
	16	13	7	9	7
2	34	22	17	12	11
	14	10	8	6	6

Species B (# Females)	Species A (# Females)				
	1	2	3	4	5
3	25	17	12	12	11
	14	9	7	6	6
4	22	18	14	11	10
	10	7	6	5	4
5	14	15	14	10	10
	10	6	5	5	6

Because two species are involved, analyses of the comparative effects of adding conspecific and allospecific females must be performed twice, first with A as the conspecific and B as the allospecific, second with B as the conspecific and A as the allospecific species.

The effect of adding one conspecific and one allospecific female from the viewpoint of species A can be tabulated as follows (the reduction in the average number of progeny per female):

Adding 1 Allospecific Female	Adding 1 Conspecific Female				
	1 → 2	2 → 3	3 → 4	4 → 5	Total
1 → 2	18	7	3	4	32
	9	2	1	3	16
2 → 3	12	5	5	1	23
	9	5	5	0	19
3 → 4	8	5	0	1	14
	3	-1	-2	1	1
4 → 5	4	4	3	1	12
	8	3	0	1	12
Total	42	21	11	7	81
	29	10	4	5	48

A similar tabulation in respect to species B appears as follows (note that conspecific relationships with B normally would appear vertically but have been arranged horizontally to correspond with the earlier table):

Adding 1 Allospecific Female	Adding 1 Conspecific Female				
	1 → 2	2 → 3	3 → 4	4 → 5	Total
1 → 2	2	0	4	0	6
	3	4	5	3	15
2 → 3	3	1	2	1	7
	6	2	2	1	11

Adding 1 Allospecific Female	Adding 1 Conspecific Female				
	1 → 2	2 → 3	3 → 4	4 → 5	Total
3 → 4	-1 / -2	1 / 2	1 / 1	1 / 1	2 / 2
4 → 5	3 / 2	0 / 0	0 / 0	0 / 1	4 / 3
Total	7 / 9	2 / 8	8 / 8	2 / 6	19 / 31

When these data are combined, the overall effect of adding a conspecific and an allospecific female equals

		Conspecific	Allospecific
A	1 → 2	42	16
	2 → 3	21	19
	3 → 4	11	1
	4 → 5	7	12
B	1 → 2	7	15
	2 → 3	2	11
	3 → 4	8	2
	4 → 5	2	3
		100	79

The averages through all 8 instances when one female was added would be 12.5 (adding 1 conspecific female) and 9.9 (adding 1 allospecific female).

In the special case of going from 1 female to 2 (where the impact of the additional female is greatest), adding 1 conspecific female lowers the progeny per female for species A by 42 and for B by 7—an average of 24.5. Adding 1 allospecific female lowers the progeny per female for species A by 16 and for B for 15—an average of 15.5.

The calculations shown here are symmetrical in the sense that B was allowed to be allospecific to A, and that A was allowed to be allospecific to B. In table 17.3b where many species were involved, the data are symmetric in this same sense: for every species pair included, each was given the opportunity to be the allospecific competitor for the other.

FOR YOUR EXTRA ATTENTION

1. Collins and Tuskes (1979) studied how two sympatric species of moths, *Hemileuca eglanterina* and *H. nuttalli*, split the day for mating purposes; some of their data are summarized below:

Calling Female	Responding Male	Time	
		10:00–13:00	*13:00–18:00*
eleganterina	*eleganterina*	286	106
	nuttalli	3	22
nuttalli	*eleganterina*	0	0
	nuttalli	26	70

How might these species come to divide the day in this manner?
HINT: See figure 11.7.

2. The following account, cited from Halkka et al. (1977), illustrates both the simplicity and the complexity of interspecific competition: "The nymphs of these four species [of spittlebugs (Homoptera)] differ in their food plant choice. Between some of the species, there is considerable niche overlap in the utilization of the various food plant species. Nymphs of these species differ in their vertical distribution on the food plants. The nymphs produce spittle masses for their protection, and nymphs belonging to different species are frequently found within a single mass. Depending on conditions, this extreme form of coexistence may lead either to fruitful protocooperation or to the most intense competition possible."

3. The grasslands of the Central Valley of California often contain "vernal" pools, small pools of water 4–10 meters in diameter and about 35 cm deep at their deepest points. *Veronica peregrina* is an annual plant exhibiting considerable phenotypic plasticity that grows in these pools. Growth is most dense at the centers of the pool and rather sparse at the edges where *V. peregrina* must compete with surrounding grasses.

Looking at the populations of *V. peregrina* as populations involved primarily with intraspecific competition at the centers of the vernal pools and interspecific competition at the pools' edges, what sorts of phenotypic (and genotypic?) differences might be expected between "center" and "border" plants? Check your predictions with the findings reported by Linhart (1974) and Keeler (1978).

18

Darwinian Fitness and the Fitness of Populations

PREVIEW: The struggle to measure fitness was described in chapter 13. Behind this effort on the part of many persons was a belief that variation in Darwinian fitness represents a load or burden for a population, and that high average Darwinian fitness corresponds to high population fitness. An important component of population fitness, however, is *persistence*: the ability of the population to continue its existence through time. In the present chapter it will be shown that persistence is enhanced by variation in the competitive component of Darwinian fitness and, consequently, that high population fitness requires some sort of phenotypic (if not genetic) load.

THE STRUGGLE BY experimentalists to measure fitness was introduced in chapter 13. This struggle had several aspects: in part, it represented an effort to associate various facets of the life cycle (birth, longevity, reproduction) to the abstract mathematical symbols $(1-s,\ 1-t,\ W_{ij},\ \overline{W})$; it was, in this sense, an effort to identify the actual operation of natural selection (see chapter 10). In part, it attempted to rank different populations according to some concept of well-being—to rank them so that one population could be identified as being "better" or "more desirable" than another. The task was reasonably simple when it concerned the question of exposing human populations to mutagenic, high-energy radiation; the smaller the exposure to radiation, the lower the rate of mutation, and the lower the incidence of genetic illness among the members of the population. From a concern over the possible genetic damage in human populations (Muller 1950b, for example), it was but a short step to the speculation that natural selection acts to minimize genetic load (Kimura 1960).

The struggle in estimating the fitness of populations revolved about the question, "what should be measured?" Experimenters quickly found that, as a rule, circumstances and logic forced them to measure the same components of fitness that were measured when estimating the relative (= Darwinian) fitnesses of various genotypes *within* populations. In ranking the relative fitnesses of populations, population geneticists accepted absolute measures of fecundity, population size, and intrinsic rate of increase under the assumption that the greater (or the more, or the larger) the better. If one teaspoonful is good, two teaspoonfuls must be better. Unfortunately, the estimated fitnesses of populations were not always correlated with expectations based on the radiation histories of these populations. The results of different sorts of experiments were discordant.

The question of population fitness was made immensely more complex by Thoday (1953), who suggested that the *persistence* of the population should be included in the concept of population fitness. The complexity arose from a need to foretell the future if persistence were to be included. Slobodkin (1968b) carried the matter much farther by comparing a population (or a species) to a participant in an existentialist poker game: the player can never win in the sense of pocketing his winnings (or cutting his losses) and departing; the best he can do is to avoid losing, to avoid extinction. Slobodkin's colorful analogy permits one to consider alternative strategies that might be employed in lessening the probability of extinction; clearly, large population size is not necessarily a good strategy. The flush-crash sequences reviewed in chapter 16 would, in that case, suggest that periods of high fitness may immediately be followed by periods of greatest uncertainty.

The sorts of problems faced by populations were reviewed by Slobodkin (1968b) together with means by which populations might cope with each of several kinds. He suggests that adaptive responses occur at several levels: behavioral responses (the individual withdraws from an unpleasant situation), physiological responses (unable to withdraw, the individual undergoes appropriate physiological changes), and long-term quasi-evolutionary responses (forced to live permanently under what are unpleasant circumstances, natural selection alters the genetic composition of the population; the population *adapts* to the prevailing conditions).

A complementary classification of the problems confronted by a population of individuals might be built as follows:

1. *Cyclic problems that recur on a set schedule.* Genetic solutions for these problems can be built into DNA or, in the case of animals, the central nervous system can be entrusted to develop appropriate ad hoc solutions. The cycle of seasons in the temperate zones of the Earth can serve as an example. Plants in these zones have developed a series of gene-mediated responses to the coming of winter and the reoccurrence of warm weather in the spring; each of these responses contributes to the survival of the plants involved. Animals also respond to the coming of winter by producing overwintering eggs and larvae, by deep hibernation, by fitful hibernation that is interrupted by each warm spell, or by continuous activity throughout the winter season. To ascribe the solution of a problem to DNA may be an unfamiliar way of speech, but it is accurate, nevertheless. That is what natural selection is all about. The case was expressed exceedingly well by Dr. Lewis Thomas (1974) upon contemplating the possibility that he might operate his liver by conscious thought; he admitted that he is constitutionally unsuited for making hepatic decisions. The solutions to internal problems are best left to the cybernetics and feedback loops of cellular genetics and biochemistry.

2. *Problems that recur at irregular intervals*, poorly correlated with previous environmental conditions. Winter arrives annually on the twenty-first of December only to be displaced by spring on March 21. Although calendar dates are of little use to a maple tree in regulating the annual flow of sap, the increasing length of daylight and increasing average daily temperatures give hints that spring is on its way. Suppose, however, that serious problems occur at irregular intervals in an entirely capricious manner (see page 19). What devices are available to the population in this case? The bulk of the chapter will be devoted to one such problem—extreme overcrowding—and the means by which the dangers of overcrowding can be lessened. The discussion of this point is, consequently, deferred for the moment.

3. *Severe problems that arise sporadically*, but at extremely rare intervals. The distinction between these problems and those that might fall within the previous category hinges on the frequency of their occurrence. Unpredictable chance events may, despite

their uncertainty, occur at frequent intervals; any existing population must have dealt with such problems repeatedly. Unpredictable events of extreme rarity may never have occurred during the history of a population (or species). We can only admit, then, that such an event when it occurs may be devastating and may even exterminate the population. This conclusion parallels Schmalhausen's (1949) claim that alterations in phenotype that are induced by normal variations in the environment are adaptive modifications; those that occur in response to unusual or artificial environments are nonadaptive morphoses. "Modifications," to cite Dobzhansky (1970:38), "are forged in the evolutionary history of a species; morphoses are 'new reactions which have not yet attained a historical basis.'"

Overcrowding

Overcrowding has been mentioned several times as a threat to a population's existence. It was mentioned in conjunction with the flush-crash "cycle." It was also mentioned in conjunction with the enormous numbers of seedlings and larvae that many plants and animals regularly produce; offspring that must be reduced to a number corresponding to that which the *prevailing* (not the *last*) environment can support. Finally, overcrowding was identified in the preceding section as a problem that can arise with no warning and at unpredictable intervals. Two questions must now be considered: Is overcrowding a serious problem for a population? If it is, how best can a population handle the matter when it arises?

Nicholson (1957) reported on the size fluctuations that occurred in a population of sheep-blowfly, *Lucilia cuprina*, that was maintained in a laboratory cage which received a daily quota of 50 grams of beef liver. The numbers of adults in these cages would increase rapidly from very few to 2,500–3,500 individuals only to crash almost at once to nearly zero, after which the numbers rose rapidly once more. Nicholson found that when the number of adults was high, eggs were laid in such quantities that *no* larvae of those periods survived. Only when the adults had crashed to a low point could a few larvae survive, only to restart the cycle once more. Had the 50 grams of food been introduced into the popu-

Table 18.1 The yield in terms of numbers of pupae or of adults of the house fly, *Musca domestica*, obtained from cultures differing in initial larval densities. (After Bøggild and Keiding 1958.)

Number of eggs:	40	40	80	80	160	160	320	320	640	640
Initial larval density:	37	38	73	68	132	141	283	279	580	565
Number of pupae:	36	35	65	64	111	113	177	181	163	154
Number of adults:	33	35	64	59	105	107	163	155	124	131
Proportion surviving										
pupae	.973	.921	.890	.941	.841	.801	.625	.649	.281	.273
adults	.892	.921	.877	.868	.795	.759	.576	.556	.214	.232

lation at irregular intervals determined by chance, the observed cycles would not have recurred at regular intervals, and extinction of these populations may easily have been a common occurrence.

The housefly has been used by Bøggild and Keiding (1958) to illustrate the same point. Eggs were transferred to fresh cultures in the numbers shown in table 18.1. Because of the occasional failure of an egg to hatch, the initial larval densities in these cultures are also listed in the table. The proportion (p) of pupae and successfully emerging adults declined linearly with increasing larval density. The equation in the case of adults could be written.

$$p = \frac{i}{X} = 0.952 - 0.00129\,X$$

where i = the number of imagoes (adults) and X = the starting number of competing larvae. Multiplying both sides by X, we obtain

$$i = 0.952X - 0.00129\,X^2.$$

The parabola that fits this equation is shown in figure 18.1. The number of adults under these experimental conditions is expected to be zero when $X = 0$ and when $X = 0.952 \div 0.00129$ or, approximately, 740.

The authors refer to the unchanging proportion of pupae that successfully transform into adult flies; such a constant proportion could serve as an example of density-independent selection standing in sharp contrast to larval survival which is clearly density dependent. However, there is reason to believe (and I leave it to the reader to confirm this claim) that the proportion of pupae from which adult flies successfully emerge declines from an intercept of about 0.976 by 0.00030 for each starting larva so that it falls to 0.806 at the highest densities studied. I regard this decline as a

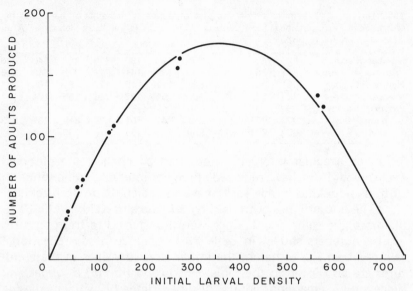

Figure 18.1 The parabolic rise and fall in the number of adult house flies yielded by cultures in which larvae were placed at different initial densities. (After Bøggild and Keiding 1958.)

measure of the imperfect judgment of larvae which, when under competitive stress in overcrowded cultures, pupate at the earliest practical moment; at times, they pupate too early and fail to emerge as adults.

On several occasions, the claim that overcrowding threatens a population's existence has been coupled to the further idea that such overcrowding may occur with little or no advance warning. That is, for many natural populations of all kinds there are good "years" and bad "years," and that a bad year may unexpectedly follow immediately after a good one. That would, according to the argument being developed here, be the worst of all possible circumstances.

The role of unexpected occurrences of bad "years" has been subjected to test in the laboratory (Wallace 1979b). One hundred cultures of wildtype *D. melanogaster* were started using a culture medium consisting of agar, dead brewer's yeast, and sucrose. Chloramphenicol and proprionic acid were added to inhibit the growth of microorganisms; the growing larvae, consequently, obtained their carbon from the sugar for the most part, and the remainder of their nutrients from the brewers yeast (of which there was always 20 grams per 500 cc of medium). These cultures were maintained by transferring *all* of the progeny of one generation into fresh vials as parents of the next generation; for each vial in each generation, then, the number of parent flies is known as well as the number of progeny these parents produced (a number that equals the number of parents in the succeeding generation).

In fifty of the 100 vials, the amount of sucrose was held constant at 50 grams per 500 cc of medium. In the other 50 vials the amount of sucrose was allowed to vary haphazardly according to random digits generated by a small pocket calculator. The results obtained from these two sets of vials (each maintained through 10 generations but sustaining losses in the meanwhile) are summarized in tables 18.2 and 18.3, and in figure 18.2. In vials within which the amount of sugar was constant, a fairly linear relationship holds between numbers of parents and number of progeny they produce. Perhaps the number of progeny decreases when that of parents exceeds 200 (recall that these would be exceedingly small, semistarved parental flies capable of laying only a few eggs each). Above 130–150 parents, the average number of progeny recovered is smaller than the number of parents themselves. In contrast, in vials within which the amount of sucrose varied erratically from generation to generation, the average number of progeny falls rapidly after the number of parents exceeds 130–150. Table 18.4 shows the history of 14 vials that failed to produce any progeny and, consequently, were lost in the sixth generation: in generation number 4, the grandparents of these vials had all numbered 130 or more. In generation number 5, the quota of sucrose was halved (see table 18.3) and the number of progeny produced by these numerous parental flies was very low.

Table 18.2 The mean number of progeny flies (*D. melanogaster*) produced per vial on constant food where all progeny flies of the previous generation are transferred to fresh vials as parents. (After Wallace 1979b.)

Parental Generation	Number of Parental Flies (midpoint)												
	10	30	50	70	90	110	130	150	170	190	210	230	250
2	45	84	48	60	72	67	102	94	—	—	—	—	—
3	112	117	135	129	133	131	155	—	—	—	—	—	—
4	—	—	—	103	92	138	147	107	47	—	—	—	—
5	3	23	144	112	146	147	160	162	113	178	161	157	78
6	46	—	156	116	123	160	157	156	199	207	193	152	—
7	26	92	153	147	149	157	182	171	164	143	160	173	186
8	—	51	—	—	134	133	128	129	142	142	129	135	—
9	—	—	99	86	116	142	141	139	146	140	—	—	—
10	—	—	—	—	106	133	143	147	137	172	—	—	—
Unweighted Average	46	73	123	108	119	134	146	138	135	164	161	154	132

Table 18.3 The mean number of progeny flies (*D. melanogaster*) produced per vial on variable food where all progeny flies of the previous generation were transferred to fresh vials as parents. (After Wallace 1979b.)

Parental Generation	Grams of Sugar		Number of Parental Flies (midpoint)												
	Parent Vial	Progeny Vial	10	30	50	70	90	110	130	150	170	190	210	230	250+
2	40	50	76	90	63	71	113	132	113	113	—	—	—	—	—
3	50	40	88	107	133	141	144	134	141	149	138	—	—	—	—
4	40	20	—	—	116	72	8	25	30	16	3	3	—	—	—
5	20	40	12	84	—	119	98	—	—	—	—	—	—	—
6	40	20	66	83	—	22	32	57	22	20	3	—	—	—	—
7	20	50	26	38	—	136	106	157	—	137	—	—	—	—	—
8	50	0	7	34	44	64	100	75	81	57	60	[a]	—	—	—
9	0	0	42	10	50	60	38	—	—	—	—	—	—	—	—
10	0	40	13	79	51	97	105	—	—	—	—	—	—	—	—
Unweighted Average			41	66	76	87	81	94	77	82	51	3	—	—	—

[a] Vial lost by accident.

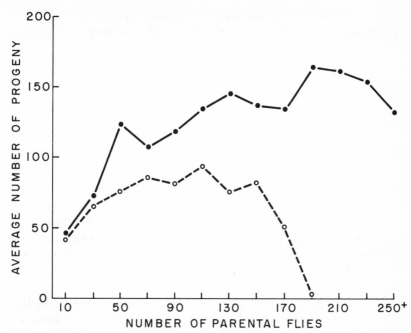

Figure 18.2 The average number of progeny flies (*D. melanogaster*) produced in vials containing various numbers of parental flies (see tables 18.2 and 18.3). Solid line and circles; constant food; dashed line and open circles: variable food as explained in text. (After Wallace 1979b.)

Table 18.4 The average number of flies (*D. melanogaster*) produced per vial in the fifth and sixth generations by vials which, in the fifth generation, had received various numbers of fourth-generation progeny as parental flies. "Zero" is the number of vials yielding no progeny at all; it can be seen that the few progeny flies hatching in some vials of the fifth generation were unable to reproduce. (After Wallace 1979b.)

Progeny in Parental Vials (Gen. 4)	Number of Vials	Generation 5		Generation 6	
		Number of Progeny	Zero	Number of Progeny	Zero
50	2	116	0	51	0
70	1	72	0	145	0
90	1	8	0	12	0
110	6	25	0	46	0
130	13	22	0	44	2
150	13	16	0	20	7
170	4	3	0	3	3
190	2	3	0	0	2

Then, in the following generation (number 6) despite an increase in the sucrose allotment, 14 of the 45 vials present in generation number 4 were lost. Clearly, overcrowding combined with subnormal culture conditions caused the loss of these overcrowded populations.

Self-thinning: The Solution to Overcrowding

Overcrowded cultures (laboratory) or areas (natural populations) may or may not thin themselves successfully. The maple seedlings shown in figures 16.1 and 16.2 did thin themselves successfully. The genetically dissimilar, outcrossed cabbages shown in figure 1.6, thinned themselves successfully. The sheep-blowflies (page 472) did not, nor did the fourteen fly populations described in tables 18.3 and 18.4.

Periodic censuses of plant populations have been made by a number of workers in an effort to reconstruct their life tables. The results of one such study (Sarukhan and Harper 1973) involving three species of buttercups are summarized in table 18.5. Here we have the number of plants in one April, that of newcomers, and the sum of the two numbers. The number remaining after two years is also given, thus allowing a calculation of the number of plants lost. As the table shows, about two plants in three are lost to the population. Unless these plants produce remarkably few seeds, other losses occurred without having been recorded.

A more ambitious attempt at reconstructing the life-table statistics of a plant has been attempted by Van Valen (1975) for a grove of palms, *Euterpe globosa* (table 18.6; figure 18.3). In Van Valen's study we sense the tremendous thinning that occurs

Table 18.5 A summary of detailed censuses of three buttercup (*Ranunculus*) species inhabiting one square meter areas over the course of two years. (After Sarukhan and Harper 1973.)

	R. repens	R. bulbosus	R. acris
Number of plants: April 1969	1,431	548	272
Number arriving in area	2,364	1,509	928
Total plants recorded	3,795	2,057	1,200
Number of plants lost	2,600	1,229	767
Number remaining: April 1971	1,195	828	433
Number 1969/Number 1971	1.20	0.66	0.63
Average '69-'71/Total recorded	0.35	0.33	0.29

Table 18.6   Life-table statistics for the palm *Euterpe globosa*. (After Van Valen 1975.)

Stage	Height (m)	Age (years)	Number per 5,926 m^2 per Year-Class	Surviving Proportion of Initial Cohort
Seed on tree	—	0	170,000	1
Seed on ground	—	0.1	46,000	2.7×10^{-1}
Germinated seed	—	1	2,200	1.3×10^{-2}
End of seedling	0.5	9	205	1.2×10^{-3}
Immature tree	1.0	15	80	4.7×10^{-4}
	2.0	24	40	2.4×10^{-4}
	3.0	30	18	1.1×10^{-4}
	4.0	36	7	4×10^{-5}
	5.5	45	3	2×10^{-5}
Reproductive maturity	6.5	51	0.5	3×10^{-6}
	9.0	66	0.2	1×10^{-6}
In canopy	12.0	88	0.2	1×10^{-6}
	14.0	104	0.5	3×10^{-6}
	16.0	130	0.3	2×10^{-6}
Senescent	18.0	156	0.1	6×10^{-7}
	20.0	182	0	0

in a stand of trees whose more or less stable number of mature trees represents only one-millionth of the seeds initially formed. The curve shown in figure 18.3 can be interpreted as illustrating three periods in the palm's existence during which survival drops exponentially with time but at very different rates: An early period of about one year's duration when upward of 99 of every 100 seeds are destroyed, an intermediate period of some 70 years during which young plants decrease exponentially after which only about one seedling remains for every 1,000 present at the outset; and a third period during which mature trees decline in number very gradually or not at all until senescence sets in after some 150 years.

Phenotypic Variation and Self-Thinning

An important point made first in describing the single-unit model (page 17) and again following the discussion of flush-and-crash sequences (page 432) is that a population of organisms can thin itself efficiently only if the individuals involved differ *phenotypically* in respect to the competitive component of fitness. In the case of the maple seedlings (figure 16.2), time of germination may have given one seedling of each of the competing pairs an advantage over the other. Among the seedlings studied, there was one that showed extraordinarily rapid growth; consequently differen-

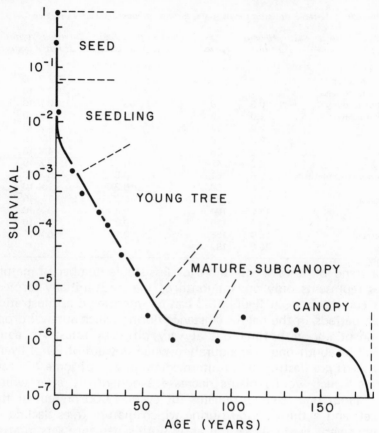

Figure 18.3 Survivorship curve for the palm *Euterpe globosa* extending over seven orders of magnitude, thus illustrating the extent to which initial cohorts of this species are thinned (self-thinning or otherwise) before the quasi-stable population of mature plants is arrived at. (After Van Valen 1975.)

tial growth rates of young seedlings might at times offset slight differences in times of germination.

Among animals, phenotypic variation associated with age provides an efficient procedure leading to a reduction in population number during periods of overcrowding. Figure 18.4 illustrates the role of age in deciding the outcome of selection between wildtype and *sepia D. melanogaster*. In pure cultures, wildtype and *sepia* flies produce about the same number of progeny per vial, about 120 flies. If (as table 17.6 shows) females of the two strains are placed in culture vials simultaneously (four females of each),

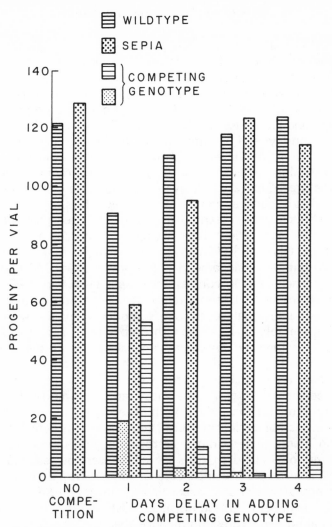

Figure 18.4 A demonstration of the phenotypic variation in competitive fitness resulting from age differences within *D. melanogaster*. The two leftmost bars represent the average number of progeny per vial produced by 4 nonvirgin wild-type (crosshatch) or *sepia* (stippled) females. The remaining graphs reveal the effects of adding *sepia* females as competitors to vials containing wildtype ones, and of adding wild-type females as competitors to vials containing *sepia* ones; the competing females are added after delays of 1–4 days. Cultures (not shown in this figure) in which 4 *sepia* and 4 wild-type nonvirgin females are introduced simultaneously could be expected to yield about 20 percent *sepia* progeny (see table 17.6).

about 20 percent of all progeny produced are *sepia*; the remainder are wildtype. Roughly the same result is observed if *sepia* females are added to the vials some 24 hours after the wildtype ones. On the other hand, if *sepia* females are placed in the vials first and the introduction of wildtype is delayed 24 hours, the proportions of wildtype and *sepia* flies are about equal. Delays of 2, 3, and 4 days in adding one type of parental females to the vials virtually guarantees that larvae of the earlier strain control the utilization of the culture medium. The production of *sepia* flies from vials in which wildtype are added after delays of 3 and 4 days is at the same level as that of cultures containing only *sepia* flies.

Although data presented in figure 17.1 (see page 443) suggested that seed weight is an extremely constant aspect of the phenotype (Harper 1977:201 reports a mere 1.04 variation in mean weight per grain among plants sown at an 816-fold range of sowing densities; in contrast, seeds per plant varied 833-fold), the constancy seems to apply to *average* weights rather than to the weights of individual seeds. Seeds produced by the same plant vary considerably in size; even the proverbial peas in a pod, renowned for their sameness, vary considerably as the data presented in table 18.7 and figure 18.5 show. Variation in seed size offers two types of phenotypic variation in respect to fitness: small seeds tend to germinate most rapidly; large seeds contain more nutrients for the early growth of seedlings.

Considerable evidence exists suggesting that dense stands of uniform plants suffer from overcrowding. Foresters refer to the *stagnation* of a stand of trees which is overcrowded and whose individual members are incapable of expressing "dominance". A wide range of size classes, according to Hawley (1946:222), indicates an expression of dominance. Eastern pine, he continues, has the ability to express dominance (because of unequal growth rates); the black waddle (used for tanbark) is poor at expressing dominance and so stands must be thinned manually from 40,000 seedlings per acre to 200 trees (each 40-feet tall) over a period of three years.

Table 18.7 The proportions of vestigial (aborted), small, and normal-sized garden peas at various positions in the pod (position 1 is at the peduncle end). (After Linck 1961.)

Size	Position in Pod						
	1	2	3	4	5	6	7
Vestigial	59	22	0	0	19	41	54
Small	41	59	41	12	31	44	46
Normal	0	19	59	88	50	16	0

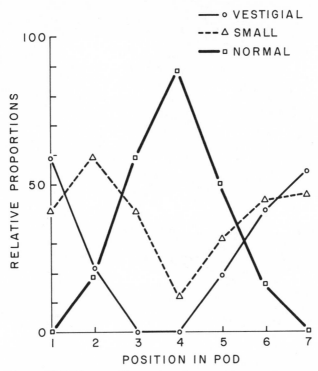

Figure 18.5 Diagram illustrating that peas in a pod are not necessarily alike. Solid line and circles: vestigial or aborted peas; dashed line and triangles: small peas; heavy line and squares: normal-sized peas. (After Linck 1961.)

A second illustration of the fatal outcome of gross over-crowding is the so-called "Grigg effect" (Grigg 1958) of micro-biologists. Reverse mutations in bacteria and fungi are detected by plating mutant auxotrophs in large numbers on minimal media where they are unable to grow; rare reverse mutations that restore prototrophy to their carriers are detected by the occurrence of rare colonies. Grigg discovered that these rare revertants often fail to grow, however, even though they are present; the carpet of auxotrophic individuals, merely by their presence and despite their inability to grow significantly and divide, may remove limiting substances from the medium so that even "normal" individuals are unable to grow.

Figure 18.6 illustrates an effect of overcrowding in holy-hocks. Seeds densely planted in a short furrow in late summer, 1973, produced numerous small plants in 1974. In each of the fol-lowing years, the number of plants at the site of the original fur-

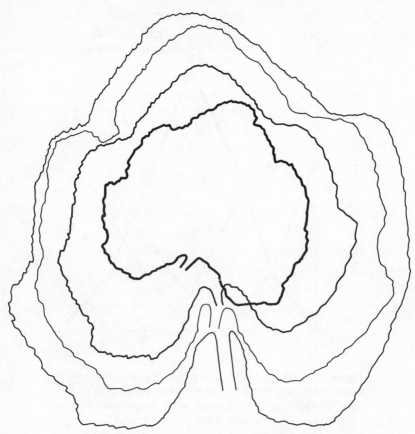

Figure 18.6 The relative sizes of leaves from hollyhock (*Althaea rosea*) plants, one of which (center leaf) was the largest from one of three plants that survived four years of self-thinning in an initially overcrowded bed; the other leaves were taken from nearby, single plants.

row dwindled until, in 1978, there were only three plants remaining. The small size of these plants is illustrated in the figure by comparing a large leaf of one with more or less average leaves from nearby, isolated plants. Thinning in the case of these hollyhocks was not accomplished easily, or without lasting harm. The point made here and in the preceding anecdotal accounts is well-known to farmers and gardeners who appreciate the harm that accompanies overcrowding and the benefit that results from thinning growing seedlings by hand.

Self-thinning can be accomplished by a variety of means including, for predatory animals, cannibalism. In an article reviewing the role of cannibalism in natural populations, Fox (1975) cites a number of examples that bear on the present discussion. Only one of four species of sheep blowfly, *Chrysomyia albiceps*, studied by Ullyett (1950) was cannibalistic, and that was the only species that persisted when food was scarce (compare with the study described on page 472). According to Paine (1965) cannibalism in the mollusc, *Navanax inermis*, is density-dependent, and may serve as the major factor regulating intertidal populations. Fox also reports that cannibalism among back swimmers, *Notonecta hoffmanni*, increases with decreasing food; under dire conditions, cannibalism ensures that the oldest nymphs are the ones most likely to survive and reproduce. A similar cannibalism of young siblings ("cainism") is practiced by the progeny of hawks and other rapacious birds; by laying eggs at three-day intervals, the parental female promotes a size difference among the young which eases decisions when and if self-thinning becomes necessary. Combat between parasitic Ichneumon flies infecting moth larvae assures that each host larva will contain but one of the larval parasites. Older and larger parasites suppress younger ones by combat during which the smaller one is punctured and thus exposed to encystment by the host larva. Older and larger parasitic larvae also secrete inhibitory substances into the host's endolymph; in this case, the young parasitic larvae simply become quiescent and die (Fisher 1959).

Self-thinning and Heterosis

The discussions presented in this and the preceding chapters have now converged so that they can be combined into one account— an account that bodes well for the persistence of a population, an account that describes the attainment of high population fitness.

In earlier chapters, especially in chapters 14 and 15, evidence was presented (an an appropriate model was described) suggesting that, for many gene loci, heterozygosity is selectively favored over homozygosity. The model developed by Wallace and Kass (1974) is essentially a model of gene action leading to the selective advantage of "heteozygosity per se." It is a model, however, that is in terms which generate testable predictions. Opposing a model that is based on the ubiquitous advantage of heterozygotes at many loci are the genetic load calculations (chapter 12) purporting to prove that the segregational load imposed by such an array of selectively inferior, segregating genotypes would be too great for any population to bear.

The present chapter has dealt with the frequent need for a population to thin itself of individuals too numerous to be supported by the environment. At times, this need extends to the weeding of adult individuals that have survived under a previously permissive set of favorable circumstances; more often it applies to the recurrent need to reduce an enormous number of larvae, juveniles, or germinating seedlings to a much smaller number of functional, reproducing adults. In the case of palm trees (figure 18.3), this reduction of young zygotes extended over a range of six orders of magnitude.

Self-thinning, consequently, depends upon phenotypic variation in fitness. The superiority of heterozygotes leads to segregating genotypes whose corresponding phenotypes are variable in respect to fitness. Consequently, rather than generating an unwanted (and unbearable) genetic load, the ubiquitous advantage of multiply heterozygous individuals generates the phenotypic variation which is required for self-thinning—that is, for the successful persistence of a population under normally varying conditions. Defining a load (genotypic or phenotypic) as the proportional amount by which the average fitness of a population is depressed relative to the optimum genotype (or phenotype) makes little sense in a world in which the elimination of nearly all zygotes is a requisite for a population's continued existence. There is, however, an important point that must be remembered here; namely, by common agreement, human beings do not accept the elimination of their young as a "natural" way of life and, consequently, genetic load calculations are not entirely inappropriate when applied to the public health aspects of genetic disease.

A Concluding Speculation

The case that has been built in this chapter rests on two substantiated claims: (1) that the potentially terminal consequences of overcrowding can be avoided by a population whose members vary in respect to fitness (specifically, in respect to competitive ability) and (2) that the selective superiority of largely heterozygous individuals provides (a) the array of individuals that differ in fitnesses and (b) the survivors who, upon reproducing, are capable of regenerating once more this necessary array.

In concluding this chapter, I wish to extend the case outlined above: if the *proximate* genetic mechanism does not provide the necessary array of variation in fitness, *ultimate* genetic mechanisms capable of providing it will be established by natural selection. Such ultimate causes result in the variation among peas in each pod (figure 18.5), variation that plant breeders have to a large extent eliminated by artificial selection; today's food industry prefers uniformity of agricultural produce. The same ultimate causes lead to the decreasing gradient of milk production extending from a sow's anterior teats to posterior ones (E. O. Wilson 1975:288); aggressive piglets claim the more productive anterior teats while runts are shunted to the rear. Should the level of the sow's milk production be lowered for any reason, the larger, aggressive piglets are more likely to survive than would any piglet if each had received an inadequate amount. The same argument applies to birds of prey that lay their eggs at intervals, while incubating them continuously; this behavior (with its ultimate genetic basis) provides the variation in fitness that our argument demands.

Can the argument developed in this chapter be extended in the manner just described without invoking group selection (selection that occurs by the differential extinction of entire populations)? I believe that it can be. First, a reminder concerning the argument that has been developed throughout the past six chapters: the variation in fitness found in a population was said to result from the superior average fitnesses of heterozygous individuals; *by lucky happenstance*, this variation is useful in assuring the persistence of the population through periods when self-thinning is necessary. In that connection, it was useful to show that overcrowding is primarily an intraspecific affair.

In order to extend the case for self-thinning to phenotypic variation not of (proximate) genetic origin, the contrast between intraspecific and interspecific competition is essential. If the two types of competition do not differ, those genotypes will be preserved whose carriers produce the greatest number of offspring possessing uniformly high fitness. Given that intraspecific and interspecific competition do differ (as the data presented in chapter 17 suggested), and given that the carrying capacity of the environment fluctuates capriciously, then those individuals who produce progeny of varying fitnesses (in respect to competitive ability) will succeed better in the long run than those who produce progeny of uniform fitnesses, provided that the *average* fitnesses of the two batches of progeny are equal. The advantage residing in the production of variable (in respect to competitive ability) offspring is revealed by considering the following extreme cases: During times of plenty (on the "up" side of a population flush), nearly all individuals survive whether variable or uniform; during stable periods, or, even more importantly, during periods when the population crashes, only the most competitive progeny survive. *The surviving progeny during these periods are the better ones of variable broods.* The resemblance between these extreme cases and the illustration of soft selection given on page 421 is not accidental; soft selection is a sine qua non of much that has been written in this chapter.

Summarizing Remarks

In introducing part 2, I said that its first six chapters (chapters 13–18) would constitute the third movement if this were a symphonic score rather than a textbook of population genetics. All six chapters dealt, each in its own way, with a question first raised in chapter 12: Is the neutralist theory necessary?

Two assumptions appeared to demand the neutrality of most observable genetic variation (see Kimura 1968): (1) the magnitude of the substitutional load as calculated by Haldane (1957, 1960) as the cost of evolution and (2) the magnitude of the segregational load that would result if observed allozyme (and other protein) variation were to be maintained by balancing selection (Kimura and Crow 1964).

Neither assumption is compelling. In chapter 12, Haldane's cost of evolution was shown to be a spurious cost, a computational artifact. During the past six chapters, the segregational load has been shown not only to be no threat to a population's existence but, on the contrary, to contribute to a variation that is essential for the population's persistence through time. Genetic load has little or no bearing on the ecological aspects (that is, on numbers or kinds of individuals) of natural populations of most organisms. The notable exception is, of course, the human population where the ideal stabilizing conditions are thought to be (1) an average of two children per couple, (2) no deaths before child-bearing age has been exceeded, and (3) a population size compatible with the resources available and the demands placed upon them.

FOR YOUR EXTRA ATTENTION

1. Consider the following statements which appeared in a paper by Mourão et al (1972): "The relationship between Darwinian fitness and adaptedness deserves further investigation. It is unfortunate that experimental studies have almost completely neglected this problem, and that theoretical investigations often assume that the average Darwinian fitness of a population is also a measure of its ability to survive and reproduce. This is clearly not so."

Has the material of the preceding six chapters helped (1) in understanding these authors' complaint, and (2) in understanding the relationship between the two kinds of "fitness?"

2. The following comments in a book review by Tinkle (1978) bear on some of the statements made in this chapter on the *persistence* of populations as an essential component of the *fitness* of populations: "One persistently distressing aspect of the book [*Reptile Ecology* by Harold Heatwole] is a tendency to explain phenomena by a group-selection mechanism. For example, there is the argument that small clutches of tropical reptiles are perhaps compensated for by a longer reproductive season. . . . Heatwole also concludes that multiple broods in annual species are advantageous because, were they single brooded, reproductive failure could result in a disastrous decline in the population. Then there is the argument that territoriality provides a mechanism by which at least a portion of the population can obtain adequate food and that selection should favor attributes that maximize stability of populations."

How would you restate Heatwole's position in making it acceptable to Tinkle? Alternatively, how would you respond to Tinkle's objections?

HINT: Recall that persistence is a multiplicative concept; a zero introduced into the series of terms that are to be multiplied reduces the product to zero.

What attribute of a genotype (an individual, not a population) is most likely to assure that at least some of its offspring will survive to reproduce in a temporally and spatially erratic environment?

3. The "self-thinning" and "automatic culling" of this chapter has been referred to by others as "truncation selection." Kimura and Ohta (1971: 155) have suggested that natural selection does not mimic artificial selection, and that it does not count the number of heterozygous loci and cull an individual if its number is below some critical value. Nei (1975:62) argues that truncation selection is possible only when competition occurs just once in life for a single limiting resource. Crow and Kimura (1979), while conceding that truncation selection might absorb the mutational load, are unwilling to extend its influence farther. To them, rank-order (i.e., truncation) selection assumes that genes affecting fitness contribute additively ($aa < Aa < AA$; contrast this view with the spectulations of chapter 15), and they point out that various components of fitness, especially among individuals of high fitness, are negatively correlated (recall figure 14.6).

Any consideration of the arguments of Crow and Kimura should include reference to Robertson (1955) who pointed out that if additive genetic variation exists in a population in respect to a component of fitness, either the population is not at equilibrium or various components of fitness must be negatively correlated. (Why?)

The material in this chapter and elsewhere in this book has stressed nonadditive genetic interactions in the determination of fitness (in this regard, refer once more to figure 9.8 and its suggestion that both high and low viabilities may be the result of ephemeral genetic interactions).

To defend the position that self-thinning provides a means by which segregational loads are absorbed and rendered unimportant in respect to population ecology, one must be able to respond to the opposing view that truncation selection is ineffectual and, therefore, of little consequence in affecting the genetics of populations.

Persistent Polymorphisms

PREVIEW: A number of stable equilibria were discussed in Part I: forward *vs.* reverse mutations, mutation *vs.* selection, rare genotype advantage, and selection in favor of heterozygotes. In addition, mutation can maintain a stable (though dynamic) number of segregating neutral alleles in a finite population of sufficient size. In the present chapter, naturally occurring polymorphisms whose maintenance is the result of natural selection will be described. Of special interest are those instances in which the polymorphisms, themselves, have caused secondary genetic changes.

DURING THE PRECEDING six chapters, an argument has been developed claiming, in effect, that the standard multiplicative model by which genetic loads are calculated has little or no bearing on the persistence of populations through time—that is, on the survival of populations. The polymorphisms that prompted this critical review of genetic load theory (specifically, of segregational load theory) were the presumably ubiquitous ones said to be associated with the efficient control of gene action. The model proposed by Wallace and Kass (1974) claims that at each locus whose structural gene is controlled by many transcription-initiation sites (promoters, or sensors), the structural gene operates most efficiently if the linear sequences of sensors for the two alleles are not identical; even random differences in the placement of sensors within these sequences are more efficient on the average than an identity of the two sequences.

In large part, the ubiquitous polymorphisms of these earlier chapters were speculative. The existence of genetic variation is not speculative, however; the high levels of protein variation are revealed unambiguously by electrophoretic and similar molecular techniques. The present chapter deals with the "large," easily observed polymorphisms of many organisms. These are the polymorphisms that, depending upon one's point of view, serve to illustrate the advantage of heterozygous individuals (chapter 10; see Ford 1965) or the indifference of natural selection in respect to genetic variation. The latter point of view, championed recently by "neutralists" such as King and Jukes (1969) and Kimura (1968) was the prevailing point of view throughout the 1940s (see, for example, Dobzhansky and Epling 1944). Here, however, attention is paid deliberately to those polymorphisms that either are, or appear to be, maintained by natural selection. The purpose is to see what sorts of pressures (both kind and magnitude) operate to maintain persistent polymorphisms, to estimate the ages of some polymorphisms, and to note the secondary and tertiary effects that existing polymorphisms may exert on the genetics of natural populations.

Selection and Polymorphisms

Despite the expenditure of considerable effort on the part of many persons, the nature of the selective forces acting on polymorphisms is known, even imperfectly, in surprisingly few instances. Such a pessimistic statement can be made, of course, because the boundary between the known and unknown recedes as knowledge is gained. What constitutes an "understanding" at one level of inquiry merely raises, in turn, new problems that demand explanation at another level.

A clear-cut polymorphism that arose in a laboratory population of *D. willistoni* was studied by Souza and his colleagues (Souza et al. 1970). Following an initial period during which a newly started population contained relatively few flies, the numbers of adults in the population cage suddenly increased dramatically. Accompanying this increase in size, large numbers of larvae

Table 19.1 The formal genetics of a phenotypic variant in *D. willistoni.* "In" signifies flies that pupate within the food cups of population cages; "out," flies that pupate on the floor of the cage. (After Souza et al. 1970.)

Cross		Percent Pupating on	Number of Pupae
Males	Females	Floor of Cage	Counted
In	Out	98.5	1,000
Out	In	98.8	860
Out/In	Out/In	73.4	1,130
In/Out	In/Out	76.6	1,754
Out	Out/In	98.6	728
Out/In	Out	98.9	726
In	In/Out	50.3	1,369
Out/In	In	45.0	724
Out	In/Out	98.7	758
Out/In	Out	98.9	809
In	Out/In	46.6	749
In/Out	In	50.6	1,214

began pupating on the floor of the population cage, outside the food cups in which larval development normally occurs. Having isolated strains of flies from pupae recovered from within the food cups and from the floor of the cage, Souza and his colleagues made genetic analyses with the results shown in table 19.1. The results are consistent with the hypothesis that "inside" and "outside" pupation are determined by two alleles at a single locus; the allele predisposing larvae to pupate outside the food cup is clearly dominant to its homologue.

The effect of the inside/outside polymorphism on the size of laboratory populations of *D. willistoni* is shown in table 19.2. Cages B and C, which were started with only one type of fly or the other, had equilibrium sizes of about 4,000–4,500 flies. A cage (E) started with recently wild-caught flies grew to approximately the same size. Cages A and D that were started either with a mixture of the inside and outside strains or with hybrids formed by crossing the two strains grew to equilibrium sizes more than half again as large.

The advantage accompanying the polymorphism for pupation-

Table 19.2 The effect of a single-gene (inside and outside pupators) polymorphism on the equilibrium size of laboratory populations of *D. willistoni.* (After Souza et al. 1970.)

Population	Started with	Equilibrium Size
A	750 Out; 750 In	7,060
B	1,500 Out	4,000
C	1,500 In	4,280
D	1,500 Out/In hybrids	7,070
E	1,500 wild-caught flies	4,250

site preference lies in the exploitation of an otherwise unoccupied territory within the laboratory cage. Those larvae pupating on the floor of the cage, had they instead remained within the food cup, would have either died or caused the death of other larvae. Clearly, this polymorphism is maintained by soft selection; that is, it is both frequency- and density-dependent.

Among the proteins for which vertebrates are frequently polymorphic are the transferrins, blood-serum proteins that serve to transport iron in the plasma. Data presented in table 19.3 (Frelinger 1972) show that the proportions of pigeon eggs which hatch successfully is related to the genotype of the females that produce them. Furthermore, the growth of yeast in the presence either of purified transferrins recovered from the birds themselves or of egg whites is considerably less in the case of *AB* heterozygotes than of the two homozygotes. Assuming that transferrins in the egg white increase the likelihood of hatching by means of a myocidal or bacteriocidal action, this *component* of fitness could be both density- and frequency-*in*dependent—that is, this might be an example of hard selection. Overall Darwinian fitness is *soft*, of course, if *any* major component of fitness is frequency- and density-dependent; this claim is true even though specific individual components of fitness (such as the antibioticlike action of egg-white proteins) may prove to be density- or frequency-*in*dependent.

In recent years (see Kojima 1971, Powell and Taylor 1979), frequency-dependent selection has attracted considerable interest because it seems to offer a means by which natural selection can maintain genetic variation in a population without incurring a tremendous genetic (segregational) load. The means by which the load is avoided has been seen on page 230: the values assigned to the Darwinian fitnesses of different genotypes are adjusted so that at equilibrium $\bar{w}$ equals 1.00.

In my opinion, reliance on frequency-dependent selection as a means of avoiding genetic load is both unrealistic and unneces-

Table 19.3 Egg hatch in pigeons in relation to a geographically widespread (Europe and the United States) polymorphism for transferrins. Also shown is the inhibitory effect of the whites of eggs laid by *AB* individuals on the growth of yeast. (After Frelinger 1972.)

	Egg Hatch		
Genotype	Tested	Hatched (%)	Growth of Yeast
AA	128	46	+++
BB	144	52	+++
AB	267	67	+

sary. The previous chapters of part 2, to repeat what was said earlier, were devoted to a demonstration that genetic load has little or no bearing on a population's ability to survive; indeed, a load (genetic or phenotypic) was shown in chapter 18 to be essential in resolving problems of survival or elimination of individuals during the self-thinning of a population.

My claim that frequency-dependent selection is an unrealistic response to the challenge of genetic load calculations is based on the following reasoning: If the multiplicative model is to be retained, selection must operate so that $\overline{w}$ becomes *exactly* 1.00 at equilibrium. Selection, that is, must operate in such a manner that $\overline{w}$ is *exactly* 1.00 at each of thousands of segregating loci. If the equilibrium fitnesses deviate even the slightest amount from 1.00, the product obtained under the multiplicative model will be exceedingly small. Kimura and Crow (1964) rested their case concerning the improbability that genetic variation is maintained by selection in favor of heterozygotes on the answer obtained by raising 0.99875 to the 5000th power (= 0.002). There is no reason to expect that 5,000 loci responsible for specifying the structure of 5,000 polypeptide chains would be subjected to natural frequency-dependent selection such that *at each locus* the equilibrium value of w ($\overline{w}$) approaches 1.00 more closely than 0.99875.

The discussion of polymorphisms and their maintenance has often focused on hybrid superiority and frequency-dependent selection as if these are mutually exclusive alternatives. No reason exists for insisting on such a clear-cut dichotomy except for the mathematical calculations that have been developed under the simplifying assumptions of density- and frequency-independent (or, hard) selection. Tables 19.4 and 19.5 illustrate the interrelation of heterozygous advantage and density. Cultures of *D. pavani* were set up with larval densities of 10, 50, and 100 larvae per vial. The proportion of inversion heterozygotes among the *adult* survivors increased from 48 percent to 58 percent for

Table 19.4 The relation between inversion polymorphism and survival in variously crowded cultures of *D. pavani*. (After Budnik et al. 1971.)

Larvae Per Vial	Total		4-R		4-L	
	Eggs	Larvae	% Homo- zygotes	% Hetero- zygotes	% Homo- zygotes	% Hetero- zygotes
10	1,000	626	52.5	47.5	50.3	49.7
50	1,000	519	49.7	50.3	48.5	51.5
100	800	292	42.3	57.7	38.0	62.0

Table 19.5 The role of competition between *D. gaucha* and *D. pavani* on selection for inversion polymorphisms in *D. pavani*. (After Budnik et al. 1971.)

Eggs per Vial	Total Eggs	Total Survival	pavani	gaucha	D. pavani Homozygotes 4-R	4-L
10	4,400	63.3%	51.5%	48.5%	58.7%	53.3%
50	4,250	65.7	50.2	49.8	51.7	39.8
100	4,300	53.7	52.2	47.8	46.7	45.5
200	3,400	34.8	56.7	43.3	33.3	27.8

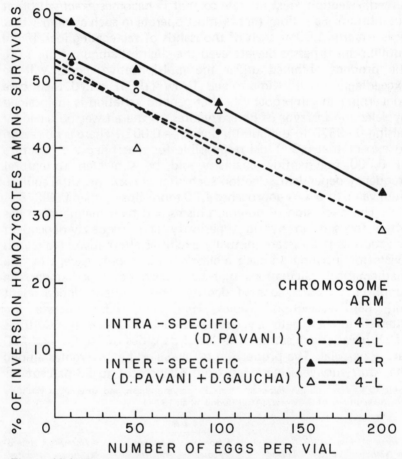

Figure 19.1 The regression of proportions of inversion homozygotes (chromosome arms 4L and 4R of *D. pavani*) on larval density for both one-species (*D. pavani*, alone) and two-species (*D. pavani* plus *D. gaucha*) vial cultures. (After Budnik et al. 1971.)

the right arm of the fourth chromosome, and from 50 percent to 62 percent for the left arm. Clearly, individuals that were inversion heterozygotes attained increased selective advantage under increased crowding (see table 19.4). Table 19.5 shows that increasing interspecific competition also exaggerates the selective advantage of inversion heterozygotes in *D. pavani*. In these cultures, *D. pavani* survives in increasing proportion as the total density increases, and as overall survival decreases. But, as the right-most columns show, inversion homozygotes constitute a smaller and smaller fraction of the *D. pavani* survivors as density increases.

Figure 19.1 illustrates the decrease in the proportions of chromosome 4-L and 4-R homozygous survivors under increasing levels of crowding, both in pure and in mixed species cultures. If the actual levels are ignored (because the starting frequencies for different sorts of experiments were not necessarily identical), the rates at which homozygotes decline with increasing larval densities are remarkably similar. The main point has already been made, however: the selective advantage of heterozygous individuals is not a constant, independent of gene frequencies or population density. A demonstration, therefore, that a particular component of selection is either frequency-dependent or density-dependent is not sufficient in itself to justify a claim that heterozygotes are not generally favored over homozygotes.

Self-Sterility Alleles: Selection in Favor of Rare Alleles

In many plant species, among them various species of the genus *Oenothera*, pollen carrying a given allele at a particular locus (the self-incompatibility locus) is unable to germinate on the stigma, or to grow down the style, of a plant whose genotype includes the same allele. Thus no a_1 pollen can fertilize eggs of $a_1 a_x$ plants, where a_x stands for any other allele at the a locus.

A plant species cannot begin to utilize a system such as this unless there are at least three alleles in the population; it is only then that pollen of a given sort (a_2, for example) encounters a plant ($a_1 a_3$) upon which it can function. It seems likely that the advantage of self-incompatibility systems of this sort lies in the protection they afford against inbreeding.

It is not difficult to see that new self-incompatibility alleles have an advantage over well-established ones in freely breeding populations. Imagine a population in which three alleles—a_1, a_2, and a_3—exist in equal frequencies. Each type of pollen can function in one-third of all plants since a_1 can function only on $a_2 a_3$, a_2 on $a_1 a_3$, and a_3 on $a_1 a_2$. Suppose, however, that an allele a_4 arises but is very rare. It can function on all three types of common plants—$a_1 a_2$, $a_1 a_3$, and $a_2 a_3$. Pollen carrying the allele a_4 is three times as likely to encounter a receptive stigma as is pollen carrying any one of the original alleles. Consequently, the new allele has a tremendous advantage over the old ones.

The same argument can be used to show that at equilibrium the various alleles have equal frequencies. Suppose a_3 is rare in the original three-allele system. Pollen of types a_1 or a_2 will rarely encounter $a_2 a_3$ or $a_1 a_3$ plants. Pollen bearing a_3 will frequently encounter the more common $a_1 a_2$ heterozygotes. Thus the number of $a_1 a_3$ and $a_2 a_3$ plants will tend to increase. The advantage of the a_3 allele will vanish when the frequencies of a_1, a_2, and a_3 are equal. Similarly, with the new allele a_4; its advantage will vanish when it is as common as the original alleles (25 percent each).

Each new allele as it arises has an advantage over those which have already been incorporated into the population. Let the new allele be a_{n+1} and let the n previous alleles be at equilibrium frequencies ($1/n$ each). Only heterozygous plants exist and with n alleles there are $[n(n-1)]/2$ different heterozygous genotypes. Pollen carrying a given allele can fertilize eggs carried by heterozygous genotypes in which it is not involved; there are $n-1$ other alleles so there are $[(n-1)(n-2)]/2$ genotypes that are receptive to pollen carrying any one allele. Within the original population, then, pollen of each type could function on $[(n-1)(n-2)/2] \div [n(n-1)/2]$ or $(n-2)/n$ of all plants. Pollen carrying the new allele, a_{n+1}, can function on all plants when this allele is rare. Consequently, if the rare allele is assigned fitness 1.00, $(n-2)/n$ represents the fitness of the other alleles. The selection coefficient s of these older alleles equals $2/n$; as the number of existing alleles increases, the disadvantage of the older ones (and therefore the advantage of the new one) becomes less and less.

The mathematical analysis of the number of self-incompatibility alleles that one would expect to find in a population of limited size is not a simple one (see Wright 1964 and references listed by him). Nevertheless, we can outline the main parts of the problem:

1. The advantage of new alleles becomes smaller as the number of previously retained ones increases.

2. Although the "old" alleles should occur in equal frequencies, these frequencies will vary as the result of chance events in a finite population; rare alleles will then enjoy an advantage while common ones will be at a disadvantage.

3. Some alleles will be lost by chance each generation; the smaller their frequency, the easier they are lost—despite their advantage.

4. If the population consists of N individuals, $2N$ gametes will be "chosen" to form the surviving zygotes of each generation. If the new alleles arise at a rate u, then $2Nu$ new alleles are present in the population each generation.

The number of alleles becomes constant when the number lost each generation equals the number gained through mutation. In his paper, Wright (1964) cites a study by Emerson (1939 and unpublished) in which forty-five self-incompatibility alleles were found in a population of *Oenothera organensis* numbering no more than 1,000 and probably fewer than 500 individuals. The system we have described here is capable, then, of maintaining a considerable number of alleles in a population.

The Age of Certain Polymorphisms

Simply because polymorphisms are often maintained at *stable* equilibria (whether because of the advantage enjoyed by heterozygous individuals or the advantage enjoyed by rare ones), they persist for long times in populations—exceedingly so in some instances. They need not persist forever, though, in order to qualify as stable polymorphisms. Nei (1975: 72), for example, remarks that the inversion polymorphisms in natural populations of *D. pseudoobscura*, once thought to be stable, appear to be transient because some chromosome frequencies are changing slowly (see Dobzhansky 1966). This remark, made to bolster the case for the neutralist hypothesis by arguing that most observable genetic variation represents transient rather than stable polymorphisms, is based on a misconception. Inversion frequencies in *D. pseudoobscura* have been known from the earliest studies (see Sturtevant and Dobzhansky 1936) to differ in different geographic regions (even on a microgeographic scale); such geographic differences can arise *only* if chromosome frequencies change with time. Indeed, Stone (1955) and Stone et al. (1960) have reviewed the everchanging patterns of chromosomal polymorphisms within the genus *Drosophila*. Qualitative and quantitative changes of inversion frequencies do not negate the observation that in many cases and under many conditions, inversion heterozygotes have been shown to be selectively superior to homozygotes.

Many species of snails are polymorphic in respect to the background colors and banding patterns of their shells. E. B. Ford and his colleagues (see Ford 1971: 178–203) in Great Britain and Lamotte (1951, 1959, 1966) in France have carried out extensive studies on populations of the garden (edible) snail, *Cepaea nemoralis*. The color polymorphism in *Cepaea* has been correlated with patterns of lights and shadows in fields and woods, and along hedgerows, as well as to the manner in which these patterns change with the seasons. Seasonally changing selection patterns can be detected in the color patterns of the debris that accumulates near rocks ("anvils") where thrushes carry live snails in order to crack open their shells. Despite the matrix of selective forces that are based on the season of the year in respect to both concealment and thermoregulation, the color and banding polymorphisms have persisted in *Cepaea* since the Pleistocene, as many as one million years (Diver 1929).

The ancient color and banding polymorphisms of another snail, *Limicolaria martensiana*, are shown in table 19.6. The color morphs in this species of African snail have been found among fossils 8,000–10,000 years old. The frequencies of the different forms vary among living snails from population to population,

Table 19.6 Frequency (percent) of color forms in a fossil and several living populations of the snail, *Limicolaria martensiana*. (After Owen 1966.)

Color Form	Fossil, Kabazimu Island (N = 1,277)	Living Kayanja (N = 2,840)	Ishasha Road (N = 882)	Rwenshama (N = 841)
Streaked	61.0	54.6	40.6	33.7
Broken-streaked	5.2	8.9	4.2	9.8
Pallid 1	3.9	24.0	7.0	1.9
Pallid 2	28.3	11.9	43.2	37.3
Pallid 3	1.6	0.6	5.0	17.3

and no present-day population has an array of frequencies identical to that recovered among fossil shells; nevertheless, the same polymorphism can be said to have existed in this species throughout these ten millennia.

Several chemical and radiological techniques exist which yield a direct measure of the age of fossilized material; consequently, in attempting to assign an age to a polymorphism, evidence surpassing that obtained from the fossil record itself is difficult to find. Nevertheless, present-day geographic distribution patterns can be used for this purpose, if the data are interpreted with care. An allele possessing a modest selective advantage over its homologues cannot be found in many more individuals than the number of generations that have elapsed since it first arose. If a newly arisen mutant allele came to be carried by a great many individuals within a short time, that allele would not be considered neutral or nearly so; on the contrary, it would be said to possess a tremendous selective advantage.

Figure 19.2 shows the geographical distributions of eight gene arrangements in *D. pseudoobscura*. Each of these arrangements almost certainly originated as a single altered chromosome following the restructuring of the gene order of a preexisting one. That single altered chromosome, borne originally by a single heterozygous fly, then spread first within and then between populations, to occupy the areas outlined in the figure, or even larger ones. The qualifying phrase has been added to the previous sentence because the large, *geographically disjunct* areas occupied by Tree Line and Olympic (areas in many of which neither gene arrangement is particularly common) probably represent relic distributions of a once far-greater geographic area occupied by these two gene arrangements. Despite many uncertainties, the individuals now carrying each of these eight once-unique gene arrangements must now number in the hundreds of millions. These gene arrangements, consequently, did not arise yesterday, or last year. Estimates of the ages of these various gene arrangements range from a conservative quarter of a million to several million years. Nevertheless, within today's distribution patterns, one can discern the balanced polymorphisms of the past (Wallace 1954b).

The retention of different gene arrangements in populations is subject to rates of origin and extinction just as are gene mutations themselves. MacArthur and Wilson (1967) popularized a geometric model for calculating equilibrium frequencies of species on islands (or other isolated regions); this method was illustrated

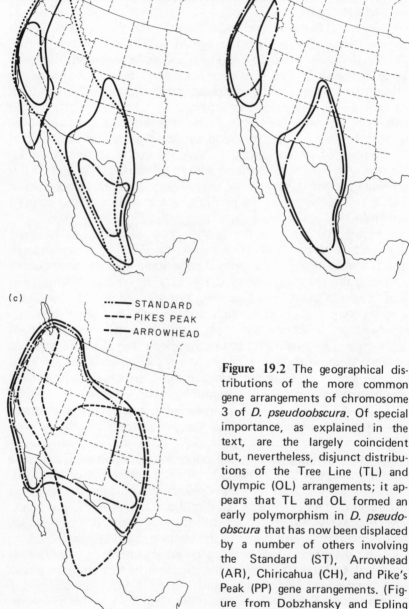

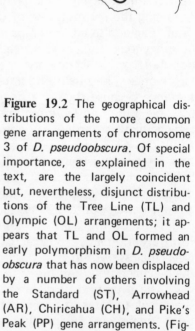

Figure 19.2 The geographical distributions of the more common gene arrangements of chromosome 3 of *D. pseudoobscura*. Of special importance, as explained in the text, are the largely coincident but, nevertheless, disjunct distributions of the Tree Line (TL) and Olympic (OL) arrangements; it appears that TL and OL formed an early polymorphism in *D. pseudoobscura* that has now been displaced by a number of others involving the Standard (ST), Arrowhead (AR), Chiricahua (CH), and Pike's Peak (PP) gene arrangements. (Figure from Dobzhansky and Epling 1944.)

(figure 9.10) in connection with forward and reverse mutations. This same model has been used by Jaenike (1973) in analyzing the polymorphisms maintained by island populations (figure 19.3). Far from being immune to extrinsic factors, polymorphisms (even those maintained by the average superior fitness of hetero- zygous individuals) respond to the rate at which new alleles arrive from outside and the rate at which they are lost (because of popu- lation size and fluctuations in relative fitnesses).

Yoon and Richardson (1976) have examined the gene ar-

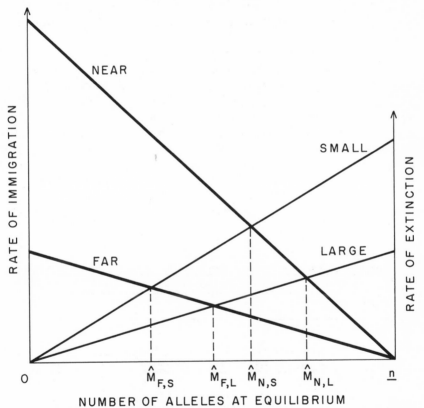

Figure 19.3 The interplay between rates of immigration of alleles to islands, and of extinction on near and far, and large and small islands, in establishing stable (but dynamic) polymorphisms on oceanic islands. The total number of alleles equals *n*. The four equilibria shown (from right to left) are $(M_{N,L})$ for near, large islands, $(M_{N,S})$ near, small islands, $(M_{F,L})$ far, large islands, and $(M_{F,S})$ far, small ones. Compare with figure 9.10. Reprinted by per- mission of the Chicago University Press from J. R. Jaenike (1973), *Amer. Natur.* (1973), 107:793–95; © 1973 University of Chicago.

rangements in seven species of Drosophilid flies inhabiting the Hawaiian Islands. In all, the chromosomes of the six species other than *Antopocerus longiseta* differ from those of *A. longiseta* by 16 paracentric inversions (figure 19.4). The distribution of these inversions among the species is tabulated in table 19.7. Clearly, polymorphisms for the various inversions have existed through the origins of several species. The data do not allow an ordering of the origins of *A. diamphidiopodus* and *A. cognatus*. However, species number 1 adds inversions B, G, and K to the previously existing ones; *A. tanythrix* then adds inversions C, J, and L in turn. The polymorphism for inversions E and F extends through four of the seven species.

The extreme age of certain genetic polymorphisms is not *proof* that they are subject to natural selection; Nei (1975: 104) emphasizes the immense time required on the average for the chance replacement of one allele by another, selectively neutral one; this time equals (approximately) $4N_e$ generations; that is, four times as many generations as the effective size of the population. Kimura and Ohta (1971) have argued that in the case of neutral alleles, migration renders the entire species one large, essentially panmictic population. Consequently, neutral alleles are

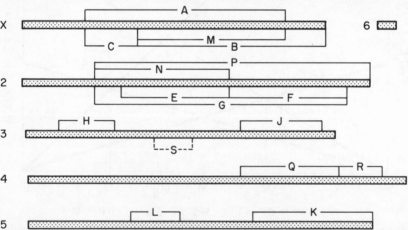

Figure 19.4 Diagrammatic representation of the salivary-gland chromosomes of *Antopocerus longiseta* on which are shown the limits of 16 paracentric inversions found in one or another of six closely related species. The distribution of these inversions among related species, the distribution of the species among the islands of Hawaii, and the known geologic history of these islands permit inferences regarding the evolutionary past of these inversion polymorphisms. (After Yoon and Richardson 1976.)

Table 19.7 The distribution of the 16 paracentric inversions shown in figure 19.4 among *Antopocerus longiseta* and six other closely related species. (After Yoon and Richardson 1976.)

Species	Chromosome Element					
	X	2	3	4	5	6
A. arcuatus	+	+	S/+	+	+	+
A. aduncus	M	N,P	+	Q,R	+	+
A. longiseta	+	+	+	+	+	+
A. diamphidiopodus	A	E,F	H	+	+	+
A. cognatus	A	E,F	H	+	+	+
A. species #1	A,B	E,F,G	H	+	K	+
A. tanythrix	A,B,C,	E,F,G	H,J	+	K,L	+

expected to exist at intermediate gene frequencies for tens or hundreds of millions of years, a period of time much longer than that required for species formation according to many estimates. A demonstration that any component of an existing polymorphism is subject to selective pressures of any sort, however, diminishes the probability that neutrality underlies polymorphisms which are older than the present-day Hawaiian Islands, the Yosemite Valley in California, or the Mediterranean Sea. Randomly fluctuating selective pressures in different generations cause a gene to *behave* as if it were neutral (Nei 1975: 165); these fluctuations do not, however, *confer* neutrality upon such genes. My feeling is that, because selection pressures neither are nor can be constant, many non-neutral genes have been claimed by neutralists because their calculations are blind to fluctuating selective pressures.

The Genetic Consequences of Persistent Polymorphisms

Within a given environment the interaction of alleles at the same and at different loci determine the phenotype of the individual. Because of the concentration by geneticists on mutant genes with major, easily detected effects, interactions between genes have frequently been treated as anomalies (as leading, for example, to the numerous modifications of the traditional 2-locus $9:3:3:1$ ratio). Lewontin (1974a: 318) has attempted to place gene interactions in their proper perspective relative to the genetics of populations in these words: "The fitness at a single locus ripped from its interactive context is about as relevant to real problems of evolutionary genetics as the study of the psychology of individuals isolated from their social context is to an understanding of man's sociopolitical evolution. In both cases context and inter-action are not simply second-order effects to be superimposed on a primary monadic analysis. *Context and interaction are of the essence''* (emphasis added).

The material to be covered in this section deals with genetic changes in populations that are evoked by certain genetic features of these same populations. These changes, for the purpose of analysis, must be looked upon as ripple effects caused by a pri-mary event. It is not difficult to imagine, however, ripple effects caused by ripples themselves. The context within which alleles exist and the interactions between these numerous alleles are, as Lewontin says, the essence of the genetics of populations. The beginning made here will be expanded later into discussions of coadaptation, of the evolution of dominance, and of the co-herence of gene pools. Disruption of the latter, following loss of cohesion, leads to the formation of new and separate species.

The extent to which "context and interaction" may be of the essence was well known to early geneticists, who then promptly sought simpler problems for analysis. Figure 19.5, for example, shows the phenotypic variation in the wings of *Beaded* flies; this illustration is from an early publication of the Morgan group (Morgan et al. 1925). Wings of flies taken from the stock bottle labeled ''Beaded'' are shown at the top of the figure. That these wing shapes are the result not only of the mutation known as

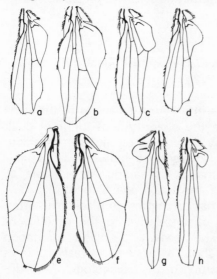

Figure 19.5 The modification of the mutant phenotype, *Beaded*, in *D. melanogaster. a-d:* Beaded wings in a standard laboratory strain; *e* and *f*: slightly Beaded phenotype obtained by outcrossing to wild-type flies; *g* and *h*: extreme Beaded phenotype obtained by deliberate selection. (After Morgan et al. 1925.)

Beaded (whose locus is at 93.8 on the genetic map of chromosome 3) but also of alleles at other loci is shown in the bottom portion of the figure: outcrossing the *Beaded* stock to (an unspecified) wildtype strain led to nearly normal wing shapes; selection for reduced wings led to thin, bladelike wings.

Morphological mutations are not the only ones that are subject to obvious genetic modification by the background genotype; enzymes and other proteins which are synthesized according to the DNA structure of one locus may have their amounts controlled by other loci. Figure 19.6 shows that the amount of alcohol dehydrogenase synthesized by the *Adh* locus in *D. melanogaster* can be readily modified by artificial selection for higher and lower activities.

Kettlewell (1958) has reviewed the experimental data obtained over 50 to 60 years from crosses involving heterozygous melanic forms of *Biston betularia* and homozygous (recessive) wildtype individuals; the results are summarized in table 19.8. Now, the oldest melanic individual known in Britain is a specimen captured in Manchester in 1850. This specimen is believed to be heterozygous because of its markings, and because the incompletely dominant gene for melanism was rare at that time. Among the progeny of melanic heterozygotes in crosses made 50 to 55 years later, melanics were ostensibly less frequent than homozy-

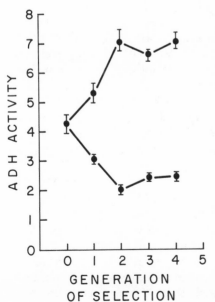

Figure 19.6 Evidence that the level of enzyme activity at a given locus (*Adh*) in *D. melanogaster* is subject to modification by artificial selection. The strain subjected to high and low selection in this study was homozygous for the *Adh-S* allele. (After Ward and Hebert 1972.)

Table 19.8 An apparent change in the relative viability of melanic (heterozygous) and type (homozygous) backcross individuals in *Biston betularia*. (After Kettlewell 1958.)

Tests Made between 1900 and 1905

Melanic	Type	Total	
109	123	232	
47	57	104	
11	18	29	
50	57	107	
217	255	472	46.6% melanics

Summary of Results Obtained in 1953–1956

22	14	36	
10	7	17	
30	28	58	
39	14	53	
5	1	6	
2	1	3	
108	65	173	62.4% melanics

gous (recessive) wildtype individuals; the data are only suggestive, however ($X^2 = 3.1$, ld.f., $p = 0.15–0.20$). Similar crosses made during the mid-1950s, however, gave clearly different results: the number of melanic progeny exceeds that of their wildtype sibs in these later crosses. The obvious suggestion is that natural selection, because of the frequency of melanic heterozygotes (and, possibly of homozygotes as well), has resulted in an improvement in the overall health (and consequent survival) of these dark forms. Because wildtype individuals today are rather rare in the industrialized areas of Britain where the melanic form occurs, selection for maintaining their health has waned. Alternatively, however, one can imagine that heterozygous melanic individuals do not form functional gametes in the expected 50:50 proportions. Genes causing such segregational distortions (SD genes) are not unknown (see a review by Crow, 1979); any such distorting element that increased the proportion of cryptic melanic offspring born of heterozygous parents would have been favored in natural populations of *B. betularia*.

Genetic interactions affecting the viability of mutants can also be illustrated by data on the sweet pea that have been summarized by Haldane (1957); these data are reproduced in table 19.9. The more recently the mutant forms of the sweet pea have been discovered by floricultural specialists, the poorer the viability of the mutant form. Mutants discovered during the eighteenth century are some 10 percent more viable (measured as deviations from expected Mendelian ratios) than those discovered since

Table 19.9 Evidence for the accumulation of modifiers that enhance the viability of homozygous mutants of the sweet pea, *Lathyrus odoratus*. In F_2 crosses, older mutants approximate the expected ratio more closely than do more recently discovered ones. (After Haldane 1957.)

Found	Mutant	Viability	S.E.	Average
1700–1800	g_1 White	1.037	0.024	
	a_1 Red	1.021	0.017	
	b_1 Light axil	1.011	0.017	1.009
	f_1 White	0.996	0.038	
	d_5 Picotee	0.996	0.022	
	a_2 Round Pollen	0.990	0.024	
1880–1899	b_2 Sterile	0.988	0.017	
	a_3 Hooded	0.977	0.021	
	e Cupid	0.976	0.032	0.954
	f_2 Bush	0.936	0.030	
	d_2 Blue	0.931	0.024	
	g_3 Mauve	0.917	0.031	
1900–1912	d_1 Acacia	0.964	0.020	
	d_4 Smooth	0.940	0.020	
	d_2 Copper	0.909	0.040	0.903
	h Spencer	0.897	0.030	
	b_3 Cretin	0.886	0.048	
	f_3 Marbled	0.821	0.023	

1900; mutants dating from the late nineteenth century are precisely intermediate.

The final example illustrating that context and interactions are of the essence, rather than being of little importance to the genetics of populations, is that of the platyfish, *Platypoecilus maculatus* (figures 19.7 and 19.8).

Platyfish inhabit various river basins in Mexico. All populations are polymorphic in respect to black spots whose shapes and positions are determined by a series of dominant alleles governing the distribution of macromelanophores. Figure 19.7 shows a number of characteristic patterns whose relative proportions remain reasonably constant within localities, but differ from river basin to river basin. Figure 19.8 shows that happens when platyfish that are captured at different localities are crossed. The characteristic patterns are obliterated as if the genetic control over the distribution of melanophores were lost. Backcross individuals, as the figure shows, may have large black growths covering one-half or more of the fish's body.

Clearly, the polymorphisms for macromelanophore spots in the platyfish are ancient ones that affect numerous, widespread, and largely isolated modern populations of these fish. The phenotype associated with each of the dominant alleles at the poly-

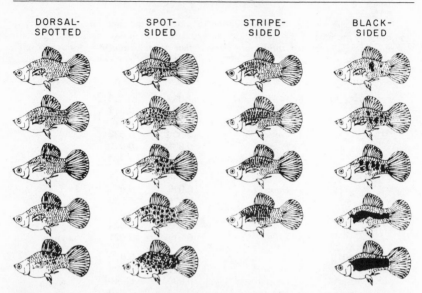

Figure 19.7 Variation in the expression of the macromelanophore pattern genes in the Rio Jamapa and Rio Coatzacoalcos populations of the platyfish, *Platypoecilus maculatus*. (After Gordon and Gordon 1957.)

morphic locus is determined not by the particular allele alone but by each of these alleles functioning *within the context* of the gene pool of the population in which it is found. That these contexts differ is revealed by interpopulation hybridization and subsequent backcrossing. Under hybridization, control of macro-

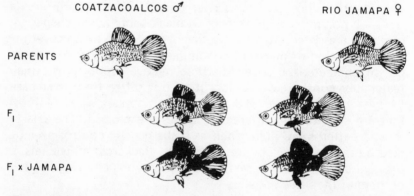

Figure 19.8 Parents, F_1 hybrids, and backcross hybrids of platyfish from two isolated river basins, Rio Jamapa and Rio Coatzacoalcos. (After Gordon and Gordon 1957.)

melanophore production and distribution breaks down with drastic results; the potential danger associated with the use by this fish of these pigment cells for creating pattern polymorphisms is revealed when they form hugh cancerlike growths on hybrid individuals. (See Borowsky 1973 for a report on a non-hybrid melanoma, one of two known cases.)

Summarizing Comment

The notion that the "normal" phenotype arises through the harmonious interaction of the environment and a harmoniously functioning collection of genes is not new. Nor is the notion that in isolation from one another different populations come to possess different collections of harmoniously acting genes. Muller (1940), for example, postulated that the chance establishment of some mutations in isolated populations provides a different genetic background for further mutations. In his words, "These earlier mutations thereby have their role changed from that of superfluous or merely advantageous deviations to necessary parts of the system." Muller developed this view as early as 1918.

When the alleles that are found at different loci are selected in response to one another's presence or absence in populations because of the differential survival and fertility of their carriers, it is appropriate to say that these alleles are coadapted. To some persons, as we shall see in chapter 20, the idea that the genes found at various loci (that is, the particular *alleles* that represent these genes at those loci) are there only because they have interacted harmoniously with other genes at other loci borders on mysticism.

To learn that the coadaptation of the various parts of the genome was suspected not long after the rediscovery of Mendel's papers is satisfying. To that early view we have, in this chapter, added the further point that heterozygosity at various loci can evoke selective changes at other loci (which are often heterozygous themselves). The effect is an enormous speeding up of the rate at which harmonious interactions evolve. Heterozygous individuals appear immediately following mutation, but new homozygotes do not appear in the population for many generations.

FOR YOUR EXTRA ATTENTION

Populations of the sea blush, *Plectritis congesta*, are generally polymorphic in respect to winged and wingless fruit. Individual plants produce either winged or wingless fruit; the frequency of wingless fruited plants varies in different populations from 1 percent or less to 46 percent on small islands in British Columbia.

Pretend that you are to study this polymorphism. What would you want to know about the inheritance (if any) of the two types of fruit? How would you proceed to get answers for this problem? How would you proceed in understanding the polymorphism, itself? Check your answers against the study reported by Ganders et al. (1977).

The Genetic Integration of Mendelian Populations: Coadaptation

PREVIEW: Persistent polymorphisms, as we saw in chapter 19, may act as the basis for additional selective change in a population; in essence, they produce ripple effects. In this chapter, evidence will be presented suggesting that ripple effects are important in determining the retention or loss of alleles at many gene loci; thus, the gene pool of a population becomes integrated through the coadapted interactions of genes at many loci.

> In order that an animal should acquire some structure specially and largely developed, it is almost indispensable that several other parts should be modified and co-adapted.
>
> Darwin; *The Origin of Species* (Second Edition)

> Each new mutant in turn must have derived its survival value from the effect which it produced upon the "reaction system" that had been brought into being by the many previously formed factors in cooperation . . . many of *the characters and factors which, when new, were originally merely an asset finally became necessary* because other necessary characters and factors had subsequently become so as to be dependent on the former.
>
> H. J. Muller (1918)

INDIVIDUAL ORGANISMS ARE not haphazard collections of anatomical structures nor are populations haphazard collections of genes. In Darwin's case, coadaptation was treated as a miscellaneous objection to his theory of natural selection. The need for structural harmony would dampen the rate of evolutionary change but, as Darwin argued, that does not make change impossible. (A modern, analogous example would be the co-evolution of certain insects and plants; see Gilbert and Ravin, 1975.) Muller's argument, outlined in the second of the two quotations, moves the need for harmony from the anatomical structures of each individual to the genetic factors that are shared by all members of any population of a cross-fertilizing species. The process leading from an asset to a necessity was, in Muller's estimation, a slow one dependent primarily upon interactions between homozygous loci. That normalizing and harmonizing selection may occur at and among heterozygous loci increases the need for, and the origin of, "co-operation" tremendously. Nevertheless, whether it comes about rapidly or slowly, cooperation between genes (coadaptation) remains a consequence of continuous selection for high Darwinian fitness within populations.

Mendelian inheritance provides more than ample material on which natural selection can act. Table 20.1 provides some numbers substantiating this claim. Only if there were but one allele per locus at all gene loci would a population of individuals be free of genetic variance. Two alleles at each of ten loci are able to generate 59,000 different genotypes. Ten alleles at each of six loci could generate gene combinations in numbers sufficient to provide a choice of six unique genotypes for every one of the four billion persons now alive. From earlier discussions (chapter 4), we know that the numbers of alleles per locus often approaches ten and that the proportion of polymorphic loci frequently exceeds 25 percent

Table 20.1 The relation between the number of gene loci, the number of alleles per locus, and the number of possible genotypes.

Number of Alleles	Number of Loci				
	1	2	$\cdots$ 10	$\cdots$	n
1	1	1	$\cdots$ 1	$\cdots$	1
2	3	9	$\cdots$ 59,000	$\cdots$	3^n
10	55	3,025	2.5×10^{17}		55^n
k	$\dfrac{k(k+1)}{2}$	$\left[\dfrac{k(k+1)}{2}\right]^2$	$\cdots$ $\left[\dfrac{k(k+1)}{2}\right]^{10}$	$\cdots$	$\left[\dfrac{k(k+1)}{2}\right]^n$

of those tested; the number of genotypes available, then, must be astronomically large. In order that selection *not* operate in retaining favorable constellations of alleles would, under such bountiful conditions, require a selective neutrality for these alleles of extraordinary precision.

The modern use of the term "coadaptation" stems from Dobzhanksy's studies on the inversion polymorphisms in natural populations of *D. pseudoobscura*; these studies will be discussed shortly. For the moment, I prefer to ignore history and continue with an example of coadaptation based on what are, in effect, epistatic interactions among genes at different loci. This example comes from work by Cavalli and Maccacaro (1952) on the development of resistance to chloramphenicol in *E. coli*. Using strains of bacteria which, if given the opportunity, could undergo sexual mating and recombination, Cavalli and Maccacaro selected these organisms for resistance to ever higher concentrations of chloramphenicol within asexually reproducing cultures. Bacteria were plated on near-lethal levels of chloramphenicol, surviving colonies were rescued, and these were re-plated either on the same or an even higher concentration of the antibiotic. The parental strains developed resistance to as much as 320 γ of chloramphenicol per ml or, later, to 1280 γ/ml.

Having obtained resistance independently in these different lines, Cavalli and Maccacaro allowed the bacteria of different selected lines to mate and undergo genetic recombination. The resistances of clones obtained from the recombinant progeny are illustrated in figure 20.1. Without exception, the average level of resistance among the recombinant lines was unexpectedly low. The most resistant recombinant lines were, at best, equal to the poorer of the two parental lines.

The interpretation of the events occurring in this study is as follows: Each colony that survived chloramphenicol exposure, did so because of a rare mutational event. In the absence of recombination, however, each new mutation was tested on a particular genetic background which remained intact through subsequent (asexual) generations. Consequently, in each selected line the successful mutations may have conferred high resistance largely through their interactions with the rest of the genome. As long as asexual reproduction continued, these favorable epistatic interactions persisted as well. Recombination between lines following the sexual cross would tend to destroy those gene combinations that generated resistance through epistasis thus revealing that different

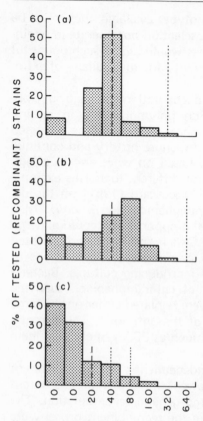

Figure 20.1 The distribution of strains of *Escherichia coli* resistant to various levels of chloramphenicol following the sexual cross (and recombination) of two different asexually selected resistant strains (*a* and *b*), or of an unselected sensitive strain and an asexually resistant one (*c*). Note that the mean level of resistance among the recombinant strains (dashed line) is consistently lower than the average of the parental strains (dotted line). (After Cavalli and Maccacaro 1952.)

lines had achieved resistance to chloramphenicol by different genetic routes. The interline gene combinations, having lost the selected epistatically interacting gene combinations, would possess a lower resistance to the antibiotic.

This example has been cited immediately because it illustrates—even though by means of halpoid organisms—the sorts of intergenic dependencies and interactions that can lead to a coadapted "reaction system."

Inversion Polymorphism in *D. pseudoobscura*

The third chromosome of *D. pseudoobscura* possesses a variety of naturally occurring gene arrangements. We have referred repeatedly to these gene arrangements in this book. They served to illustrate certain features of the Hardy-Weinberg equilibrium; they served, too, to illustrate that a selective superiority can be a property of heterozygous individuals (page 297).

That the gene arrangements found in populations of *D. pseudoobscura* do in fact possess selective significance was shown by Wright and Dobzhansky (1946); proof that these arrangements were subject to selective pressures was given by the systematic changes in their frequencies which occurred in cage populations. In figure 20.2 we see the changes in the frequencies of *ST* which occurred in four populations that contained initially 20 percent Standard and 80 percent Chiricahua (*CH*) gene arrangements from Piñon Flats, Calif. From these data, the relative fitnesses of *ST/ST*, *ST/CH*, and *CH/CH* individuals were estimated as 0.90, 1.00, and 0.41.

Fairly large numbers of combinations of several gene arrange-

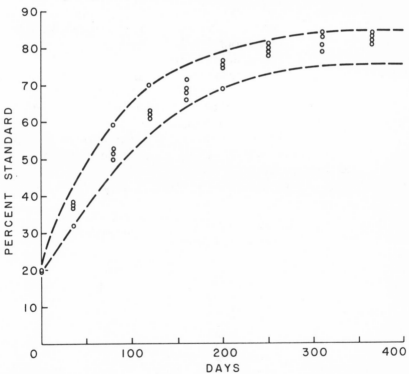

Figure 20.2 The frequency of *ST* chromosomes in four replicate populations of *D. pseudoobscura* that were started initially with 20 percent *ST* and 80 percent *CH* obtained originally from Piñon Flats, Calif. (After Dobzhansky and Pavlovsky 1953.)

ments have been studied in experimental populations. Among these tests, there have been many that involved two alternative arrangements—*AR* versus *ST*, *AR* versus *CH*, and *ST* versus *CH*, for example. Furthermore, a number of these tests have involved the same gene arrangements obtained from different geographic localities. Such tests were made to determine whether the same two gene arrangements would go through identical sequences of frequency changes even though they had come from geographically remote areas.

The results of a number of such "intralocality" populations are listed in table 20.2. With the single exception of the population containing *ST* and *TL* chromosomes from Mather, the heterozygotes in these tests consistently exhibit a higher fitness than the two corresponding homozygotes. In each of these populations, therefore, inversion frequencies reached stable equilibria. Furthermore, they approached these equilibria at rates consistent with the estimated fitnesses. Not all equilibrium frequencies were identical even for populations involving the same pair of gene arrangements. Thus, the equilibrium frequencies of *ST* in different *ST* versus *AR* populations range from 73 to 54 percent, depending upon the geographic origin of the chromosomes involved. The gene content of chromosomes of the same gene arrangement but from different localities is not identical.

As a further test of the geographic differentiation in the gene content of chromosomes with various gene arrangements, a number of "interlocality" laboratory populations were set up. In these one gene arrangement was taken from one locality while the othe

Table 20.2 The relative fitnesses of a number of homozygous and heterozygous geno types involving gene arrangements of *D. pseudoobscura* obtained from a number o localities within California. (Dobzhansky 1948b.)

Locality	1	2	Genotype 1/1	1/2	2/2
Piñon	ST	CH	0.85	1.00	0.58
Keen	ST	CH	0.91	1.00	0.42
Mather	ST	CH	0.78	1.00	0.28
Piñon	ST	AR	0.81	1.00	0.50
Keen	ST	AR	0.79	1.00	0.58
Mather	ST	AR	0.64	1.00	0.58
Piñon	AR	CH	0.86	1.00	0.48
Keen	AR	CH	0.54	1.00	0.60
Mather	AR	CH	0.81	1.00	0.60
Mather	AR	TL	0.69	1.00	0.12
Mather	ST	TL	1.12	1.00	0.33

was taken from a second, geographically remote locality. The results of these tests differ markedly from those in which the gene arrangements were obtained from the same locality.

The contrast between intralocality and interlocality populations can be seen by comparing the data listed in table 20.2 with those listed in table 20.3. Because changes of inversion frequencies in the mixed-locality populations are extremely erratic, estimates of relative fitnesses in these populations cannot be calculated from observed frequency changes. Instead, the estimates of fitness are based upon the deviations from the $1:2:1$ ratio expected among the F_2 generation offspring of interpopulation hybrid parents. Flies for these one-generation tests are raised, of course, under population-cage conditions. These one-generation tests measure but one component of fitness: the differential survival of larvae under extremely crowded conditions. One-generation tests of intralocality populations give results similar to but not numerically identical to those listed in table 20.2. The results listed in table 20.3 are altogether different. In every experimental population without exception, one homozygote (in one case, both homozygotes) exceeds the heterozygote in fitness. In the interlocality populations one gene arrangement frequently displaces the other; one attaining a frequency of 100 percent the other dropping to 0 percent. The data in table 20.3 reveal why this is so: stable intermediate frequencies depend for the most part on the superior fitness of inversion heterozygotes.

The results reported in tables 20.2 and 20.3 led Dobzhansky (1948b, 1950) to suggest that chromosomes with different gene arrangements which occur together within populations

Table 20.3 The relative fitnesses of individuals homozygous and heterozygous for various gene arrangements in laboratory populations of *D. pseudoobscura*. Unlike those listed in table 20.2, these are geographically mixed populations; the superscript identifies the geographic origin of each gene arrangement as Mather, Piñon, or Mexico. (After Dobzhansky 1950.)

	Arrangement		Fitnesses		
Localities	1	2	1/1	1/2	2/2
Mather vs. Piñon					
45	AR^M	CH^P	1.28	1.00	0.47
46	ST^P	AR^M	0.63	1.00	1.51
47	ST^M	AR^P	1.31	1.00	0.57
48	ST^M	CH^P	1.18	1.00	0.48
49	ST^P	CH^M	1.38	1.00	0.43
Mexico vs. Piñon					
55	ST^P	CH^{Mex}	1.26	1.00	0.87
56	AR^P	CH^{Mex}	1.53	1.00	1.16

inhabiting the same geographic area are *coadapted* through natural selection and, as a result, inversion heterozygotes tend to exhibit high fitness. Gene arrangements from geographically isolated populations cannot be subjected to selection for coadaptation, and hence the heterozygotes of mixed-locality laboratory populations have fitnesses intermediate (as a rule) to those of the two corresponding inversion homozygotes.

An alternative, testable explanation can be suggested to account for the observations described in the preceding paragraphs. One can argue (see Sturtevant and Mather 1938) that dissimilarity in origin decreases the probability that two alleles at the same locus will be identical. Alleles found at various loci in a given Standard chromosome from Piñon Flats, for example, are more likely to be identical to those found in a second Standard chromosome from Piñon Flats than in one from Mather, a locality several hundred miles away. Similarly, alleles found at loci in a given Standard chromosome are more likely to be identical to those in a second Standard chromosome from the same locality than to those in an Arrowhead chromosome from that locality. This statement applies particularly to loci within the inverted segment where gene exchange between chromosomes of different gene arrangements does not occur.

According to the alternative explanation, gene arrangements A and B should show the following sequence of fitnesses in intra-locality populations: $A/B > A/A \geqslant B/B$. This sequence is that which is actually observed, although the observed differences in fitness are much greater than those that would be expected by the mere covering up of deleterious recessive alleles. The latter expectations, based on genetic-load-type calculations, would be of the order of $\frac{1}{3}$ (one of three autosomes) $\times \frac{1}{3}$ (the average length of an inverted segment) $\times 0.10$ (total mutation rate per gamete) or about 1 percent. In contrast to this expected small advantage of inversion heterozygotes, two-, four-, and even eightfold differences in relative fitness are listed in table 20.2.

The results of the mixed locality populations can be written as follows: $A/A > A/B > B/B$. An explanation based on the covering up of deleterious mutations now encounters difficulties. To retain it, one must make the unlikely assumption that arrangement B of one locality is more likely to carry alleles identical to those of arrangement A of a second locality than are two different A's from the same locality. No logic can support such an assumption.

There is, however, a third experiment, not as well known as those already described, which eliminates "greater hybridity" as an alternative explanation. Through the use of an ingenious series of crosses, Dobzhansky (1950) compared the fitnesses of

$$ST^M/ST^P: \quad \frac{AR^P/ST^P}{+ \atop AR^M/ST^M} : AR^M/AR^P.$$

The superscripts M and P refer to the two localities Mather and Piñon Flats. It can be seen that the two structural homozygotes are both interlocality heterozygotes, but the structural heterozygotes carry chromosomes obtained from the same locality. The fitnesses estimated for these flies were 0.87, 1.00, and 0.46. The intrapopulation structural heterozygotes exceeded the two structural homozygotes in fitness despite the hybridity of the latter arising from their remote geographic origins. The results of this test, together with those obtained by studying single and mixed locality populations, are enough to dispose of the "greater hybridity" explanation for the observed coadaptation of gene arrangements coexisting within given geographic localities.

The Wallace-Kass model of gene control (see page 396) provides a (speculative) explanation for the origin of coadaptation between different gene arrangements within local populations. According to the model, structural genes that are controlled by numerous promoters or transcription-initiation sites function most efficiently in diploid organisms if the sequences of these promoters are not identical for the two homologous alleles. Even random differences in the two homologous sequences were said to improve the functioning of their associated alleles (figure 15.14). The *most* efficient arrangement for the promoters of one sequence, however, is that which is *precisely the reverse* of that of the other; two contrasting sequences of that sort reduce the effective size of the gene control region by one-half. No promoter would be farther than one-half the length of the control region from one or the other of the homologous structural genes.

The arrangements of promoters in precisely reverse orders within the homologous gene loci, however, maximizes the probability that a crossover will occur within the control region and give rise to recombinant chromosomes that carry duplications of some and deficiencies for other promoters:

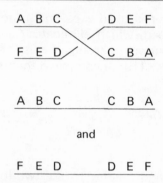

Thus, under the Wallace-Kass model a population is subjected to conflicting selection pressures: dissimilar sequences of promoters increase the efficiency of gene action but give rise to defective control regions through recombination; similar sequences decrease the efficiency of gene action but remove the danger that otherwise accompanies recombination. Paracentric inversions (by far the most common type of chromosomal rearrangement found in natural populations) remove the danger which recombination poses for dissimilar homologous promoter sequences; the recombinant strands that arise through recombination within a paracentric inversion are relegated to nonfunctioning polar bodies in Drosophila females (there is little or no recombination in the males of most Drosophila species) and tend to be discarded as well in the megaspores of plants (see McClintock 1941). Because paracentric inversions cause the elimination of recombinant strands in *Drosophila* (at little or not cost in terms of female fertility), the sequences of transcription-initiation sites at homologous loci within the inverted segments can take on the most efficient orders; in the extreme case, the two sequences of each homologous pair could be the exact reverse of one another. The protection against the generation of duplications and deficiencies provided by the inversion would explain the degree of hybrid vigor that accompanies inversion heterozygosity.

The coadaptation of inversions within populations, and the lack of it between them, would be explained by the large number of possible sequences that transcription initiation sites might possibly take. The contrasting sequences in one population may be (extreme cases are used here as illustrative examples) A B C D E F and F E D C B A. In another locality, the pair of sequences settled upon may be C D E F A B and B A F E D C. In still another local-

ity, the two contrasting sequences might be D C B E F A and
A F E B C D. Coadaptation, under this scheme, resides in the
(postulated) tendency for the sequences at homologous loci in
local populations to take on reverse orders. There is no "between-
population" selection (other than that arising occasionally from
an exchange of migrant individuals) that modifies the sequential
arrangement of promoters at a given locus in one population
relative to that which exists in another, isolated population. Consequently, interpopulation inversion heterozygotes need not
have as efficient pairs of control regions as intrapopulation
heterozygotes.

Finally, in this connection, it must be recognized that whatever pair of more or less reversed sequential promoter arrangements that may be established within the inverted blocks of genes
of one geographic locality (that is, whether A B C D E F and
F E D C B A or C D B A E F and F E A B D C), these sequences
in large measure determine within the local population what are or
are not efficient arrangements of promoters at other gene loci
throughout the genome, beyond the limits of the inversion itself.
These are the ripples that spread from the persistent inversion
polymorphism through the gene pool as described in the preceding
chapter. Furthermore, because the control of genes throughout
the genome has been largely determined within each population
by those of others located within the inverted regions, interpopulation hybridization can give rise to a bewildering array of gene
combinations, among which the interpopulation structural heterozygotes may not possess a consistent selective advantage.

Coadaptation at the Gene Level

Dobzhansky's term "coadaptation" referred specifically to chromosomal polymorphisms and to their maintenance in local populations owing to the superior fitness of structural heterozygotes. The
term, however, has historical implications that go beyond this
special use; Darwin used the term in this more general sense when
he referred to the reciprocal adjustment of anatomical features to
one another. Extending this general concept to genetic features
of populations, one can argue that the alleles (by "allele" I refer to
both structural genes and their associated control regions, if any)

at one locus owe their existence in a population to the harmonious
way in which they interact with alleles at other loci. This, of
course, leads to the position expressed by Muller (1918, 1940); it
is also the position I have expressed (Wallace and Vetukhiv 1955)
as follows: ". . . genes are found within populations because they
have been retained as a result of selective properties they exhibit
in heterozygous individuals . . . . Any particular allele at any locus
will be retained or eliminated by virtue of its average adaptive
value when in combination with various other alleles at that or
other loci . . . ." The second point of view differs from that expressed by Muller only in stressing that the harmonious interactions arise in response to alleles while carried by heterozygous
individuals; this stipulation permits the coadaptation of gene pools
to be a relatively speedy process.

An early experiment designed to reveal the intrapopulation
coadaptation of alleles of one chromosome (chromosome 3, AR
gene arrangement, of *D. pseudoobscura*) was performed by Brncic
(1954; see table 20.4). Using the modified *CIB*-techniques that are
available to Drosophila geneticists, Brncic compared the relative
viabilities of flies carrying two different chromosomes of one
geographic origin with those of flies carrying chromosomes obtained from different geographic localities; the latter exceeded the
former in viability by approximately 6 percent.

The superiority of F_1 interpopulation heterozygous combinations was confirmed in a second experiment. In this second experiment, additional tests were made on the viabilities of flies carrying
interpopulation recombinant chromosomes. These tests reveal that
the superiority of interpopulation heterozygosity resides largely
in the possession of nonrecombinant chromosomes; flies carrying
recombinant ones have average viabilities lower than those of the
intrapopulation heterozygotes.

Data similar to those obtained by Brncic were obtained by

Table 20.4 The effect of recombination between wildtype chromosomes of different
geographic origins on viability in *D. pseudoobscura*. (After Brncic 1954.)

First experiment
Parental   0.94
Hybrid (F_1)   1.00
Second experiment
Parental   0.92–0.93
Hybrid (F_1)   1.00
Hybrid (F_2)   0.91
Hybrid (F_3A)   0.91
Hybrid (F_3B)   0.86

Table 20.5 A comparison of the viability, fecundity, and longevity of intra- and inter-population hybrids of *D. pseudoobscura* and *D. willistoni*. (After Vetukhiv 1953, 1954, 1956.)

	F_1	F_2
Survival		
D. pseudoobscura		
Intrapopulation	1.00	1.00
Interpopulation	1.18	0.83
D. willistoni		
Intrapopulation	1.00	1.00
Interpopulation	1.14	0.90
Eggs/♀/day.		
D. pseudoobscura		
Intrapopulation	1.00	1.00
Interpopulation	1.27	0.94
Longevity		
D. pseudoobscura		
Intrapopulation	1.00	1.00
Interpopulation (16°)	1.25	0.94
Interpopulation (25°)	1.13	0.78–0.95

Vetukhiv (1953, 1954, 1956), whose results are summarized in table 20.5. In the F_1 generation tests, whatever the species (*D. pseudoobscura* or *D. willistoni*) and whatever the measure of Darwinian fitness (survival, eggs/female/day, or longevity) inter-population hybrids exceed heterozygotes formed by mating strains of single localities. Contrasting sharply with the F_1 results, those of the F_2—a generation that has undergone both assortment and recombination—reveal the consistent inferiority of interpopulation gene combinations. Interpopulation hybridity represents an improvement over intrapopulation heterozygosity, but only if the genomes have not been disrupted by recombination.

King (1955), during extensive studies on the development of DDT resistance in *D. melanogaster*, made observations paralleling those of Cavalli and Maccacaro (1952) for the development of chloramphenicol resistance in *E. coli*. In eleven crosses between two resistant strains of flies or between flies of resistant and sensitive strains, the resistance of F_1 hybrid flies consistently approximated the mean (or, even more closely, the geometrical mean) of the two parental strains (table 20.6). Without exception, however, the DDT resistance of the F_2 progeny raised from un-treated F_1 individuals was lower than that of the F_1 flies; no comparable decrease was encountered when resistant populations were allowed to reproduce several generations without selection. Resistance in these lines dropped in the presence of interstrain recombination.

Table 20.6 DDT resistance of F_1 and F_2 hybrids of *D. melanogaster*; the parental flies were either from resistant laboratory populations or from a resistant population and its control. Figures represent minutes exposure needed to kill one-half the exposed individuals (LD_{50}). (After King 1955.)

Experiment	Parental	Populations	Average	F_1	F_2
X1	10.5	2.5	6.5	6.5	3.5
X1R	2.5	10.5	6.5	6.8	3.5
X2	8.8	10.8	9.8	10.9	5.1
X2R	10.8	8.8	9.8	10.1	5.5
X3	13.3	12.6	13.0	13.6	7.4
X3R	12.6	13.3	13.0	12.5	7.1
X4	21.3	13.5	17.4	20.0	11.6
X5	17.6	2.5	10.1	7.8	4.8
X5R	2.5	17.6	10.1	7.4	4.8
X6	(13.0)	2.5	7.8	6.4	2.6
X8	(19.0)	(18.0)	18.5	19.8	13.0

The account of the coadaptation of gene pools in the general sense (as opposed to Dobzhansky's coadaptation of inverted gene arrangements) can be extended to the matter of sex determination. An individual's sex may seem to be one of nature's "givens" but it is not. The maleness of males and femaleness of females is in many organisms determined by the environment, by age, by interactions between individuals, or by genetic factors of opposing influences— factors very often associated with sex-chromosomes or autosomes.

The control of sex often resides in the proper balance of male- and female-determining factors—genetic or cytoplasmic. Hybridization of individuals from different geographic localities sometimes reveals the presence of "strong" and "weak" sex-determining factors. The opposition of strong factors for maleness and femaleness results in normal males and females; similarily, the opposition of weak factors results in normal sex determination. The opposition of strong and weak factors, on the other hand, produces intersexual individuals.

Table 20.7 summarizes Goldschmidt's (1934) interpretation of many data obtained by crossing Gypsy moths, *Lymantria dispar*, of different geographic origins. According to this interpretation, the egg cytoplasm is important in determining femaleness and the sex-chromosome (for which male moths, unlike *Drosophila*, are homogamic) in determining maleness. These opposing tendencies are both strong in Japanese populations but weak (still balanced, however) in European ones. Notice that the model

Table 20.7 A summary of some of the results obtained by intercrossing the gypsy moth, *Lymantria dispar*, from different localities (left), and a symbolic interpretation of these results (right) in which (F) represents maternally inherited cytoplasm which is said to promote femaleness, M represents the sex-chromosome (males of this species are homogamic, females heterogamic) which is said to promote maleness, and the subscripts s and w signify strong and weak. (After Goldschmidt 1934.)

Observation	Interpretation
1. Japanese females × European males	$(F_s)M_s \times M_wM_w$
Males: normal	$(F_s)M_sM_w$
Females: normal	$(F_s)M_w$
2. European females × Japanese males	$(F_w)M_w \times M_sM_s$
Males: normal	$(F_w)M_wM_s$
Females: intersexual	$(F_w)M_s$
3. F$_2$ from (1)	$(F_s)M_w \times M_sM_w$
Males: $\frac{1}{2}$ normal	$(F_s)M_wM_s$
$\frac{1}{2}$ intersexual	$(F_s)M_wM_w$
Females: Normal	$(F_s)M_s$ and $(F_s)M_w$
4. F$_2$ from (2), using fertile	
intersexual females	$(F_w)M_s \times M_wM_s$
Males: normal	$(F_w)M_sM_s$ and $(F_w)M_sM_w$
Females: $\frac{1}{2}$ normal	$(F_w)M_w$
$\frac{1}{2}$ intersexual	$(F_w)M_s$

developed by Goldschmidt matches the empirical results extremely well.

A point of some interest might be interjected here. The term "intersex" suggests an individual whose sexual characteristics fall uniformly somewhere between those of normal males and normal females. An examination of the illustrative plates in Goldschmidt's 1934 publication reveals, however, that the Gypsy moth intersexes are actually mosaics of male and female sectors as if some cellular control device is too easily switched from one pathway to another in these hybrids (see page 386); Ernst Caspari (personal communication) has also recognized the mosaic nature of Goldschmidt's intersexual moths.

Polymorphisms for chromosomes that interfere with the hatching of eggs in *D. melanogaster*, and which exist in two widely separated populations (Austin, Texas and Inhaca, Mozambique) have been described by Kearsey et al. (1977). The data listed in table 20.8 reveal that both the Mozambique and Texas populations contain second and third chromosomes which, when heterozygous in a background derived from a mutant laboratory strain (*yellow*; *brown*; *scarlet*: mutant alleles that mark the X, second,

Table 20.8 The mean hatchability (%) of eggs produced by *D. melanogaster* females carrying second and third wildtype chromosomes (+) from Texas (T) or Mozambique (M). The mutants listed are *y* (yellow body color), *bw* (brown eye color), and *st* (scarlet eye color); *bw/bw st/st* flies have white eyes. Number: The number of second and third chromosome combinations extracted from each locality (37 from Texas and 54 from Mozambique in all) that yielded each class of results. (After Kearsey et al. 1977.)

		Eye Color				
Class	Source	White *y bw st* / *y bw st*	Brown *y bw st* / *y bw +*	Scarlet *y bw st* / *y + st*	Wildtype *y bw st* / *y + +*	Number
I	M	69	67	78	68	8
	T	77	58	57	38	4
II	M	69	6	72	8	20
	T	72	8	68	12	6
III	M	68	70	3	3	9
	T	86	44	1	5	2
IV	M	67	5	3	4	17
	T	76	2	7	5	25

and third chromosomes), reduce egg hatch considerably. The authors of this report point out, however, that they have no reason to believe that the "low" chromosomes (i.e., chromosomes causing poor egg hatch) have an effect on hatchability in the cytoplasm of the populations from which they were taken. Inbred lines homozygous for second and third "low" chromosomes, it seems, have normal hatchability. It may now be time to repeat Lewontin's phrase once more: Context and interaction are of the essence.

A final example of coadaptation may be cited in respect to a specific, second chromosomal factor (*Segregation-Distorter*, *SD*) of *D. melanogaster*. *SD* is a genetic factor which tends to promote its own spread in a population by appearing in more than one half of the functional sperm of heterozygous (+/*SD*) males. The detection of *SD* generally depends upon the excess nonmutant second chromosomes recovered from males heterozygous for a wildtype and a genetically marked second chromosome of a laboratory stock. As the diagram in figure 20.3c shows, *SD* chromosomes are recovered from these crosses in proportions ranging from 0.70 to 1.00. These same chromosomes are recovered for the most part, at least, in expected proportions from males carrying an X-chromosome and autosomes (second, third, and fourth) from the wild population (figure 20.3a). In figure 20.3b, the males were similar to those used to generate figure 20.3a except that the Y- rather than the X-chromosome came from the *SD*-containing

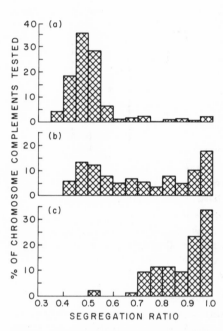

Figure 20.3 Diagrams revealing the distribution of genes that are capable of suppressing the action of *Segregation-Distorter* (*SD*), a chromosomal element in *D. melanogaster* which favors its own appearance among the gametes produced by heterozygous males. (a) The proportions of *SD*-bearing sperm recovered from males heterozygous for wildtype X, 2d, 3d, and 4th chromosomes. (b) The porportions recovered from males heterozygous for wildtype Y, 2d, 3d, and 4th chromosomes. (c) Proportions recovered from males carrying genetically marked chromosomes from a laboratory strain of flies; these are the control crosses that normally reveal the presence of *SD*. Notice that the suppressors occur both on the X-chromosome and on the autosomes. (After Hartl 1970.) Reproduced by permission of the Genetics Society of Canada.

population. The three diagrams in figure 20.3 reveal that the self-promoting tendency of *SD* is clearly suppressed in this wild population by factors located on virtually all X-chromosomes and on a third or more of all autosomal sets tested. The existence of the dominant suppressors has presumably been evoked as a coadaptational response to *SD*.

Mayr (1954) has probably expressed the most all-encompassing view of the coadaptation of gene pools of local populations. Virtually as a matter of definition in local populations, the nature and frequencies of alleles at every locus are determined by natural selection based on their interactions with the alleles (and combinations of alleles) at all other loci. Mayr's view, expressed a decade before allozyme studies had revealed the existence of numerous alleles at many loci, was based on multiple alleles and innumerable gene combinations.

An important consequence that emerged from Mayr's concept was the divergent paths that populations might take if they were established by small numbers of "founder" individuals.

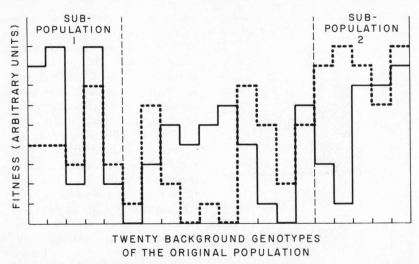

Figure 20.4 Model illustrating that the average fitnesses (i.e., tendencies to increase or decrease in frequency) of two alleles, a_1 (dashed line) and a_2 (solid line), need not be identical in two small subpopulations, even though these averages were identical in the original, larger population from which the smaller subpopulations were drawn. (After Mayr 1954.)

Figure 20.4 represents how such divergence might occur rapidly in isolation. The entire diagram represents the average fitnesses of two alleles in a series of twenty background genotypes; averaged over all backgrounds, the fitnesses of the two alleles are equal. (The "fitness" of an allele refers only to its tendency to increase or decrease in frequency; in a population at equilibrium where gene frequencies do not change, the fitnesses of the various alleles are said to be equal.) Two small samples of individuals are shown as being founders of geographically isolated subpopulations 1 and 2. Inspection of the diagram reveals that the average fitnesses of the alternative alleles in the sampled backgrounds are not equal: a_1 will be favored in subpopulation 2; a_2 in subpopulation 1.

Supergenes

The concept of the *supergene* is related to that of coadaptation but is not to be confused with it. As originally defined (and, more importantly, discussed) by Darlington and Mather (1949), a supergene is merely a group of genes that are held together mechanically (by an inversion or otherwise) so that they are transmitted as a unit. E. B. Ford (1971: 110) modified the simple block concept of Darlington and Mather into a cooperative-block concept: "If two major genes co-operate in an advantageous way, selection will favour rare structural interchanges, as well as translocations, bringing them on the same chromosome and then the means of checking crossing-over between them . . . . This will continue until the two genes so seldom break apart that they act effectively as a single switch-mechanism; that is to say, *until they have become a super-gene*" (emphasis added).

Ford's supergene, consequently, encompasses the notion of cooperation (Muller's term, too) or coadaptation. In addition, Ford's account proposes a means by which supergenes can arise. In the succeeding paragraphs, examples of supergenes (in Ford's sense) will be presented; molecular, microbial, and other laboratory geneticists generally are unaware that a problem exists which demands supergenes as an explanation. Next, we shall discuss an alternative means by which cooperating genes can be assembled into a single unit.

Mimicry is a common phenomenon among butterflies, a familiar example being the edible Viceroy (*Basilarchia archippus*) and the distasteful, inedible Monarch (*Danaus plexippus*) of the eastern United States. The pattern of the Viceroy is an almost perfect copy of the Monarch; the one difference that is obvious to even neophyte butterfly collectors (the possession of a thin, curved black line on the hind wing of the Viceroy) has disappeared in some specimens (Roderick Clayton, personal communication). The existence of mimic patterns and their remarkable perfection results from natural selection, from the greater chances of survival of an edible insect if its predators commonly mistake it for something distasteful.

E. B. Ford, P. M. Sheppard, and their colleagues have carried out extensive studies on the genetics of mimetic forms (see Ford 1971: chapters 12 and 13, and Turner 1977). Among the most complex of the species studied is *Papilio memnon*, an Asiatic butterfly. As in many *Papilio* species, the males of *P. memnon* are nonmimetic; females are polymorphic: one form is nonmimetic like the males, the remaining ones mimic other sympatric, distasteful butterfly species. What is inherited in *P. memnon* as a single, closely linked unit (supergene) are at least five "conventional" genes: one controlling the presence or absence of tails on the hind wings, a second controlling the color of the abdomen, a third controlling the color pattern of the hind wings, a fourth controlling the color pattern of the forewing, and the fifth controlling the presence or absence of "epaulettes"—spots resembling the "bleeding points" of some poisonous butterfly models. A successful mimic must simultaneously copy many features of its model including such behavioral traits as the time of day chosen for flight and the obvious characteristics of the model's flight pattern. The responsible genes must be present or absent as a single unit.

Ford suggests that chromosomal insertions and translocations are the means by which cooperating genes are moved about within the genome so that they eventually assemble at one complex locus, otherwise known as a supergene. Alternative means by which supergenes can be assembled exist (see Turner 1977). One of these can be understood properly only if one understands genetic data and interprets them properly—only if one understands the meaning of the words "a gene for . . . ."

From the earliest days of genetic research, a large number of persons have been confused by terminology such as "*the gene for* white eye color (in *D. melanogaster*) is sex-linked." These persons picture at a given spot (locus) on the X-chromosome a small amount of material that causes the eyes of its homozygous or hemizygous carriers to be white. In contrast, they also picture at this same locus other, normal bits of material that cause the eyes to be red, the normal wildtype color. But these pictures are wrong. At the *white*-locus on the X-chromosome of *D. melanogaster* is some material (now known to be a segment of DNA) that is *necessary* for the formation of normal red-colored eyes but which is by no means *sufficient* for the formation of such eyes. If the material at this locus is altered, white instead of red eyes may result. On the other hand, alterations at scores of other loci also disrupt the normal development of the fly's eye and eye pigmentation. *The gene for* red eyes is not located at any single locus; virtually the entire genome is required for the production of a normal eye.

The importance of genetics has grown and the practitioners of more and more biological disciplines now resort to genetic arguments. Consequently, one must keep in mind the difference between *necessary* and *sufficient*. Each of many genes may be necessary for the development of a certain phenotype (in saying so, we have already assumed that the environment is suitable); consequently, a change at any one of them can upset normal development. None of these many genes, however, is sufficient in itself to bring about normal development. "Sufficiency" is virtually nonexistent in genetics. One might say that the base composition of DNA in the region of a structural gene is sufficient to *specify* a given sequence of amino acids; even so, the DNA of no structural gene is sufficient to guarantee that the polypeptide it specifies will in fact be synthesized. The synthesis of a protein requires a great deal of metabolic machinery whose many parts are under the control and specifications of many gene loci.

An appreciation of the preceding paragraphs should make one wary about a great deal of terminology now used by sociologists, anthropologists, and sociobiologists in speaking of genes for altruism, genes for aggression, genes hor homosexuality, and genes for other complex behavioral traits. In the 1930s, a corresponding caution would have been to beware of terms such as genes for nomadism, genes for the love of the sea (*thalassophilia*; see Davenport and Scudder 1919), genes for alcoholism, and genes for other complex traits including poverty.

As an example of a complex trait, I shall use the maintenance of double-stranded (ds) RNA particles by wildtype yeast cells (see Wickner 1978). A normal yeast cell contains many dsRNA particles in its cytoplasm; mutant cells are known which cannot maintain these particles. Wickner collected 28 such mutants and found that the mutant genes were located at 26 loci: 24 loci had one mutant each, 2 had two mutants. Now, if genes at 150 loci are required to maintain dsRNA particles (and if mutations at each are equally probable), 28 random mutations should be distributed as 23.3 singlets and 2.2 doublets; if genes at 200 loci are required, the 28 mutations should be distributed as 24.3 singlets and 1.7 doublets. (Why?) Thus, genes at some 150 to 200 loci may be needed to maintain dsRNA particles in yeast.

Looking at the maintenance of dsRNA particles now simply as a complex phenotype (such as the formation and retention of a "tail" on an insect's wing, or as a complex behavioral trait such as flying lazily at tree-branch height during the midafternoon), and transposing the 150–200 loci needed for their maintenance onto the *D. melanogaster* genome (about 300 map units in length), what do we find? On the average, any gene in the entire genome could be less than one map unit from a locus at which is located a gene essential for the maintenance of the complex trait. As figure 20.5 shows, any one of the 150–200 loci can be converted into the switch gene controlling the presence or absence of the

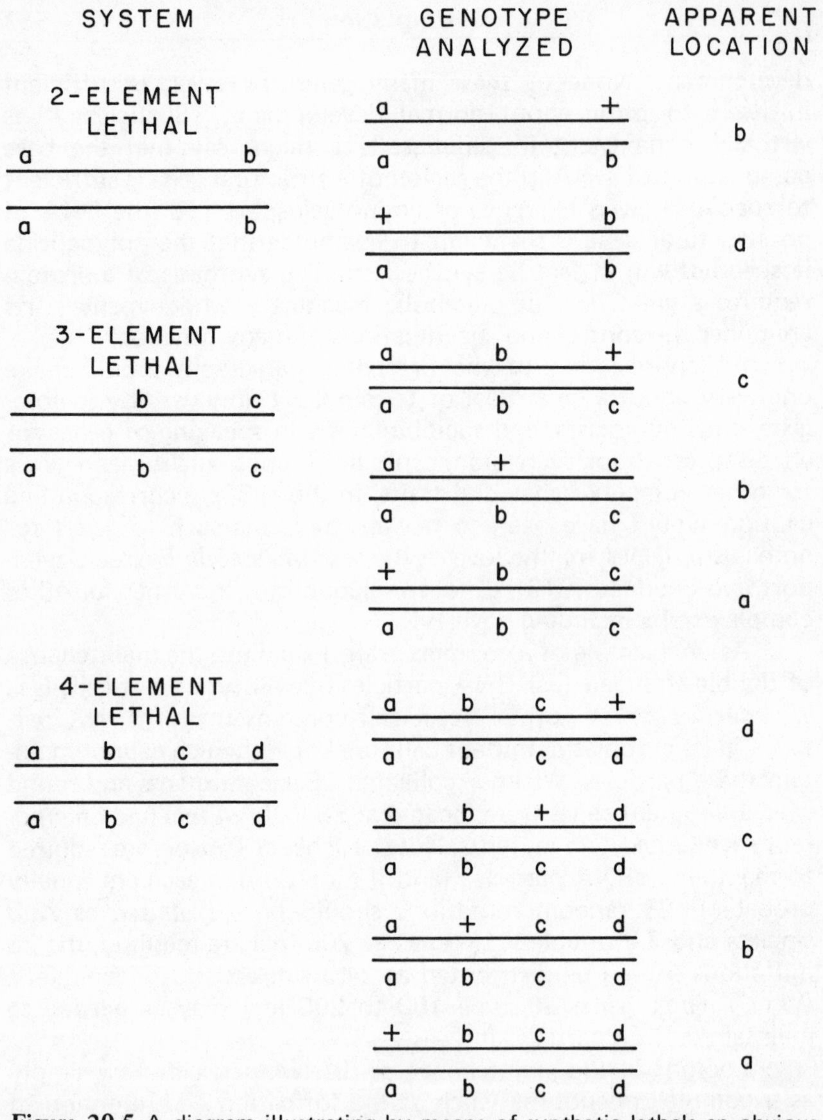

SYSTEM   GENOTYPE   APPARENT
        ANALYZED   LOCATION

2-ELEMENT LETHAL

a ———————— b

a ———————— b

 a   +
 ——————————   b
 a   b

 +   b
 ——————————   a
 a   b

3-ELEMENT LETHAL

a b c

a b c

 a b +
 ——————————   c
 a b c

 a + c
 ——————————   b
 a b c

 + b c
 ——————————   a
 a b c

4-ELEMENT LETHAL

a b c d

a b c d

 a b c +
 ——————————   d
 a b c d

 a b + d
 ——————————   c
 a b c d

 a + c d
 ——————————   b
 a b c d

 + b c d
 ——————————   a
 a b c d

Figure 20.5 A diagram illustrating by means of synthetic lethals an obvious fact: genetic analyses detect and locate gene loci at which alleles with dissimilar phenotypic effects are segregating. Each of the genotypes diagrammed for analysis leads to the conclusion that the lethal "gene" is located at a particular locus. An extension of the systems illustrated here suggests that nearly any one of the hundreds of genes controlling a complex trait could, in effect, become the "switch" gene for that trait; furthermore, if it were a matter of importance in respect to fitness, the switch gene could be chosen from any one of many possible ones in respect to its linkage with other genetically controlled traits.

trait. Consequently, it is unnecessary to postulate that a gene for this, and a gene for that, and a gene for something else were *assembled* into a supergene by means of chromosomal inversions, insertions, and translocations. In the case of complex traits, it is highly likely that at numerous places throughout the genome, genes that are necessary for the development of various aspects of a complex trait are closely linked to one another by chance; consequently, no need for shifting bits of chromatin exists. It is only necessary, as figure 20.5 demonstrates, that the effective segregation be restricted to one of these clusters. Turner (1977: 180) refers to the events described here as an example of "the largesse of the genome." As Wickner's (1978) results with yeast reveal, the genome is not miserly in respect to genes that are *necessary* for the development and expression of a complex trait— it merely lacks any that is *sufficient*.

Concluding Comments

The contrast expressed by Lewontin (page 505) between (1) regarding gene interactions as second-order effects and (2) recognizing them as the essence of heredity is nowhere illustrated as well as in the treatment accorded coadaptation, the molding within Mendelian populations of collections of harmoniously acting genes. Ecological and other field geneticists recognize the complexities of the genetic mechanisms that underlie the segregating morphological patterns and different behaviors characteristic of individuals in wild populations. Laboratory geneticists, working under a different philosophy, deliberately homogenize and purify their experimental material in order that one (or, at best, few) variables can be successfully isolated for intensive study. Interactions, under laboratory procedures, are often the essence of a poorly designed experiment.

In a review of Spiess' (1977) *Genes in Populations*, Simmons (1978) speaks of the paucity of evidence for epistasis and over-dominance, and of the vagueness of coadaptation and the integration of gene pools. Selection, he adds, is building an organism, not a monument. Selection does not build *an* organism in the sense that Volkswagon once built *the* "Beetle." Selection, at least in cross-fertilizing species, fashions a collection of alleles at various gene loci such that in the course of each generation, following their recombination and reassortment, a variety of individuals are produced some of which may survive and reproduce, thus perpetuating the population. The patchiness of the environment in which these individuals develop and the need for phenotypic variation among them assure that these individuals are not best cut from identical genetic patterns.

Coadaptation, supergenes, and polygenic variation have been discussed by Hedrick, et al. (1978) in a review of multilocus systems in evolution. Unfortunately, in an otherwise useful review, some confusion concerning these terminologies managed to intrude. In a paragraph dealing with *supergene*, Mayr's (1963) ostensible view has been contrasted with the original "mechanical unit" of Darlington and Mather (1949). A check of Mayr (1963) reveals that each of his references to supergenes is a reference to an inverted gene sequence or to a block of genes otherwise protected against recombination. Furthermore, the view expressed by Mayr in reference to the integration of genotypes which the authors contrast with the "mechanical unit" (a quotation to be found in Mayr—1963: 278) clearly refers to coadaptation (not to supergenes); in the cited quotation, Mayr merely argues that *position effect* probably plays a minor role in evolution.

Again, for the benefit of those who might refer to the review by Hedrick et al., some additional points of possible confusion might be identified here: (1) In the study by Brncic (page 524) coadaptation between chromosomes of different gene arrangements (Dobzhansky's "coadaptation") was not involved; the third chromosomes of different origins in this study were allowed to recombine freely. (2) To suggest, as the reviewers do, that differences in viability should have been independent of the environment is misleading; that viability differences are both frequency- and density-dependent has been a consistent theme in the present as well as in earlier chapters. (3) Contrary to the review, a specific (though speculative) model *has* been proposed to account for the interactions between different gene arrangements from the same and different geographic localities (Wallace 1976, see page 521). (4) Coadaptation in the senses of both Dobzhansky (inversion polymorphisms) and Mayr (coadaptation or integration of gene pools) applies specifically to fitness and components of fitness; body size (MacFarquhar and Robertson 1962) need not follow the same rules. Wallace et al. (1953) in a study involving recombination between chromosomes of the same or of different populations, showed that whereas numerous interactions were involved in the determination of viability, few, if any, were involved in the determination of the number of sternopleural bristles.

These concluding remarks are not intended to be a critique of the review by Hedrick and his colleagues. On the other hand, the bases of a discussion of complex matters should be made as precise as possible. E. O. Wilson (1975: 64) claims that "some of the apparent new understanding of the Modern Synthesis was false illumination created by the too-facile use of a bastardized genetic lexicon: "fitness,' 'genetic drift,' 'gene migration,' 'mutation pressure,' and the like." Words unleashed tend to masquarade as explanations. Much of the skepticism surrounding the word *coadaptation*, for example, arises from its use as an *explanation* in situations where its *existence* has yet to be demonstrated. Matters are only worsened if the necessarily complex studies that are needed for that demonstration are subsequently misconstrued.

FOR YOUR EXTRA ATTENTION

1. In searching for evidence of coadaptation within local populations of *D. subobscura*, MacFarquhar and Robertson (1963) studied the size of flies from various populations and of their hybrids. Size need not provide a measure of Darwinian fitness although it is one of the attributes that have been measured in the struggle to measure fitness (chapter 13).

The following sizes (mgs) and fitnesses were obtained for eight (of a possible nine) genotypes of an Australian grasshopper (White and Andrew 1962; see figure 3.2). Using a pocket calculator, calculate the regression of viability (Y) on size (X):

	X	Y		X	Y
1.	34.28	1.015	5.	32.52	0.853
2.	33.18	1.000	6.	31.75	1.052
3.	32.75	0.929	7.	32.63	1.048
4.	35.00	0.641	8.	29.25	0.621

2. Norman and Prakash (1980) have studied the activities of various amylase allozymes during development, and the association of these allozymes with different inversions and different (closely related) species: *D. pseudo-obscura*, *D. persimilis*, and *D. miranda*.

Among their findings were these: *D. pseudoobscura* and *D. persimilis* have different patterns of amylase activity; activity patterns of alleles carried by the shared Standard (*ST*) gene arrangement in contrast, are similar. The levels of activity of these allozymes depends upon the gene arrangements in which they are found. The time of activity of the allozymes during develop-ment also depends upon the gene arrangements in which they are found.

The authors conclude from the above observations that the lack of recombination between inverted chromosomal segments has allowed genic divergence of amylase activity.

Consider these observations in the context of coadaptation (especially in reference to Dobzhansky's original use of the term), and in relation to some of the speculations on gene control presented in chapter 15.

3. Wilson et al. (1974) suggest that the inability of two species to hy-bridize may arise from an accumulation of incompatibilities between two systems for regulating the expression of genes during embryonic develop-ment. Mammals, they point out, have undergone both rapid regulatory evolution and a rapid rearrangement of gene orders. This correlation leads them to suggest that gene arrangement provides an important means of achieving new patterns of regulation.

The means by which rearrangements might modify gene regulation are not specified by these authors. What are some of the possibilities?
HINT: Do not overlook position effect.

4. Sved (1979) has reviewed a topic of recent interest to Drosophila geneticists, *hybrid dysgenesis*. Crosses between wild-caught *D. melanogaster* and appropriately marked laboratory strains have led to the disclosure of a variety of genetic anomalies: recombination in males, increased mutation rates, chromosomal aberrations, and sterility.

Consider these effects from the point of view of the coadaptation of local populations (including laboratory strains). Design experiments capable of revealing whether crosses between wild-caught flies of different popula-tions also exhibit dysgenesis.

5. Sheppard (1969) has suggested that the course of resistence to in-secticides on the part of many insect pests is often a two-step process: Shortly after the introduction of a new insecticide (commonly a lag of three or four years), a low-level, partially dominant (and often single locus) resistant form is detected. The second stage, possibly involving selection for modifier alleles, is characterized by a marked increase in the level of resistance. Consider this suggestion in the light of material discussed in this chapter.

6. In considering the origin and structure of supergenes, the model of gene control proposed by Britten and Davidson should not be overlooked. Using much of the material illustrated in figures 15.6, 15.7, and 15.8, develop a model for *Papilio memnon* in which the supergene that is said to be a cluster of five conventional genes is, instead, a sensor gene (in the Britten-Davidson

sense) that controls the production of signals, which, in turn, activate numerous genes scattered throughout the genome.

7. Powell (1978) forced laboratory populations of *D. pseudoobscura* to undergo "flush-and-crash" cycles in which the sizes of the populations increased from single pairs to 10,000 flies or more, only to start anew as single pairs. Upon testing eventually for mating preference, Powell found that certain of these populations showed excesses of homogamic matings. These experiments were performed as a test for Carson's (1968, 1975) views on the relation between population flushes and crashes and speciation. Relate the same results, however, to Mayr's (1954) views as illustrated in figure 20.4.

8. In summarizing their studies, Cavalli and Maccacaro (1952) say: "Repeated selection tends to build a polygenic system rich in positive interactions. Recombination is likely to break down such positively interacting systems . . ." In these words they describe the essence of coadaptation.

Without denying Cavalli and Maccacaro's conclusion, some words of caution have been made necessary because of advances in our knowledge of bacterial inheritance, especially in reference to antibiotic resistance. Many such genes in *E. coli* are located on plasmids rather than on the bacterial chromosome. Certain procedures carried out by Cavalli and Maccacaro seem to rule out the presence of plasmids (unknown in 1952) in their material.

Seldom during an *E. coli* K-12 F^+-mediated "mating" of two bacteria do both entire parental chromosomes come to lie in a single cell; this is in sharp contrast to events that occur in the mating of higher organisms. The male (or donor) bacterium generally furnishes only a fragment of its chromosome, starting at an arbitrarily fixed point and ending at a second point that is determined by a chance break in the DNA strand. Thus, unlike a mating of resistant and sensitive strains of *D. melanogaster*, a chloramphenicol-resistant donor bacterium is unlikely to transmit all of its genes for resistance in any cross; consequently, the average resistance of the progeny of a resistant donor times a sensitive recipient bacterium should be lower than a calculation based on parental averages would suggest. Despite this precautionary note, Cavalli and Maccacaro's conclusions seem to be correct. (Joseph Falkinham provided invaluable assistance in making this assessment of the 1952 experimental procedures possible.)

21

The Genetic Integration
of Mendelian Populations:
Evolution of Dominance

PREVIEW: The genetic means by which the favorable allele at a locus gains dominance over other alternative alleles in respect to some feature of the phenotype constitutes a special case of coadaptation. Both theoretical calculations and experimental observations are discussed in this chapter. The matter is much more complex than the obvious explanation: normal alleles make functional enzymes and, therefore, are dominant to alleles which make nonfunctional ones, or none at all. On the other hand, the mathematically analyzed models are not based on assumptions consistent with experimentally detected dominance modification.

DOMINANCE IS AN attribute of gene expression. In many ways, the genetic bases of dominance and of coadaptation are similar; consequently, an understanding of one helps in the understanding of the other. The advent of molecular and microbial genetics has led many to equate dominance and recessiveness to the presence or absence of a primary gene product (of a functional enzyme, for example); this view resembles Bateson's "presence and absence" hypothesis of the early 1900s.

In the present chapter we shall comment on the old and new uses of the term "dominance." Following that, we shall discuss two early views of the evolution of dominance. Departing from theory at that point, the biochemical framework within which dominance occurs will be described, as well as experimental data bearing on dominance and its modification by selection. We shall then return once more to a theoretical treatment of dominance because, despite the fifty or more years that have elapsed since Fisher (1928) published a paper entitled "The Possible Modification of the Response of the Wildtype to Recurrent Mutations," the theoretical analysis of dominance modification is still a matter of considerable interest to mathematicians (see Feldman and Karlin 1971).

Dominance: The Old and the New Concepts

Dominance is a term that refers to the phenotype, not to the gene. It refers, as Mendel used the term, to the "roundish form" of the pea or to its "yellow colouring" (Sinnott et al. 1958:423). The curly wing that is caused by the mutant gene, Cy, is dominant to the normal, straight wing—especially at high temperatures. The lethality that is also caused by the mutant gene, Cy, is recessive to the nonlethality of the normal allele. Either the terms "dominant" and "recessive" apply to the phenotype rather than to the responsible allele or they apply to the allele *in respect to* specified aspects of the phenotype. *Curly* (Cy) is a dominant allele *in respect to* wing shape but is recessive *in respect to* lethality. To refer to *Curly* as either a dominant or a recessive allele with no further information is meaningless.

Experimental procedures in genetics have become exceedingly refined in recent years. Molecular geneticists assay routinely for the presence of enzymes in tissues or cell cultures. The normal allele at a given locus usually makes an enzyme; mutant alleles frequently do not. The assay determines the presence or absence of the enzyme; consequently, it is tempting to talk of the allele as being dominant or recessive. The presence of the functional enzyme is directly related to the presence of the normal allele. The presence of the functional enzyme is also directly related to the observed "phenotype" (growth in minimal medium, for example). In this case, can dominance be safely used in reference to the allele itself?

It might seem in these very basic analyses that dominance can be used in reference to the allele but, even here, its use in this way is erroneous. Many allelic forms of the same gene have been revealed by electrophoretic analyses of gene-enzyme variants. Individuals heterozygous for two dissimilar alleles frequently have enzyme variants that show up on starch gels as two distinct bands; such alleles are said to be "co-dominant." The starch gel becomes, in this case, the medium in which the phenotype is revealed; the presence of two colored bands means that both enzymes are present and, so it seems, the genes have equal weight in determining the phenotype. Even at this level, the use of dominance in reference to the allele itself can be misleading. One allele may be active in one tissue but inactive in another. Consequently, even at the gene-product level, the term "co-dominant" demands qualification (see figure 15.2).

Crosby (1963) has recommended that "dominance" be reserved for use in reference to the competition of messenger RNA molecules for attachment to ribosomes immediately prior to protein synthesis. This is an interesting suggestion for this competition has important consequences in respect to gene action. But to identify this one problem with a term that has meant something quite different for a half-century or more will surely lead to confusion. The great debates on the evolution of dominance—debates that form one of the exciting chapters of evolutionary genetics—were not concerned with a single step of the many that lie between gene action and phenotypic expression. On the contrary, these debates dealt with the control of the phenotype. To a large extent they dealt not with the ability or inability to make a certain protein but with the control of time and place at which this protein would be made.

The Evolution of Dominance: Fisher's View

Fisher (1930:chap. 3) was one of the first to attempt to give an evolutionary interpretation of the genetic facts concerning dominance. In making this attempt he clearly saw that evolutionary genetics must deal effectively with two major questions: Can evolution be explained in terms of genetic causes? Can genetical phenomena be explained in terms of known evolutionary causes? Molecular biology, through amino acid sequence analyses and the breaking of the genetic code, has made tremendous contributions toward the solution of the first of these questions; some molecular biologists, by failing to appreciate the historical aspect of life, are seemingly unaware that the second question exists.

Fisher's account begins with an enumeration of respects by which one allele can be distinguished from another:

1. One allele can be rare while another is common.

2. One allele can be selectively advantageous while the other can be disadvantageous.

3. One allele can be a mutant that has arisen from the other, earlier one.

4. One allele can be dominant in respect to some aspect of the phenotype while the other is recessive.

Associations between the four categories listed above are not random. Although evolution in the long run consists of the replacement of common alleles by once-rare alleles, the common allele tends in the vast majority of cases to be the advantageous one as well. In studying mutations, common alleles generally serve as the starting point and so mutant alleles (3) are also disadvantageous (2), as a rule.

Although it is not absolutely so, one allele of a series is generally dominant to all other alleles in respect to selectively important aspects of the phenotype. Of the dozen or more alleles at the *white* locus in *D. melanogaster*, for example, the wildtype allele produces heterozygotes in combination with the other ones that are indistinguishable from the wildtype homozygote; all other heterozygous combinations are intermediate to the two corresponding homozygotes. The need to restrict this point to "selectively important aspects of the phenotype" is illustrated by work of Dobzhansky and Holz (1943) on the effects of various recessive mutations on the shape of spermathecae in female flies (table 21.1). The wildtype allele, in respect to the shape of the spermathecae, is not at all dominant; dominance, as we emphasized earlier, is an attribute of an allele in respect to a specified aspect of the phenotype.

Dominance is not determined by the mutational origin of a new allele. That one allele has arisen from another does not necessarily mean that the newly arisen one is recessive nor that

Table 21.1 The effect of various "recessive" mutations on the shape of the spermatheca in female *D. melanogaster*. The numbers represent mean ratios obtained by dividing the height of the spermatheca by its diameter; "mid-parent" represents the mean of (1) the ratio observed for the mutant homozygote and (2) the ratio (1.56) observed in homozygous wildtype females. Note that the ratio for heterozygotes resembles the mid-parent value; the recessivity of these mutants, then, applies to the morphological effect for which they are named (e.g., white eye color). (After Dobzhansky and Holz 1943.)

Mutation	Homozygote	Heterozygote	Mid-Parent
white-1	1.41	1.51	1.49
white-2	1.43	1.49	1.50
white-3	1.41	1.49	1.49
white-4	1.42	1.50	1.49
yellow-1	1.41	1.49	1.49
yellow-2	1.41	1.48	1.49
yellow-3	1.42	1.49	1.49
yellow-4	1.42	1.49	1.49
yellow-5	1.42	1.49	1.49
yellow-6	1.49	1.51	1.53
yellow-7	1.47	1.51	1.52
yellow-8	1.45	1.52	1.51
yellow-9	1.47	1.51	1.52
yellow-10	1.42	1.51	1.49
forked	1.44	1.46	1.50
vermillion	1.42	1.50	1.49
dusky	1.50	1.47	1.53
ruby-1	1.50	1.49	1.53
ruby-2	1.56	1.54	1.56

the original one is dominant. The experimental evidence bearing on this point comes from studies of back mutations. Dominant wildtype alleles can be obtained as rare mutations from mutant alleles.

If evolution occurs by the substitution of new alleles for old ones within populations (Fisher argued), and if this substitution can occur at any or all loci, then we must be prepared to admit that new genes must often *acquire* the dominance that we associate with common, advantageous alleles. Furthermore, the dominance of new alleles must include dominance in respect to those phenotypic traits for which the earlier wildtype allele was dominant. Conversely, as they are displaced from populations, old wildtype genes acquire recessiveness.

In seeking an explanation for the acquisition of dominance by the "incoming" allele and the loss of it by the "outgoing" one, Fisher turned to the modification of gene action by the background genotype. That gene action can be modified is known. Mutations that are maintained in homozygous cultures tend to revert toward normal appearance or viability (see figure 19.5 and table 19.9); the strong expression of the mutant phenotype can generally be regained by outcrossing it to an unrelated wildtype stock and reextracting the homozygous mutant anew.

Fisher's argument for the modification of the dominance of a gene by selection for its background genotype in *heterozygous individuals* depends, in part, on the great excess of heterozygous over homozygous mutant individuals. This preponderance we have seen in earlier chapters. If a gene is extremely rare, the ratio $2pq/q^2$ is very nearly $2/q$. Thus, for a gene whose frequency is 0.001, there are about 2,000 heterozygotes for every homozygote. More to the point, if the mutation lowers the fitness of heterozygotes by an amount hs, its frequency at equilibrium will be u/hs (see page 271). The ratio of heterozygotes to mutant homozygotes will then be $2hs/u$ to 1.00. Modification of the heterozygotes will be subject to a great deal more selection than will the modification of the mutant homozygote.

The crucial point for Fisher's argument is the extent to which selection is able to affect heterozygotes. The heterozygotes may greatly outnumber mutant homozygotes but still be but a minuscule fraction of the population generally. To evaluate the amount of selection acting on heterozygotes, we assume that the ratio of descendants of a heterozygote to those of a homozygous normal is $x:1$; the sum of these descendants is $1 + x$. The fitness of the heterozygotes has been said to equal $1 - hs$; as a rule, in the progenies of heterozygotes, one-half of the individuals are heterozygous in turn. Thus we can say that $x/(1 + x)$ equals $\frac{1}{2}(1 - hs)$ or that x equals $(1 - hs)/(1 + hs)$. The proportion of heterozygotes within an equilibrium population equals $2u/hs$. The proportionate contribution of heterozygotes to some future generation, therefore, is the product of their frequency times the proportion ascribable to a single heterozygote:

$$\frac{2u(1 - hs)}{hs(1 + hs)} \; .$$

If u equals 10^{-6} and hs equals 10^{-2}, the proportionate contribution equals 1/5,000. Thus the rate of progress achieved by selection for dominance of the wildtype gene, selection acting within the heterozygous individuals alone, is about 1/5,000 of that which would be achieved if the entire population consisted of heterozygotes.

With this demonstration Fisher rested his case. Using data of the sort we shall present later in this chapter, Fisher could argue that artificial selection for gene expression in heterozygous individuals is in fact capable of modifying dominance. Furthermore, the calculations we have just made show that an identical but much less intense selection operates in natural populations. Thus Fisher concluded that the dominance exhibited by wildtype alleles is achieved by the accumulation of genetic modifiers through selection acting on heterozygous individuals.

The Evolution of Dominance: Wright's View

Criticism of Fisher's theory of the evolution of dominance through the selection of modifiers that tend to make heterozygotes resemble homozygous normals centered primarily upon the "second-order" selection involved. Selection, as we usually think of it, acts directly upon the relative frequencies of alternative alleles in a population. Fisher's scheme postulated that the relative frequencies of alleles at one locus are modified in response to an advantage accompanying the modification of heterozygotes at a second locus—heterozygotes whose frequency in the population is determined in turn by natural selection.

The intensity of selection in respect to the modifier under Fisher's model was calculated by Wright (1934). His calculations took the following form:

Let A and a be the alleles at the primary locus. Let M be the modifier that causes Aa to approach AA in respect to fitness but which has no effect on AA. Let M be completely dominant to m. Let q_M be the frequency of M and $1 - q_M$ be the frequency of m. Finally, let the $u/h''s$ be the frequency of a, where u is the mutation rate of A to a and $h''s$ is the net disadvantage of Aa relative to AA averaged over all combinations of M and m. The disadvantage of Aa relative to AA in mm individuals is hs, while that in MM and Mm individuals is $h's$. The relation between h, h', and h'' is given by

$$h'' = (1 - q_M)^2 h + [2q_M (1 - q_M) + q_M^2] h'.$$

The distribution of genotypes in respect to A and M is given by

$$\left[\frac{u}{h''s} a + \left(1 - \frac{u}{h''s} \right) A \right]^2 [(1 - q_M) m + q_M M]^2.$$

The selective disadvantage of *mm* individuals as a group (relative to the average of *AA*) equals $2uh/h''$, ignoring second-order terms. The disadvantage of *M-* individuals is $2uh'/h''$. Thus the advantage of *M-* over *mm* individuals, s_M, equals

$$\frac{2u(h - h')}{h''} .$$

The most favorable case for the selective modification of dominance exists when *M* makes *Aa* equal in fitness to *AA* ($h' = 0$). In this case, h'' equals $(1 - q_M)^2 h$ and s_M equals $2u/(1 - q_M)^2$. The rate of change of *M* equals

$$\frac{s_M q_M}{(1 - q_M)^2}$$

or

$$2uq_M .$$

The value $2uq_M$ is the rate of change per generation of the dominant allele, *M*, brought about by the favorable selection it obtains as a modifier of the dominance of the allele *A*. The rate of change is so small that it would be completely erased by a mutation rate of *M* to *m* equal to $2u$ (twice the rate at which *A* mutates to *a*).

The calculations given above (based on Wright 1934) make it seem unlikely that selection of dominance modifiers through their action on rare heterozygotes is an effective factor. As an alternative possibility, Wright suggested that there may be no need for a new mutation to *acquire* dominance. There may, indeed, be largely recessive genes whose heterozygous expression can be modified by selection; there is no reason to believe, however, that these are the ones that are seized upon in the evolution of populations. On the contrary, successful alleles, according to Wright, may not differ greatly in activity from their predecessors.

The Biochemical Framework of Dominance

A genuine understanding of dominance must be based on facts that are consonant with results obtained (1) by the genetic analyses of wild-caught organisms (as opposed to highly inbred, laboratory strains), (2) by experiments in which the phenotype of heterozygous individuals is altered by artificial selection, (3) by the analyses of the control of gene action at the molecular level, and (4) by the theoretical analyses of gene-frequency changes that have been made by mathematical geneticists. An understanding that neglects any of these four sources of information risks being superficial rather than genuine.

A free-ranging discussion of dominance tends to reach back farther and farther in time, eventually to the origin of life itself. The discussion can be restricted, however, to life as it now exists. In that case, we must acknowledge that life consists of a network of interacting biochemical reactions, each of which requires certain enzymes (gene products) but each of which in turn is regulated by a number of feedback controls. These feedback controls can affect both the enzymatically controlled processes and the production of the enzymes themselves. If we admit these things, and if we limit our attention momentarily to haploid organisms so that dominance will not enter our discussion, we can see that alleles favored by natural selection will be those that tend to overproduce their respective products.

The tendency to overproduce is a necessary property of any gene that is regulated by feedback controls (see figure 9.2). Just as a thermostat is useless if a furnace is incapable of overheating a house, so a feedback control is useless if a gene is incapable of performing more than adequately. On the other hand, in the presence of efficient controls, it is extremely difficult for a gene to actually overproduce. It simply gets turned off. Muller (1932b) described selection for this type of gene action although he had in mind homozygous diploid organisms. By this argument he circumvented the need to rely on selection in rare heterozygotes. In our above comments we have pointed out that overproduction of gene product is a feature "built into" a gene by selection early in the evolutionary history of life—even in haploids where there is no question of dominance.

With the admission that single functional alleles do indeed have a tendency toward dominance over inactive ones—a tendency based on enzyme kinetics and control mechanisms established by selection under conditions where dominance need not exist—we can now turn our attention to the problem that really concerns us: Given that individuals of genotype Aa have a phenotype intermediate between the (favored) wildtype (AA) and the homozygous mutant (*aa*), what avenues for rectification are amenable to selection? Our discussion is limited, consequently, to those cases in which complete dominance within diploid organisms is lacking.

For purposes of discussion, it is assumed that the attainment of a particular phenotype (phenotype$_{AA}$ of figure 21.1) requires the presence of a substance E (for end product) in a critical concentration at a given time and place within the organism. In the absence (or with insufficient amounts) of E, phenotype$_{AA}$ is not formed; instead, an aberrant phenotype (resembling, for example,

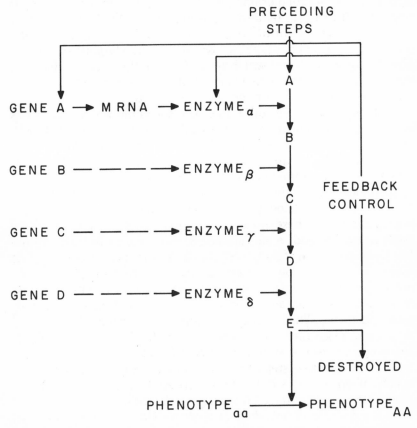

Figure 21.1 The biochemical steps and interactions that are concerned with the control of a hypothetical phenotype (phenotype$_{AA}$) by a gene A whose function is to specify enzyme α that transforms substance A into substance B; these steps of necessity are involved in the degree and the modification of the dominance of gene A over its allele a.

phenotype$_{aa}$ of the homozygote aa) results. This generalized statement covers phenotypic variation based on either quantities of phenotypic substances per cell or spatial distributions of such substances in gross patterns. It merely recognizes that the difference between the appearance of one genotype and that of another is to be explained ultimately in terms of material substances.

Substance E is but one substance of many in a complex network of biochemical reactions. In figure 21.1, E has been represented as the final step of a series leading from A through B, C, and D. As the figure shows, however, many steps precede the formation of A; furthermore, E itself is the start of a series of steps that lead to its removal or destruction. (We might note that should E be an inhibitor so that its *removal* is necessary for the development of phenotype$_{AA}$, we would merely shift our attention to a substance E' which is needed for the destruction of E. Whether the phenotype depends upon the presence or absence of a substance, the discussion remains unchanged.)

Gene A, shown in the figure as responsible for the synthesis of enzyme α which in turn controls the transformation of A to B, occasionally mutates to an allele a. Homozygous aa individuals are phenotypically unlike AA individuals, as noted above. Our interest centers on the phenotype of Aa individuals.

In figure 21.1 the presence of feedback controls connecting E and gene$_A$ and its enzyme have been illustrated. We have, however, agreed that these are not under discussion; such controls are essential to life and, once established, automatically tend to confer dominance on functional alleles in diploid organisms. The interesting case for our present discussion is that in which the Aa heterozygote fails to develop phenotype$_{AA}$, presumably because it is unable to maintain an adequate supply of E at some critical time and place. What mechanisms exist for adjusting phenotype$_{Aa}$ so that it resembles phenotype$_{AA}$? Whatever these mechanisms are, they control the modification of dominance.

The corrective steps that are available to the organism and which when taken would be interpreted as the evolution of dominance can be inferred from an examination of figure 21.1.

1. *The rate at which E is destroyed or otherwise removed can be lowered.* The concentration of E can be effectively increased by reducing the rate at which it is destroyed. Presumably this solution for dominance modification would require the modification of some gene in the destruction pathway or in still others that regulate the milieu within which this destruction is carried

out. Upon genetic analysis, a corrective measure of this sort would be found to depend upon genes other than A—that is, upon modifier genes.

2. *The rate at which E is synthesized can be increased.* There are several means by which this can be achieved. If any one of the enzymes b through d limits the overall rate of production of E, its quantity or its activity per molecule can be increased. Such alterations, whether caused by mutations at locus B (or C or D) or by still other gene changes, would represent changes in genetic modifiers of gene A.

If steps preceding A were limiting in the sense that more E could be produced in heterozygous individuals if more A were present, corrective measures could involve these "early" reactions. Once more, whatever the precise enzymatic change responsible for the adjustment, it would be looked upon as a change in a genetic modifier of gene A.

Finally, the transformation of A into B might be the limiting reaction in respect to the shortage of E in heterozygotes. In this case, a number of adjustments are possible. Upon genetic analysis, some would prove to be the work of modifiers: changes in the cellular milieu that make the enzyme more efficient, changes rendering the enzyme less sensitive to feedback control, or a reduction in the rate of destruction of enzyme α. On the other hand, some of these alterations could be met by a change in the amino acid composition of the enzyme itself, by a change at locus A. Alternatively, the rate of transcription of the locus could be altered by mutational changes in the control region (see, for example, Dickson, et al. 1977, Miozzari and Yanofsky 1978). Only changes such as these would represent an alteration of dominance by the substitution of one allele for another.

The details listed above would surely differ for different loci. Nevertheless, the main point is clear: A change in the relationship between gene dosage and phenotype is basically a change in the relationship between a given level of enzyme activity and the quantity of an often remote end product. There are a multitude of associated gene-controlled processes that influence this relationship. Consequently, when the need for change arises, the solution will often be found in the modification of physiologically associated processes that are under the control of genes at other loci.

Feedback mechanisms promote dominance. But where sufficient control is lacking, where dominance does not exist, the corrective steps—the evolution of dominance—must frequently involve the evolution of modifier systems. Furthermore, the probability that a change in a single modifier gene will provide the entire solution to a given instance of incomplete dominance appears to be no greater than that the corrective measure will arise within the original locus. For this reason dominance will often appear to depend largely on many loci, on the "genetic milieu" of the locus in question.

Experimental Observations on Dominance

Industrial melanism

Industrial melanism in various species of moths has given us one of our best opportunities to follow the evolution of dominance. "Industrial melanism" refers to the blackening of certain moths that has occurred since the onset of the industrial revolution in the early nineteenth century. Ford (1971) and, especially, Kettlewell (1973) have given excellent accounts of this evolutionary phenomenon; in this section we shall cite only those details that are especially pertinent to the discussion of the evolution of dominance.

The natural history of industrial melanics is well known. The moths that have been affected are those with cryptic markings; warningly colored ones or those that conceal themselves in crevices when at rest have been unaffected. The selective agent in the case of these melanics is differential predation by birds; Kettlewell and Tinbergen have excellent films showing birds of several species finding and eating moths taken at rest on tree trunks. Kettlewell has also shown that the amount of predation is considerable and that the protection offered by cryptic coloration is appreciable. Melanics are difficult to see on soot-blackened tree trunks; normals are difficult to see on the lichen-covered tree trunks of nonindustrial areas.

During the spread of melanic forms, dominance has evolved. The early heterozygotes of *Biston betularia*, while markedly different from typical wildtype specimens, had several white lines on their wings as well as pale patches; these have disappeared so that present-day heterozygotes are (with the exception of minute white spots on some) indistinguishable from homozygous melanics. The modification of the appearance of heterozygotes, an evolutionary intensification of melanism, that can be followed by an examination of pinned museum specimens, is important. The industrialization of Britain has resulted in the spread of melanism in one hundred or more species of moths. That dominance evolved in these cases would show that fully dominant alleles are not the only ones seized upon in evolution to the exclusion of those whose heterozygotes are intermediate.

Granted that the heterozygotes of *B. betularia* have changed in appearance during the past century, can we decide whether the change came through selection for modifiers or by selection for new and more dominant alleles? As the data summarized in figure 21.2 show, at least part of the effect has been brought about by modifier genes. A backcross involving homozygous wildtypes (*B. b. typica*) and heterozygous melanics (*B. b. carbonaria*) from Birmingham, England gives an approximately 1:1 ratio of wildtype and melanic forms (see table 19.8, however for evidence of a slight excess of heterozygous progeny in this type of cross). If however, melanic heterozygotes from Britain are backcrossed to a wildtype form from Canada (known formerly as *B. cognataria*, but now also referred to as *B. betularia*: see Rindge 1975) and the heterozygous progeny are backcrossed for three or four additional generations, few if any melanic forms are obtained, non-wildtype (non-*typica*) individuals have clearly intermediate phenotypes. "Breakdown" heterozygotes of the previous matings, when crossed with one another, do generate some full melanics; many of the heterozygous progeny have intermediate phenotypes, however. On the other hand, if artificial selection is applied to the progeny of these breakdown heterozygotes, rather complete dominance is rapidly built up once more.

Data similar to those presented for *Biston betularia* have also been obtained in respect to the dominance of a melanic (not *industrial* melanic) form of *Triphaena comes* (Ford, 1955). Crosses between melanic heterozygotes from either the Hebrides or Orkney Islands produce standard monohydrid F_2 ratios—3 melanics to 1 typical. Only 1 percent or 2 percent of all F_2 progeny are intermediate in phenotype. In crosses of heterozygotes that come from the different island groups, about 15 percent of all progeny are intermediate in phenotype. These intermediate forms are heterozygotes, presumably, because melanics in these crosses fall considerably below their expected proportions. In short, the melanic phenotype of the heterozygous individuals depends upon

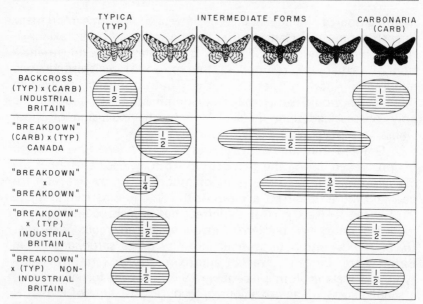

Figure 21.2 The distribution of phenotypes of *Biston betularia* ("typ" = *typica* or light form; "carb" = *carbonaria* or dark form) into six categories within broods, thus illustrating the "breakdown" and "buildup" of dominance in respect to industrial melanism in this species. The most pronounced "breakdown" involves the use of typical individuals (*B. cognataria*) from Canada, a country within which few dark individuals occur. Note that crosses between "breakdown" heterozygotes and typical individuals from either industrialized or nonindustrialized regions of Britain result in the same distributions of categories among progeny phenotypes; the genetic background of British moths seems to be uniform in this respect. (After Kettlewell 1965.) Reproduced with permission (and modification): from *Science* (1965) 148:1290–96; original figure © 1965 American Association for the Advancement of Science.

genetic backgrounds that are unique to the isolated geographic localities.

The final point to be made here in reference to industrial melanism can be expressed in the following form: whatever allele possesses the greatest immediate advantage will spread most rapidly in a population. The greatest immediate advantage lies with the rare allele that confers an advantage on heterozygous individuals. These alleles may be displaced ultimately by even more efficient but slower-spreading ones. Nevertheless, at any moment the rapidly spreading alleles on the average will be the more common ones in the genome. This statement applies to the succession of melanic alleles that have been observed in moth species (see Kettlewell 1965). More spectacularly, however, it applies to the relative proportions of partially dominant and recessive alleles used in the development of industrial melanisms. Ford (1971: 314-16) mentions only three instances among over one hundred in which recessive alleles have been utilized in the attainment of protective melanism.

Dominance in mimicry

Mimicry is a fascinating problem for population geneticists. In the case of Batesian mimicry, in which harmless organisms masquerade as others that are poisonous or otherwise harmful, the mimic must not become too common. If it does, predators may associate its appearance with "good" taste rather than with "bad." The problems raised by this need, and the answers hit upon by mimic species, underlie some of the classic studies of populations. *Ecological Genetics* by E. B. Ford (1971) has several chapters devoted to mimicry and polymorphism; these are recommended to those who are unfamiliar with the subject.

Here we are interested in the evidence mimicry presents us regarding the genetic basis of dominance. Some of this evidence is given in table 21.2. *Papilio dardanus*, an African butterfly related to one of our own swallowtails, mimics a number of Danaids, butterflies related to our Monarch (an ill-tasting, ill-smelling insect that in the United States is mimicked by the Viceroy). In the area near Entebbe, on the northern edge of Lake Victoria, models are extremely common relative to mimics; in this locality, imperfect mimics are extremely rare. Some five hundred kilometers southeast near Nairobi, models are much rarer than their mimics. In this area, nearly one third of all mimics are imperfect in design. There is no reason to suspect that the dominant alleles controlling the mimic patterns found in the two geographic localities are different; it is much more likely that natural selec-

Table 21.2 The dependence of the frequency of imperfect mimics in local populations of *Papilio dardanus* upon the number of models in the same locality. (Ford 1971: table 12.)

| Locality | Total Models | Mimics | |
		Total	% Imperfect
Entebbe	1949	111	4
Nairobi	32	133	32

tion, in an area where models are rare, is not sufficiently strong to maintain the proper modifiers for the main alleles.

Dominance modifiers of *Scute* in *D. pseudoobscura*

A shrewd guess about the path dominance modification will take—selection for modifiers or substitution of new alleles at the locus involved—can be made by studying the variation exhibited by partially dominant mutant genes. Helfer (1939) has made such a study using the "dominant" mutation *Scute* in *D. pseudoobscura*. The phenotypic manifestation of *Scute* is a lack of certain large bristles on the head, thorax, and scutellum of its homozygous and heterozygous carriers. Of the four dorsocentral and four scutellar bristles found on normal flies (eight bristles in all), only an average of 1.08 bristles is found on homozygous *Scute* females. The average number of bristles found on females heterozygous for *Scute* depends upon the source of the non-*Scute* chromosome ("genome" in F_1 hybrid females obtained by mating Sc/Sc females by wild-type males); values obtained by Helfer in his study ranged from four or more to about one. *Scute*, in other words, behaved as a complete dominant in some crosses and as a semidominant (four bristles rather than eight) in others.

The matings Helfer made in studying the variation in the dominance of *Scute* involved strains of *D. pseudoobscura* (and the sibling species *D. persimilis*, known at that time as Race B) from localities in the western United States, Mexico, and Canada. Three possibilities might explain this variation: (1) The wildtype alleles of *Scute* found in different localities may differ (isoalleles); (2) modifying genes in the rest of the genome may differ from locality to locality (dominance modifiers); or (3) both of these possibilities may be true.

The genetic analysis used by Helfer involved some rather elaborate mating schemes using genetic markers and crossover suppressors wherever needed. The essential features of his experiments are the following: Two wildtype strains (one from British Columbia and the other from Southern California) that differed considerably in their effect on the number of bristles on *Scute* heterozygotes were chosen for detailed analysis. The wildtype allele of each of these strains was transferred to an unrelated ("neutral") genetic background. Taken out of their original backgrounds, the two alleles appear to be identical (table 21.3). In fact, in a follow-up study based on different material, it appeared that the wildtype alleles at the *Scute* locus in the two

Table 21.3 A summary of a genetic analysis of the modification of the dominance of *Scute* in respect to bristle numbers in *D. pseudoobscura.* The numbers refer to the average number of bristles (out of a possible eight) possessed by different types of flies. (After Helfer 1939.)

Origin of Wildtype	F_1 Hybrid
Wildtype males X *Sc/Sc* females	
Merritt, B. C.	2.5
Henshaw, Calif.	1.2
Sc/Sc homozygotes	1.1
Wildtype allele of *Scute* in neutral background	
+Merritt/*Sc*	1.2
+Henshaw/*Sc*	1.2
+*pseudoobscura*/*Sc*	2.6
+*persimilis*/*Sc*	2.5

Amount of background from given locality:

Geographic Origin of Wildtype Chromosomes	Number of Autosomes Heterozygous			
	0	1	2	3
Merritt	1.58	1.93	2.40	3.12
Henshaw	1.61	1.28	1.07	1.21
Grand Canyon	1.52	1.74	2.03	2.82
Big Horn	1.86	1.56	1.72	1.43

sibling species, *pseudoobscura* and *persimilis*, are identical as well.

The analysis of background chromosomes was made by mating F_1 hybrid males (carrying genetically marked autosomes) with homozygous *Scute* females. Female flies carrying one (second, third, or fourth), two, or three major autosomes from the test locality could be identified by their phenotypes. The results of this test are also summarized in table 21.3. Wildtype autosomes of Merritt origin carry suppressors of *Scute*; the more "Merritt" autosomes an individual carries, the greater the number of bristles. Wildtype chromosomes from Henshaw appear to enhance *Scute* relative to the "neutral" autosomes of this particular analysis. Two additional localities, Grand Canyon (Arizona) and Big Horn (Wyoming), were also subjected to this chromosomal analysis, the wildtype chromosomes of one of these suppressed and the other enhanced the expression of *Scute*.

The implications of this fine analysis of dominance modification are clear. If the suppression of *Scute* were to become a major evolutionary problem in populations of *D. pseudoobscura*, the response would involve alterations in the frequencies of modifier genes at many different loci. There is no evidence that different wildtype alleles at the *Scute* locus exist; at least, if different wildtype alleles do exist, they are not markedly different in the extent to which they exhibit dominance over the mutant allele, *Scute*.

Artificial selection for dominance modification

Studies on both the Peppered moth, *B. betularia* (figure 21.1) and on *D. pseudoobscura* (table 21.3) have shown that genetic variation for the modification of a heterozygote's phenotype preexists in natural populations. This should not be a startling observation because it is known that the response of most populations to novel challenges comes about by virtue of pre-existing rather than brand-new genetic variation (see Luria and Delbruck 1943, and Lederberg and Lederberg 1952, for selection in bacterial cultures, and Bennett 1960 for selection of *D. melanogaster* in respect to DDT-resistance). A number of experiments have been carried out in which the variable expression of heterozygous individuals has been utilized to select artificially for both the enhanced dominance and recessivity of mutant genes. Ford (1940) succeeded in rendering a *lutea* mutant in the Currant moth (*Abraxas grossulariata*) virtually a complete recessive in one selected line, and a complete dominant in another—all within four generations of artificial selection. Fisher (1958: 63–68) reviews this and a similar experiment which he performed on the short-tail (*Sd*) mutation in mice; within two years, two selected lines were obtained in which heterozygotes had tails of average lengths of 50 mm and 12 mm.

A more recent study has been carried out by Srb and his colleagues (Russell and Srb 1972; Srb and Basl 1972). *Peak*, a dominant mutation in *Neurospora crassa*, causes colonial growth of the vegetative mycelium and the formation of spherical rather than linear asci. Wildtype *N. crassa* exhibits colonial growth when grown on sorbose; the normal growth pattern on sorbose can be restored by sorbose-resistant mutations. Srb and his colleagues screened for possible *Peak* modifiers by isolating mutations that restored normal growth of a wild-type strain on sorbose-containing medium; these mutations were further screened by testing them for the ability to restore at least some linear asci in the *Peak*-1/+ strain. Table 21.4 summarizes data bearing on the interactions of five dominant *Peak* mutations with various wildtype backgrounds (*N. crassa*, *N. sitophila*, and *N. tetrasperma*) and five modifier mutations. A number of points are illustrated by these data: First, the wildtype backgrounds of the three *Neurospora* species modify the dominance of the mutant *Peak*-4; nearly 80 percent of the asci are linear in the *sitophila* background, but only 2 percent are linear in that of *tetrasperma*. Second, although the modifiers (including a sorbose-resistant mutation, *sor*, that was obtained from a Neurospora stock center) were chosen by virtue of the modification of *Peak*-1, this proven ability does not guarantee a similar

Table 21.4 Various modifications of the phenotypic effect of five dominant alleles (*Pk*-1 through *Pk*-5) in *Neurospora crassa* by (1) wildtype alleles of the *peak* locus that were transferred into *N. crassa* from other *Neurospora* species and (2) five modifier mutants at other gene loci (*mod-A* through *mod-D*, and *sor*), some of which were tested for their effects in both heterozygous and homozygous condition. Wildtype (+/+) *N. crassa* has nearly 100 percent linear asci; *Pk*-1/+ produces asci only 4 percent of which are normally linear. None of the modifiers listed here, incidentally, restores linearity to the asci of *N. crassa* that is homozygous for the recessive allele of *peak* (*pk/pk*). (After Russell and Srb 1972; Srb and Basl 1972.)

Genotype	(Species)	Pk-1	Pk-2	Pk-3	Pk-4	Pk-5
+/+	crassa	4	20	77	27	31
	sitophila	—	—	—	79[a]	—
	tetrasperma	—	—	—	2[a]	—
mod-A/+		10	20	92	58	57
mod-A/mod-A		6	—	—	45	—
mod-B/+		12	18	96	48	52
mod-B/mod-B		—	—	45	—	—
mod-C/+		11	37	83	12	55
mod-C/mod-C		—	—	—	—	25
mod-D/+		10	22	81	42	43
sor/+		10	16	16	31	24

[a]Value unchanged through five generations of backcrossing.

modification of the remaining four dominant *Peak* mutations. Nevertheless, an inspection of the table reveals that, on the average, *Peak*-4 and *Peak*-5 produce more linear asci in the presence of the modifiers than in the normal (wildtype) *crassa* background. None of the modifiers, incidentally, had any effect on the proportion of linear asci in the perithecia of homozygous *Peak/Peak* strains nor of those in perithecia of homozygous, recessive *peak/peak* strains. A final point, one of special importance to the development of a theory of dominance modification, is that the selected modifiers themselves have their greatest effect (in the four cases tested) in the *heterozygous* condition; homozygotes for mod-A, mod-B, and mod-C produce fewer linear asci (a greater recessivity of the dominant *Peak*) than the corresponding heterozygotes.

Linkage and the Selection of Dominance Modifiers

A number of studies, several of which have been cited above, have shown that the dominance exhibited by one allele over another is indeed subject to modification by alleles at other loci; the dominance of a given allele in a given locality may at times be the result of special modifiers accumulated in that one, and to a large extent only in that one, population. On the other hand, studies such as that of Kettlewell (1965) on *Biston betularia* have shown that successive alleles, each exhibiting greater melanism than its predecessor, may sweep through populations in response to selection for dominance modification. In this section we shall consider dominance modification in algebraic terms. Following that, we shall return once more to reconsider the calculations of Fisher and Wright as given earlier in this chapter.

Polymorphism and dominance

Suppose that a population, homozygous for gene m, is polymorphic for alleles A and a. The relative fitnesses of AA, Aa, and aa individuals are $1 - s$, 1, and $1 - t$, respectively. Consequently, the equilibrium frequencies of A and a are, in that order, $t/(s + t)$ and $s/(s + t)$.

Let the relative fitnesses listed above $(1 - s : 1 : 1 - t)$ be based in part upon the *partial* relative fitnesses of AA, Aa, and aa of 1, $1 - r$, and $1 - r$, respectively. These are fitnesses ascribable to a single cause (such as predation by one particular predator) of all those events that enter into and determine the overall selection coefficients s and t. By *partial* fitnesses 1, $1 - r$, and $1 - r$ we mean that each of the overall adaptive values $1 - s$, 1, and $1 - t$ can be written $(1 - s)(1) : (1 + r)(1 - r)$ (approximately): $[(1 - t)/(1 - r)](1 - r)$. (It is immaterial whether the partial coefficient r is associated with a pleiotropic effect of a or with an allele at a neighboring locus that has been caught up by accident in the polymorphism at the A locus.)

Finally, let gene M, a modifier that increases the partial fitness of Aa individuals from $1 - r$ to 1.00 (that is, equal in this one respect to AA individuals), arise within the otherwise homozygous mm population. What will be the selective advantage of M over the earlier allele, m?

We restrict our calculations to the case where M is very rare and, consequently, is found only in Mm heterozygotes. Because M is rare, the average fitness of mm homozygotes is very nearly the same as that of the entire population, $1 - st/(s + t)$. The fitness of Mm individuals can be calculated as the average of the following:

Genotype	Frequency	Fitness
$AAMm$	$t^2/(s + t)^2$	$1 - s$
$AaMm$	$2st/(s + t)^2$	$1 + r$ (very nearly)
$aaMm$	$s^2/(s + t)^2$	$1 - t$

The average fitness of *Mm* individuals is $1 - st/(s + t) + 2rst/(s + t)^2$. The increase in fitness of *Mm* individuals relative to the average fitness of the population (and hence of *mm* individuals) is $2rst/[(s + t)(s + t - st)]$. Since only one-half of the gametes of *Mm* individuals carry the gene *M*, the selective advantage of this gene relative to its allele *m* is $rst/[(s + t)(s + t - st)]$. This, in turn, can be written as the product of an equilibrium gene frequency and a second term:

$$\left(\frac{s}{s + t}\right)\left(\frac{rt}{s + t - st}\right) \text{ or } \left(\frac{t}{s + t}\right)\left(\frac{rs}{s + t - st}\right).$$

In the case just described, the alleles at the *m* locus were combined at random with those at the *a* locus. If, on the contrary, one assumes that *M* is tightly linked with either *A* or *a*, calculations similar to those given above yield the following advantages of *M* over *m*:

Linked with *A*	$rs/(s + t - st)$
Linked with *a*	$rt/(s + t - st)$.

Since both $s/(s + t)$ and $t/(s + t)$ are less than 1.00, it is obvious that linkage of *M* with either *A* or *a* confers a greater advantage on *M* than this allele enjoys when its locus is independent of the *a* locus. Furthermore, the advantage of *M* is greatest when it is linked to the rarer (*a*) of the two alleles. This point is self-evident because linkage with the rare allele virtually guarantees that *M* will be found in the heterozygous *Aa* individuals whose fitness it modifies; Crosby (1963) refers to this as the evolution of recessiveness.

A numerical example, using values of *s* and *t* commonly encountered in the case of inversion polymorphisms in *D. pseudoobscura*, shows that the selective advantage of *M* over *m* need not be negligible. Let *s* equal 0.20, *t* equal 0.70, and *r* equal 0.05. The advantages of *M* over *m* under the three linkage possibilities are as follows:

M independent of *A*	0.010
M linked to *A*	0.013
M linked to *a*	0.046

These values, we must admit, apply to extremely favorable assumptions regarding dominance modification. If, for example, the existence of *Aa* individuals were to depend upon recurrent mutation alone as in the calculations made by Fisher and Wright, the advantage of M over m under otherwise equally favorable assumptions would be thousands of times smaller. Selection for dominance modifiers need not be negligible if heterozygotes are at all common, as they are in the case of a persistent polymorphism.

Another Look at the Theoretical Calculations

At the beginning of this chapter, I said that I would present two early views of the evolution of dominance (Fisher's and Wright's) and only later would I return to consider the formal mathematical treatment of this problem once more. A number of modern workers have reconsidered this problem (Ewens 1967 and references given in that paper; Feldman and Karlin 1971; Roughgarden 1979: 197) with conclusions differing from those of both Fisher and Wright, as well as from those reached by Bodmer and Parsons (1960), Crosby (1963), and myself (pages 560–562). I shall not attempt to reproduce the mathematical treatment here; I shall present only the array of fitness coefficients, identify the initial conditions, and list some of the conclusions.

Let the "primary" locus be occupied by two alleles, A and a, and the modifier locus be occupied originally by only m, but later (following the mutation of m to M) by M and m. Further, let the array of fitness coefficients be as follows:

	AA	Aa	aa
MM	1	1	$1 - s$
Mm	1	$1 - ks$	$1 - s$
mm	1	$1 - hs$	$1 - s$

Inspection of this array reveals that at the outset (in an mm background), a is dominant (or partially so) in respect to fitness. The allele M decreases this dominance in Mm heterozygotes ($k < h$), and causes a to be recessive in MM homozygotes. The frequency of a at the outset (assumed to be at equilibrium between mutation, u, and selection) equals u/hs; after M has displaced m in the population (and the calculations reveal that M will indeed displace m), the frequency of a rises to $\sqrt{u/s}$ (see chapter 10). The effect of the replacement of m by M has been to increase the mean fitness of the population, in respect to the A locus, from $1 - 2u$ to $1 - u$ (see chapter 12).

Ewens (1967) stresses that his interest in the problem of the evolution of dominance lies within the mathematical treatment— that is, with the correct method of calculating the outcome of selection—given the array of fitness coefficients listed above. He then lists a number of fallacies that have arisen from the earlier calculations (fallacies that apply, then, to the discussion on page 547): (1) that Δq_M is of the order of u (when q_M is nearly 1.00, Δq_M approaches $\sqrt{u}$); (2) that $\Delta q_M = 2uq_M$ (if $\Delta q_M = iq_M$ (1 − q_M), i is always small and approaches zero as q_M approaches 1.00); (3) that the action of the modifiers is most effective if $k = 0$ (not so because, as q_M approaches 1.00, m is sheltered and becomes difficult to displace from the population); (4) that linkage between the modifying and primary loci is essential; and (5) that sampling errors are important in the outcome (random effects tend to cancel out, whereas selection provides for a steady increase). Feldman and Karlin (1971) are even more precise about the role of linkage: "... the looser the linkage the faster evolution starts and finishes." As Roughgarden (1979: 197) points out, however, evolution under this model is extremely slow; with $u = 10^{-6}$, M increases initially by a factor of 1.000001.

There are two features of these recent calculations that are of some interest relative to the theme of the present chapter. The first is the apparent contradiction concerning the role of linkage. This contradiction may reside in the contrasting models that were analyzed: on page 560 we discussed the modification of one aspect of the phenotype of heterozygous individuals which existed in the population at high frequency by virtue of an already existing, persistent polymorphism; the model presented on page 562 deals with the modification of the dominance of A in heterozygous individuals who form only $2u/hs$ of the population.

Both models may be quite useless in attempting to understand events occurring in natural populations. The recent extensions of Fisher's and Wright's calculations constitute manipulations within the framework that led to genetic load calculations, a framework shown to be grossly inadequate in discussions extending from chapter 13 to chapter 18. The impetus for the selection of the modifier allele, M, is said to be the reduction of the genetic load at the A locus from $2u$ to u, from 2×10^{-6} to 10^{-6} if $u = 10^{-6}$.

Furthermore, both the recent calculations and those presented on page 545 that dealt with the modification of the phenotype of numerically common heterozygotes assumed that the modifying allele, M, is rare in the population at the outset—newly mutated, for example. These assumptions do not coincide with the observation that heterozygous individuals are often variable in appearance and that *either* allele can be rendered dominant to the other by artificial (Ford 1940) or natural (Ford 1971: 144) selection within three or four generations. Nor do they coincide with observations, such as those of Helfer (see page 556), revealing a host of dominance modifiers present and waiting in populations. Finally, the (perhaps) intuitively pleasing and logical array of fitness coefficients tabulated on page 562 do not coincide with the findings in *Neurospora* where, in each of the four instances tested, the dominance modifier proved to be more effective when *it* was heterozygous than when homozygous. It seems as if new models are needed—models that take into account both experimental observations and the underlying molecular mechanisms of dominance. Otherwise, as Ewens (1967) suggests, the analysis of the evolution of dominance becomes merely an exercise in mathematics.

The Evolution of Dominance and the Genetic Load

The evolution of dominance offers once more an opportunity to demonstrate that the genetic load of a population is not identical with the loss of individuals the population suffers as the result of natural selection. Further, it offers an opportunity to demonstrate that "genetic deaths" and "nongenetic deaths" (environmentally caused deaths, ostensibly) are not independent compartments in an evolving population; as a matter of mathematical convenience they can be treated separately but only under rigidly defined, static conditions.

In the section that dealt with polymorphism and the evolution of dominance, we described an allele M that corrected a slight flaw in what were generally superior Aa heterozygotes. The fitnesses of AA, Aa, and aa individuals before M arose and after it had spread in the population can be tabulated as follows:

Genotype	Original (mm) Background	New (MM)	New (MM) Background (Adjusted)
AA	$1 - s$	$1 - s$	$1 - S$
Aa	1	$1 + r$	1
aa	$1 - t$	$1 - t$	$1 - T$

The values $1 - S$ and $1 - T$ are obtained by dividing $1 - s$ and $1 - t$ by $1 + r$. S and T are obviously greater than s and t. The genetic load, which has changed from $st/(s + t)$ to $ST/(S + T)$, has become larger. But the advantage conferred by M upon Aa individuals according to the earlier discussion was the result of *decreased* predation. We find ourselves, then, claiming that lessened predation has increased the genetic load. This seeming paradox has a simple explanation: Predation in the earlier population would have been ascribed to environmental causes; the population was uniformly mm, so no genetic deaths could be ascribed to that locus. The postulated advent of the allele M has changed all this. Deaths have been retrieved from those formerly regarded as accidental and have been relabeled "genetic deaths"; following the origin of M, these deaths have become part of the genetic load. If we recall that an average of only one daughter can survive per mother (page 423), we can see how artificial this apportionment of deaths can be. The allele M in our example does not affect progeny size; therefore, the number of offspring that must die before reaching maturity is precisely the same after the origin of M as before. The enlarged genetic load brought on by the allele M is an enlarged load in name only; some other, unrecognized load borne by the initial population has diminished correspondingly.

FOR YOUR EXTRA ATTENTION

1. What is wrong with the following sentences? The gene causing melanism in *Biston betularia* today behaves as a dominant. The original gene was recessive; its present dominant condition arose through the selection of modifier genes.

2. West (1977) has repeated experiments of the sort performed by Kettlewell (see page 553 and figure 21.2); his results differ considerably from Kettlewell's: West found no breakdown in the dominance of the *car-*

bonaria phenotype following repeated crosses of *B. betularia* from Britain with *B. b. cognataria* of Virginia.

West suggests a possible explanation for the divergent results of these two studies: "I used a melanic *female* in every generation, and Kettlewell used males in at least some generations . . . . If there is no crossing over in *Biston* females . . . and if dominance modifiers are on the same chromosome as the locus for melanism, then my melanic females would have been passing on the whole British chromosome whereas Kettlewell's males would have been producing gametes recombinant for British and Canadian alleles at many loci."

Consider West's remarks in the light of comments on linkage and dominance modification presented in this chapter. Design an experiment capable of resolving the present contradiction.

The Genetic Integration of Mendelian Populations: Genetic Assimilation

PREVIEW: Adaptive modification of individual behavior or physiology in response to changing environmental conditions is a characteristic of nearly all forms of life. Numerous instances are known, however, in which individuals exhibit special structures before they have encountered appropriate inducing stimuli. The means by which responsibility for adaptive responses is transferred from the realm of physiology (where genetic responsibility is ultimate rather than proximal) to that of genetics (where genetic responsibility is proximal) will be the topic of the present chapter. As we shall see, the transfer can be accomplished without invoking mystical phenomena.

LIVING THINGS HAVE developed two mechanisms for assessing and appropriately responding to the environment. The first of these is synonymous with life itself: DNA—the genetic material which controls biochemical reactions at the cellular level. The second, unique to animals, is the nervous system. Both DNA and the nervous system, in time, arrive at logical solutions to problems that are posed by the environment. The amount of time required is, of course, immensely different. Hesitate while swatting a fly, and you miss; the fly's nervous system has analyzed the situation in a fraction of a second, and—presto!—the fly has flown away.

On the other hand, let the environment vary in a manner that can be predicted from various warning signals—let spring arrive with the lengthening of daylight hours, let desert flowers bloom after seasonal rains, or let forest fires signal a devastation of the local environment—and adaptive (that is, appropriate) responses on the part of various organisms can be brought into play by genetic mechanisms alone, by genetic mechanisms responding, that is, to diverse environmental signals. This point is often missed by physical scientists who, seeing the logic and the seeming purpose in the functioning of biological structures, conclude that life originated under the guidance of a Master Engineer. These persons fail to see that DNA, by virtue of both its variation and its ability to faithfully reproduce these variations—and because its successful carriers generally produce many more zygotes than can possibly survive—accumulates the information needed to construct organisms able to cope successfully with predictable environmental changes. Variation, replication, and amplification are the features that enable DNA to accumulate the information needed for evoking appropriate responses on the part of its carrier organisms.

Schmalhausen (1949) refers to appropriate responses of individuals to environments that have been encountered repeatedly during the history of the species as *modifications*; modifications reflect the proper working of an information-laden genetic system. In contrast, individual reactions to unique or novel environmental conditions—to massive doses of ionizing radiation, to man-made toxic chemicals, or to being struck by lightning—are not necessarily adaptive or appropriate; these reactions Schmalhausen calls *morphoses.* In his words, morphoses are reactions which have not attained a historical basis. DNA, of course, is the chemical that records history.

The response of genetically controlled, adaptive mechanisms to environmental signals is understandable to most persons. Increasing numbers of daylight hours stimulate the pituitary glands of birds; pituitary hormones then set in motion a train of events that lead to the maturation of sexual characteristics, mating behavior, the production and care of offspring. What, though, does one make of adaptive responses that occur without an appropriate signal? Responses that seemingly arise, that is, as if in anticipation of a proper signal.

A commonly cited example of a precocious adaptive response —a response, that is, that seems to anticipate its own need—is that of the callosities of the skin on the underside of the ostrich, one at the anterior and the other at the posterior end of the bird's brestbone (see Waddington 1975:29). These thickened growths of skin occur where the breast of the adult bird rubs against the ground as it squats. What startles many persons is the appearance of these thickenings on the breast of the embryo ostrich before it has hatched from the egg. Why should a callous form before the bird's skin has been abraded? A little thought would reveal that ostriches are not unique: human embryos long before birth have thickened skin on the soles of their feet and the palms of their hands even though their feet have not yet borne weight, nor their hands grasped any object.

It seems to me that little is gained by posing questions in the form of exotic morphological traits and their possible origins. A simpler example than the embryonic development of callosities might be the tanning of skin. Large numbers of persons respond to an exposure to sunlight by the darkening of their skin. The tanning process is more complex than we need describe here (see Wurtman 1975) but it can be visualized in terms of inducible enzyme systems of bacteria: sunlight is the trigger which induces pigment formation by cells in the skin. In addition to those persons who tan, there are those who are unable to develop any pigmentation despite exposure to sunlight, and those who are permanently pigmented, sunlight or no. The latter, continuing the bacterial analogy, are constituitive in respect to pigment formation while the former are unable to synthesize pigment. Permanent pigmentation, like the precocious development of callouses, can be viewed as an anticipatory adaptive response. The pigment-forming system in this case either needs no inducing signals or responds to signals other than those released by sunlight.

In ending these preliminary remarks, I might point out that all potentially devastating, one-time challenges must be anticipated by the individual prior to their actual occurrence. The appropriate responses to these challenges must be induced by events other than the challenge itself. The unhatched chick develops a temporary horny point on its beak, an *egg tooth*, that is needed for piercing its shell; once out of the shell the egg tooth is lost. This structure, like many others that might be cited in the case of other organisms, must be formed before a need for it arises. The development of the egg tooth must be controlled by genetic mechanisms that respond to signals other than the immediate need to escape from an egg shell.

Certain adaptive traits are developed before the circumstances requiring them arise. These traits are then said to be under genetic control (a more precise phrase might read "under proximate genetic control"; callouses that develop as the result of strenuous manual labor even though environmentally induced are under ultimate genetic control because the *ability* to develop callouses resides in the genotype). How is the control for such traits transferred to the genotype?

Body Size in *Drosophila*

The size of adult Drosophila flies is greatly affected by the temperature at which they are grown. Provided that stressful extremes are excluded, flies grown at high temperatures (23°C and above) are smaller than those grown at lower ones (say, 21° and below). Developmental rates are also affected: development is faster at high than at low temperatures.

Individuals of various wild Drosophila species captured at different localities frequently differ systematically in size. That is, some localities possess characteristically large individuals, others small ones. Studies of this sort have been carried out by Stalker and Carson (1947, 1948, 1949; see Carson 1958b) on *D. robusta*, and by Sokoloff (1965) on *D. pseudoobscura*. The results compare with observations that might be made among human populations in which, for example, the average heights of Greeks and Frieslanders would be found to differ considerably.

An exceptionally thorough study of *D. subobscura* was carried out by Prevosti (1955) with the beautifully clear results shown in Figure 22.1. Flies sampled from localities throughout Great Britain from northern Scotland to southern England not only differed in size but also exhibited a clear-cut gradient: the flies captured in the cooler regions were larger than those captured in warmer ones. Nor was this merely a reflection of the immediate temperature under which they developed in their natural habitat. Flies from the different regions, when allowed to produce progeny in the laboratory under uniform conditions, produced progenies of different average sizes. The larger flies of cool regions produce

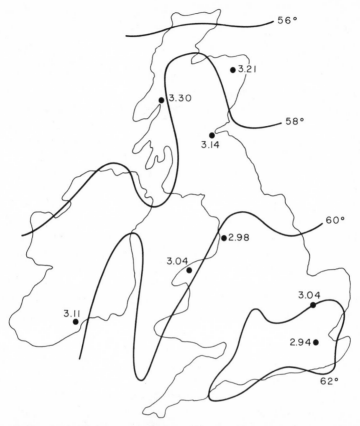

Figure 22.1 Variation in the size of *D. subobscura* flies in the British Isles. Size is given in terms of the relative areas of wings. Superimposed on the map are July isotherms. Flies from Barcelona, Spain, that were measured together with those shown here had a wing-area index of 2.58. (Redrawn from Prevosti 1955.)

larger-sized progeny at a given temperature than do the smaller flies of warmer regions. It may be difficult to see an advantage accruing to this or that body size in an organism that is small under any circumstance, but there is no question that in these flies body size is not only influenced by the temperature at which development occurs, but also by the genotypes of the flies inhabiting different geographical regions. Furthermore, the relative influence that the average genotype of these flies has on size parallels the influence of the prevailing temperature of the locality in which they are found: cool areas are characterized by large flies and "large" genotypes, warm areas by small flies and "small" genotypes.

The correspondence between body size and temperature as revealed by laboratory studies on the one hand, and between body size and genotype as detected among wild-caught individuals on the other, suggests that differences in body size represent adaptive modifications in Schmalhausen's terminology. That this suggestion is probably correct has been shown by events that occurred in a number of laboratory populations of *D. pseudoobscura*, populations referred to as "Vetukhiv's cages" (Anderson 1973).

In 1958, Vetukhiv, one of Th. Dobzhansky's collaborators, started six laboratory populations at Columbia University: two were to be maintained at 16°C, two at 25°C, and two at 27°C. The parental flies in these cages were not genetically uniform, but the same heterogeneous mixture was used in starting all populations. Vetukhiv's intent was to follow the genetic divergence of isolated populations in the laboratory. Unfortunately, he died shortly after starting these populations; his colleagues, however, maintained them and periodically carried out various tests in an attempt to reveal the genetic divergence of these laboratory populations.

Anderson (1973) analyzed the body sizes of the flies from the Vetukhiv cages and, in doing so, summarized earlier observations made by himself and others. His results are presented in table 22.1. Anderson and his predecessors asked the same question: What are the sizes of the flies from each of the populations when egg samples from the cages are allowed to develop at two different temperatures, 16° and 25°?

Examination of the data in table 22.1 reveals the basis of an

Table 22.1 The mean wing length of female *D. pseudoobscura* flies from the six "Vetukhiv" laboratory populations. Notice that tests were made $1\frac{1}{2}$, 6, and 12 years after these populations were begun during May 1958. Populations A and B were maintained at 16°C, C and D at 25°C, and E and F at 27°C. (After Anderson 1973.)

	Years from Origin	Population						S.E.[a]
		A	B	C	D	E	F	
Tests at 16°C	1.5	78.52	78.61	78.96	80.83	79.72	79.32	0.71
	6	83.07	84.03	82.15	81.41	81.29	80.63	0.51
	12	—	83.85	81.57	82.38	80.33	77.04	0.25
Tests at 25°C	1.5	69.99	69.94	70.21	71.05	71.01	70.17	0.62
	6	72.97	74.09	70.90	70.91	71.88	68.31	0.31
	12	—	71.63	69.02	68.82	67.59	65.88	0.20

[a]Standard error for all population means, from error mean square in analysis of variance.

earlier claim: comparable samples of flies grown at the two temperatures differ considerably in size. Averaged throughout the table, tests run at 16° yield flies that are some 15 percent larger than those obtained at 25°.

Three tests of body size were made of the flies in Vetukhiv's cages. The first was carried out a year and a half after the populations were first set up. As the data in table 22.1 show, there were no differences in the sizes of flies from the different populations when these were grown at either 16° or 25°. (Between 16° and 25°, yes; between the different populations at a given temperature, no.)

Six years after the populations were set up, tests reveal that at both 16° and 25° egg samples from cages A and B (cages maintained at 16°) gave rise to larger adults than did those from the remaining four populations. This differentiation in size continued so that after twelve years, each pair of populations produced flies at a given temperature whose size differed from that of the others. The one remaining population maintained at 16° produced flies substantially larger than those obtained from populations maintained at 25°, and those maintained at 25° gave rise to flies larger than those obtained from populations maintained at 27°.

The events revealed by Anderson twelve years after Vetukhiv's populations were set up confirm that differences in body size which at the outset existed only between tests made at the two temperatures (16° and 25°) eventually assumed a hereditary basis such that populations maintained at lower temperatures are "genetically" larger than those maintained at higher temperatures. In a term to be used in the following section, large size has been *assimilated* into the genetic composition of the laboratory populations. In every way, Anderson's analysis parallel the observations made by Prevosti (1955) in respect to natural populations of *D. subobscura*.

Genetic Assimilation

The transfer of the control of a phenotypic trait from physiological processes to genetic ones has been the subject of a number of experimental studies carried out by Waddington (for a summary, see Waddington 1975:16–98). One of his earliest studies involved

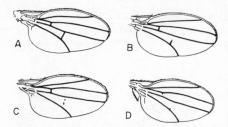

Figure 22.2 The crossveinless phenotype that was the basis of selection
in the experiment described in table
22.2 and figure 22.3. A. The wing
of a normal *D. melanogaster*. B–D.
Various expressions of the crossveinless phenotype. (Modified from
Waddington 1953.)

the gradual modification of the crossveinless phenotype from an abnormality (see figure 22.2) induced in a small proportion of flies exposed as pupae to a brief heat shock to the characteristic phenotype of most flies in unshocked cultures.

The experiment performed by Waddington consisted of exposing pupae (*D. melanogaster*) to a 40°C heat shock some 20 hours after puparium formation. Following this brief heat shock, a few flies of most wildtype strains will exhibit imperfect wing venation as shown in figure 22.2. That these imperfections are somatic and, therefore, not hereditary is demonstrated ordinarily by raising progeny from the affected flies under standard culture conditions; such progeny have normal wing venation. A great deal of genetic research during the 1930s involved the production of "phenocopies" by heat shock and other insults to normal development; Goldschmidt (1938) believed that any change of the phenotype caused by a mutant gene could be reproduced during development if a suitable experimental technique were found. Phenocopies, however, were somatic, not germinal, modifications.

The experimental procedure used by Waddington was not the ordinary one, however. Each generation he exposed the progeny of the phenotypically crossveinless flies to the four hour, 40° heat shock. And, for the most part, the proportion of flies possessing aberrant wing veins increased in successive generations (table 22.2 and figure 22.3). Eventually, nearly all heat-shocked flies exhibited crossveinless wings. Alternatively, by deliberately selecting heat-shocked flies which did *not* develop crossveinless wings, Waddington succeeded (with some difficulty, it seems) in obtaining a strain of flies more resistant to the effects of heat shock than were his original flies.

As the experiment has been described, it resembles the artificial selection for any quantitative trait. *The capacity to respond to the heat shock*, whether the response is the crossveinless or the normal phenotype, *has a genetic basis.* Individuals that develop abnormal wings are most likely to have offspring which also

Table 22.2 The number of crossveinless flies in two selected lines of *D. melanogaster*. The crossveinless phenotype was induced each generation by the subjection of pupae to a 40° temperature shock. The "upward" line was maintained by selecting crossveinless flies (phenocopies) for parents; the "downward" line was maintained by choosing normal flies from among the treated ones as parents. (After Waddington 1953.)

		Upward Line			Downward Line		
	Age treated	++	cve	% cve	++	cve	% cve
P1.	17–23	1466	747	33.8			
F1.	17–23	1582	470	22.9			
F2.	17–23	1727	502	22.5			
F3.	17–23	1784	521	22.6	1812	337	15.7
F4.	17–23	1312	738	36.0	1884	425	18.4
F5.	21–23	649	1070	62.2	1136	587	34.1
F6.	21–23	438	599	57.8	584	314	35.0
F7.	21–23	450	1009	68.5	1355	525	27.9
F8.	21–23	566	1618	74.1	1355	665	32.9
F9.	21–23	475	1308	73.4	1727	544	24.0
F10.	21–23	694	1674	70.7	2131	653	23.5
F11.	21–23	456	1808	79.9	1884	452	19.3
F12.	21–23	558	1583	73.9	1541	242	13.6
F13.	21–23	484	1510	75.7	1812	170	8.6
F14.	21–23	221	822	78.8	1082	90	7.7
F15.	21–23	165	630	79.2	1233	116	8.6
F16.	21–23	463	2598	84.9	2478	179	6.7
F17.	21–23	235	3521	93.7	2423	311	11.4
F18.	21–23	101	2567	96.2	2522	677	21.2
F19.	21–23	97	2387	96.1	2326	382	14.1
F20.	21–23	63	2564	97.6	2168	645	22.6
F21.	21–23	58	1804	96.9	1788	664	27.0
F22.	21–23	58	2277	97.5	1873	406	17.8
F23.	21–23	72	2133	96.7	2037	321	13.6

develop abnormal wings when exposed to the 40° shock; alternatively, those that do *not* respond are most likely to leave progeny who do not respond.

The experiment was carried one step farther. As the proportion of crossveinless individuals in the treated lines increased, Waddington began raising other progeny without exposing them to the heat shock; he found eventually that a small proportion of these untreated flies also showed abnormal wing veins. These flies were then selected as parents. In successive generations the proportion of crossveinless flies increased in these selected lines even though no further exposures to the 40° heat shock occurred. By applying artificial selection, Waddington was able to transform a particular phenotype (crossveinless) from a rather rare, environmentally induced abnormality (with no *obvious* genetic basis) to the prevalent phenotype of a strain of flies—a phenotype no longer dependent upon the formerly essential temperature shock.

The experiment on the genetic assimilation of the crossvein-

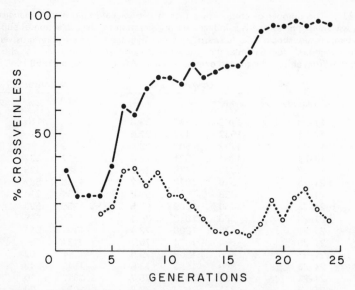

Figure 22.3 Changes in the frequencies of crossveinless pheno-copies in two selected lines of *D. melanogaster*, members of which were subjected as pupae to a 40° temperature shock each generation. Phenocopies were chosen as parents each generation in the upward line; nonreacting (i.e., normal) flies were chosen each generation in the downward line. (After Waddington 1953.)

less phenotype was subsequently buttressed by another one on the genetic assimilation of an environmentally induced bithorax phenotype (Waddington 1956). If eggs of *D. melanogaster* are exposed to ether vapor for a few minutes some 2 to 3 hours after being laid, many of the adults that eventually hatch will possess winglike structures instead of halteres (figure 22.4). Once more, the use of flies exhibiting the ether-induced "bithorax" phenotype as parents resulted in still higher proportions of bithorax flies when the eggs of the subsequent generation were also exposed to ether vapor. Furthermore, lines of flies were eventually obtained in which an exposure to ether was no longer necessary to induce the bithorax phenotype; the abnormal phenotype became the characteristic phenotype of the selected strains.

An apparent genetic assimilation of shell shape in freshwater snails is illustrated in figure 22.5. *Limnaea stagnalis* has an elongated shell compared to those of *L. lacustris* and *L. bodamica*. If, however, *L. stagnalis* is forced to grow in agitated water, the tension exerted by various of the snail's muscles shortens and distorts the shell; its shape under these circumstances resembles those of

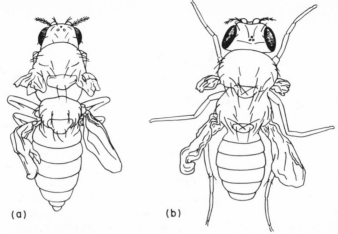

(a)   (b)

Figure 22.4 A. A bithoraxlike *phenocopy* of *D. melanogaster* that
may be obtained in relatively high frequencies by exposing 2–3
hour old eggs to ether vapor. B. The bithoraxlike *phenotype* that
occurs commonly in a strain of flies (*D. melanogaster*) that was
obtained by selecting as parents the bithorax phenocopies developing each generation from ether-treated eggs. The selected
strain, after 20 generations, no longer required the exposure to
ether in order to produce bithorax-like flies. The wings of both
flies illustrated here have been removed to better reveal the
halteres. (A. Modified from Gloor 1947. B. After Waddington
1956. Figure *A* reproduced with permission: from the *Revue
Suisse de Zoologie* (1947), 54:637–712.

the other two species. The other two species, *L. lacustris* and *L.
bodamica*, do, in fact, live in water that is agitated by waves;
individuals of these species that are raised in laboratory aquaria,
however, do not develop elongated shells. The compact phenotype
in the case of these species is determined genetically.

The role of environmental shocks, whether they are temperature shocks, exposure to ether fumes, or physical agitation, in
enabling the selection for particular phenotypes, and the eventual
development of genetic control over these phenotypes, is illustrated by analogy in figure 22.6. The shock, whatever its form,
opens a window (*B*) that reveals otherwise concealed genetic
variation. Once genetic variation has been revealed, it can be
subjected to either artificial or natural selection. Under these
circumstances, selection can eventually remove the need for the
initially essential shock; genetically caused phenotypic variation
becomes visible through the original window (*A*).

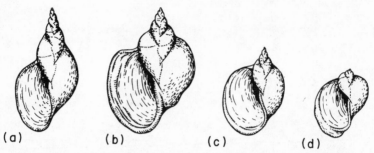

Figure 22.5 Drawings of the shells of three species of *Limnaea. L. stagnalis* (a) will develop shells (b) resembling those of the other two species if after hatching from the egg it is forced to grow in agitated water. The other species, *L. lacustris* (c) and *L. bodamica* (d) live in water that is agitated by waves; they retain their characteristic shapes, however, when raised in aquaria. The contorted shape has apparently been "assimilated" by the latter two species. (From Piaget.) Reprinted from C. H. Waddington, *The Evolution of an Evolutionist*; used by permission of the publisher, Cornell University Press; © 1975 C. H. Waddington.

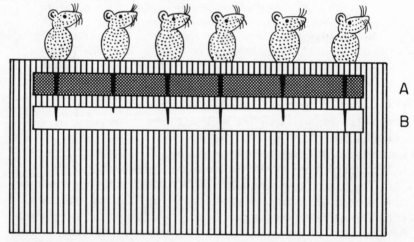

Figure 22.6 A hypothetical experiment involving artificial selection for short tails in mice. Window *A* does not permit the experimenter to see the variation in the lengths of tails; consequently, he performs what must of necessity be random matings. Window *B* reveals the existing variation in tail length, and allows the experimenter to select short-tailed parents. Eventually, the shortness of some tails may be discernible even in window *A*.

To believe that genetic assimilation is *explained* by transforming it into a standard selection format would be wrong because even progress made under artificial selection for quantitative traits is "understood" only in the formal terms of statistical genetics. Increases in the yields of food crops, improvement in egg production in poultry, increases in milk production, and changes in the conformation of beef, hogs, and other farm animals —all of these changes are understood only in the most superficial sense. Virtually nothing is known in any of these examples of the genetic mechanisms that underlie the selectively modified phenotypes—that is, of the basic changes in gene action resulting from artificial selection.

A hint regarding the genetic control of body size in *Drosophila* is provided by results obtained by Druger (1962). Using *D. pseudoobscura*, the species used by Vetukhiv in setting up his six populations, Druger selected individuals for large and small body sizes at two temperatures, 25°C and 16°C (see table 22.3). Having achieved divergence in the sizes of flies in the lines maintained according to the four regimes (large size, low temperature; large size, high temperature; small size, low temperature; and small size, high temperature), Druger determined the sizes of these flies (and unselected controls) when they were raised at both high and low temperatures.

Inspection of the data in table 22.3 shows that in six of Druger's eight tests, the selected lines deviate (upward or downward) some 9 percent to 10 percent from their corresponding controls. In two instances, however, the deviation is closer to 5 percent. These exceptional instances are those in which the selection imposed upon the line corresponded in direction to the in-

Table 22.3 An analysis of the wing lengths (= body size) of flies (*D. pseudoobscura*) that were selected for long and short wings at two temperatures (16° and 25°), and which were then tested at each of these temperatures. Numbers in parentheses are percent deviations from the corresponding unselected control. (After Druger 1962.)

Selected at	Tested at	Long	Short	Control
25°	16°	194.8 (10.3)	167.3 (5.2)[a]	176.5
25°	25°	170.0 (9.5)	139.3 (10.3)	155.3
16°	16°	189.5 (8.9)	159.0 (8.6)	174.0
16°	25°	157.3 (5.4)[a]	136.3 (8.7)	149.3

[a]These classes represent flies that were subjected to selection in accord with the prevailing environmental influence (small size at high temperature, large size at low), but were then tested under the temperature possessing the contrasting influence; the deviations of these classes from their corresponding controls are the smallest recorded.

fluence of the temperature under which the selection was carried out: selection for large body size under low temperature, and for small body size under high temperature. When these lines are tested at the *other* temperature, about half of the gains previously accomplished by selection are lost.

One can imagine that temperature is a trigger normally activating certain gene loci in *D. pseudoobscura*, thus bringing about an adaptive response in body size. (That body size in *Drosophila* is adaptive in respect to temperature and is genetically controlled has been suggested by the studies of Prevosti and Anderson described earlier.) Considering only one of Druger's two atypical results, one can argue that gene loci concerned with diminishing body size respond to high temperatures but are inactive at low ones; consequently, they are unaffected by selection for large size that is carried out at low temperature. (To paraphrase King and Jukes [1969], selection cannot change that which it canot perceive.) However, when the selected line is eventually exposed to high temperature, these "silent" loci for diminished body size are activated once more, thus negating a great deal of the otherwise accomplished gains. The other atypical case can be explained by a precisely converse argument.

When viewed as suggested here, Druger's study emphasizes the role of gene activation or lack of it (that is, of the control of gene action) in the response of organisms to natural selection. Beneath the formal terminology (consisting largely of "plus" and "minus" modifiers) of quantitative genetics lies a network of reactions: signals of various efficiencies that control transcription, molecular structures that modify rates of translation, and cellular milieus that affect the fates of various protein molecules. These are the factors which speed up or delay the commitment of cells to one developmental pathway or another. The fates of cells, after all, must underlie all anatomical and physiological adaptations.

For the genetic assimilation of a particular phenotype to succeed, genetic variation in respect to that phenotype must exist, and a means for revealing that variation must also exist. Waddington used environmental shocks—temperature and ether fumes—to reveal otherwise concealed genetic variation. Alleles at one locus can be used to reveal variation at still other loci.

Dun and Fraser (1958, 1959) have reported on the selection for high and low vibrissa ("whisker") number in mice despite the almost invariate number of these hairs on normal individuals.

The second window in this case was provided by a semidominant, sex-linked mutant gene, *Tabby*. *Tabby* males have an average number of 8.7 vibrissae while heterozygous *Tabby* females have an average of 15.1. Normal mice have 19 vibrissae. While the variance in the number of vibrissae in normal mice is virtually zero (nearly all individuals have exactly 19 bristles), considerable variation exists in the case of *Tabby* males and heterozygous *Tabby* females.

Dun and Fraser used the variation among *Ta/+* females to start lines of mice in which the mothers were chosen for low ("low" lines) and high ("high" lines) numbers of vibrissae. On alternate generations, the mating involved *Tabby* males and heterozygous *Tabby* females. This cross produced wildtype males. By the sixth generation of selection for high and low bristle numbers in *Ta/+* females, wildtype males were found that were aberrant: vibrissae were missing in the low lines, extra vibrissae were present in the high lines. Thus, by shifting the window through which a given trait was viewed (as in figure 22.6), it was possible to see variation that was not visible before. This variation proved to be genetic, at least in part, because selection was effective in changing vibrissae number. The effectiveness of selection extended to an alteration of vibrissae number even as viewed through the window provide by normal mice. (As an important corollary, we learn from this experiment that a highly uniform phenotype cannot be accepted as evidence for the absence of genetic variation. We also learned this in chapter 3!)

Genetic Assimilation and Behavior

"The history of behavioral biology," according to Mayr (1974b; see Mayr 1976), "is a history of controversies. . . ." Anyone who has observed the controversy swirling about E. O. Wilson following the publication of his book, *Sociobiology* (1975), must agree with Mayr's assessment. Fortunately, there is no need here to become embroiled in each of the many disputes that plague behavioral biology, or even those to be found in behavioral genetics. Nevertheless, a discussion of genetic assimilation which has touched on the formation of callouses "in anticipation of" future skin abrasions and constitutive skin pigmentation that is independent of exposure to ultraviolet radiation might be extended to touch on the claim: if a particular behavior is adaptive, it will have a genetic basis.

One difficulty with the claim as stated is an ambiguity in the term "genetic." Behavioral traits are complex traits and, in their studies, behavioral geneticists detect alleles which are *necessary* for the expression of these traits but which are by no means *sufficient* for bringing about their expression. This contrast between *necessary* and *sufficient* was illustrated earlier (page 533), when it was shown that dsRNA particles in yeast seem to be maintained by some 150–200 different genes. This contrast obscures the meaning of a statement such as, "I now believe that [the points of similarity between human and chimpanzee social behavior] are based in part on the possession of identical genes" (Wilson 1978:31). Does this identity apply to the few loci at which experimentalists might, with luck, eventually find segregating alleles, or does it apply to the bulk of all the DNA of these two species whose identify can be verified equally well by examining the amino acid sequences in various proteins or by DNA–DNA-hybridization experiments? The ambiguity grows when we read that "we must expect lies and deceit, and selfish exploitation of communication to arise whenever the interests of the genes of different individuals diverge" (Dawkins 1976:70).

Certain behaviors are essential to survival or reproduction and, furthermore, must be precisely executed without prior dress rehearsals. Such behaviors are among the *closed programs* of Mayr (1974b) because experience has no opportunity to bring about an improvement; perfection is demanded from the first. It seems to me that genetic assimilation as defined and demonstrated by Waddington, and described in the present chapter, provides a reasonable explanation for the existence of such behaviors. If an unhatched chick's egg tooth is cited as an example of genetic assimilation, is it reasonable not to extend the same notion to the pecking and scratching behavior that brings the tooth into play, tearing egg membranes and breaking the egg's shell?

In contrast to closed behavioral programs, Mayr describes open ones to which experience contributes greatly. Here, as Dawkins (1976) explains extremely well, the role of genes is restricted to the construction of the best possible brain and nervous system. Here again, certain circuits within the nervous system must be correct from the outset; consequently, their patterns are as genetically controlled as are the batteries of detoxifying enzymes that are manufactured as needed by mammalian liver cells. Reflex actions—the snatching of one's hand from a hot surface—are examples of such prewired patterns. Even these, because their role in survival is not all or none from the beginning, can be rewired, however, At least one Roman general (and, more recently, E. Gordon Liddy of Watergate fame) demonstrated his self-control to doubting onlookers by allowing his hand to be broiled in an open flame.

Granted that the behavioral program is an open one, there exists no obvious basis for arguing that adaptive behaviors should be placed under genetic control. The correct response to a recurrent challenge might be transformed into a "conditioned reflex" but that in no sense suggests that the control is genetic. An argument can easily be made that among social animals having individual life-spans of several years or more, the direct genetic control of behavior would be the least efficient. Lorenz (1952) pointed out that experienced jackdaws recognize "enemies" on sight and by vocal signals identify them for younger birds in the flock; having learned in this way, young ones subsequently also give the proper warning cry. One need not assume (worse, one would be foolish in assuming) that this system, which is extremely efficient within social flocks, tends to change into the innate, seemingly genetically controlled behavior of the European robin which is fearful of most other animals.

Concluding Remarks

During the 1940s and 1950s when Waddington was doing considerable experimentation on genetic assimilation, he used the unfortunate descriptive term "inheritance of acquired characters" in reference to his results. Inheritance of acquired characters is, of course, a term that arouses strong emotions among biologists. It represents, at least among English-speaking biologists, the chief distinction between Lamarckian and Darwinian evolution. Its use by Waddington has enabled anti-Darwinists such as Koestler (1971) and Grassé (1977) to refer uncritically to his work as if his results confirmed their own views. What Waddington did, in truth, was to show that individuals which respond to an environmental shock by producing an aberrant phenotype may differ genetically from those that do not respond in that manner. The environmental shock then becomes a tool which allows the experimentalist (or Nature) to perform a standard selection experiment. To the extent that the genetics of artificial and natural selection is understood, so is the phenomenon of genetic assimilation. It remains for the future to reveal whether the genetic assimilation of heat-shock-induced phenocopies in *D. melanogaster* has any bearing on the production and fate of heat-shock proteins that are produced in large quantity by stressed cells of these flies (Moran et al. 1978).

FOR YOUR EXTRA ATTENTION

1. Powell (1974) has tested *D. willistoni* for the possible genetic assimilation of body size within populations maintained for 2.5 years at $19°$ and $25°$C. His results can be summarized as follows:

| | | Population maintained at | |
		$19°$	$25°$
	$19°$ ♀♀	64.5	62.9
	♂♂	58.4	57.2
Population tested at	$25°$ ♀♀	57.8	57.2
	♂♂	51.3	50.5

Interpret these results, and compare them with those of Anderson (1973; see table 22.1).

2. Utida (1972) has described two forms of the southern cowpea weevil, *Callosobruchus maculatus*: a flight form and a flightless one. The two forms differ in a variety of physiological and behavioral traits. The flight form is induced by a number of environmental factors: crowding, high temperature, lack of water, very short or very long photoperiods. He has also reported that in maintaining the beetles in the laboratory over a period of 30 generations, the effectiveness of the inducing conditions declined markedly.

How would you explain the origin of the induction of the flight "syndrome" by the environment, and the loss of this ability in laboratory cultures?

3. Douglas and Grula (1978) have studied the pigmentation of the butterfly, *Nathalis iole*, in relation to photoperiod. Their data suggest that the photoperiod regime experienced by larvae regulates the deposition of thermo-regulatory melanin. Short-day photoperiods result in dark adult forms; long-day photoperiods induce the development of adults ("immaculate") with few melanic scales.

Suggest alternative means by which a population might arrange to possess the appropriate pigmentation in respect to warm and cool seasons. How does the control developed by *N. iole* differe from tanning in human beings?

4. Williams (1966:71–83) has discussed Waddington's account of genetic assimilation at length. Regarding the bithorax phenotype he says, "Selection of chance differences between individuals was the evolutionary force that produced the bithorax stock from the normal stock. . . . the ether did not, in a Lamarckian sense, produce the genetic variation that was selected by the experimenter but it certainly did produce the expression of that variation."

Later (having switched from bithorax to callosities), however, he has this to say: "The process [of selection] starts with a germ plasm that says: 'thicken the sole of it is mechanically stimulated; do not thicken it if this stimulus is absent,' and ends with one that says, 'thicken the soles.' I fail to see how anyone could regard this as the origin of an evolutionary adaptation. It represents merely a degeneration of a part of the original adaptation. . . . It must, as a general rule, take more information to specify a facultative adaptation than a fixed one".

Consider Williams' comments carefully in respect (1) to the notion that evolutionary adaptations should not represent "degenerations" and (2) to the suggestion that facultative and (anatomically localized) fixed gene actions differ markedly in the complexity of their control mechanisms.

Hint: Can localized gene action occur in the absence of control? Suggest a phrase that might be more precise than Williams' "chance differences."

23

Challenges to Integration: Polymorphism or Speciation?

PREVIEW: Natural environments are often heterogeneous. If this heterogeneity is discerned by the individuals of a population (and, of course, members of different species differ in the levels of heterogeneity that they can detect), it may result in contradictory selective pressures.

    The present chapter considers alternative courses that are available to a population under such circumstances: ignore the heterogeneity, develop genetic polymorphisms to match the grainy environment, or erect mating barriers which subdivide the original population into two or more new species.

THE PRECEDING THREE chapters were devoted to a discussion of forces that tend to mold gene pools of populations into coherent and harmonious collections of alleles. This coadaptation, if the term is used in its broadest sense, is a subtle task in the case of cross-fertilizing species where pairs of individuals of different genotypes mate and produce progenies consisting of arrays of recombinant genotypes. Each offspring then proceeds to search in turn for a mate possessing still a different genotype. The collection of alleles that is eventually consolidated within such a population is one from which more or less randomly generated genotypes of high fitness are put together in sufficient numbers to fill the available environment. The dominance of alleles at various loci is modified by those at others, the timing of these and other alleles is adjusted by genetic assimilation—all such adjustments might be viewed as the coadaptation or integration of the local gene pool.

Challenges to the integration of gene pools exist and are to be the subject for the present chapter. These challenges arise primarily from inconsistencies of the environment: environments are grainy, and the different grains, by making diverse demands of organisms, impose conflicting, even contrasting, pressures on the gene pool. Certain components of the genotype may also interact badly and, for that reason, act disruptively; these can be mentioned briefly before the larger matter of environmental challenges is considered.

Genetic Challenges to Integration

Genetic challenges to integration are those which lead to unstable equilibria; one allele, or one genetic alternative, tends to displace the other. In this respect, these challenges represent the converse of coadaptation. In the latter, two states are retained in the population by virtue of their selectively beneficial interactions; in the former, selection tends to cleanse the population of one or the other alternative.

Perhaps the best-known (supposedly) unstable condition in human genetics is that involving the Rhesus (*Rh*) alleles (see Li 1955:262). Without entering into the matter of multiplicity of alleles, their nomenclature, or their symbolic representation (whether, for instance, they should be represented as eight alleles at one locus or as two alleles at each of three closely linked loci), we can claim that there are Rh-positive and Rh-negative persons. The responsible genotypes can be represented as *rhrh* (Rh-negative) and *RhRh* or *Rhrh* (Rh-positive). This representation greatly simplifies a complex situation.

Mothers who are *rhrh* (Rh-negative) will form antibodies against red blood cells of Rh-positive persons. The exposure to such cells may arise through blood transfusions or, because of an accidental exchange of fetal and maternal blood through a defective placenta, during pregnancy.

Restricting our attention to pregnancies, we can see that Rh-positive embryos carried by Rh-negative (*rhrh*) mothers must of necessity be *Rhrh* in genotype. These embryos, especially the second or later ones born by the mother, are in danger of having their red blood cells lysed by the anti-Rh antibodies of the mother. Babies affected in this way are said to suffer from *erythroblastosis foetalis*, an affliction that may be lethal if a total transfusion is not available to the affected child at birth.

Genetically, the potential incompatibility of *Rhrh* embryos and their sensitized *rhrh* mothers leads to an unstable polymorphism. Because each affected newborn is heterozygous for the *Rh* and *rh* alleles, each lethality removes one allele of each sort. The *proportionate* loss is greater for the rarer than for the more common allele. Therefore, theoretical calculations reveal that the commoner allele should eventually reach 100 percent while the rarer one dwindles to 0 percent. Because most human populations are polymorphic for the Rh-alleles, it appears that other forces are involved in maintaining these alleles in various populations (see Li 1953 for suggestions concerning these other forces).

An even more drastic example of genetic incompatibility can be found in the recently synthesized attached-autosomes of *D. melanogaster*. The major autosomes of this fly are both V-shaped; that is, the centromere of each is located in the center of the chromosome. By exposing females to a moderate dose of X-irradiation, it is possible to break and rearrange these chromosomes so that instead of the normal 2L · 2R/2L · 2R composition (using chromosome 2 as an example), the altered flies have the composition 2L · 2L/2R · 2R.

Obtaining attached-autosome strains requires some skill (see Rasmussen 1960); maintaining them after they have been established requires none. Because the gametes of an attached-second-chromosome strain of flies carry either 2L · 2L or 2R · 2R chromosomes, only half of all fertilizations produce viable offspring (2L · 2L sperm X 2R · 2R egg, or the reverse). A mutant strain of flies of which only half of all progeny survive is not difficult to maintain; recall that the *CyL/Pm* stock is also maintained despite the loss of one half of all zygotes—the *CyL/CyL* and *Pm/Pm* homozygotes.

Matings between normal *D. melanogaster* and flies from an attached-autosome strain cannot produce viable offspring; progeny flies carry, for example, either 3 2L's and 1 2R or 1 2L and 3 2R's. Consequently, a population consisting of a mixture of normal and attached-autosome flies is unstable; "heterozygotes" are lethal and, consequently, either the normal chromosomes should displace the attached ones, or vice versa.

Because the chromosomal manipulation of wild populations of insect pests offers one possibility for their control (for example, chromosomes carrying genes for sensitivity to an insecticide can be introduced into a population as a replacement for those carrying established resistant alleles), a number of experiments have been carried out on the fate of attached-autosomes in laboratory populations of *D. melanogaster*; figure 23.1 summarizes the results of such experiments. It appears that the wildtype (normal) chromosomes have a tremendous advantage over the radiation-induced attached ones. The critical point in these unstable populations seems to be about 95 percent attached-autosomes: 5 percent normal ones. That is, if the attached autosomes have a frequency of 95 percent or more, they tend to eliminate the normal chromosomes from the laboratory populations; otherwise, they are instead eliminated by the normal chromosomes.

As a rule, it would seem that the mechanical consequences of chromosomal rearrangements are more likely to produce unstable situations in which heterozygotes (or hybrids) are at a disadvantage than are gene mutations alone. Exempt from this claim in the case of mammals are mutations (such as the Rhesus-factor discussed above) involving antigens which can leak through defects in the placenta. Also exempt would be gross incompatibilities of gene control mechanisms of the sort described in chapter 15. In the case of chromosomal alterations, (1) translocation heterozygotes are generally at a severe disadvantage, (2) female Drosophila flies heterozygous for pericentric inversions are at a selective disadvantage (see Vann 1966), (3) female flies heterozygous for certain combinations of overlapping inversions may be at a disadvantage (see Sturtevant 1938, Wallace 1954b), and (4) triploid hybrids produced by matings between diploids and their tetraploid relatives are generally quite sterile. These, then, are the sorts of genetic polymorphisms that are unstable and for which populations tend to become monomorphic (without, it seems, anyone regarding the resulting monomorphism as an example of coadaptation); later we shall see that these same sorts of cytological alterations may have a role in the splitting of one species from another.

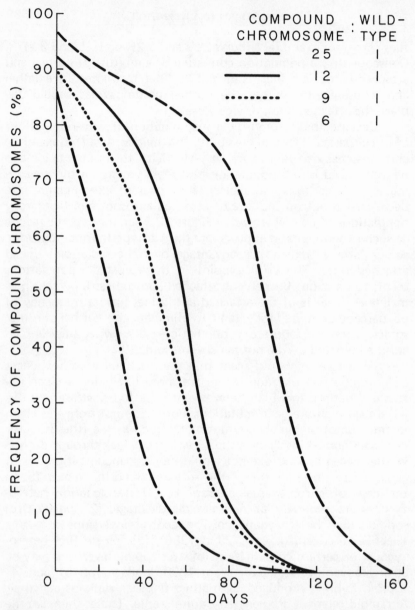

Figure 23.1 The elimination of compound-chromosome-carrying flies (*D. melanogaster*) from populations in which they and normal flies competed. (The two types produce no viable hybrids.) Starting proportions are listed in the key. Not shown in the figure are the few cases in which the carriers of compound-chromosomes eliminated their normal competitors; the unstable "switch point" is well above starting proportions of 95 percent compound-chromosome carriers (fewer than 5 percent normal flies). (After Fitz-Earle et al. 1973.)

Environmental Challenges

The integration of a population is challenged whenever natural (or artificial) selection operates in conflicting directions. This must occur, of course, in any nonuniform environment—and few environments are, in fact, uniform. It is useful in this case to speak of a grainy environment as if—like a salt-and-pepper fabric—it were composed of only two elements. As in the case of photographic film, to extend our analogy, we can speak of the size of the grains: fine-grained and coarse-grained. We can borrow a concept from colloidal chemistry and speak as well of the matrix and the dispersed phase (cream is butterfat dispersed in a matrix of water; butter is water dispersed in a matrix of butterfat; whipped cream is a delicate balance between the two).

The graininess of the environment does not depend upon the environment alone; it depends as well upon the organism. Wolves and bison do not perceive the environment in the same manner as snails and beetles. Given that the environment is grainy, three (perhaps five) organism-environment relationships can be distinguished:

1. The organism is so sedentary relative to the size of the grains that a given individual seldom encounters both types of grains.

2. The organism may be extremely mobile so that it flits from one grain to another, never recognizing that different grains exist—only an average "grain."

3. Organisms move (in relation to the grains) sufficiently slowly that they must respond as individuals first to one, then to the other, type of environment.

The fourth and fifth cases would resemble the third except that one type of grain or the other represents the major challenge (perhaps because it is the matrix within which the other is dispersed); the other is too large to be ignored but too small to constitute a major challenge.

Now, when the environment "pulls" in two directions, populations have a number of options:

1. They can settle on one genotype and remain in one environment while ignoring the other—in effect, leaving the second environment to another species.

2. They can settle on a single genotype that to some extent manages to cope with both environments, but coping with neither one as well as it might.

3. The population can establish a polymorphism such that

each of two homozygotes is fitted for one environment, and the heterozygote to both (or to one).

    4. The population can split into two species—one living in one environment, the other in the second. This solution, of course, resembles the first option described above except that the other species is now said to be derived from the original one.

Patterns of Selection

The above paragraphs constitute a catalogue of the interactions that might occur between organisms and a grainy or patchy environment. Figure 23.2 recalls the various types of selection that can act on a population; these were also described in chapter 11. Stabilizing selection is that which would be expected in a constant environment; it would also be that to which organisms would be subjected if they were restricted to a single portion of a grainy environment.

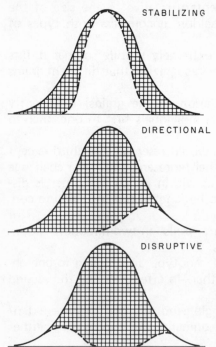

STABILIZING

DIRECTIONAL

DISRUPTIVE

Figure 23.2 Three types of selection. The solid line indicates the initial collection of individuals; the shaded portion represents those eliminated by selection; and the clear area under the broken line represents surviving individuals. (See earlier discussion in chapter 11.)

Directional selection is that which occurs in most instances of artificial selection (selection aimed at higher yields in farm crops and at more meat or milk from farm animals), in environments which are themselves changing, and (in the case of a grainy environment) following the successful invasion of the second portion of a grainy environment.

Disruptive selection is that to which any population must be subjected if it tries to occupy dissimilar portions of the surrounding environment simultaneously. The graphic representation of disruptive selection shown in figure 23.2 vividly illustrates the chapter title: challenges to integration.

Environments which are detected by human beings as being grainy may not be so to other organisms. Imagine a laboratory population of *Drosophila melanogaster* that is maintained in the presence of (1) standard medium and (2) a special medium that is highly toxic to developing larvae. Here, however, the toxicity is so great that no surviving flies emerge from the treated food. In such a case, the presence of the second type of food medium has no effect on the population except, possibly, for favoring flies which avoid laying eggs on the toxic medium (what is the source of this selection?) and (to the extent that the poison affects adult flies that visit the treated food cups) either an increasing resistance to the poison among adult flies or a growing tendency to avoid the poisonous food.

One of the options open to a population that is exposed to divergent environments is to settle on one genotype that functions reasonably well in either. Because the actions of genes are subject to various controls, this solution may be extended to include individual phenotypic plasticity. An example is shown in figure 23.3. The plant (*Ranunculus aquatilis*, the water crowfoot) shown here exhibits one phenotype when growing submerged in water (an adaptive phenotype, incidentally, with leaves that are not dangerously buffeted by water currents) and another when growing on land (in which case the leaves are broad and rigid). As the figure shows, the same individual can exhibit the two phenotypes if it grows partially submerged.

Although the water crowfoot possesses a remarkable plasticity, it should not be regarded as something unique in the world of plants and animals. All insects that pass through larval, pupal, and adult stages are equally remarkable. Two nonidentical sets of genes (some loci may function in both) are responsible for the larval and adult morphologies and behaviors; in the case of insects,

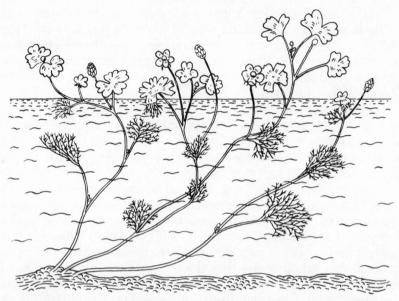

Figure 23.3 An illustration of the divergent phenotypes which the same genotype can produce in different environments. The water crowfoot (*Ranunculus aquatilis*) can grow on land, submerged in water, or, as the illustration shows, one individual can grow both below and above water.

hormones are known to trigger genes that switch development from one pattern to the other.

Even tanning can be viewed as an instance of individual plasticity that enables a person to adapt to two (cloudy and sunny) environments. Because tanning does represent such phenotypic flexibility, it also illustrates how one genotype can give rise to two or more distinct phenotypes, depending upon the environmental stimuli to which the individual is subjected.

Many of the points being made here in an anecdotal way are encompassed in the theory of fitness sets that has been developed largely by Levins (1968 and earlier references). Figure 23.4a illustrates the set of fitnesses corresponding to all genotypes in a population; for each genotype, W_1 is its fitness in environment number 1 and W_2 is its fitness in environment number 2. Each genotype can be represented as a point corresponding to (W_1, W_2). The numerous genotypes result in a collection or cloud of points as shown in the figure. Assuming that the environments are encountered sequentially either by movement through a grainy environment or during the passage of time, the long-term or overall fitness of an individual is given by the product $W_1 \times W_2$; if

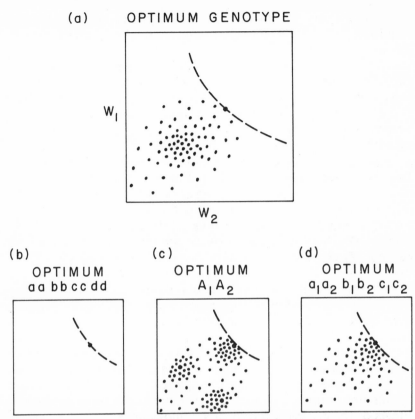

Figure 23.4 Given two environments (E_1 and E_2) and many possible genotypes, the two fitnesses (W_1 and W_2) of each genotype in the two environments can be plotted as shown in the top diagram (a). Of the entire set of points, that representing an optimal genotype can be identified (maximum value for the product, $W_1 \times W_2$). This genotype, depending upon whether it is homozygous, heterozygous for a simple polymorphism, or heterozygous at many gene loci will generate populations of offspring shown in the lower diagrams, b, c, and d. Under the classic view of populations with its stress on genetic uniformity, the optimum genotype would tend to recreate itself as in b. Simple polymorphisms generate genotypes clustered locally around the homozygous and heterozygous individuals. An optimal genotype that is heterozygous at many loci would tend to recreate the pattern shown in a.

either drops to zero, for example, the individual either dies or becomes sterile. For an overall fitness of a given value, say $\overline{W}$, the points representing $W_1 \times W_2 = \overline{W}$ form a hyperbolic curve. Of all such curves, many will miss the fitness set completely, others will pass well inside the set, but *one* curve will pass through one (perhaps more if the margin of the set is concave) point which, by definition, represents the genotype with the greatest overall fitness.

The optimum genotype has, under Levins's model, now been identified. There remain, however, several possibilities which are illustrated in the remaining three diagrams (figure 23.4b, c, and d). The classical view of population structure with which we were largely concerned in the early chapters of this book held that, except for rare deleterious mutant alleles, all individual members of a population are (or should be) genetically identical. Under ideal conditions (see figure 14.1), the fitness set would, in this case, consist of a single point lying as far from the origin ($W_1 = 0$, $W_2 = 0$) as is biologically possible (figure 23.4b).

Chromosomal polymorphisms such as those studied by Dobzhansky and his colleagues in *D. pseudoobscura* (see chapters 13 and 19) could easily give rise to a complex fitness set such as that shown in (c); here the inversion heterozygotes possess the highest fitness and, therefore, the population is polymorphic. The two types of inversion homozygotes could differ in the fitnesses they exhibit in the two environments (cyclic seasonal changes in inversion frequencies suggest that they do; see Dobzhansky 1947a). The fitness set in this case would consist of three dense clusters of points representing the inversion heterozygote and the two types of homozygotes; the fuzziness of each of these centers reveals only that the rest of the genome has some effect on the fitness of the inversion hetero- and homozygotes.

The last diagram (figure 23.4d) illustrates the case where the optimal genotype tends to be a complex heterozygote (see figure 14.2 and chapter 15); in each generation, the entire fitness set would be regenerated by the mating of those individuals possessing the optimal genotype(s). This case corresponds to that which was discussed in chapter 18.

Not shown in figure 23.4 but easily imagined nevertheless is the case where the "northeast" portion of the fitness set is concave, possessing two prongs, with the hyperbolic curve tangent to two distinctly different sets of "optimal" genotypes. That would be a case corresponding to disruptive selection (figure 23.2). Such a case can easily lead to a balanced polymorphism within the population (Levene 1953; Maynard-Smith 1962) even though heterozygous (A_1A_2) individuals are not necessarily superior in fitness to the two homozygotes (A_1A_1 and A_2A_2); in this sense the polymorphism differs from the one illustrated in figure 23.4c. A stable polymorphism can be maintained by a heterogeneous environment even though the average fitness of heterozygous individuals is less than that of the homozygotes; a sufficient but seemingly not necessary condition for this equilibrium is that the average fitness of the heterozygote exceed the geometric mean of those of the corresponding homozygotes.

Two polymorphisms that are directly dependent upon the complexity of the environment can be cited as illustrative examples. The first is the polymorphism discovered by Souza et al. (1970) in a laboratory population of *D. willistoni* (see tables 19.1 and 19.2); this polymorphism, it may be recalled, was based on pupation site preferences: one morph pupated within food cups, the other preferred the floor of the population cage.

The second example, involving easily discerned environmental differences, is a case of gene substitution in Tasmanian populations of Eucalyptus (Barber 1955, 1965). In a region where the elevation changes abruptly within a short distance, Eucalyptus trees with waxy (glaucous) stems displace those with nonwaxy (green) ones. It seems that trees protected by wax tolerate the cold of higher elevations better than do the others. Within the region where the two types of trees grow in mixed stands there is no reduction in the density of the stands, thus prompting Barber to invoke density-dependent selection ("soft" selection in my terminology) as the mode of natural selection. Because the two environments—high and low elevation—are physically separated (albeit nearby), one can (as Barber does) look upon this example as illustrating a *cline* rather than a grainy environment. In this regard, however, the example serves to remind us that a grainy environment is a degenerate or pathological cline in which the distance between alternative environments is reduced to zero.

Animals, because of their sense organs and mobility, add complications to the more simplistic models of grainy environments. Anyone who has watched a cat seek out a patch of sunlight for warmth, or the shade of a chair or bush during the heat of a summer day, knows that the "grains" of the environment need not be encountered by chance alone. Earlier (page 110) we saw evidence that individual *D. pseudoobscura* with various genotypes (both chromosomal and allozymic) sought different environments when released in the mountains at Mather, California (Taylor and Powell 1977). Table 23.1 shows the preference of individual *Asellus aquaticus* of differing amylase genotypes for two different sorts of rotting leaves. Clearly, these isopods do not make random choices in going to one or the other type of food.

The extreme response to a pattern of disruptive environ-

Table 23.1 The observed distribution of amylase genotypes in experiments in which 2,183 individuals of *Asellus aquaticus* were given a choice of beech and willow leaves. (After Christensen 1977.)

Genotype	Beech	Willow	Total
A_1A_1	330 (43.6 percent)	427 (56.4 percent)	757
A_1A_2	410 (41.4 percent)	581 (58.6 percent)	991
A_2A_2	127 (29.2 percent)	308 (70.8 percent)	435
Totals	867	1,216	2,183

Chi-square, 2 d.f., = 26.0; $p \ll 0.001$.

mental challenges would be the splitting of the population into two—one meeting one challenge, the second meeting the other. Whether or not a single population, living in a restricted area, can split in this manner has been the subject of discussion among evolutionary biologists for many years. The discussion dwells on sympatric versus allopatric speciation. Under the influence of Mayr (1942, 1963, and see 1976) most biologists believe that truly adaptive responses to diverse environmental challenges occur only in spatially (geographically) isolated populations, each local population meeting one challenge. Subsequent dispersal and alterations in distribution patterns then lead to the coexistence of the two forms in a single locality; at that time, three events can be imagined: (1) the patchy environment itself keeps the two adaptively different groups apart; (2) individuals of the two groups meet and interbreed, but the inability of hybrids to survive on either patch leads through natural selection to sexual isolation; or (3) individuals of the two groups meet and interbreed with the resultant production of a highly variable, polymorphic, introgressive population (see figure 6.7).

Not all persons accept the idea that geographic isolation (that is, a spatial isolation brought about by nongenetic means) is a *necessary* condition for speciation; many of these persons would admit, however, that such isolation is *commonly* involved in the origin of species. In effect, they maintain only that under the right circumstances a population *may* split into two separate and isolated subpopulations all within a single locality (sympatric speciation). Maynard-Smith (1966) provided a theoretical account of the problem of sympatric speciation; Bush (1975a, 1975b, and earlier references) has reviewed the evidence for sympatric speciation, including that obtained during his own extensive studies of the true fruit flies (*Rhagoletis* spp.).

Experimental data suggesting that a program of disruptive selection can successfully subdivide a single population (*Dro-*

sophila melanogaster) into two largely isolated subgroups have been obtained by Thoday and his colleagues (Thoday and Gibson 1962; Thoday and Boam 1959; Thoday 1958, 1964). Figure 23.5, for example, shows the gradual divergence in the numbers of sternoplural bristles among the progeny of parents that were chosen each generation because they possessed either the highest or the lowest number of bristles among the flies of their generation. The divergence occurred even though the parental males and females were identified while virgin, and an opportunity for random mating was provided by placing these virgin flies into a common mating chamber.

That the divergence between the high and low lines in Thoday's experiment involved the origin of a sexual isolation between the high and low selected lines is shown in figure 23.6 and table 23.2. The figure shows the histograms representing the distributions of bristle numbers among the flies of high and low lines during an advanced generation of selection. Beneath these histograms are others obtained by deliberately mating low males and females, high males and females, high males with low females,

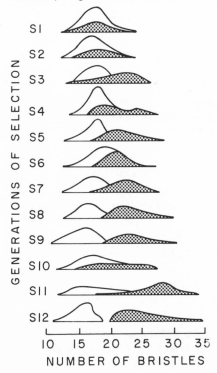

Figure 23.5 The divergence of lines of *D. melanogaster* that were selected for high and low numbers of sternopleural bristles. This divergence was obtained even though the opportunity for random mating was provided each generation by placing males and virgin females of the two selected lines in a common mating chamber. (After Thoday and Gibson 1962.)

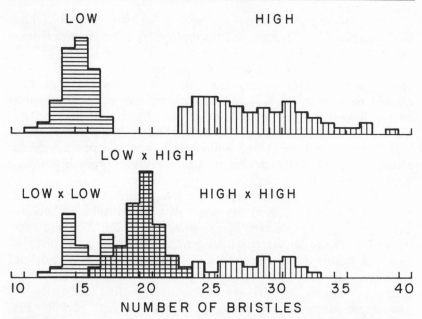

Figure 23.6 Evidence for sexual isolation between lines of *D. melanogaster* maintained by disruptive selection. The upper histograms represent the bristle numbers observed among the progeny of the two selected (high and low) lines. The lower ones are those of progeny of high X high, high X low, and low X low matings. Few, if any, offspring of high X low matings are found within the selected lines even though the opportunity for such matings was provided. (After Thoday 1964, 1972). Reproduced with permission: from *Insect Reproduction* (Entomological Society of London) and the Proceedings of the Royal Society (1972) B 182:109-43.

and low males with high females. The distributions of high X high and low X low progeny coincide with those observed in the selected lines; the high X low crosses (crosshatched histogram in the figure) produce progeny with intermediate bristle numbers that are not observed in the selection experiment itself. This observation suggests that matings between high and low flies in the lines subjected to disruptive selection do not occur. The suggestion of sexual isolation has been borne out by direct observation (table 23.2); the males and females of the high and low lines exhibit a marked mating preference for their own kind.

Figure 23.7 illustrates the bimodal distribution of bristle numbers obtained in another line of *D. melanogaster* that was subjected to disruptive selection. The selection scheme in this case was complex, involving a round of matings each generation in

Table 23.2 Evidence for preferential mating of males with high and low numbers of sternopleural bristles following a number of generations of disruptive selection. (After Thoday 1964.)

	Female:	High		Low	
Generation	Successful male:	High	Low	High	Low
7		12	3	4	12
8		14	2	6	10
9		10	4	6	7
10		8	4	3	13
19		27	2	8	20
		71	15	27	62

which the flies possessing the lowest bristle numbers that were found in the high lines were mated with those possessing the lowest number among the progeny of the low lines and, conversely, the flies possessing the highest bristle numbers among the progeny of the designated low lines were mated each generation with the highest of the high lines. As the figure shows, the flies in these selected lines came to possess two quite different distributions of bristle numbers: low lines had low numbers while high lines had high ones. A genetic analysis of these flies revealed, however, that the distributions resulted from the presence of chromosomes characterized by a complex combination of recessive and dominant multi-locus lethals. In brief, the solution was an expensive one in terms of the survival of offspring each generation.

Species exhibiting sexual dimorphisms can be regarded as species in which the two sexes have been (or, are) subjected to disruptive forces. One can, in fact, claim that males of many species must be conspicuously marked in order to attract mates, whereas females must be inconspicuous in order to remain with and rear their offspring successfully; this conflict of interest is especially obvious among birds. Butterflies in which the females mimic one or another of several models also represent species subjected to disruptive forces.

A laboratory experiment in which an attempt was made to obtain a strain of *D. melanogaster* in which males and females

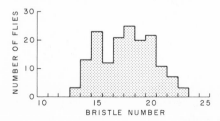

Figure 23.7 The bimodal distribution of sternopleural bristle numbers in one line (D$^+$) of *D. melanogaster* that was subjected to a complex disruptive selection (see page 261). (After Thoday and Boam 1959.)

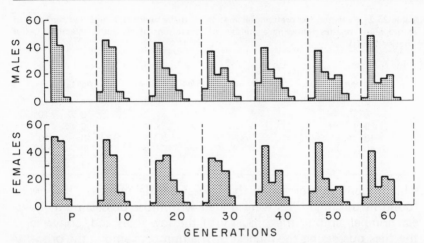

Figure 23.8 The progress at 10-generation intervals of selection (line ALM) for sexual dimorphism in *D. melanogaster*; males with the lowest and females with the highest numbers of bristles were chosen each generation as parents. (The bars of each histogram represent 5–6, 7–8, 9–10, 11–12, 13–14, 15–16, 17–18 and 19+ sternopleural bristles.) (Wallace unpublished.)

possessed different numbers of sternopleural bristles partially succeeded (one "success" in four tries) but revealed that this success, like that described immediately above, was obtained at a price.

Figure 23.8 illustrates the gradual development during 60 generations of selection of a bimodal distribution of sternopleural bristle numbers in a selected line in which males possessing the lowest bristle numbers were mated in each generation with females possessing the highest numbers. Neither sex succeeded in eliminating individuals possessing the "wrong" modal numbers. Males with high numbers were sterile, however. Figure 23.9 and table

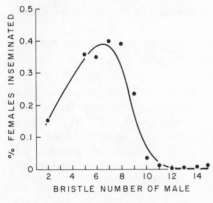

Figure 23.9 The relation between the number of sternopleural bristles of males in line ALM and the ability of these males to fertilize females. Note the opportunism of selection: only males with low numbers of sternopleural bristles were chosen to be parents each generation in this line. (Wallace unpublished.)

Table 23.3 Sterility of males associated with high numbers of sternopleural bristles in an artificially selected line of *D. melanogaster*. (Wallace unpublished.)

Number of Bristles on Male	Number of Females Examined	Number of Females Inseminated	Percent Inseminated
2	27	4	14.8
5	67	24	35.8
6	293	101	34.5
7	284	112	39.4
8	292	113	38.7
9	280	65	23.2
10	287	9	3.1
11	282	3	1.1
12	287	0	0
13	271	0	0
14	233	1	0.4
15	52	1	1.9

23.3 present data on the percentages of females that were inseminated when exposed to males from the selected line that possessed different bristle numbers. Recall that the disruptive selection procedure normally called for the mating of males with low bristle numbers to females with high ones. The figure and table reveal that males with low numbers were highly successful at mating. Males with high bristle numbers, those that normally were destined to be discarded, proved to be virtually sterile; males with ten or more bristles inseminated only 1 percent of the females to which they were exposed whereas males with lower numbers mated successfully with 30 percent to 40 percent of the available females.

Modes of Speciation

Thoday and others have interpreted the results obtained by experiments on disruptive selection as proof that sympatric speciation can occur. Even the negative results obtained by Scharloo et al. (1967) and Chabora (1968) do not alter this interpretation (see Thoday and Gibson 1970) because no one claims that disruptive selection must succeed in splitting *every* population subjected to such selection. Evolutionary time is long; therefore, the proponents of sympatric speciation need find only that disruptive selection has succeeded in splitting a population (as in figure 23.5) at least once. What happens once in a human lifetime could happen countless times during the history of a species.

Advances in our understanding of the behavior and reproductive biology of many species have made many aspects of the allopatric-sympatric dispute obsolete. For example, early on it became obvious that what constitutes geographical isolation varies by orders of magnitude for different species; the movement of ground snails is normally much more restricted than that of mammals or of flying insects. Furthermore, the aggregation of birds from many separate breeding areas within a common winter feeding area does not constitute sympatry if each spring every individual returns to the locality in which it was born.

Mayr (1963:449, 451), in an attempt to avoid misunderstandings and circular arguments, refers to "the normal cruising range of the individuals" and to the origin of isolating mechanisms "within the dispersal area of the offspring of a single deme." Even these precautions are inadequate for coping with species with specialized feeding and mating areas; sometimes one spot serves for both feeding and mating, in other cases (Carson et al. 1979:483) the two functions are carried out at different places. In either case, however, courtship and mating need not occur at every point within an individual's normal cruising range. The mating behavior of human beings is not this haphazard nor is that of many other species.

A number of complicating restrictions on the mating behavior of true fruit flies are shown in Figure 23.10. These flies (*Rhagoletis*) are subdivided into a number of species, each of which infests one or more host plants. The figure shows (1) that flies of both sexes identify their proper host by both visual and olfactory cues; they then recheck their initial decision through chemical and physical clues before oviposition or mating. The restrictions noted in the figure imply that upon leaving one host plant, an individual may fly past any number of host plants infested by other *Rhagoletis* species before sighting and alighting on its own particular host species; here it once again encounters individuals of its own kind. Maynard-Smith (1966) has pointed out that "if in a heterogeneous environment mating takes place within the 'niche' . . . then the populations from two niches, whether genetically different or not, will be largely isolated." He goes on to say that "this should perhaps be regarded as a form of allopatric speciation in which isolation is behavioral rather than geographic."

Behavioral isolation, in Maynard-Smith's sense, is nongenetic in origin, thus resembling geographic isolation. Allopatry and sympatry lose their normal meanings once nongenetic "niche"

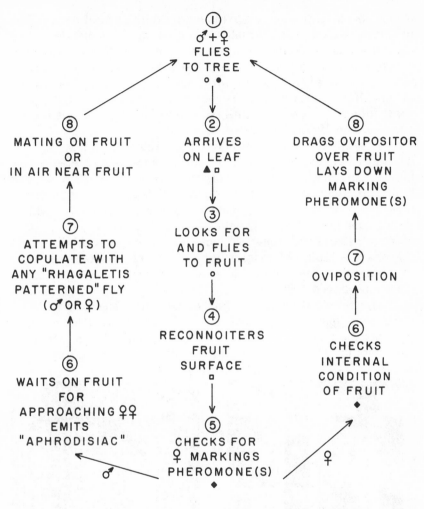

①
♂+♀
FLIES
TO TREE
o •

⑧
MATING ON FRUIT
OR
IN AIR NEAR FRUIT

②
ARRIVES
ON LEAF
▲ □

⑧
DRAGS OVIPOSITOR
OVER FRUIT
LAYS DOWN
MARKING
PHEROMONE(S)

⑦
ATTEMPTS TO
COPULATE WITH
ANY "RHAGALETIS
PATTERNED" FLY
(♂OR♀)

③
LOOKS FOR
AND FLIES
TO FRUIT
o

⑦
OVIPOSITION

⑥
WAITS ON FRUIT
FOR
APPROACHING ♀♀
EMITS
"APHRODISIAC"

④
RECONNOITERS
FRUIT
SURFACE
□

⑥
CHECKS
INTERNAL
CONDITION
OF FRUIT
♦

⑤
CHECKS FOR
♀ MARKINGS
PHEROMONE(S)
♦

♂

♀

TYPE OF CUE :
o VISUAL • OLFACTORY
▲ TACTILE ♦ CHEMOTACTILE
□ PHYSICAL & CHEMICAL

Figure 23.10 A diagram illustrating some of the steps that occur in the selection of a host tree by *Rhagoletis* species and in mating. Steps 1 through 5 apply to both males and females. Note that the restriction of mating to the immediate area of host plants (or even to these plants themselves) introduces an element of allopatry into what would otherwise be regarded as sympatry. (After Bush 1975b.)

choice (behavioral isolation in Maynard-Smith's sense) occurs. The boundary which one normally visualizes as lying between two regions (thus separating that which is in one region from that which is in the other) becomes intricately convoluted when it separates two niches that are interspersed in the same geographic region. Ultimately, one can visualize this boundary as pathological, space-filling curves such as those shown in figure 23.11 (see Kasner and Newman 1940:352–53).

The simplest model of allopatric speciation, that which was developed in the 1940s and is still represented in elementary biology textbooks (see, for example, Keeton 1972:614; also figure 23.14, type 1a), is the "dumbbell" model: a population is divided into two approximately equal-sized portions which, during isolation, diverge genetically until (when contact is once more re-established) they produce hybrids of inferior fitness. If the newly established contact persists, natural selection will tend to strengthen pre-mating isolating mechanisms of any sort— behavioral or ecological.

The simple dumbbell model of allopatric speciation has its difficulties: many species that have been widely split for millions of years show no sign of having become genetically incompatible (see Stebbins 1950:540). Although mating barriers are frequently

(a)   (b)

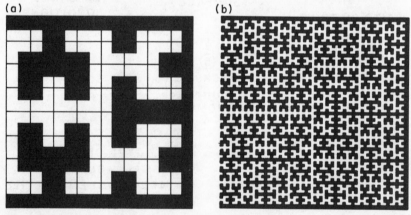

Figure 23.11 An illustration of the progressive development (a and b) of a pathological space-filling curve. One can imagine two species of host plants (white and black in the diagram) whose insect parasites tend to return as adults to, and oviposit on, the plant species on which they developed as larvae. Such plants, even though they were to grow intermingled in natural stands or in cultivated orchards, could form complex patterns defying classification in the accepted sense of allopatry or sympatry.

greater in areas of sympatry (see table 11.5), this obvious expectation is not always fulfilled (Moore 1957). Alternative models have been sought, especially models capable of accounting for the origin of a second species not from a sizable fraction of the first but, if possible, from a single local population.

Theoretical papers by Mayr (1954) and Carson (1955) serve to focus attention upon the species border as a likely locale for speciation events. In his paper, Mayr (1954) asked why the range of a species should terminate abruptly in an area seemingly lacking any abrupt environmental change. Seacoasts present no problem; land plants and animals reach the shore and go no farther. The problem arises in continental areas, sometimes accompanied by gradual elevational, temperature, or rainfall changes, where the border of the distributional area of a species is so clear-cut that it can be drawn accurately on a map. Mayr's analysis is exceedingly simple: the species' border reflects a dynamic equilibrium between the need of border populations (1) to adapt to the local environment and (2) to remain genetically compatible with individuals from interior populations who constantly move into the border region as migrants. The conflicting needs (actually a form of disruptive selection) reach an equilibrium at the species border: here the species is unable to adapt further to environmental changes because of the locally dysgenic effects of genes arriving through migration. Having been said, Mayr's scheme appears obvious. How have plant and animal breeders developed domesticated crops and farm animals? By isolating their selected lines and preventing their outcrossing with wild relatives. Populations at the borders of wild species do not have access to isolated gardens and fenced-in areas. Or do they?

Carson (1955) presented a physical view of species borders which is represented graphically in figure 23.12. The species border represents a dynamic equilibrium but, over long periods of time, it is not a stable border. On the contrary, as climatic conditions undergo cyclic changes, the species border ebbs and flows like the water in a seaside tidal flat. Refugia exist such that as the species border retracts, pockets containing small populations remain behind, beyond the receding border. These pockets correspond to the tidal pools remaining on a beach at low tide in the seaside analogy.

These small isolated populations inhabiting refugia outside the larger species distribution are the fenced-in pens and isolated experimental gardens of natural populations. In them, selection

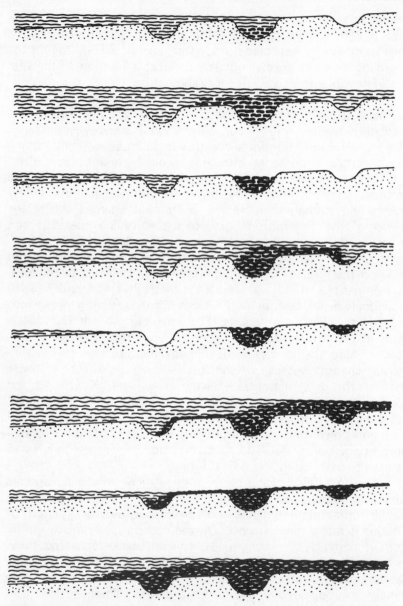

Figure 23.12 A diagram based on Carson's (1955) analogy of the ebb and flow of the tide, and the creation of tidal pools, illustrating a species border. The species as a whole fails to adapt successfully to the area at, and immediately beyond, the border even though environmental challenges need not change abruptly in this region. Within the often isolated

acts to adapt the local isolated populations to border and near-border environmental conditions. If the ebb and flow of the border occurs frequently, however, the newly developed adaptations are swamped each time the species range expands and the isolated populations are overrun. Presumably this periodic swamping rather than the arrival of chance migrants provides the dysgenic tendencies required by Mayr's model.

How can a locally adapted population that is isolated in a refugium avoid having its gene pool swamped by the alleles carried by nonadapted relatives arriving when the species distribution undergoes a temporary expansion? Wallace (1959c) suggested that if the small population during its isolation acquires a new genic or chromosomal constitution (especially the latter) that would be incompatible with, and eliminated from, the gene pools of most larger populations, the swamping can to some extent be avoided (figure 23.12 and 23.13). The chromosomal changes suggested included certain combinations of paracentric inversions, pericentric inversions, translocations, and aneuploidy. Self-fertilization was included in the list as a genetically determined phenotype that would also serve to maintain the better-adapted genotypes of refuge populations.

Interest in models of speciation other than the classical (= dumbbell) one has increased in recent years (see White 1968, 1978; Bush 1975a; and Wilson et al. 1974). A brief outline of various alternative models is presented in figure 23.14. The classical dumbbell model is illustrated as type Ia. Type Ib represents a variant of type Ia in which accidentally dispersed individuals (gravid females) serve as founders of geographically isolated populations which then expand, thus meeting the original distribution range (compare with figure 2.15).

The model referred to in figure 23.14 as parapatric (type II) corresponds to the models illustrated in figures 23.12 and 23.13 except that figure 23.14 fails to emphasize the concurrent genetic changes which Wallace (1959c) advocated as those providing solutions to the problem of maintaining a "border-adapted" geno-

refugium represented as the deepest tidal pool in the diagrams, successful adaptation is possible provided that the necessary genotypes can be kept intact. A number of genetic mechanisms and reproductive behaviors can achieve the necessary isolation. The outcome (bottom diagram) could be a clear-cut separation (speciation) as shown, or it could be a long-lasting zone of hybridization between two geographic races.

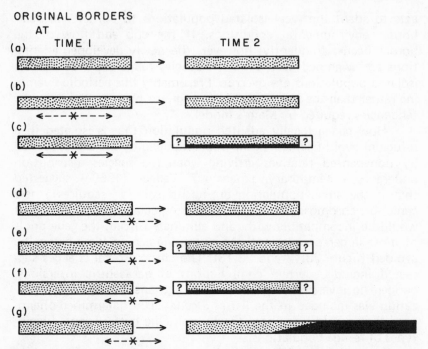

Figure 23.13 A diagram suggesting that a chromosomal or other genetic alteration which occurs at a species border and which succeeds in allowing the species to extend its range (perhaps by forming a new species) is one which, had it arisen within the original species range, would generally be eliminated from the population containing it. The origin of the genetic alteration is shown by an asterisk (*). Dashed lines indicate disfavored changes; solid lines, favored ones. Diagram *a* represents a population in which no change occurs between time 1 and time 2; hence, its range remains unchanged. Diagrams *b* and *c* suggest that whether the new change is incorporated into populations or not, if it arises well within the species range, the range remains largely unchanged. Diagrams *d* and *e* show that changes arising at the border of the range leave the range unchanged if they are unsuited for environmental conditions beyond the border; such a change may, however, be incorporated by populations within the existing range. Diagram *f* suggests that an alteration which might otherwise be suitable for areas beyond the species border may not increase the species range if it is, in fact, incorporated within populations lying within the present species range; such a change merely adds more genetic heterogeneity to what is presumably an already heterogeneous assemblage of populations. Diagram *g* represents a change that is incompatible with the genetic structures of (and, therefore, would be eliminated from) existing populations; this change, if suitable for environments beyond the species range, would lead to an expansion of this range. Compare this diagram (*g*) with the "ebb and flow" model described in figure 23.12. (After Wallace 1959) reproduced with permission: from *C. S. H. Symp. Quant. Biol.* (1959), 24:193–204; © Cold Spring Harbor Laboratory.

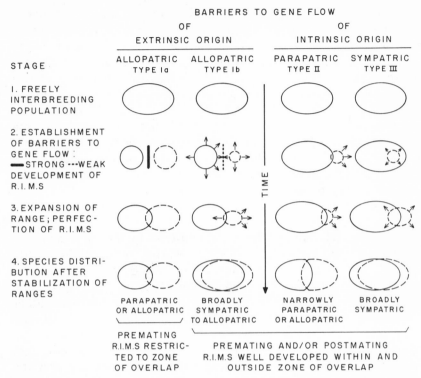

BARRIERS TO GENE FLOW

OF   OF

EXTRINSIC ORIGIN   INTRINSIC ORIGIN

STAGE	ALLOPATRIC TYPE Ia	ALLOPATRIC TYPE Ib	PARAPATRIC TYPE II	SYMPATRIC TYPE III

1. FREELY INTERBREEDING POPULATION

2. ESTABLISHMENT OF BARRIERS TO GENE FLOW :
—STRONG ---WEAK DEVELOPMENT OF R.I.M.S

3. EXPANSION OF RANGE; PERFEC-TION OF R.I.M.S

4. SPECIES DISTRI-BUTION AFTER STABILIZATION OF RANGES

PARAPATRIC OR ALLOPATRIC

BROADLY SYMPATRIC TO ALLOPATRIC

NARROWLY PARAPATRIC OR ALLOPATRIC

BROADLY SYMPATRIC

PREMATING R.I.M.S RESTRIC-TED TO ZONE OF OVERLAP

PREMATING AND/OR POSTMATING R.I.M.S WELL DEVELOPED WITHIN AND OUTSIDE ZONE OF OVERLAP

Figure 23.14 Diagrams illustrating the basic modes of speciation. Notice that Carson's tidal-pool analogy encompasses elements of both allopatric (Type Ib) and parapatric (Type II) modes. R.I.M.'s are reproductive isolating mechanisms. (Bush 1975a.) Reproduced with permission: from *Annual Review of Ecology and Systematics*, vol. 6; © 1975 Annual Reviews, Inc.

type through periods of swamping and introgression. The last model, sympatric specialization (type III), shows the new species arising well within the range of the original species. This is the sort of speciation that Mayr (see Mayr 1963) regards as improbable. Increases in our understanding of the reproductive and feeding behavior of many organisms render this type of speciation less unlikely; simultaneously, however, as figure 23.11 emphasized, these same advances have eroded the contrast originally implied by the terms "allopatric" and "sympatric." The line separating this area from that area has become convoluted and may at times even fall entirely *within* a given area.

Concluding remarks
The genetics of speciation has its roots in the challenges to integration outlined in this chapter. Of what does this evolution consist?

To Mayr (1954) it consisted of a revolution: allelic differences at one locus were thought to produce a cascade of compensatory changes at all other loci. The same view, of course, lay beneath Muller's (1940) view of the origins of reproductive isolation. Unfortunately, the term "revolution" has not been defined and so we find Dobzhansky (1970:364) writing in reference to Hubby and Throckmorton (1968): "The authors 'interpret these results to indicate that speciation does not *require* a change in a large number of loci.' I believe their findings warrant the opposite conclusion." The same data, that is, drives different persons to diametrically opposed conclusions; this can only mean that the subject of debate has not been adequately defined.

Improvements in cytological techniques have led to an appreciation of the frequency with which speciation has been accompanied by chromosomal alterations in many organisms, especially mammals (Bush 1975a; Wilson et al. 1974; White 1978). The role of these inversions, translocations, and transpositions has frequently been ascribed to the control of gene action ("position effects" are well known in *Drosophila* and other organisms; see Lewis 1950; Baker 1968 for reviews). Bush (1975a: 357) has suggested that "even the view of Goldschmidt [1940] that a 'hopeful monster,' a mutation that, in a single genetic step, simultaneously permits the occupation of a new niche and the development of reproductive isolation, no longer seems entirely unacceptable."

Chromosomal rearrangements do indeed often alter the normal functioning of alleles at loci near the points at which the chromosomes are broken. Furthermore, such altered gene function might in certain cases prove to be selectively favored. I would, however, suggest two other roles that chromosomal aberrations may play in the process of species formation. One has been discussed in this chapter (figures 23.1, 23.12, and 23.13): the origin of incompatible genetic systems such that a border population's newly consolidated and locally adapted genotype can be spared the swamping effect that accompanies periodic enlargements of the "normal" species range. The other is related to gene control as described in chapters 15 and 20: chromosomal segments such as those contained within inverted gene blocks or near the breakage points of translocations (crossing-over is substantially reduced in translocation heterozygotes for physical reasons alone because chromosome pairing is difficult) may serve to accumulate contrasting control regions that interact in selectively advantageous ways.

In short, the *content* of the inverted (or translocated)

chromosomal segment may become its selectively important feature, rather than the breakage points and their associated position effects in the classical sense of position effects. The differentiation of these two possibilities prompted much of the early work on the gene arrangements in *D. pseudoobscura* (see, for example, Dobzhansky 1948b).

FOR YOUR EXTRA ATTENTION

1. Tauber et al. (1977) and Tauber and Tauber (1977a, 1977b) analyzed the genetic differences between *Chrysopa carnea* and *C. downesi*, two species of lacewings, and found that the temporal reproductive barrier isolating them is the result of alleles at two gene loci. They suggest that reproductive isolation arose in sympatry. Subsequent to the publication of their observations, Hendrickson (1978) published a dissenting letter to which Tauber and Tauber (1978) replied.

Try to construct a three-step model by which a polymorphic species in a two-habitat environment might be transformed into two species by photoperiod-sensitive mutations. Check your model against that of the Taubers, and against the criticism they encountered.

2. Relate the following conclusions drawn by Power (1979) to material covered in chapters 2, 3, 6, and 8: As a prelude to speciation, geographically isolated populations should show relatively great phenetic variation. Populations of house finches (*Carpodacus* spp.) on California islands show greater phenetic divergence than mainland populations. The most divergent populations are found on islands farthest removed from the coast. Very small islands also have some of the most divergent populations.

3. Silander (1979) analyzed the allozymes at six loci in the perennial grass, *Spartina patens*, inhabiting a barrier island off the coast of North Carolina.

Among 346 plants analyzed, 101 genotypes were identified: 15 among 90 dune plants, 31 among 91 swale plants, 46 among 75 marsh plants, and 47 among 90 transect plants. Three genotypes were common between dune and swale populations, two between dune and marsh, six between swale and marsh, and 27 between transect and other sites.

What bearing do these observations have on the possible patchiness of the coastal environment? And, on the plant's response to its surroundings?

4. The account given in this chapter emphasized the alternatives facing populations that are "torn" between two choices: polymorphism or speciation. Moore (1977), in discussing a similar problem in the context of narrow hybrid zones at the juncture of two separate but closely related species, cites three possibilities in respect to the observed hybrids:

a. The hybrid zone is ephemeral: either isolation will arise or the populations will fuse. (Compare with the account given in this chapter.)

b. The hybrid zone represents a dynamic equilibrium, selection for isolation is thwarted by the constant arrival of naive migrants. (Compare with figure 2.15.)

c. The hybrid zone is maintained because of the superiority of hybrid individuals in a zone of overlap that falls at the limits of the suitable ranges of both parental species. (Compare with Mayr 1954.)

5. Watanabe and Kawanishi (1979) have argued that females of a derived species do not mate with males of the ancestral species, whereas females of the ancestral species readily mate with males of the derived ones.

Referring to the pattern of speciation suggested in figures 23.13 and 23.14, what is the logic behind this argument? Compare Watanabe and Kawanishi's arguments with those advanced by Kaneshiro (1976).

6. Inversion polymorphisms frequently span species so that closely related *Drosophila*, for example, either share the same polymorphisms or carry a visible record of such earlier polymorphisms in the structure of their giant chromosomes. (For a botanical example, see the account of translocations in *Datura* species given in Wallace 1966d:146.)

Often the polymorphisms of related species involve the same chromosome; those in *D. pseudoobscura* and *D. persimilis*, for example, are restricted to the third chromosome.

On the basis of what you have read in the earlier chapters of this book, suggest why polymorphisms of related species might be restricted to the same chromosome. Compare your suggestions with those of Carson (1969) who speaks of parallel evolution and Wallace (1966e) who refers to cause and effect. What about the possibility that the origin and retention of inversions is a matter of chance (an expectation based on their being selectively neutral)?

7. Consider the following statement from Lloyd (1980) in the light of ideas discussed in this chapter: "In plant and animal species which have both sexual and asexual races, the asexual forms often have a wider distribution than the sexual forms. . . . The greater frequency of asexual animals in habitats freed from continental ice sheets and on isolated islands . . . is probably assisted by the easier colonization by asexual populations."

Convince yourself that the last sentence could allude to two (possibly more) distinctly different mechanisms.

The Evolution of
Mendelian Populations

PREVIEW: Following a number of general comments related to evolutionary genetics and the genetics of speciation, this chapter deals with several items: the measurement of evolutionary change, the role of sex in evolution, and the divergent views on evolution held by neutralists and neo-Darwinians.

BIOLOGICAL EVOLUTION HAS two aspects: gradual change through time and the generation of new species. The second aspect has been treated either implicitly or explicitly in the past four chapters. In learning of the integration of gene pools, we learned of forces that oppose speciation; in discussing disruptive forces, we touched on matters that promote the origin of species.

Evolution as the gradual change of organisms through time need not involve speciation in a biological sense. Organisms that change through time do acquire new names but these denote taxonomic, not biological, species. These names are applied to morphologically distinct fossils in successive geological strata, for example, as an aid to communication among paleontologists, not because anything is known of the possible compatibility or incompatibility of the gene pools of these temporally isolated populations.

To many persons, one of the goals of modern population genetics is the development of a quantitative theory of speciation in terms of genotypic frequencies. Lewontin (1974a: 159) not only identifies that goal but adds, "we know virtually nothing about the genetic changes that occur in species formation." Indeed, we do know virtually nothing about the genetics of species formation; Lewontin is quite correct. But, by the same token, we know virtually nothing about the genetics of milk production in cattle, of egg production in chickens, of body conformation in swine, or of yield in a host of agricultural plants. In truth, we know the *genetics* of very little of the living world; those who are interested in species formation should not feel alone in their ignorance.

Quite possibly, genotypic frequencies in the sense of frequencies of alleles at loci throughout an animal's or a plant's genome will have little or nothing to do with speciation. Genotypic frequencies at some loci must, of course, be involved; alterations in the production of pheromones, for example, must involve the genetic basis of their biochemical synthesis. The need for sexual isolation, however, need not arise as the result of a potentially predictable, quantitative change in genotypic frequencies; on the contrary, it may arise because of instabilities associated with heterozygosity for a chromosomal translocation. In fact, a thorough understanding of the cohesive forces operating within the gene pools of Mendelian populations, would provide the basis needed for understanding the genetic basis of active speciation: *The genetics of species formation lies in the loss or destruction of any one or more of the genetic systems that promote the cohesive integration of gene pools.*

Active speciation as this term is used above means the generation and maintenance of two gene pools where earlier only one existed. Such speciation represents not so much the acquisition of a "species-forming" quality as the loss of a quality that is essential for the coherence of a single gene pool. There remains, however, the notion of gradual change through time, the accumulation of genetic differences, and the eventual arrival of the population at a state which, if it could only be compared with the same population of a much earlier time, would be recognized as deserving of a new *biological* species designation. This is the pattern of speciation that Muller (1918, 1940) envisioned when he wrote of gradual genetic divergence, with the substitution of each new allele calling for alterations at other loci and with alleles that were substituted by chance at one moment becoming essential alleles as other alterations occur—each made possible by, and being dependent upon, the earlier one.

Time cannot be compressed; consequently, we shall never know what intervals in the fossil record correspond to the occurrence of biological rather than taxonomic speciation. Of one thing we can be sure: the process of creating species differences through nonselected genetic divergence alone is an exceedingly slow one. Many of the plants found in the southeastern portions of the United States and of China belong to the same species (Stebbins 1950: 538ff). Carson (1978) has described the apparent genetic compatibility of inter-specific hybrids of Hawaiian *Drosophila*; species that are normally isolated by behavioral differences, when crossed in the laboratory, produce what appear to be completely viable and fertile F_1, F_2 and backcross hybrids.

The Divergence of Populations

Populations do change through time and, because the changes are often random in origin or occur in response to haphazard environmental events, isolated populations tend to become less and less similar in genetic composition. Such divergence, the simplest of all evolutionary changes, was the subject of much discussion in the early chapters of this book, beginning with chapters 3 and 4 ("Detecting and Measuring Genetic Variation"). The subsequent chapters of Part I were then devoted to understanding the origin, nature, and maintenance of observed variation.

In this, the final chapter, emphasis shifts to the matter of change itself, and to procedures that reveal how divergent two populations have become—about the genetic "distance" between populations.

DNA sequencing
The ultimate genetic comparison of any two species, or of any two populations within a species, would be that in which the base sequences of the entire DNA of individuals of both species were studied and compared. Because the base pairs in the DNA of higher organisms number in the tens of millions (*Drosophila*) or billions

(man and other vertebrates), determining the complete sequence for any species poses horrendous problems. Nevertheless, appropriate molecular techniques are available and the task of making the needed analyses (albeit on a piecemeal basis) has gotten under way. In several laboratories the DNA of *D. melanogaster* has been cut into segments and annealed into replicating plasmids of *E. coli* (see Wensink et al. 1974). The sequences of more and more DNA segments of this sort are being determined daily. A more modest evolutionary comparison has been reported by Nichols and Yanofsky (1979) on the *trpA* genes of *Salmonella typhimurium* and *E. coli*. In this one structural gene, nearly 25 percent of the base pairs differ. In contrast, the differences in the enzyme molecules (α-chain of tryptophan synthetases) involve only 15 percent of all amino acids. Over one-half of all codons in the *TrpA* genes carried by these different bacterial species contain synonymous base substitutions. Nichols and Yanofsky conclude, following a statistical analysis of the distribution of the differences between the two species, that the synonymous substitutions have been selectively neutral.

DNA quantity

Polyploidy is a well-known and well-understood mechanism by which "instant speciation" occurs in plants. Hybrids between distantly (but not *too* distantly) related plant species generally have low fertility because of poor chromosome pairing. Occasionally such hybrids undergo spontaneous doubling of their chromosomes; the result is an effective "diploid" (in terms of chromosome pairing), but a plant that contains the full chromosomal complement of each parental species (and, hence, is an allotetraploid).

Molecular and bacterial geneticists have uncovered evidence that many of the interspecific differences in amounts of DNA per cell can be explained in terms of DNA-doubling. Many of the genes carried by bacteria possess related functions—evidence that they have arisen by gene duplication. A study of the location of these genes in *E. coli's* circular chromosome (Zipkas and Riley 1975) reveals that many are located at the ends of 180° or 90° arcs (Figure 24.1). A distribution of this sort suggests that the chromosome of present-day *E. coli* has arisen through two doublings of a much smaller, simpler ancestral chromosome.

An ambitious attempt to relate the DNA content of all types of organisms—prokaryotes and eukaryotes, plants and

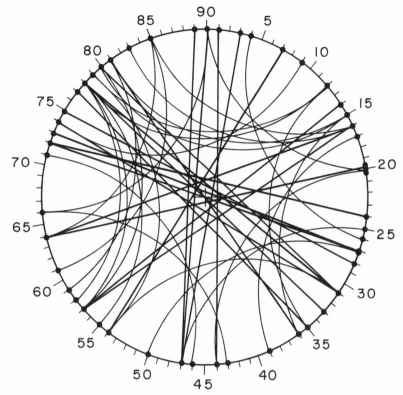

Figure 24.1 The circular genetic map of *E. coli*, 90 minutes in length. Straight lines connect known genes of related function—putative duplications—whose map positions are separated by approximately 180°; curved lines connect genes of related function whose map positions are separated by approximately 90°. The greater-than-expected number of such geometric relationships suggests that the modern *E. coli* chromosome has arisen by two duplications of an earlier, simpler bacterial chromosome. (After Zipkas and Riley 1975.)

animals—has been made by Sparrow and Nauman (1976). Because closely related species and genera often differ in the amount of DNA per cell, Sparrow and Nauman restricted their analysis to the *minimum* DNA content per genome known for each of 23 major phylogenetic groups (see figure 24.2). The interpretation of such an analysis obviously requires considerable caution, but even granting the need for caution, the evidence presented by Sparrow and Nauman suggesting that the amount of DNA has frequently doubled during the course of evolution is impressive.

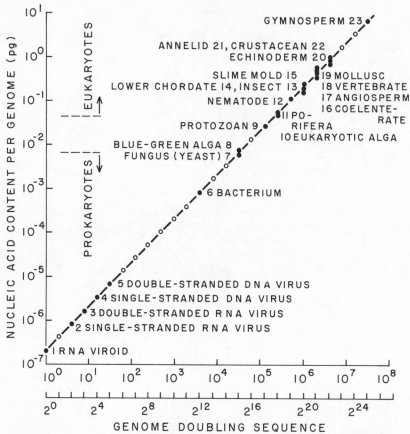

Figure 24.2 A log-log plot of the *minimum* individual DNA (or RNA) per genome for 23 major phylogenetic groups of organisms. The vertical axis gives the absolute nucleic acid content per genome in picograms. The horizontal axis represents a theoretical sequence of genome doublings. Solid circles represent the 23 designated groups; open circles represent doublings that do not correspond to any *known* minimum value, although organisms are known which possess these amounts of DNA per genome. (After Sparrow and Nauman 1976.) Reproduced with permission: from *Science* (1976), 192:524–29; © 1976 American Association for the Advancement of Science.

DNA melting

Although an analysis of evolutionary change through the total sequencing of base pairs in the DNA of different species may not yet be practical, an analysis of the similarity of DNA of different species by estimating the base-pair homologies is possible.

    The physical basis for the complementarity of the purine and pyrimidine bases of the two strands of the DNA double helix consists of hydrogen bonding—the sharing of electrons—between adenine on one chain and thymine on the other or, alternatively, between cytosine and guanine. Adenine and thymine share two electrons ($A:T$); cytosine and guanine share three ($C:G$).

Thermal energy is sufficient to break the hydrogen bonds that hold the two strands of DNA together. $A:T$ bonds separate ("melt") at a lower temperature than do the $C:G$ bonds which involve the additional shared electron. The melting points of DNA molecules having different $A:T/C:G$ ratios have proven to be linearly related to these ratios; thus, the composition of DNA can be accurately assessed by its melting point alone (see Lewin 1974: his fig. 4.6).

The melting point of DNA (that is, the temperature at which the two strands of the double helix fall apart when heated) is also capable of assessing the proportion of mismatched pairs in a DNA molecule. For example, if DNA in solution is melted, after which the solution is allowed to cool once more, the single strands of DNA reassociate into complementarily paired, double strands. If these double strands are reheated, they melt at a somewhat lower temperature than did the original material. Reassociation of DNA in a beaker is not as precise as the enzymatic synthesis of one DNA strand on the basis of the base-pair sequence of another. Chemical analysis has shown that (at least as a rough rule of thumb) the melting temperature of DNA is lowered $1°C$ for every 1 percent mismatched base pairs.

Clearly, the DNA of different species of organisms can now be compared in order to estimate the proportion of base pairs in the single strands of each which do not correspond properly in reassociated double strands. The experimental procedures involve the use of "labeled" (radioactive iodine) DNA of one species and (for the control) an excess of unlabeled DNA from the same species (to form *homologous* DNA hybrid molecules) or from a different species (to form a *heterologous* DNA hybrid). There is no need to describe these techniques here; Sibley and Ahlquist (1981) have described the techniques and their possible errors in great detail.

The quantitative aspects of figures 24.3 and 24.4 were generated by the use of random numbers generated by a pocket calculator; the simulation corresponds to the currently held view that base-pair alterations occur in DNA at a steady rate which is constant in clock time (as opposed to generation time).

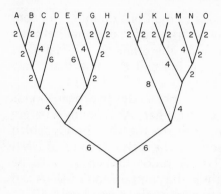

Figure 24.3 A diagram representing a phylogeny in which the relationships between 15 present-day (but hypothetical) species are depicted. The numbers refer to independent mutational events that have occurred during each time interval represented in the diagram. Notice that, under the assumption that mutation rate is at least nearly constant, the phylogeny also represents the chronology of speciation.

The simulation involved a fragment of "DNA" one hundred base-pairs long (00–99). Each site within this fragment was occupied by one of the four "nucleotide bases" (0–3). The composition of the ancestral strand at the bottom of the diagram was determined by generating 100 digits that had values 0, 1, 2, or 3. Proceeding from the ancestral strand, first to the left and then to the right, two descendant strands were generated by identifying (in each case) six sites at which changes would occur (identified by randomly generated 2-digit numbers, 00–99) and what that change would be (a random choice of the four digits, 0–3; by chance the new "base" could be identical to the original one). The newly created strands were then altered successively as shown in figure 24.3. The number of sites that were to be altered in each instance is shown in the figure; the substitution (if any) was based on a random choice of the four digits, 0 to 3. Each fork in the diagram (known officially as a *cladogram*) represents a "fossil" strand of DNA that was altered according to the next higher set of instructions. Eventually, fifteen contemporary strands of DNA were obtained; these are labeled *A* through *O* in the diagram (figure 24.3).

A pairwise comparison of the contemporary strands reveals that these existing fragments of "DNA" are not identical; the numbers of sites at which the bases found in two strands differ are given in figure 24.4a. Strands *N* and *O*, for example, differ at four sites; the two sets of two changes each that were introduced into the common ancestral strand were different. *A* and *B* differ at three sites, while *G* and *H* differ at only two. In these cases, the two mutational changes in each branch involved common sites.

Examination of figure 24.4a reveals that the matrix would suffice to reconstruct the cladogram illustrated in figure 24.3, although not necessarily in such detail that the number of muta-

Matrix A

	A	B	C	D	E	F	G	H	I	J	K	L	M	N	O
A	–														
B	3	–													
C	5	6	–												
D	8	10	7	–											
E	9	11	9	10	–										
F	11	13	11	13	6	–									
G	10	12	10	12	7	5	–								
H	11	14	12	14	7	5	2	–							
I	21	23	22	22	18	21	22	22	–						
J	21	24	23	22	19	22	23	22	3	–					
K	20	22	22	22	20	22	23	23	15	13	–				
L	22	24	23	23	20	22	23	23	13	13	2	–			
M	20	21	20	19	16	19	20	21	11	12	7	7	–		
N	24	26	24	24	21	22	23	23	15	13	9	9	6	–	
O	24	27	24	24	21	23	24	22	15	13	9	9	6	4	–

Matrix B

	A	B	C	D	E	F	G	H	I	J	K	L	M	N	O
A	–	3	5	8	9	11	10	11	31	31	20	22	20	24	24
B		–	6	10	11	13	12	14	32	32	22	24	21	26	27
C			–	7	9	11	10	12	30	31	22	23	20	24	24
D				–	10	13	12	14	29	30	22	23	19	24	24
E					–	6	7	7	28	28	20	20	16	21	21
F						–	5	5	30	29	22	22	19	22	23
G							–	2	32	31	28	23	20	23	24
H								–	31	30	23	23	21	23	22
I									–	6	23	24	21	25	25
J										–	26	26	24	26	27
K											–	2	7	9	9
L												–	7	9	9
M													–	6	6
N														–	4
O															–

Figure 24.4 Matrices of mutational differences obtained by a (pocket-calculator) simulation. The simulated "strand" of DNA consisted of 100 (00–99) sites that could be occupied by any one of four nitrogenous bases (0–3). Starting with a randomly generated strand (base of figure 24.3), numbers of mutational substitutions (random in respect to both site and base substitution, including no change) were made according to the numbers shown in figure 24.3. The result is shown at the left in the figure (matrix A). Matrix B was constructed in a manner identical to A except that the number of mutations on the branch leading to I and J were increased 2.5 times: 20, 5, and 5, rather than 8, 2, and 2.

tional changes in each branch could have been identified exactly. Clearly, however, the rectangular, lower-left portion of the matrix reveals the early split between contemporary strands A–H and the remaining ones, I–O. The smaller segment containing 16 cells reveals an early split between strands A–D and E–H; this pattern does not resemble the corresponding one in the other half of the cladogram where the two strands I and J differ systematically from strands K–O.

The logic used above to reconstruct the cladogram (figure 24.3) from observed differences (figure 24.4a) corresponds precisely with that which is used by evolutionary biologists in assessing the evolutionary relationships among living organisms by creating and then melting heterologous DNA.

The melting temperature is lower in the case of heterologous hybrid DNA than it is for the homologous hybrid molecule. Because mismatched base pairs lower the melting temperature, the difference in melting temperature can be interpreted as a measure of differences between DNA strands; that is, a difference in melting temperature can be interpreted as a base pair difference just as the differences in random digits were interpreted in figure 24.4a. Using only the unique sequences of the DNA of several species of flightless (ratite) birds (repetitive sequences that exist in thousands or millions of copies per genome are excluded from DNA analyses for technical reasons), Sibley and Ahlquist (1981) report the following: (1) The melting point of heterologous DNA molecules involving the cassowary (*Casuarius*) and the emu (*Dromaius*) is 3.4 degrees lower than that of the homologous hybrid DNA of either the cassowary or the emu. (2) The melting point of the heterologous DNA molecules involving either the cassowary and the ostrich (*Struthio*) or the cassowary and the rhea (*Rhea*) is about 13 degrees lower than that of the homologous DNA hybrid of any of these birds; the heterologous hybrid molecules involving ostriches and rheas has a melting point that is lowered by some 15 degrees.

Clearly the above results correspond to those that were obtained from the simulation experiment (figure 24.4a): the emus and cassowaries have DNA that is quite similar; the rule of thumb which says that 1 mismatched pair in 100 lowers the melting temperature by 1° suggests that some 3–4 percent of the base pairs in the DNA of these two birds are dissimilar. On the other hand, the DNA of either the emu or the cassowary differs as much from that of the ostrich as it does from that of the rhea—two birds that are themselves quite different. Clearly, these melting point changes correspond to the differences listed in the lower left portion of Figure 24.4a where species *C*, for example, differs equally from *I*, *J* . . . , and *O* even though these species differ from one another. (Compare this reasoning with that used in reference to figure 14.8.)

The tabular material presented by Sibley and Ahlquist (1981) reveals other cases corresponding to that involving the emu, cassowary, ostrich, and the rhea. The neo-tropical tinamou (*Nothoprocta*) is equally distantly related to the ostrich, rhea, emu, cassowary, and the kiwi. Heterologous DNA hybrid molecules between the tinamou and these other ratite birds have melting points that are uniformly 16 degrees lower than that of the homologous hybrid DNA of any one species.

The substitution of base pairs in DNA has been regarded (for example, Fitch 1976) as providing an evolutionary clock by which the time elapsed since two organisms have had a common ancestor can be measured. Unfortunately, the fossil record is not particularly accurate, especially when the dating of common ancestors is concerned; the bifurcation between man and the other great apes, for example, is variously placed from one to five million years ago.

In place of a clock which has been synchronized with clock time, one can regard data similar to those given in figure 24.4a as yielding evidence on the relative rates of evolution (see Wilson, et al. 1977). An external reference (say, E) provides information about A, B, C, and D as they diverged first from E, and then from one another. Sibley and Ahlquist (1981) make excellent use of their data in reviewing, for example, geologic data on the movements of continents, including the possible existence of long-vanished archipelagoes. Comparable data would be useful for identifying instances in which the displacement of morphological characteristics of two species living in sympatry have led taxonomists to overestimate the unrelatedness of the two forms. Base pair substitutions in DNA occur at rates that are virtually unaffected by the rapid changes in flowering times, plumage colors, and other physical and behavioral traits that assure the sexual isolation of closely related species living in sympatry.

Having discussed the DNA-hybridization technique and the problems to which it can be addressed, it is necessary to introduce a word of caution: arguments based on the technique may become circular; external evidence *not* dependent upon molecular techniques may, in fact, be necessary if the circularity of these arguments is to be detected. This need is unfortunate because it has been supposed by some that molecular data alone are sufficient to abolish the need for cruder taxonomic or paleontological evidence.

Figure 24.4b consists of a matrix, similar in all respects to the one in figure 24.4a except for some numerical changes. On the forked branch of the cladogram (figure 24.3) leading to I and J, 2.5 times as many base substitutions were introduced into the lineages as are indicated in the earlier figure: 20 instead of 8, and 5 instead of each 2. The result, diagramed properly according to the method by which the new matrix was generated, should create a diagram resembling figure 24.3 but with an enormously exaggerated Y-shaped set of lines leading to I and J. Such a diagram, if drawn, would have represented accurately the higher rate of base substitutions in the one branch of the cladogram.

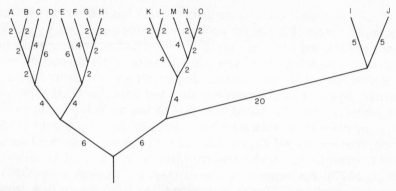

Figure 24.5 A diagram derived from the one shown in figure 24.3. The number of mutations in the last three line segments leading to species *I* and *J* has been increased 2.5-fold (8 to 20, 2 to 5) in this diagram. The phylogeny has been redrawn to represent the new mutational relationships. Notice that, under the erroneous assumption that mutation rate is nearly constant, the present phylogeny could (as was the case with figure 24.3) be interpreted as representing the chronology of speciation. Mutational distances, that is, do not measure time unlesss they are constant, but this postulated constancy cannot then be proven by a phylogeny that is itself based on mutational distances. The system requires independently arrived at points of reference.

By regarding the level at which the letters *A* to *O* are printed as representing the *present* in a temporal sequence, *and by assuming that the rate at which bases are substituted in DNA remains constant through time*, a new diagram (figure 24.5) can be constructed. Within a cladogram, time and mutation rate are interchangeable. Without independent evidence serving to lock the diagram into a fixed position, a cladogram based on dissimilarities of heterologous DNA can seemingly prove that mutation rates (rates of base substitutions) are constant through time—even though they are not.

Amino acid substitutions

Because proteins are (other than RNA) primary gene products, variation in protein structure has been extensively used in studying evolutionary changes at the molecular level. Such studies include immunological alterations (see Sibley et al. 1974; Sarich and Wilson 1967), electrophoretic differences (Lewontin 1974a: chapter 4 and included references), and differences in amino acid sequences.

A simple example of the use of amino acid sequences in the

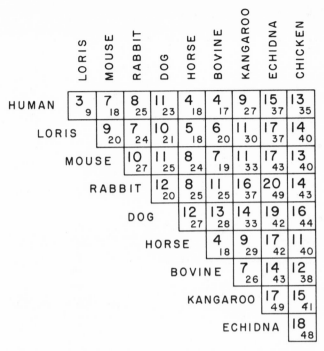

Figure 24.6 Matrices based on amino acid substitutions in the alpha-globulin of the hemoglobins of a number of vertebrates. The larger print shows the substitutions observed among the 45 amino acids in positions 3-47 (see figure 24.7); the smaller print shows the total substitution in the entire alpha-chain of 144 amino acids.

construction of a phylogeny is presented in figures 24.6 and 24.7. Table 24.1 provides a reference in which the amino acids and their 3-letter and 1-letter abbreviations are listed.

Forty-five amino acids (positions 3 to 47) in the alpha-chain of the hemoglobin molecule have been chosen as a convenient example. The amino acids of the hemoglobin molecule have been completely sequenced in a number of vertebrate species, including the nine listed in the figures: human, loris, mouse, rabbit, dog, horse, "bovine," kangaroo, echidna, and chicken. A matrix listing the number of amino acid differences between pairwise combinations of two hemoglobins is shown in figure 24.6. In this matrix, differences are shown not only for the 45 amino acids chosen for analysis (larger print) but also for the entire alpha-chain (smaller print; see Dayhoff 1972). Clearly, the differences within the entire chain and within the smaller portion chosen for illustration are

Figure 24.7 A phylogeny connecting ten vertebrate "groups" based on amino acid substitutions in a sequence of 45 positions (3–47) in the alpha-chain of the hemoglobin molecule. Although the phylogeny has been constructed with care, it may be neither the true nor even a good one; the total number of possible branching phylogenies that connect ten end points (species or other taxonomic groups) exceeds 50 billion. By the use of computers, many phylogenies of this sort can be constructed from which the most parsimonious one can be chosen as the one most likely to be correct.

more numerous in combinations involving placental mammals with monotremes and birds than it is between primates (man and loris) or among the ungulates (horse and bovine hemoglobins).

Sufficiently few amino acid substitutions have occurred among the 45 in the segment chosen for study that an attempt at recreating a phylogeny has been made (figure 24.7). Certain primi-

Table 24.1 The 20 amino acids that compose proteins and their three- and one-letter abbreviations. (From Nei 1975.)

Name	Three letter	One letter	Name	Three letter	One letter
	Abbreviations			*Abbreviations*	
1. Alanine	Ala	A	11. Leucine	Leu	L
2. Arginine	Arg	R	12. Lysine	Lys	K
3. Asparagine	Asn	N	13. Methionine	Met	M
4. Aspartic acid	Asp	D	14. Phenylalanine	Phe	F
5. Cysteine	Cys	C	15. Proline	Pro	P
6. Glutamine	Gln	Q	16. Serine	Ser	S
7. Glutamic acid	Glu	E	17. Threonine	Thr	T
8. Glycine	Gly	G	18. Tryptophan	Trp	W
9. Histidine	His	H	19. Tyrosine	Tyr	Y
10. Isoleucine	Ile	I	20. Valine	Val	V

tive sequences of ambiguous composition have been postulated; the ambiguities at various amino acid sites are shown in each instance. Wherever one amino acid has been substituted for another, unambiguous one, the substitution is identified in the figure by a dot ($\cdot$). The vast majority of these substitutions might have involved a *single* base-pair substitution within the corresponding codon.

Genetic distance

Even without the sophisticated information provided by advanced molecular techniques, early geneticists developed means by which they could express the differences between groups quantitatively, thus providing a means by which scales of similarity could be constructed. A crude method might treat *differences* in the frequencies of alleles as dimensions in a multidimensional space, and calculate a "distance" as one would calculate the distance between the diagonal corners of a room:

$$d = \sqrt{a^2 + b^2 + c^2 + \cdots}$$

Thus, suppose gene frequencies for four loci in each of three populations are as follows:

Population	a^1	a^2	b^1	b^2	c^1	c^2	d^1	d^2
1	.90	.10	.80	.20	.50	.50	.30	.70
2	.30	.70	.50	.50	.20	.80	.10	.90
3	.50	.50	.50	.50	.60	.40	.40	.60

The distance between populations 1 and 2 would be calculated as the square root of (.36 + .36 + .09 + .09 + .09 + .09 + .04 + .04) or

1.077. In the case of populations 1 and 3 the distance would be the square root of (.16 + .16 + .09 + .09 + .01 + .01 + .01 + .01) or 0.735. And, in the case of populations 2 and 3, the distance would equal the square root of (.04 + .04 + 0 + 0 + .16 + .16 + .09 + .09) or 0.762. Thus, we would conclude that population 1 differed more from population 2 than it does from 3, and that population 2 differs slightly more from 3 than does population 1. Visualized as the end points of a cladogram, 1 and 2 would be the extreme points, and 3 would lie somewhat closer to 1 than to 2. Bodmer and Cavalli-Sforza (1976: 733) develop a similar measure of genetic distance which also involves the square of the difference between gene frequencies.

Nei (1975: 175–202) has reviewed and expanded the calculation of genetic distance. Consider a locus containing several alleles whose frequencies in two populations are X_i and Y_i (i refers to the ith allele). The probability that two alleles chosen at random from population X will be identical equals the sum of $(X_i)^2$; let this be j_x. For population Y, j_y equals the sum of $(Y_i)^2$. The probability that two alleles, one of which is chosen from one population and the other from the second, will be identical equals the sum of $X_i Y_i$, or j_{xy}. In going from one locus to many, we may set J_x, J_y, and J_{xy} equal to the arithmetic means of j_x, j_y, and j_{xy} (monomorphic loci are included in calculating these means). J's (both lower and upper case) measure the identity of randomly chosen pairs of alleles. If *differences* are represented by D's, then within population X_1, $D_x = 1 - J_x$; within population Y, $D_y = 1 - J_y$; and between the two populations $D_{xy} = 1 - J_{xy}$. If the genetic distance between populations is looked upon as a distance exceeding the "distance" which exists *within* populations, we can compute

$$D = D_{xy} - (D_x + D_y)/2.$$

The distance calculated here equals Nei's *minimum* genetic distance; still other distances (estimates that are useful under given circumstances) can be computed using logs ($D = \ln J$, for example) or geometric means (J equals the geometric mean of the individual j's).

A quantity, genetic identity (I), can be defined as

$$D = \ln I$$

or

$$I = e^{-D}.$$

(Note the similarity between this expression and the zero term of the Poisson distribution discussed on page 21.)

Ayala et al. (1974) have measured the genetic similarity and genetic distance between flies belonging to the *willistoni* group of *Drosophila*. This group contains a number of different species: *D. willistoni*, *D. tropicalis*, *D. equinoxialis*, *D. paulistorum*, and *D. nebulosa*. In addition, *D. equinoxialis* consists of two distinctly different subspecies: *D.e. equinoxialis* and *D.e. caribbensis*. Furthermore, *D. paulistorum* contains a number of subgroups (known as semispecies) which generally produce sterile hybrid offspring. Finally, of course, samples of flies of each species or semispecies taken at many localities throughout South and Central America and the Caribbean were available for study.

The results of the above study are summarized in table 24.2. The genetic distance increases steadily (and genetic similarity decreases) as one moves from local populations of one species, through semispecies and subspecies, and then on to sibling and nonsibling species (the latter involves *D. nebulosa*, which is morphologically distinct from the other four species which can be identified only with difficulty).

The values for genetic similarity and genetic distance listed in table 24.2 are averages obtained by combining electrophoretic observations at many loci. Because the allozymes produced by genes at many loci were examined in this study, measures of genetic similarity can also be made at individual loci in order to determine the proportions of loci that exhibit various genetic similarities. The results of these locus-by-locus analyses are shown in figure 24.8. Most loci show a high level of genetic similarity between geographic localities of a single species. As the sibling-species and full species levels are attained, fewer loci show high levels of genetic similarity; a growing proportion show low levels of similarity and high levels of genetic distance. Ayala and his colleagues (see Ayala and Tracey 1974) use the U-shaped distribution of genetic similarities among sibling and full species of the

Table 24.2　The average genetic similarity (I) and genetic distance (D) between flies of the same or different taxa of the *D. willistoni* group. (After Ayala et al. 1974.)

Relatedness	I	D
Local populations	0.970 ± 0.006	0.031 ± 0.007
Subspecies	0.795 ± 0.013	0.230 ± 0.016
Semispecies	0.798 ± 0.026	0.226 ± 0.033
Sibling species	0.475 ± 0.037	0.740 ± 0.078
Non-sibling species	0.352 ± 0.023	1.056 ± 0.068

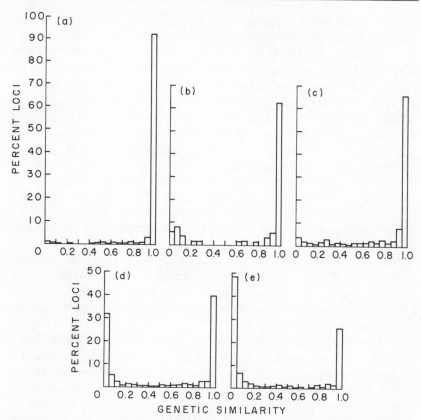

Figure 24.8 The changing proportions of loci exhibiting various genetic similarities as a function of biological relatedness: (a) geographic populations, (b) subspecies, (c) semispecies, (d) sibling species, and (e) nonsibling species. (After Ayala et al. 1974.)

willistoni group as evidence that the alleles involved are not neutral but, instead, are maintained in Drosophila populations by natural selection. Nei (1975: 165) has emphasized that the *average* time for a neutral mutant allele to reach fixation equals 4 N_e generations (where N_e equals the effective size of the population); for a population numbering in the millions [Kimura and Ohta (1971: 153) have suggested that an entire species or subspecies may be practically panmictic] or hundreds of millions, this is an enormous length of time even by geological standards. Neutral alleles, fluctuating in frequency under the influence of random events, should have created bell-shaped rather than the U-shaped distributions as shown in figure 24.8. Kimura (1979) has suggested, despite their current occupation of most of South and Central America, that flies of the *willistoni* group may have gone through a rather recent bottleneck during which their numbers were drastically smaller than those estimated to exist today. Except to adjust the observed distributions of gene frequencies to make them fit expected ones, neither evidence of, nor a biogeographical basis for, the bottleneck has been suggested.

Geological time and the age of species

Evolutionary biologists continually search for known dates which can be associated with reconstructed evolutionary events. The submersion of the Isthmus of Panama during which South America becomes an island is an important event in evolutionary studies. The advance of glaciers from the north into much of the United States and the creation of refugia in the southern Appalachian, Ozark, and Alaskan regions also identify an important era. The age of the Hawaiian Islands became important when it became apparent that an explosive outburst of speciation had occurred on those islands, virtually in insolation from the rest of the world (Zimmerman 1958; Hardy 1965; Carson et al. 1970). More than 500 drosophilid species have been described on these islands (Carson et al. 1970); the twenty-two species of Hawaiian honey creepers serve as a classical textbook example of adaptive radiation among birds, rivaling the familiar example of Darwin's finches of the Galapagos Islands (see Wilson et al. 1973: 856–858).

Because the age of the various islands of the Hawaiian archipelago are known with reasonable certainty (the oldest is estimated to be 4 million to 5.6 million years old), it appeared that in this instance time limits might be assigned to the origin of species (Carson, 1976). However, it now appears that this is not the case, and the purpose of these paragraphs is merely to mention why the *assignment of dates* has failed; the study of the evolution of Hawaiian *Drosophila* continues unabated.

The Hawaiian Islands lie in a chain far out in the Pacific Ocean. Volcanos located near a rift in the ocean's floor form undersea mountains which, as they break through the ocean's surface, give rise to the youngest islands lying at the eastern-most end of the existing chain. The island of Hawaii is the youngest of the archipelago. Islands once formed are carried north westwardly by the movement of the earth's crust; hence, the island (Kauai) at the western end of the archipelago is the oldest one. The islands,

furthermore, sink beneath the ocean's surface beyond the archipelago's present-day limit, become more and more submerged, and
eventually are recycled within the earth's crust.

The age of the oldest Hawaiian island has little or no bearing
on the length of time *Drosophila* flies and their relatives may have
lived on these islands. Figure 24.9 illustrates this claim by means
of a small figure running for his life on the exposed rim of a partially submerged wheel. That the wheel makes one revolution per
day, that the exposed arc equals one-quarter of its circumference,
and therefore, that a point (P) is above water for only six hours
each day have nothing to do with the length of time the figure
may have been running on the rim. Perhaps he has been running
for weeks. Similarly with the *Drosophila* flies on the Hawaiian
Islands: they may have lived there, hopping from island to island

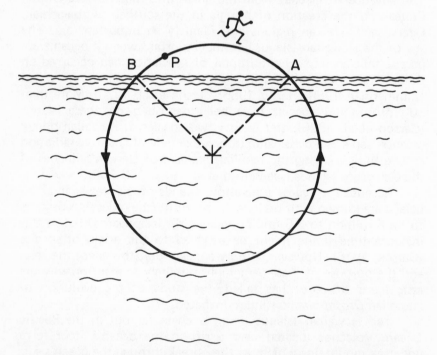

Figure 24.9 An analogy illustrating how a knowledge of island geology is
essential for the proper interpretation of evolution that has occurred on an
island archipelago. Knowledge concerning the speed of rotation of the partially
submerged wheel (and, consequently, of the time required for P to travel
from A to B) is insufficient for estimating the time during which the small
figure has been running for his life along the exposed rim.

(by lucky accident, of course) for lengths of time greatly exceeding the age of the oldest island.

Summarizing comments

Newly developed molecular techniques such as starch gel and acrylomide electrophoresis, amino acid sequencing of proteins, and base-pair sequencing of DNA have yielded and continue to yield more information on the divergence of genetic material through time than earlier students of evolution could have imagined. The results of modern studies have not cast doubt on the fact of evolution itself; all forms of life share the same genetic materials (DNA and RNA) and, with minor exceptions, the same genetic code. In the case of large taxonomic categories, molecular divergence between different forms agrees in extent with expectations based on evolutionary patterns established by the older biological disciplines: taxonomy, paleontology, morphology, and embryology. Smaller categories do not always agree; the explanation will probably lie in the unexpectedly rapid morphological changes which the members of certain genera have undergone under novel and intense selection pressures.

The alteration of DNA through base-pair substitutions has been viewed by some as a steady process which has occurred at a uniform rate throughout geologic time. If true, differences in DNA structure (or of protein composition) would provide an evolutionary clock which, if only one or more fixed reference points could be found, would accurately time all past evolutionary events. Paleontological checkpoints are not always dated accurately, nor do geological events (see figure 24.9) always provide suitable checkpoints; isotope dating, on the other hand, may provide the required information.

The seeming regularity with which protein composition has changed and DNA has undergone base-pair alterations has been offered as evidence for the selective neutrality of these changes. In my opinion, there may be extremely little connection between the apparent uniformity of genetic change over a long period of time and the selective neutrality of these changes. Stone (1955; Stone et al. 1960) attempted to estimate the number of chromosomal aberrations (fusions, translocations, pericentric and paracentric inversions) that have occurred throughout the evolution of the genus *Drosophila*. He and his colleagues estimated that as many as 40,000 paracentric inversions alone may have occurred. Now, paracentric inversions are probably not selectively neutral: they encompass a great many gene loci; many inversions possess an advantage in heterozygous individuals; upon entering a species each favored one must have at least a local advantage; and upon leaving, an average disadvantage. Fluctuations in selection intensity, however, make a gene (or inversion) "behave just like a neutral gene" (Nei 1975: 165). To *behave* as a neutral gene, however, is not to *be* a neutral gene. Consequently, the molecular clock, even if it exists as a practical matter, is no proof of selective neutrality.

The Role of Sex in Evolution

In 1958, at one of the national scientific meetings, Muller presented a short paper entitled "How Much Is Evolution Accelerated by Sexual Reproduction?" This rhetorical question was inadvertently answered by the program itself, for following the title was the parenthetic phrase (10 min.). In reality, sex has been much more influential than this in permitting and shaping evolutionary change.

Sex, in a genetic sense, means the exchange and recombination of genetic material once carried by two different individuals. Recombination can take place only if the DNA fragments of two individuals come to lie in the same cell. Sex, in a nongenetic sense, involves the myriad of procedures by which genetic exchange is accomplished. In the case of bacteriophage, recombination follows the infection of one bacterium by two or more phage particles. Bacteria conjugate in carrying out their sexual process or, alternatively, rely upon phagelike particles (or even naked DNA) to carry genetic information from one individual to another. Higher plants have cycles of sexual and asexual generations that are diagrammed in nearly every genetics text and are well know to all biology students; animals have reduced the halpoid phase of their life cycles to the germ cells themselves.

The role ascribed to sex by Muller (1932a, 1958) and, later, by Crow and Kimura (1965, 1969) was the acceleration it permits in the formation of selectively beneficial gene combinations. In the absence of sex, genes that could create superior combinations must occur in tandem throughout one line of descent (see figure

ASEXUAL

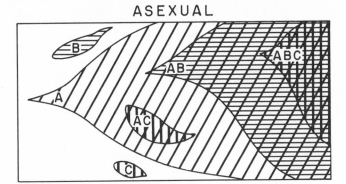

SEXUAL

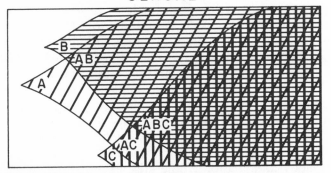

Figure 24.10 Diagrams illustrating the advantage of sexual reproduction in giving rise to advantageous gene combinations (*AB*, followed later by *ABC*) in large populations. Time flows from left to right. After their origins, the mutations *A*, *B*, and *C* either increase in number or, as represented by the small, spindle-shaped areas, disappear. (After Muller 1932a; Crow and Kimura 1965.)

24.10A). Quite possibly they would have to occur in the proper order as well. The diagram in figure 24.10A shows that *A* could succeed where *B* and *C* could not; that *AB* could succeed where *AC* could not; and that only with *AB* can *C* succeed, thus creating the combination *ABC*.

With the exchange of genes that accompanies sexual reproduction, the genes *A*, *B*, and *C* are shunted among various individuals and are tested in a variety of combinations. The successful combination *ABC* can arise very quickly, even though the individual genes arose by mutation in unrelated individuals rather than in a

Table 24.3 The relative rate of incorporation of new mutations into populations with and without recombination. Note that sexual reproduction scarcely improves the rate of incorporation in the case of small populations. (Crow and Kimura 1965.)

U/s	Population Size					
	10^3	10^4	10^5	10^6	10^7	10^8
10^{-7}	1.0007	1.01	1.12	2.38	16.7	162
10^{-6}	1.007	1.09	2.15	14.4	139	1.4×10^3
10^{-5}	1.07	1.92	12.1	116	1.2×10^3	1.2×10^4
10^{-4}	1.69	9.75	92.6	922	9.2×10^3	9.2×10^4
10^{-3}	7.50	69.6	691	6.9×10^3	6.9×10^4	6.9×10^5
10^{-2}	46.7	462	4.6×10^3	4.6×10^4	4.6×10^5	4.6×10^6
10^{-1}	240	2.4×10^3	2.4×10^4	2.4×10^5	2.4×10^6	2.4×10^7

NOTE: U/s = ratio of mutation rate to selection coefficient.

tandem series of individuals related by direct descent. Crow and Kimura (1965) calculated the role population size would play in determining the advantage of sex (table 24.3). A very small population from which newly arisen mutations would be rapidly lost by chance alone would, in a practical sense, imitate an asexual population: because little genetic variation can be maintained in small populations, the components of the advantageous combination must arise within them in tandem sequence. Hence, with mutation rates of 10^{-5} or 10^{-6}, a population of 1,000 individuals gives rise to the advantageous combinations at a rate only slightly greater than that of asexual forms. Given the same mutation rates in a population of 1,000,000 individuals, the rate at which an advantageous combination would arise exceeds that of an asexual form by one or two orders of magnitude.

The first serious challenge to the "Mullerian" point of view came from Maynard-Smith (1968b). His argument was that recombination is not necessary to promote advantageous combinations of genes; recurrent mutation and chance association are sufficient. As an example, consider the following:

Genotypes:	ab	Ab	aB	AB
Fitness:	1	$1 + K$	$1 + k$	$(1 + K)(1 + k)$
Frequency:	P_{ab}	P_{Ab}	P_{aB}	P_{AB}

If the total number of individuals in the following generation is T, the frequencies of the four genotypes in that generation will be

$$P'_{ab} = P_{ab}/T, \quad P'_{Ab} = P_{Ab}(1 + K)/T, \quad P'_{aB} = P_{aB}(1 + k)/T,$$

and

$$P'_{AB} = P_{AB}(1 + K)(1 + k)/T.$$

Consequently, if the first generation was in linkage equilibrium (P_{ab} · $P_{AB} = P_{Ab}$ · P_{aB}), so is the following one [P_{ab} · $P_{AB}(1 + K)(1 + k) = P_{Ab}(1 + K)$ · $P_{aB}(1 + k)$]. Hence, recombination (whose overall effect is to restore linkage equilibrium) is, in fact, not necessary.

Two points can be made relative to Maynard-Smith's calculations: linkage equilibrium under recurrent mutation was assumed at the outset, and the assumed fitnesses were multiplicative. Crow and Kimura, as they point out in their response (1969), assumed that selectively beneficial mutations were unique rather than recurrent events. Furthermore, as figure 24.10a shows, the fitnesses of the various gene combinations discussed by Muller were *not* multiplicative; on the contrary, C was advantageous only in a preexisting AB combination, and B was advantageous only in combination with a preexisting A.

Because of its prevalance throughout the world of living things, sex is assumed to endow its possessors with a selective advantage. As Fisher (1930) pointed out: individuals do not necessarily benefit from sexual union; therefore, the advantage of sex must be to the group possessing it. In contrast to this assumed advantage is a demonstrable disadvantage: sexually reproducing forms, because they produce males, reduce their otherwise attainable reproductive capacity by half (recall page 23). A number of persons (Williams 1975; Maynard-Smith 1970, 1978; plus their numerous references) have asked: "Where is the twofold advantage of sex?"

The consensus holds that the variability created by recombination and the random assortment of genes better prepares the sexual form for unexpected environmental vicissitudes or for rigorous environments lying beyond the range of tolerable conditions that characterizes the present habitat of any uniform clone of asexually reproducing individuals. Even in a constant environment, however, the need for self-thinning among overcrowded juvenile forms (see page 486) favors a mode of reproduction that leads to diversity among competing zygotes.

Perhaps a serious error enters into most discussions of the twofold advantage of asexual reproduction because they neglect the evolutionary history of living beings. The bulk of today's forms reproduce sexually; asexual organisms represent relatively small spots on an otherwise uniform background of sex. The switch from sexuality to asexuality may be extremely difficult because of pleiotropic effects associated with the responsible genetic changes. Carson (1961b, 1967; Carson et al. 1957) has studied parthenogenesis in a number of *Drosophila* species. In each instance, the proportions of unfertilized eggs that hatch and eventually produce surviving adults are exceedingly small, even after extensive selection for improvement. Ironically, the one species for which a parthenogenetic strain is known in nature, *D. mangabeirai*, has an overall egg hatch and survival of about 50 percent—a proportion which, when multiplied by 2, makes it "competitive" with other, sexually reproducing flies.

On the other hand, if we return to the simplest forms of life that indulge in genetic recombination, bacteria, we find that sex need not be accompanied by a tremendous cost. Transformation is an important means by which recombination is effected by microorganisms. One can then imagine that the "male" in primitive cells was merely the DNA floating free in an aqueous environment. Maleness, in effect, could have been a property of dead and disintegrating bacterial cells. The enzymatic machinery allowing the incorporation of foreign DNA into an individual's genome might have been developed under these primitive conditions. This machinery, plus the sequences in DNA now known to characterize many transposable insertional elements (Starlinger and Saedler 1972), could have been largely perfected while organisms utilized free-floating DNA. Having been adapted for the recombinational events needed for the somatic development of higher organisms as well as for sexual reproduction, turning this enzymatic machinery off may not be done with impunity today, even under selection for asexuality.

Evolution: The Contrasting Views of Neutralists and Selectionists

The classic works of Fisher (*The Genetical Theory of Natural Selection*, 1930), Haldane (*The Causes of Evolution*, 1932) and Wright (*Evolution in Mendelian Populations*, 1931), together with a series of publications based on Jesup Lectures delivered at Columbia University (Dobzhansky 1937; Mayr 1942; Stebbins 1950), form the basis for what has been called "the modern synthesis." Using genetics as their focal point, students of diverse areas of biology brought their expertise to bear upon evolution, the aspect of life without which, as Dobzhansky (1973b) has claimed, nothing else in biology makes sense.

The fundamental assumption of those who were responsible for that far-reaching synthesis was that natural selection continuously molds the content of the gene pools of populations, thus determining not only the morphological aspects of individuals but also their behaviors, their fertilities, and their reproductive isolation from individuals of other groups. Chance events were not denied, as the terms "genetic drift" and "Sewall Wright effect" testify; but the driving force for adaptive changes was identified as natural selection.

The basic assumption of the modern synthesis was challenged by Kimura (1968) and King and Jukes (1969); in its stead these three proposed that the bulk of observable genetic variation is selectively neutral. Forgetting that Charles Darwin offered only one of many early theories of evolution, King and Jukes referred to neutralism as "non-Darwinian evolution."

Many of the divergent views of neutralists and selectionists have been dealt with in earlier chapters, especially in chapters 13 through 18. This earlier material will not be repeated here. In the few remaining paragraphs it is appropriate, however, to summarize a number of the contrasting views of these two schools of modern evolutionary thought. The first and last chapters of Nei (1975) contain a generally thorough, similar account that has provided material for the following discussion.

Under the neo-Darwinian synthesis, workers assumed that natural selection operates on existing genetic variation rather than on newly arisen mutations; the decreased effectiveness of artificial selection in effecting changes on inbred or partially inbred populations confirms this assumption (Johannsen 1909; also see figure 11.4). If selection were to operate primarily on newly arisen mutations, the past history of a population exposed to selection would be unimportant. Nei (1975: 252) suggests that populations may have to await the occurrence of new favorable mutations; the evidence suggests that preexisting variation is generally sufficient.

Under the view prevailing during the 1930s and 1940s, populations were viewed as being at adaptive peaks, largely homozygous as far as genetic structure was concerned, and, hence, subject only to deleterious mutations; beneficial mutations could occur only if and when environmental conditions changes. During the 1950s this view was challenged by an alternative selectionist view which held that the adaptedness of populations rests largely on genetic heterogeneity at the population level (and heterozygosity at the individual level). This view of population structure lies beyond the comprehension of neutralists because the latter remain committed to genetic-load and cost-of-evolution calculations—calculations whose validity has been questioned earlier in this book.

Molecular evolutionists regard the rate of evolution as a constant when measured in celestial (clock) rather than generation time. They further hold that functionally less important portions of molecules evolve more rapidly than do functionally important ones, and that gene duplication precedes the emergence of new gene functions. The short-term basis for the first claim rests with experiments by Novick and Szilard (1950, 1951; also see Hartl and Dykhuizen 1979) who have shown that the accumulation of mutant bacteria in chemostat populations is proportional to clock time rather than to generation time of the dividing bacteria. The second claim (regarding functionally important molecules and portions of molecules) would not be disputed by selectionists; the proper functioning of an enzyme or any other protein is a matter impinging directly on the individual's survival and reproduction—that is, on its fitness.

Long before the neutralist theory was proposed, Bridges (1935) had suggested that gene duplication is the sole means by which a genome can increase in size. While it is true that duplication must precede the emergence of a gene having a *new* function, it is not at all clear that gene duplication does not ordinarily *follow* the establishment of an allozyme (or other protein) polymorphism (Spoffard 1969). The pre-existence of the polymorphism would provide the selective drive favoring the fixation of the duplication. Otherwise, duplications must be established (fixed) either by chance or by otherwise unattainable demands for a gene's product (see Mortlock 1981).

Several bits of evidence which Nei (1975: 246 ff.) cites in support of the neutralist theory have no obvious bearing on the neutralist-selectionist controversy. That 99 percent of all species which ever lived are now extinct, for example, is not compelling evidence for the reliance of organisms on rare advantageous mutant alleles in meeting evolutionary demands. In addition to those extinct species which existed in name only as a taxonomic convenience, other ("real") species became extinct because natural selection lacks foresight: natural selection is opportunistic.

A final point must now be made in this regard. Under the neutralist theory, Nei claims, variation is said to be the *product* of evolution, not a storage *designed for future use*. This claim unfairly ascribes to selectionists a teleology that has no place in modern evolutionary thought. To say that natural selection progresses largely *on the basis of* existing genetic variation is not to say that this variation exists *for the future use of* natural selection. The opportunism of natural selection and the absence of purpose in evolution have been acknowledged by all students of evolution except for a few confirmed vitalists (Koestler 1972; Grassé 1977).

FOR YOUR EXTRA ATTENTION

1. Evolution is essentially a matter of transmuting intrapopulation into interpopulation variation. Traditionally, the steps to evolution are thought to involve the formation of geographic races, subspecies, semispecies (as in *D. paulistorum*, see Dobzhansky and Spassky 1959), sibling species, and well-differentiated species.

What is the significance of Nei and Roychoudhury's (1972) conclusion that the number of gene differences between individuals from different races (Negro, Caucasian, and Japanese) is only slightly greater than the number between individuals from the same race?

For an extended discussion of this subject, see Lewontin (1974a: 152–57).

2. Kirkpatrick and Selander (1979), in a report on allozyme variation of normal and dwarf whitefish (*Coregonus clupeaformis*) of Maine lakes, point out that whitefish show a propensity for producing dwarf or pigmy morphotypes; normal and dwarf types occur sympatrically in lakes of the Yukon, Ontario, Alaska, and Sweden. Each dwarf form seems to have arisen independently of the others: in Maine the two forms have reached species rank; whether the others have also attained species rank or whether they represent polymorphic populations is unknown.

Kirkpatrick and Selander go on to say that dwarf forms are the only whitefish to be found in shallow lakes; dwarfs and normal sized fish exist sympatrically in deeper lakes, with the dwarfs at greater depths.

Perhaps the matter is being pushed too far here, but compare the situation in respect to dwarf and normal-sized whitefish to the discussion starting with and growing from figure 23.13.

3. Anderson and Jenkins (1979) have reported on the gene frequencies among domestic cats of Reno, Nevada. They conclude (in part) that Reno's cats are more similar to those of Boston, New York, Philadelphia, and other cities of the northeastern United States than they are to those of southwestern cities such as Dallas, Houston, Lubbock, or Denver.

What would be the nature of their evidence? What might explain their observation?

4. The olive fruitfly, *Dacus oleae*, is highly polymorphic for alleles at a locus (*A*) which produces acetylcholinesterase; more than 20 allozyme variants (in addition to a silent one) are known at this locus.

Because *D. oleae* flies are a serious pest on olive trees, they have been subjected to intensive spraying programs—especially by organophosphate, a powerful insecticide that can be detoxified by acetylcholinesterase.

Tsakas (1977) has studied many populations of *D. oleae* in Mediterranean olive-growing regions that have been subjected to heavy organophosphate treatment. The silent allele in each treated population occurs in low frequency; in addition, many flies produce three (rather than two) bands on electrophoretic gels. What is a reasonable explanation for these observations?

5. Davey and Reanney (1980) describe how gonococci tend to self-rupture and, as a result, drug resistance spreads easily by transformation—that is, by the passage of DNA molecules from the surrounding medium into living cells.

Compare this description with the "cost-free" suggestion made earlier in which maleness was equated with DNA floating free in an aqueous environment.

References

Abraham, I. and W. W. Doane. 1978. Genetic regulation of tissue-specific expression of amylase structural genes in *Drosophila melanogaster. Proc. Nat. Acad. Sci., U.S.* 75:4446–50.

Abramson, H. A. and M. H. Gorin. 1940. Skin permeability. *Cold Spring Harbor Sym. Quant. Biol.* 8:272–79.

Ahmad, M., R. K. Sakai, R. H. Baker, and R. W. Ainsley. 1978. Genetic analysis of two enzyme polymorphisms in a malaria vector mosquito. *J. Hered.* 69:155–58.

Alexander, R. D. 1974. The evolution of social behavior. *Annul. Rev. Ecol. Syst.* 5:335–83.

Allard, R. W. 1960. *Principles of Plant Breeding.* New York: Wiley.

Anderson, E. 1949. *Introgressive Hybridization.* New York: Wiley.

Anderson, M. M. and S. H. Jenkins. 1979. Gene frequencies in the domestic cats of Reno, Nevada: confirmation of a recent hypothesis. *J. Hered.* 70:267–69.

Anderson, W. W. 1969. Genetics of Natural Populations. XLI. The selection coefficients of heterozygotes for lethal chromosomes in *Drosophila* on different genetic backgrounds. *Genetics* 62:827–36.

_____ 1973. Genetic divergence in body size among experimental populations of *Drosophila pseudoobscura* kept at different temperatures. *Evolution* 27:278–84.

Anderson, W. W., L. Levine, O. Olvera, J. R. Powell, M. E. de la Rosa, V. M. Salceda, M. I. Gaso, and J. Guzman. 1979. Evidence for selection by male mating success in natural populations of *Drosophila pseudoobscura. Proc. Nat. Acad. Sci., U.S.* 76:1519–28.

Andrewartha, H. G. 1963. *Introduction to the Study of Animal Populations.* Chicago: University of Chicago Press.

Andrews, K. J. and G. D. Hegeman. 1976. Selective disadvantage of non-functional protein synthesis in *Escherichia coli. J. Mol. Evol.* 8:317-28.

Angus, R. A. and R. J. Schultz. 1979. Clonal diversity in the unisexual fish *Poeciliopsis monacha-lucida:* a tissue graft analysis. *Evolution* 33:27-40.

Atwood, K. C., L. K. Schneider, and F. J. Ryan. 1951. Selective mechanisms in bacteria. *Cold Spring Harbor Symp. Quant. Biol.* 16:345-55.

Audus, L. J. 1960. Microbiological breakdown of herbicides in soils. In E. K. Woodford and G. R. Sagar, eds., *Herbicides and the Soil*, pp. 1-19. Oxford: Blackwell Scientific Publ.

Averhoff, W. W. and R. H. Richardson. 1974. Pheromonal control of mating patterns in *Drosophila melanogaster. Behav. Genet.* 4:207-25.

Ayala, F. J. 1966. Evolution of fitness: I. Improvement in the productivity and size of irradiated populations of *Drosophila serrata and D. birchii. Genetics* 53:883-95.

—— 1968. Genotype, environment, and population numbers. *Science* 162:1453-59.

—— 1969. Genetic polymorphism and interspecific competitive ability in *Drosophila. Genet. Res. Camb.* 14:95-102.

—— 1971. Competition between species: Frequency dependence. *Science* 171:820-24.

Ayala, F. J. and C. A. Campbell. 1974. Frequency-dependent selection. *Ann. Rev. Ecol. Syst.* 5:115-38.

Ayala, F. J., M. E. Gilpin, and J. G. Ehrenfeld. 1973. Competition between species: Theoretical models and experimental tests. *Theor. Pop. Biol.* 4:331-56.

Ayala, F. J. and M. L. Tracey. 1974. Genetic differentiation within and between species of the *Drosophilia willistoni* group. *Proc. Nat. Acad. Sci., U.S.* 71:999-1003.

Ayala, F. J., M. L. Tracey, D. Hedgecock, and R. C. Richmond. 1974. Genetic differentiation during the speciation process in *Drosophila. Evolution* 28:576-92.

Baker, W. K. 1968. Position-effect variegation. *Adv. Genet.* 14:133-69.

—— 1975. Linkage disequilibrium over space and time in natural populations of *Drosophila montana. Proc. Nat. Acad. Sci., U.S.* 72:4095-99.

Banerjee, J., S. Hozra, and S. Sen. 1978. EMS-induced reversion studies in the *white* locus of *Drosophila melanogaster. Mut. Res.* 50:309-15.

Barber, H. N. 1955. Adaptive gene substitutions in Tasmanian eucalypts: I. Genes controlling the development of glaucousness. *Evolution* 9:1-14.

—— 1965. Selection in natural populations. *Heredity* 20:551-72.

Barker, J. S. F. and J. C. Mulley. 1976. Isozyme variation in natural populations of *Drosophila buzzatii. Evolution* 30:213-33.

Bateman, A. J. 1959. The viability of near-normal irradiated chromosomes. *Int. J. Rad. Biol.* 2:170-80.

Battaglia, B. 1965. Advances and problems of ecological genetics in marine animals. *Proc. 11th Inter. Cong. Gen.* 2:451-63.

Beadle, G. W. and B. Ephrussi. 1936. The differentiation of eye pigments in *Drosophila* as studied by transplantation. *Genetics* 21:225-47.

_____ 1937. Development of eye colors in *Drosophila:* diffusable substances and their interrelations. *Genetics* 22:76–86.

Beadle, G. W. and E. L. Tatum. 1941. Genetic control of biochemical reactions in *Neurospora. Proc. Nat. Acad. Sci., U.S.* 27:499–506.

Becker, H. 1957. Über Röntgenmosaikflecken und Defektmutationen am Auge von *Drosophila* und die Entwicklungsphysiologie des Auges. *Z. i. A. V.* 88:333–73.

Bennett, J. 1960. A comparison of selective methods and a test of the pre-adaptation hypothesis. *Heredity* 15:65–77.

Berberović, L. 1969. An excess of $L^M L^N$ heterozygotes in a South European population. *Heredity* 24:309–14.

Berger, F. G., G. A. M. Breen, and K. Paigen. 1979. Genetic determination of the development program for mouse liver β-galactosidase: involvement of sites proximate to and distant from the structural gene. *Genetics* 92:1187–1203.

Bijlsma, R. 1978. Polymorphism at the G6PD and 6PGD loci in *Drosophila melanogaster.* II. Evidence for interaction in fitness. *Genet. Res. Camb.* 31:227–37.

Bishop, J. A., D. J. Hartley, and G. G. Partridge. 1977. The population dynamics of genetically determined resistance to warfarin in *Rattus norvegicus* from mid Wales. *Heredity* 39:389–98.

Blair, W. F. and T. E. Kennerly, Jr. 1959. Effects of X-irradiation on a natural population of the wood-mouse (*Peromyscus leucopus*). *Texas J. Sci.* 11:137–49.

Blaylock, B. G. and H. H. Shugart, Jr. 1972. The effect of radiation-induced mutations on the fitness of Drosophilia populations. *Genetics* 72:469–74.

Bodmer, W. F. 1972. Race and IQ: The genetic background. In K. Richardson, D. Spears, and M. Richards, eds., *Race and Intelligence,* pp.83–113. Baltimore, Md.: Penguin Books.

Bodmer, W. F. and L. L. Cavalli-Sforza. 1976. *Genetics, Evolution, and Man.* San Francisco: W. H. Freeman.

Bodmer, W. F. and P. A. Parsons. 1960. The initial progress of new genes with various genetic systems. *Heredity* 15:283–99.

Bøggild, O. and J. Keiding. 1958. Competition in house fly larvae, experiments involving a DDT-resistant and a susceptible strain. *Oikos* 9:1–25.

Bonnier, G., U.-B. Jonsson, and C. Ramel. 1958. Selection pressure on irradiated populations of *Drosophila melanogaster. Hereditas* 44:378–406.

Boorman, S. A. and P. R. Levitt. 1972. Group selection on the boundary of a stable population. *Proc. Nat. Acad. Sci., U.S.* 69:2711–13.

_____ 1973. Group selection on the boundary of a stable population. *Theor. Pop. Biol.* 4:85–128.

Borowsky, R. 1973. Melanomas in *Xiphophorus variatus* (Pisces, Poeciliidae) in the absence of hybridization. *Experientia* 29:1431–33.

Boveri, T. 1903. Über die Konstitution der chromatischen Kernsubstanz. *Verh. deutsch. Zool. Ges.* 13 vers. Wurzburg 10-33.

Bridges, C. B. 1935. Salivary chromosome maps, with a key to the banding of the chromosomes of *Drosophila melanogaster. J. Heredity* 26:60–64.

Britten, R. J. and E. H. Davidson. 1969. Gene regulation for higher cells: a theory. *Science* 165:349-57.

Brncic, D. 1954. Heterosis and the integration of the genotype in geographic populations of *Drosophila pseudoobscura*. *Genetics* 39:77-88.

Brown, A. J. L. and C. H. Langley. 1979. Reevaluation of level of genic heterozygosity in natural population of *Drosophila melanogaster* by two-dimensional electrophoresis. *Proc. Nat. Acad. Sci., U.S.* 76:2381-84.

Brown, W. L., Jr. and E. O. Wilson. 1956. Character displacement. *Syst. Zool.* 5:49-64.

Brues, A. M. 1964. The cost of evolution vs. the cost of not evolving. *Evolution* 18:379-83.

Budnik, M., D. Brncic, and S. Koref-Santibanez. 1971. The effects of crowding on chromosomal polymorphism of *Drosophila pavani*. *Evolution* 25:410-19.

Bulger, A. J. and R. J. Schultz. 1979. Heterosis and interclonal variation in thermal tolerance in unisexual fishes. *Evolution* 33:848-59.

Buri, P. 1956. Gene frequency in small populations of mutant *Drosophila*. *Evolution* 10:367-402.

Burns, J. M. 1966. Preferential mating versus mimicry: Disruptive selection and sex-limited dimorphism in *Papilio glaucus*. *Science* 153:551-53.

Bush, G. L. 1975a. Modes of animal speciation. *Annul. Rev. Ecol. Syst.* 6:339-64.

—— 1975b. Sympatric speciation in phytophagous parasitic insects. In P. W. Price, ed., *Evolutionary Strategies of Parasitic Insects and Mites,* pp.187-206. New York: Plenum.

Carson, H. L. 1955. The genetic characteristics of marginal populations of *Drosophila*. *Cold Spring Harbor Symp. Quant. Biol.* 20:276-87.

—— 1957. Production of biomass as a measure of fitness of experimental populations of *Drosophila*. *Genetics* 42:363-64.

—— 1958a. Increase in fitness in experimental populations resulting from heterosis. *Proc. Nat. Acad. Sci., U.S.* 44:1136-41.

—— 1958b. The population genetics of *Drosophila robusta*. *Advances in Genetics* 9:1-40.

—— 1961a. Heterosis and fitness in experimental populations of *Drosophila melanogaster*. *Evolution* 15:496-509.

—— 1961b. Rare parthenogenesis in *Drosophila robusta*. *Amer. Natur.* 95:81-86.

—— 1967. Selection for parthenogenesis in *Drosophila mercatorum*. *Genetics* 55:157-71.

—— 1968. The population flush and its genetic consequences. In R. C. Lewontin, ed., *Population Biology and Evolution,* pp. 123-37. Syracuse, N.Y.: Syracuse University Press.

—— 1969. Parallel polymorphisms in different species of Hawaiian *Drosophila*. *Amer. Natur.* 103:323-29.

—— 1975. The genetics of speciation at the diploid level. *Amer. Natur.* 109:83-92.

_____ 1976. Inference of the time of origin of some *Drosophila* species. *Nature* 259:395–96.

_____ 1978. Speciation and sexual selection in Hawaiian *Drosophila*. In P. F. Brussard, ed., *Ecological Genetics: The Interface,* pp. 93–107. New York: Springer-Verlag.

Carson, H. L., D. E. Hardy, H. T. Spieth, and W. S. Stone. 1970. The evolutionary biology of the Hawaiian Drosophilidae. In M. K. Hecht and W. C. Steere, eds., *Essays in Evolution and Genetics,* pp. 437–543. New York: Appleton-Century-Crofts.

Carson, H. L., M. R. Wheeler, and W. B. Heed. 1957. A parthenogenetic strain of *Drosophila mangabeirai* Malogolowkin. University of Texas Publ. 5721: 115–22.

Castle, W. E. 1903. The laws of heredity of Galton and Mendel and some laws governing race improvement by selection. *Proc. Am. Acad. Arts, Sci.* 39: 223–42.

Cavalli, L. L. and G. A. Maccacaro. 1952. Polygenic inheritance of drug-resistance in the bacterium *Escherichia coli. Heredity* 6:311–31.

Cavalli-Sforza, L. L. 1959. Some data on the genetic structure of human populations. *Proc. 10th Inter. Congr. Genet.* 1:389–407.

Cesnola, A. P. di. 1904. Preliminary note on the protective value of colour in *Mantis religiosa. Biometrika* 3:58–59.

Chabora, A. J. 1968. Disruptive selection for sternopleural chaeta number in various strains of *Drosophila melanogaster. Amer. Natur.* 102:525–32.

Champion, M. J., J. B. Shaklee, and G. S. Whitt. 1975. Developmental genetics of teleost isozymes. In C. L. Markert, ed., *Isozymes. 3. Developmental Biology,* pp. 417–37. New York: Academic Press.

Chao, C. Y. 1959. Heterotic effects of a chromosome segment in maize. *Genetics* 44:657–77.

Chargaff, E. 1950. Chemical specificity of nucleic acids and mechanism of their enzymatic degradation. *Experientia* 6:201–9.

Chetverikov, S. S. 1926. On certain features of the evolutionary process from the point of view of modern genetics. *J. Exper. Biol.* (Moscow) 2:3–54.

Christensen, B. 1977. Habitat preference among amylase genotypes in *Asellus aquaticus* (Isopoda, Crustacea). *Hereditas* 87:21–26.

Christiansen, F. B. 1975. Hard and soft selection in a subdivided population. *Amer. Natur.* 109:11–16.

Clarke, B. 1973a. The effect of mutation on population size. *Nature* 242:196–97.

_____ 1973b. Mutation and population size. *Heredity* 31:367–79.

Collins, M. M. and P. M. Tuskes. 1979. Reproductive isolation in sympatric species of dayflying moths (*Hemileuca:* Saturnidae). *Evolution* 33:728–33.

Colwell, R. K. and E. R. Fuentes. 1975. Experimental studies of the niche. *Ann. Rev. Ecol. Syst.* 6:281–310.

Commoner, B. 1964. DNA and the chemistry of inheritance. *Amer. Sci.* 52:365–88.

Connor, J. L. and M. J. Bellucci. 1979. Natural selection resisting inbreeding depression in captive wild house mice (*Mus musculus*). *Evolution* 33:929–40.

Creed, E. R. 1971. Melanism in the two-spot lady bird, *Adalia bipunctata*, in Great Britain. In E. R. Creed, ed., *Ecological Genetics and Evolution*, pp. 134–51. Oxford: Blackwell Scientific Publications.

Crenshaw, J. W. 1965. Radiation-induced increases in fitness in the flour beetle *Tribolium confusum*. *Science* 149:426–27.

Crew, F. A. E. 1966. *The Foundations of Genetics*. Oxford: Pergamon Press.

Crick, F. 1971. General model for the chromosomes of higher organisms. *Nature* 234:25–27.

Crosby, J. L. 1963. The evolution and nature of dominance. *J. Theor. Biol.* 5:35–51.

Crow, J. F. 1948. Alternate hypotheses of hybrid vigor. *Genetics* 33:477–87.

—— 1952. Dominance and overdominance. In J. W. Gowen, ed., *Heterosis*, pp. 282–97. Ames: Iowa State College Press.

—— 1956. Genetics of DDT resistance in *Drosophila*. *Proc. Inter. Genet. Symp. Tokyo* (Cytologia Suppl.):408–9.

—— 1957. Possible consequences of an increased mutation rate. *Eugen. Quart.* 4:67–80.

—— 1958. Some possibilities for measuring selection intensities in man. *Human Biol.* 30:1–13.

—— 1979. Genes that violate Mendel's rules. *Sci. Amer.* 240:134–46.

Crow, J. F. and M. Kimura. 1965. Evolution in sexual and asexual populations. *Amer. Natur.* 99:439–50.

—— 1969. Evolution in sexual and asexual populations. *Amer. Natur.* 103:89–91.

—— 1970. *An Introduction to Population Genetics Theory*. New York: Harper and Row.

—— 1979. Efficiency of truncation selection. *Proc. Nat. Acad. Sci., U.S.* 76:396–99.

Crow, J. F. and R. G. Temin. 1964. Evidence for the partial dominance of recessive lethal genes in natural populations of *Drosophila*. *Amer. Nat.* 98:21–33.

Crumpacker, D. W. and J. S. Williams. 1974. Rigid and flexible chromosomal polymorphisms in neighboring populations of *Drosophila pseudoobscura*. *Evolution* 28:57–66.

Cunha, A. B. da, Th. Dobzhansky, O. Pavlovsky, and B. Spassky. 1959. Genetics of Natural Populations. xxviii. Supplementary data on the chromosomal polymorphism in *Drosophila willistoni* in its relation to its environment. *Evolution* 13:389–404.

Curtsinger, J. W. 1976a. Stabilizing selection in *Drosophila melanogaster*. *J. Hered.* 67:59–60.

—— 1976b. Stabilizing or directional selection on egg lengths: a rejoinder. *J. Hered.* 67:246–47.

Darlington, C. D. and K. Mather. 1949. *The Elements of Genetics* London: Allen and Unwin.

Darwin, C. 1859. *The Origin of Species.* New York: The Modern Library.

Davenport, C. B. and M. T. Scudder. 1919. Naval officers: their heredity and development. Publ. 259. Washington, D.C.: Carnegie Inst. Wash.

Davey, R. B. and D. C. Reanney. 1980. Extrachromosomal genetic elements and the adaptive evolution of bacteria. *Evol. Biol.* 13:113–47.

Davidson, E. H. and R. J. Britten. 1979. Regulation of gene expression: possible role of repetitive sequences. *Science* 204:1052–59.

Dawkins, R. 1976. *The Selfish Gene.* New York: Oxford University Press.

Dayhoff, M. O. 1972. *Atlas of Protein Sequence and Structure.* Vol. 5. Washington, D.C.: Nat. Biomed. Res. Foundation.

Diamond, J. M. 1978. Niche shifts and the rediscovery of interspecific competition. *Amer. Sci.* 66:322–31.

Dickson, R. C., J. Abelson, W. M. Barnes, and W. S. Reznikoff. 1975. Genetic regulation: The *lac* control region. *Science* 187:27–35.

Dickson, R. C., J. Abelson, P. Johnson, W. S. Reznikoff, and W. M. Barnes. 1977. Nucleotide sequence changes produced by mutations in the *lac* promoter of *Escherichia coli. J. Molec. Biol.* 3:65–75.

Diver, C. 1929. Fossil records of Mendelian mutants. *Nature* 123:183.

Dixon, W. J. and F. J. Massey, Jr. 1951. *Introduction to Statistical Analysis.* New York: McGraw-Hill.

Doane, W. W. 1977. Further evidence for temporal genes controlling *Amy* expression in *Drosophila melanogaster.* (Abstract) *Genetics* 86:s15–s16.

Dobzhansky, Th. 1937. *Genetics and the Origin of Species.* New York: Columbia University Press.

—— 1947a. A directional change in the genetic constitution of a natural population of *Drosophila pseudoobscura. Heredity* 1:53–64.

—— 1947b. Genetics of Natural Populations. xiv. A response of certain gene arrangements in the third chromosome of *Drosophila pseudoobscura* to natural selection. *Genetics* 32:142–60.

—— 1948a. Genetics of Natural Populations. xvi. Altitudinal and seasonal changes produced by natural selection in certain populations of *Drosophila pseudoobscura and Drosophila persimilis. Genetics* 33:158–76.

—— 1948b. Genetics of Natural Populations. xviii. Experiments on chromosomes of *Drosophila pseudoobscura* from different geographic regions. *Genetics* 33:588–602.

—— 1950. Genetics of Natural Populations. xix. Origin of heterosis through natural selection in populations of *Drosophila pseudoobscura. Genetics* 35:288–302.

—— 1952. Nature and origin of heterosis. In J. W. Gowen, ed., *Heterosis,* pp. 218–23. Ames: Iowa State College Press.

—— 1962. Rigid vs. flexible chromosomal polymorphisms in *Drosophila. Amer. Natur.* 96:321–28.

—— 1964. How do the genetic loads affect the fitness of their carriers in *Drosophila* populations? *Amer. Natur.* 98:151–66.

—— 1968. Adaptedness and fitness. In. R. C. Lewontin, ed., *Population Biology and Evolution,* pp. 109–21. Syracuse, New York: Syracuse University Press.

652   References

—— 1970. *Genetics of the Evolutionary Process.* New York: Columbia University Press.

—— 1973a. Active dispersal and passive transport in *Drosophila. Evolution* 27:565–75.

—— 1973b. Nothing in biology makes sense except in the light of evolution. *Amer. Biol. Teach.* (March), pp. 125–29.

Dobzhansky, Th., W. W. Anderson, and O. Pavlovsky. 1966. Genetics of Natural Populations. xxxviii. Continuity and change in populations of *Drosophila pseudoobscura* in western United States. *Evolution* 20:418–27.

Dobzhansky, Th. and C. Epling. 1944. Contributions to the genetics, taxonomy, and ecology of *Drosophila pseudoobscura* and its relatives. Publ. 554. Washington, D.C.: Carnegie Inst. Wash.

Dobzhansky, Th. and A. M. Holz. 1943. A re-examination of the problem of manifold effects of genes in *Drosophila melanogaster. Genetics* 28:295–303.

Dobzhansky, Th., A. M. Holz, and B. Spassky. 1942. Genetics of Natural Populations. viii. Concealed variability in the second and the fourth chromosomes of *D. pseudoobscura* and its bearing on the problem of heterosis. *Genetics* 27:463–90.

Dobzhansky, Th. and P. Koller. 1938. An experimental study of sexual isolation in *Drosophila. Biol. Zentr.* 58:589–607.

Dobzhansky, Th., C. Krimbas, and M. G. Krimbas. 1960. Genetics of Natural Populations. xxx. Is the genetic load in *Drosophila pseudoobscura* a mutational or a balanced load? *Genetics* 45:741–53.

Dobzhansky, Th., R. C. Lewontin, and O. Pavlovsky. 1964. The capacity for increase in chromosomally polymorphic and monomorphic populations of *Drosophila pseudoobscura. Heredity* 19:597–614.

Dobzhansky, Th. and C. Pavan. 1950. Local and seasonal variations in relative frequencies of species of *Drosophila* in Brazil. *J. Animal Ecol.* 19:1–14.

Dobzhansky, Th. and O. Pavlovsky. 1953. Indeterminate outcome of certain experiments on Drosophila populations. *Evolution* 7:198–210.

—— 1955. An extreme case of heterosis. *Proc. Nat. Acad. Sci., U.S.* 41:289–95.

—— 1961. A further study of fitness of chromosomally polymorphic and monomorphic populations of *Drosophila pseudoobscura. Heredity* 16:169–79.

Dobzhansky, Th. and M. L. Queal. 1938. Genetics of Natural Populations. ii. Genic variation in populations of *Drosophila pseudoobscura* inhabiting isolated mountain ranges. *Genetics* 23:463–84.

Dobzhansky, Th. and M. M. Rhoades. 1938. A possible method for locating favorable genes in maize. *J. Am. Soc. Agron.* 30:668–75.

Dobzhansky, Th. and B. Spassky. 1944. Genetics of Natural Populations. xi. Manifestation of genetic variants in *Drosophila pseudoobscura* in different environments. *Genetics* 29:270–90.

—— 1953. Genetics of Natural Populations. xxi. Concealed variability in two sympatric species of *Drosophila. Genetics* 38:471–84.

—— 1959. *Drosophila paulistorum,* a cluster of species in *statu nascendi. Proc. Nat. Acad. Sci., U.S.* 45:419–28.

_____ 1963. Genetics of Natural Populations. xxxiv. Adaptive norm, genetic load, and genetic elite in *Drosophila pseudoobscura. Genetics* 48:1467-85.

_____ 1968. Genetics of Natural Populations. xl. Heterotic and deleterious effect of recessive lethals in populations of *Drosophila pseudoobscura. Genetics* 59:411-25.

_____ 1969. Artificial and natural selection for two behavioral traits in *Drosophilia pseudoobscura. Proc. Nat. Acad. Sci., U.S.* 62:75-80.

Dobzhansky, Th. and B. Wallace. 1953. The genetics of homeostasis in *Drosophilia. Proc. Nat. Acad. Sci., U.S.* 39:162-71.

Dobzhansky, Th. and S. Wright. 1941. Genetics of Natural Populations. v. Relations between mutation rate and accumulation of lethals in populations of *Drosophila pseudoobscura. Genetics* 26:23-51.

_____ 1943. Genetics of Natural Populations. x. Dispersion rates in *Drosophila pseudoobscura. Genetics* 28:304-40.

_____ 1947. Genetics of Natural Populations. xv. Rate of diffusion of a mutant gene through a population of *Drosophila pseudoobscura. Genetics* 32:303-24.

Douglas, M. M. and J. W. Grula. 1978. Thermoregulatory adaptations allowing ecological range expansion by the Pierid butterfly, *Nathalis iole* Boisduval. *Evolution* 32:776-83.

Druger, M. 1962. Selection and body size in *Drosophila pseudoobscura* at different temperatures. *Genetics* 47:209-22.

Dun, R. B. and A. S. Fraser. 1958. Selection for an invariant character—"vibrissa number"—in the house mouse. *Nature* 181:1018-19.

_____ 1959. Selection for an invariant character, vibrissa number, in the house mouse. *Aust. J. Biol. Sci.* 12:506-23.

Dunn, J. J. and F. W. Studier. 1973. T7 early RNAs and *E. coli* RNAs are cut from long precursor RNAs in vivo by ribonuclease III. *Proc. Nat. Acad. Sci., U.S.* 70:3296-300.

East, E. M. 1936. Heterosis. *Genetics* 21:375-97.

Ehrlich, P. R. 1965. The population biology of the butterfly, *Euphydryas editha.* ii. The structure of the Jasper Ridge colony. *Evolution* 19:327-36.

Ehrman, L. 1965. Direct observation of sexual isolation between allopatric and between sympatric strains of the different *Drosophila paulistorum* races. *Evolution* 19:459-64.

_____ 1966. Mating success and genotype frequency in *Drosophila. Anim. Behav.* 14:332-39.

_____ 1967. Further studies on genotype frequency and mating success in *Drosophila. Amer. Natur.* 101:415-24.

_____ 1970. The mating advantage of rare males in *Drosophila. Proc. Nat. Acad. Sci., U.S.* 65:345-48.

Ehrman, L. and J. Probber. 1978. Rare Drosophila males: The mysterious matter of choice. *Amer. Sci.* 66:216-22.

Ehrman, L., B. Spassky, O. Pavlovsky, and Th. Dobzhansky. 1965. Sexual selection, geotaxis, and chromosomal polymorphism in experimental populations of *Drosophila pseudoobscura. Evolution* 19:337-46.

Emerson, S. 1939. A preliminary survey of the *Oenothera organensis* population. *Genetics* 24:524-37.

Emlen, J. M. 1973. *Ecology: An Evolutionary Approach.* Reading, Mass.: Addison-Wesley.

Endler, J. A. 1973. Gene flow and population differentiation. *Science* 179: 243-50.

_____ 1979. Gene flow and life history patterns. *Genetics* 93:263-84.

Endler, J. A. and J. S. Johnston. 1980. The influence of breeding site distribution on population structure and differentiation in *Drosophila.* (in manuscript).

Epp, M. D. 1974. The homozygous genetic load in mutagenized populations of *Oenothera hookeri* T. and G. *Mutat. Res.* 22:39-46.

Erickson, R. O. 1945. The *Clematis fremontii* var. *riehlii* population in the Ozarks. *Ann. Mo. Bot. Garden* 32:413-60.

Ewens, W. J. 1967. A note on the mathematical theory of the evolution of dominance. *Amer. Natur.* 101:35-40.

Falconer, D. S. 1960. *Introduction to Quantitative Genetics.* Edinburgh: Oliver and Boyd.

Falk, R. 1961. Are induced mutations in *Drosophila* overdominant? II. Experimental results. *Genetics* 46:737-57.

_____ 1967a. Viability of heterozygotes for induced mutations in *Drosophila melanogaster. Mutat. Res.* 4:59-72.

_____ 1967b. Fitness of heterozygotes for irradiated chromosomes in *Drosophila. Mutat. Res.* 4:805-19.

Falk, R. and N. Ben-Zeev. 1966. Viability of heterozygotes for induced mutations in *Drosophila melanogaster.* II. Mean effects in irradiated autosomes. *Genetics* 53:65-77.

Falk, R., A. Rahat, and N. Ben-Zeev. 1965. Viability of heterozygotes for induced mutations in *Drosophila melanogaster.* I. Irradiated X-chromosome. *Mutat. Res.* 2:438-51.

Feldman, M. W. and S. Karlin. 1971. The evolution of dominance: a direct approach through the theory of linkage and selection. *Theor. Pop. Biol.* 2:482-92.

Feller, W. 1967. On fitness and the cost of natural selection. *Genet. Res.* 9:1-15.

Felsenstein, J. 1975. A pain in the torus: some difficulties with models of isolation by distance. *Amer. Natur.* 109:359-68.

Fisher, R. A. 1928. The possible modification of the response of the wildtype to recurrent mutations. *Amer. Natur.* 62:115-26.

_____ 1930. *The Genetical Theory of Natural Selection.* Oxford: Clarendon Press.

_____ 1958. *The Genetical Theory of Natural Selection,* 2d ed. New York: Dover.

Fisher, R. C. 1959. Life history and ecology of *Horogenes chrysostictos gmelin* (Hymenoptera, ichneumonidae), a parasite of *Ephestia sericarium* Scott (Lepidoptera, phycitidae). *Can. J. Zool.* 37:429-46.

Fitch, W. M. 1976. Molecular evolutionary clocks. In F. J. Ayala, ed., *Molecular Evolution,* pp. 160-78. Sunderland, Mass.: Sinauer Assocs.

Fitz-Earle, M., D. G. Holm, and D. T. Suzuki. 1973. Genetic Control of Insect Populations. I. Cage studies of chromosome replacement by compound autosomes in *Drosophila melanogaster*. *Genetics* 74:461-75.

Fontdevila, A. and J. Mindez. 1979. Frequency-dependent mating in a modified allozyme locus of *Drosophila pseudoobscura*. *Evolution* 33: 634-40.

Ford, E. B. 1940. Genetic research in the Lepidoptera. *Ann Eugenics* 10:227-52.

——— 1945. *Butterflies.* London: Collins.

——— 1955. Polymorphism and taxonomy. *Heredity* 9:255-64.

——— 1965. *Genetic Polymorphism.* Cambridge, Mass.: MIT Press.

——— 1971. *Ecological Genetics,* 3d ed. London: Chapman and Hall.

Ford, E. D. 1975. Competition and stand structure in some even-aged plant monocultures. *J. Ecol.* 63:311-33.

Ford, H. D. and E. B. Ford. 1930. Fluctuation in numbers, and its influence on variation, in *Melitaea aurinia,* Rott. (Lepidoptera). *Trans. Roy. ent. Soc. London,* 78:345-51.

Fox, L. R. 1975. Cannibalism in natural populations. *Annul. Rev. Ecol. Syst.* 6:87-106.

Frelinger, J. A. 1972. The maintenance of transferrin polymorphism in pigeons. *Proc. Nat. Acad. Sci., U.S.* 69:326-29.

Frydenberg, O. and K. Sick. 1960. The selection of *st* and *cn* alleles on different genetical backgrounds in *Drosophila melanogaster*. *Hereditas* 46:601-21.

——— 1962. The selective value of a cinnebar mutant of *Drosophila melanogaster* in populations homozygous for an unsuppressible vermillion mutant. *Hereditas* 48:313-23.

Fujio, Y., Y. Nakamura, and M. Sugita. 1979. Selective advantage of heterozygotes at catalase locus in the Pacific oyster, *Crassostrea gigas. Japan J. Genet.* 54:359-66.

Gallo, A. J. 1978. Genetic load of the X-chromosome in natural populations of *Drosophila melanogaster. Genetica* 49:153-57.

Ganders, F. R., K. Carey, and A. F. J. Griffiths. 1977. Natural selection for a fruit dimorphism in *Plectritis congesta* (Valerianaceae). *Evolution* 31:873-81.

Garrod, A. E. 1909. *Inborn Errors of Metabolism.* Oxford: Frowde Hodder and Stoughton, Oxford University Press.

Gentry, A. H. 1974. Flowering phenology and diversity in tropical *Bignoniaceae. Biotropica* 6:64-68.

Gershenson, S. 1928. A new sex-ratio abnormality in *Drosophila obscura. Genetics* 13:488-507.

Gilbert, L. and P. Ravin. 1975. *Co-evolution of Animals and Plants.* Austin: University of Texas Press.

Gilbert, W. and A. Maxam. 1973. The nucleotide sequence of the *lac* operator. *Proc. Nat. Acad. Sci., U.S.* 70:3581-84.

Glass, H. B. and C. C. Li. 1953. The dynamics of racial intermixture—an analysis based on the American Negro. *Am. J. Human. Genet.* 5:1-20.

Gloor, H. 1947. Phaenokopie-Versuche mit Aether an *Drosophila. Rev. suisse Zool.* 54:637–712.

Goldschmidt, R. 1934. Lymantria. *Biblio. Genetica* 11:1–186.

―――― 1938. *Physiological Genetics.* New York: McGraw-Hill.

―――― 1940. *The Material Basis of Evolution.* New Haven, Conn.: Yale University Press.

Gooch, J. L. and R. J. M. Schopf. 1970. Population genetics of marine species of the phylum ectoprocta. *Biol. Bull.* 138:138–56.

Goodenough, U. and R. P. Levine. 1974. *Genetics.* New York: Holt, Rinehart and Winston.

Gordon, H. and M. Gordon. 1957. Maintenance of polymorphism by potentially injurious genes in eight natural populations of the platyfish, *Xiphophorus maculatus. J. Genet.* 55:1–44.

Gottlieb, L. D. 1977. Genotypic similarity of large and small individuals in a natural population of the annual plant *Stephanomeria exigua* ssp. *coronaria* (Compositae). *J. Ecol.* 65:127–34.

Graham, J. B. and C. A. Istock. 1979. Gene exchange and natural selection cause *Bacillus subtilis* to evolve in soil culture. *Science* 204:637–39.

Grassé, P-P. 1977. *Evolution of Living Organisms.* New York: Academic Press.

Greene, E. L. 1968. Genetic effects of radiation on mammalian populations. *Ann. Rev. Genet.* 2:87–120.

Green, M. M. 1969. Controlling element mediated transpositions of the *white* gene in *Drosophila melanogaster. Genetics* 61:429–41.

Green, M. M. and S. H. Y. Shepherd. 1979. Genetic instability in *Drosophila melanogaster:* The induction of specific chromosome 2 deletions by MR elements. *Genetics* 92:823–32.

Grigg, G. W. 1958. Competitive suppression and the detection of mutations in microbial populations. *Aust. J. Biol. Sci.* 11:69–84.

Gromko, M. H. 1977. What is frequency-dependent selection? *Evolution* 31:438–42.

Haldane, J. B. S. 1924. A mathematical theory of natural and artificial selection. Part 1. *Trans. Camb. Phil. Soc.* 23:19–41.

―――― 1932. *The Causes of Evolution.* New York: Harper and Row. (Reprinted 1966, Cornell University Press, Ithaca, N.Y.)

―――― 1937. The effect of variation on fitness. *Amer. Natur.* 71:337–49.

―――― 1953. Animal populations and their regulation. *New Biol.* 15:9–24.

―――― 1954. The measurement of natural selection. *Proc. 9th Inter. Cong. Genet.* 1:480–87.

―――― 1957. The cost of natural selection. *J. Genet.* 55:511–24.

―――― 1960. More precise expressions for the cost of natural selection. *J. Genet.* 57:351–60.

―――― 1964. A defense of beanbag genetics. *Persp. Biol. Med.* 7:343–59.

―――― 1965. The implications of genetics for human society. *Proc. 11th Inter. Cong. Genet.* 2:xci–cii.

Halkka, O., L. Halkka, and M. Raatikainen. 1975. Transfer of individuals as a means of investigating natural selection in operation. *Hereditas.* 80:27–34.

Halkka, O., M. Raatikainen, and L. Halkka. 1976. Conditions requisite for stability of polymorphic balance in *Philaenus spumarius* (L) (Homoptera). *Genetica* 46:67-76.

Halkka, O., M. Raatikainen, L. Halkka, and T. Raatikainen. 1977. Coexistence of four species of spittle-producing Homoptera. *Ann. Zool. Fennici* 14:228-31.

Hamilton, W. D. 1964a. The genetical evolution of social behavior, I. *J. Theor. Biol.* 7:1-16.

―――― 1964 b. The genetical evolution of social behaviour, II. *J. Theor. Biol.* 7:17-52.

Hamilton, W. D., P. A. Henderson, and N. A. Moran. 1980. Fluctuation of environment and coevolved antagonist polymorphism as factors in the maintenance of sex. In R. D. Alexander and D. W. Tinkle, eds., *Natural Selection and Social Behavior* (in press). New York: Chiron Press.

Handford, P. T. 1971. An esterase polymorphism in the bleak, *Alburnus alburnus,* Pisces. In E. R. Creed, ed., *Ecological Genetics and Evolution,* pp. 289-97. Oxford: Blackwell Scientific Publ.

Hardin, G. 1971. Nobody ever dies of overpopulation (editorial). *Science* 171:527.

Hardy, D. E. 1965. *Insects of Hawaii,* Vol. 12. Honolulu: University of Hawaii Press.

Hardy, G. H. 1908. Mendelian proportions in mixed population. *Science* 28: 49-50.

Harper, J. L. 1961. Approaches to the study of plant competition. *Sym. Soc. Exp. Biol.* 15:1-39.

―――― 1977. *Population Biology of Plants.* New York: Academic Press.

Harper, J. L. and I. H. McNaughton. 1962. The comparative biology of closely related species living in the same area. VII. Interference between individuals in pure and mixed populations of *Papaver* species. *New Phytology* 61:175-88.

Harris, H. 1966. Enzyme polymorphisms in man. *Proc. Roy. Soc. Ser. B.* 164:298-310.

Hartl, D. L. 1970. Meiotic drive in natural populations of *Drosophila melanogaster.* IX. Suppressors of *segregation distorter* in wild populations. *Can. J. Genet. Cytol.* 12:594-600.

Hartl, D. L. 1980. *Principles of Population Genetics.* Sunderland, Mass.: Sinauer.

Hartl, D. L. and D. Dykhuizen. 1979. A selectively driven molecular clock. *Nature* 281:230-31.

Hartl, D. L. and H. Jungen. 1979. Estimation of average fitness of populations of *Drosophila melanogaster* and the evolution of fitness in experimental populations. *Evolution* 33:371-80.

Hartman, P. E. and P. B. Hulbert. 1975. Genetic activity spectra of some antischistosomal compounds, with particular emphasis on thioxanthenous and benzothiopyranoindazoles. *J. Tox. Env. Health* 1:243-70.

Hawley, R. C. 1946. *The Practice of Silviculture.* New York: Wiley.

Herbert, P. D. N. 1974. Enzyme variability in natural populations of *Daphnia magna*. 2. Genotypic frequencies in permanent populations. *Genetics* 77:323-34.

Hedrick, P., S. Jain, and L. Holden. 1978. Multilocus systems in evolution. *Evol. Biol.* 11:104-84.

Helfer, R. G. 1939. Dominance modifiers of *Scute* in *Drosophila pseudoobscura*. *Genetics* 24:278-301.

Hendrickson, H. T. 1978. Sympatric speciation: Evidence? *Science* 200:345-46.

Hickey, D. A. 1977. Selection for amylase allozymes in *Drosophila melanogaster*. *Evolution* 31:800-04.

Hicks, J. B., A. Hinnen, and G. R. Fink. 1979. Properties of yeast transformation. *Cold Spring Harbor Symp. Quant. Biol.* 43:1305-13.

Highton, R. 1977. Comparison of microgeographic variation in morphological and electrophoretic traits. *Evol. Biol.* 10:397-436.

Hiraizumi, Y. 1965. Effects of X-ray induced mutations on several components of fitness. *Ann. Report Nat. Inst. Genet. (Japan)* 15:109-11.

Hiraizumi, Y. and J. F. Crow. 1960. Heterozygous effects on viability, fertility, rate of development, and longevity of Drosophila chromosomes that are lethal when homozygous. *Genetics* 45:1071-83.

Hirsch, J. and L. Erlenmeyer-Kimling. 1961. Sign of taxis as a property of the genotype. *Science* 134:835-36.

_____ 1962. Studies in experimental behavior genetics. 4. Chromosome analyses for geotaxis. *J. Comp. Physiol. Psychol.* 55:732-39.

Hood, L. 1976. Genetics and biological evolution: Introduction. *Fed. Proc.* 35:2077-78.

Hubby, J. L. and L. H. Throckmorton. 1968. Protein differences in *Drosophila*. 4. A study of sibling species. *Amer. Natur.* 102:193-205.

Hunter, R. L. and C. L. Markert. 1957. Histochemical demostration of enzymes separated by zone electrophoresis in starch gels. *Science* 125:1294-95.

Ives, P. T. 1949. Nonrandom production of visible mutations by the Florida high mutation rate gene in *Drosophila*. *Rec. Gen. Soc. Am.* 18:96.

_____ 1970. Further genetic studies of the south Amherst population of *Drosophila melanogaster*. *Evolution* 24:507-18.

Jacob, F. 1977. Evolution and tinkering. *Science* 196:1161-66.

Jacob, F. and J. Monod. 1961. On the regulation of gene activity. *Cold Spring Harbor Symp. Quant. Biol.* 26:193-211.

Jaenike, J. R. 1973. A steady state model of genetic polymorphism on islands. *Amer. Natur.* 107:793-95.

James, A. P. 1959. The spectrum of severity of mutant effects. 1. Haploid effects in yeast. *Genetics* 44:1309-24.

Jelinek, W. and J. E. Darnell. 1972. Double-stranded regions in heterogeneous nuclear RNA from Hela cells. *Proc. Nat. Acad. Sci., U.S.* 69:2537-41.

Jensen, A. R. 1969. How much can we boost IQ and scholastic achievement? *Harvard Ed. Rev.* 39:1-123.

Johannsen, W. 1909. *Elemente der exakten Erblichkeitslehre.* Jena:Gustav Fisher.

Johnson, F. M., C. G. Kanapi, R. H. Richardson, M. R. Wheeler, and W. S. Stone. 1966. An analysis of polymorphisms among isozyme loci in dark and light *Drosophila ananassae* strains from American and Western Samoa. *Proc. Nat. Acad. Sci., U.S.* 56:119-25.

Johnson, N. K. 1963. Biosystematics of sibling species of flycatchers in the *Empidonax hammondii-oberholseri-wrighti* complex. *Univ. Cal. Publ. Zool.* 66:79-238.

Johnson, T. K. and B. H. Judd. 1979. Analysis of the cut locus of *Drosophila melanogaster. Genetics* 92:485-502.

Johnston, J. S. and W. B. Heed. 1975. Dispersal of *Drosophila:* The effect of baiting on the behavior and distribution of natural populations. *Amer. Natur.* 109:207-16.

———— 1976. Dispersal of desert-adapted *Drosophila:* The saguaro-breeding *D. nigrospiracula. Amer. Natur.* 110:629-51.

Jong, G. de, A. J. W. Hoorn, G. E. W. Thorig, and W. Scharloo. 1972. Frequencies of amylase variants in *Drosophila melanogaster. Nature* 238:453-54.

Jungen, H. and D. L. Hartl. 1979. Average fitness of populations of *Drosophila melanogaster* as estimated using compound-autosome strains. *Evolution* 33:359-70.

Kaneshiro, K. Y. 1976. Ethological isolation and phylogeny in the *Plantitibia* subgroup of Hawaiian *Drosophila. Evolution* 30:740-45.

Karlin, S. 1976. Population subdivision and selection migration interaction. In S. Karlin and E. Nevo, eds., *Population Genetics and Ecology,* pp. 617-57. New York: Academic Press.

Kasner, E. and J. Newman. 1940. *Mathematics and the Imagination.* New York: Simon and Schuster.

Kearsey, M. J., W. R. Williams, P. Allen, and F. Coulter. 1977 Polymorphism for chromosomes capable of inducing female sterility in *Drosophila. Heredity* 38:109-15.

Keeler, K. H. 1978. Intra-population differentiation in annual plants. 2. Electropheretic variation in *Veronica peregrina. Evolution* 32:638-45.

Keeton, W. T. 1972. *Biological Science,* 2d ed. New York: Norton.

Keiding, J. 1977. Resistance in the housefly in Denmark and elsewhere. In D. L. Watson and A. W. A. Brown, eds., *Pesticide Management and Insecticide Resistance,* pp. 261-302. New York: Academic Press.

Kerr, W. E. and S. Wright. 1954. Experimental studies of the distribution of gene frequencies in very small populations of *Drosophila melanogaster.* 1. Forked. *Evolution* 8:172-77.

Kettlewell, H. B. D. 1958. A survey of the frequencies of *Biston betularia* (L) (Lep) and its melanic forms in Great Britain. *Heredity* 12:51-72.

———— 1965. Insect survival and selection for pattern. *Science* 148:1290-96.

———— 1973. *The Evolution of Melanism.* New York: Oxford University Press.

Kidwell, J. F. and M. G. Kidwell. 1979. Dynamics of natural selection on a fourth chromosome lethal of *Drosophila:* Twelve-generation study of experimental populations of *D. melanogaster. J. Hered.* 70:123-26.

Kimura, M. 1960. Optimum mutation rate and degree of dominance as determined by the principle of minimum genetic load. *J. Genet.* 57:21-34.

—— 1968. Evolutionary rate at the molecular level. *Nature* 217:624-26.

—— 1979. The neutral theory of molecular evolution. *Sci. Amer.* 241:98-126.

Kimura, M. and J. F. Crow. 1964. The number of alleles that can be maintained in a finite population. *Genetics* 49:725-38.

—— 1969. Natural selection and gene substitution. *Genet. Res.* 13:127-41.

Kimura, M. and T. Ohta. 1971. *Theoretical Aspects of Population Genetics.* Princeton, N.J.: Princeton University Press.

King, J. C. 1955. Evidence for the integration of the gene pool from studies of DDT resistance in *Drosophila. Cold Spring Harbor Symp. Quant. Biol.* 20:311-17.

—— 1971. *The Biology of Race.* New York: Harcourt Brace Jovanovich.

King. J. L. and T. H. Jukes. 1969. Non-Darwinian evolution: Random fixation of selectively neutral mutations. *Science* 164:788-98.

Kirkpatrick, M. and R. K. Selander. 1979. Genetics of speciation in lake whitefishes in the Allegash basin. *Evolution* 33:478-85.

Koehn, R. K. 1969. Esterase heterogeneity: Dynamics of a polymorphism. *Science* 163:943-44.

Koestler, A. 1972. *The Case of the Midwife Toad.* London: Picador, Pan Books.

Kojima, K. 1971. Is there a constant fitness for a given genotype? No! *Evolution* 25:281-85.

Kramer, F. R., D. R. Mills, P. E. Cole, T. Nishihara, and S. Spiegelman. 1974. Evolution in vitro: Sequence and phenotype of a mutant RNA resistant to ethidium bromide. *J. Mol. Biol.* 89:719-36.

Lamotte, M. 1951. Recherches sur la structure genetique des populations naturelles de *Cepaea nemoralis. Bull. Biol. de France et de Belgique* (Suppl. 35):1-238.

—— 1959. Polymorphism of natural populations of *Cepaea nemoralis. Cold Spring Harbor Symp. Quant. Biol.* 24:65-86.

—— 1966. Les fracteurs de la diversité du polymorphisme dans les populations naturelles de *Cepaea nemoralis. Lavori della Soc. Malacologica Italiana* 3:33-73.

Latter, B. D. H. 1962. Changes in reproductive fitness under artificial selection. In G. W. Leeper, ed., *The Evolution of Living Organisms,* pp. 191-202. Melbourne, Australia: Melbourne University Press.

—— 1964. Selection for a threshold character in *Drosophila.* 1. An analysis of the phenotypic variance on the underlying scale. *Gen. Res.* 5:198-210.

Laycock, G. 1970. *The Alien Animals: The Story of Imported Wildlife.* New York: Ballentine Books.

Lederberg, J. 1951. Genetic studies with bacteria. In L. C. Dunn, ed., *Genetics in the 20th Century,* pp. 263-89. New York: Macmillan.

Lederberg, J. and E. M. Lederberg. 1952. Replica plating and indirect selection of bacterial mutants. *J. Bact.* 63:399–406.

Lehninger, A. 1970. *Biochemistry.* New York: Worth.

Lerner, I. M. 1954. *Genetic Homeostasis.* Edinburgh: Oliver and Boyd.

_____ 1958. *The Genetic Basis of Selection.* New York: Wiley.

Lerner, I. M. and H. P. Donald. 1966. *Modern Developments in Animal Breeding.* New York: Academic Press.

Lerner, I. M. and C. A. Gunns, 1952. Egg size and reproductive fitness. *Poul. Sci.* 31:537–44.

Lerner, S. A., T. T. Wu, and E. C. C. Lin. 1964. Evolution of a catabolic pathway in bacteria. *Science* 146:1313–15.

Leslie, J. F. and R. C. Vrijenhoek. 1978. Genetic dissection of clonally inherited genomes of *Poeciliopsis.* 1. Linkage analysis and preliminary assessment of deleterious gene loads. *Genetics* 90:801–11.

Levene, H. 1953. Genetic equilibrium when more than one ecological niche is available. *Amer. Natur.* 87:331–33.

Levin, B. R., F. M. Stewart, and L. Chao. 1977. Resource-limited growth, competition, and predation: A model and experimental studies with bacteria and bacteriophage. *Amer. Natur.* 111:3–24.

Levin, D. A. 1972. Low frequency disadvantage in the exploitation of pollinators by corolla variants in *Phlox. Amer. Natur.* 106:453–60.

_____ 1978a. The origin of isolating mechanisms in flowering plants. *Evol. Biol.* 11:185–317.

_____ 1978b. Genetic variation in annual phlox: Self-compatible versus self-incompatible species. *Evolution* 32:245–63.

Levin, D. A. and H. W. Kerster. 1969. Density-dependent gene dispersal in *Liatris. Amer. Natur.* 103:61–74.

Levin, S. A. 1975. On the care and use of mathematical models. *Amer. Natur.* 109:785–86.

Levins, R. 1968. *Evolution in Changing Environments.* Princeton, N.J.: Princeton University Press.

_____ 1970. Extinction. In M. Gerstenhaber, ed., *Some Mathematical Questions in Biology.* Lectures on Mathematics in the Life Sciences, 2:77–107. Providence, R.I.: Amer. Math Soc.

Lewin, B. 1974. *Gene Expression.* Vol. 2: *Eucaryotic Chromosomes.* New York: Wiley.

Lewis, E. B. 1950. The phenomenon of position effect. *Adv. Genetics* 3:73–116.

Lewis, H. W. 1954. Studies on a melanoma-producing lethal in *Drosophila. J. Exp. Zool.* 126:235–75.

Lewontin, R. C. 1955. The effects of population density and composition on viability in *Drosophila melanogaster. Evolution* 9:27–41.

_____ 1966. Is nature probable or capricious. *Bioscience* 16:25–27.

_____ 1974a. *The Genetic Basis of Evolutionary Change.* New York: Columbia University Press.

_____ 1974b. The analysis of variance and the analysis of causes. *Amer. J. Hum. Genet.* 26:400–11.

Lewontin, R. C. and L. C. Dunn. 1960. The evolutionary dynamics of a polymorphism in the house mouse. *Genetics* 45:705-22.

Lewontin, R. C. and J. L. Hubby. 1966. A molecular approach to the study of genic heterozygosity in natural populations. 2. Amount of variation and degree of heterozygosity in natural populations of *Drosophila pseudoobscura*. *Genetics* 54:595-609.

Lewontin, R. C. and M. J. D. White. 1960. Interaction between inversion polymorphisms of two chromosome pairs in the grasshopper, *Moraba scurra*. *Evolution* 14:116-29.

Li, C. C. 1953. Is RH facing a crossroad? A critique of the compensation effect. *Amer. Natur.* 87:257-61.

———— 1955. *Population Genetics*. Chicago: University of Chicago Press.

———— 1963. The way the load ratio works. *Am. J. Hum. Genet.* 15:316-21.

Linck, A. J. 1961. The morphological development of the fruit of *Pisum sativum*, var. *Alaska*. *Phytomorphology* 11:79-84.

Linhart, Y. B. 1973. Ecological and behavioral determinants of pollen dispersal in hummingbird-pollinated *Heliconia*. *Amer. Natur.* 107:511-23.

———— 1974. Intra-population differentiation in annual plants. 1. *Veronica peregrina* L. raised under non-competitive conditions. *Evolution* 28:232-43.

Lloyd, D. G. 1980. Benefits and handicaps of sexual reproduction. *Evol. Biol.* 13:69-111.

Lomnicki, A. 1969. Individual differences among adult members of a snail population. *Nature* 223:1073-74.

Lorenz, K. Z. 1952. *King Solomon's Ring*. New York: Crowell.

———— 1958. The evolution of behavior. *Sci. Am.* (December) 199:67-78.

Luria, S. E. and M. Delbruck. 1943. Mutations of bacteria from virus sensitivity to virus resistance. *Genetics* 28:491-511.

MacArthur, R. H. and E. O. Wilson. 1967. *The Theory of Island Biogeography*. Princeton, N. J.: Princeton University Press.

McCarron, M., J. O'Donnell, A. Chovnick, B. S. Bhullar, J. Hewitt, and E. P. M. Candido. 1979. Organization of the rosy locus in *Drosophila melanogaster:* further evidence in support of a CIS-acting control element adjacent to the xanthine dehydrogenase structural element. *Genetics* 91:275-93.

McClintock, B. 1941. The stability of broken ends of chromosomes in *Zea mays*. *Genetics* 26:234-82.

———— 1951. Chromosome organization and genic expression. *Cold Spring Harbor Symp. Quant. Biol.* 16:13-47.

———— 1956. Controlling elements and the gene. *Cold Spring Harbor Symp. Quant. Biol.* 21:197-216.

———— 1965. The control of gene action in maize. *Brookhaven Sym. in Biol.* 18:162-84.

McConkey, E. H., B. J. Taylor, and Duc Phan. 1979. Human heterozygosity: a new estimate. *Proc. Nat. Acad. Sci., U.S.* 76:6500-04.

MacFarquhar, A. M. and F. W. Robertson. 1963. The lack of evidence of coadaptation in crosses between geographical races of *Drosophila subobscura* Coll. *Genet. Res.* 4:104-31.

MacIntyre, R. J. 1976. Evolution and ecological value of duplicate genes. *Ann. Rev. Ecol. Syst.* 7:421-68.

Malécot, G. 1969. *The Mathematics of Heredity.* San Francisco: Freeman.

Marçallo, F. A., N. Freire-Maia, J. B. C. Azevedo, and I. A. Simões. 1964. Inbreeding effect on mortality and morbidity in South Brazilian populations. *Ann. Hum. Genet., London* 27:203-18. ·

Marshall, J. T. 1960. Interrelations of Abert and brown towhees. *Condor* 62: 49-64.

Maruyama, T. and J. F. Crow. 1974. Heterozygous effects of X-ray induced mutations on viability of *Drosophila melanogaster, Mutat. Res.* 27:241-48.

Mather, K. 1953. Genetical control of stability in development. *Heredity* 7:297-336.

Mather, K. and L. Jinks. 1971. *Biometrical Genetics.* Ithaca, N.Y.: Cornell University Press.

Maynard-Smith, J. 1962. Disruptive selection, polymorphism, and sympatric speciation. *Nature* 195:60-62.

_____ 1964. Group selection and kin selection. *Nature* 201:1145-47.

_____ 1966. Sympatric speciation. *Amer. Natur.* 100:637-50.

_____ 1968a. "Haldane's dilemma" and the rate of evolution. *Nature* 219: 1114-16.

_____ 1968b. Evolution in sexual and asexual populations. *Amer. Natur.* 102:469-73.

_____ 1970. What use is sex? *J. Theor. Biol.* 30:319-35.

_____ 1976. Group selection. *Q. Rev. Biol.* 51:227-83.

_____ 1978. *The Evolution of Sex.* Cambridge: Cambridge University Press.

Mayr, E. 1942. *Systematics and the Origin of Species.* New York: Columbia University Press.

_____ 1954. Change of genetic environment and evolution. In J. Huxley, A. C. Hardy, and E. B. Ford, eds., *Evolution as a Process,* pp. 157-80. London: Allen and Unwin.

_____ 1959. Where are we? *Cold Spring Harbor Symp. Quant. Biol.* 24:1-14.

_____ 1963. *Animal Species and Evolution.* Cambridge, Mass.: Harvard University Press.

_____ 1974a. The definition of the term disruptive selection. *Heredity* 32: 404-6.

_____ 1974b. Behavior programs and evolutionary strategies. *Am. Sci.* 62: 650-59.

_____ 1976. *Evolution and the Diversity of Life.* Cambridge, Mass.: Belknap Press.

Merritt, R. B. 1972. Geographic distribution and enzymatic properties of lactate dehydrogenase allozymes in the fathead minnow, *Pimephales promelas. Amer. Natur.* 106:173-84.

Milkman, R. D. 1967. Heterosis as a major cause of heterozygosity in nature. *Genetics* 55:493-95.

Miller, R. S. 1967. Pattern and process in competition. *Adv. Ecol. Res.* 4:1-74.

Miozzari, G. and C. Yanofsky. 1978. Naturally occurring promoter down mutation: Nucleotide sequence of the *trp* promoter/operator/leader region of *Shigella dysenteriae* 16. *Proc. Nat. Acad. Sci., U.S.* 75:5580–84.

Mitton, J. B., Y. B. Linhart, K. B. Sturgeon, and J. L. Hamrick. 1979. Allozyme polymorphisms detected in mature needle tissue of ponderosa pine. *J. Hered.* 70:86–89.

Monod, J. and F. Jacob. 1961. Teleonomic mechanism in cellular metabolism, growth, and differentiation. *Cold Spring Harbor Symp. Quant. Biol.* 26: 389–401.

Moodie, G. E. E., J. D. McPhail, and D. W. Hagen. 1973. Experimental demonstration of selective predation on *Gasterosteus aculeatus. Behaviour* 47:95–105.

Moore, J. A. 1957. An embryologist's view of the species concept. In E. Mayr, ed., *The Species Problem*, pp. 325–38. Publ. 50. Amer. Assoc. Adv. Sci. Washington. D.C.

Moore, N. W. 1964. Intra- and interspecific competition among dragonflies (Odonata). *J. Anim. Ecol.* 33:49–71.

Moore, W. S. 1977. An evaluation of narrow hybrid zones in vertebrates. *Quart. Rev. Biol.* 52:263–77.

Moran, L., M.-E. Mirault, A. P. Arrigo, M. Goldschmidt-Clermont, and A. Tissieres. 1978. Heat shock in *D. melanogaster* induces synthesis of new mRNAs and proteins. *Trans. Roy. Soc.* 283:391–406.

Moree, R. and J. R. King. 1961. Experimental studies on relative viability in *Drosophila melanogaster. Genetics* 46:1735–52.

Morgan, T. H., C. B. Bridges, and A. H. Sturtevant. 1925. The genetics of *Drosophila. Biblio. Genet.* 2:1–262.

Mortlock, R. P. 1981. Regulatory mutations and the development of new metabolic pathways by bacteria. *Evol. Biol.* 14:(in press)

Morton, N. E. 1979. Fisher, the Life of a Scientist (Review). *Amer. J. Hum. Genet.* 31:392–93.

Morton, N. E., J. F. Crow, and H. J. Muller. 1956. An estimate of the mutational damage in man from data on consanguineous marriages. *Proc. Nat. Acad. Sci., U.S.* 42:855–63.

Mouraõ, C. A., F. J. Ayala, and W. W. Anderson. 1972. Darwinian fitness and adaptedness in experimental populations of *Drosophila willistoni. Genetica* 43:552–74.

Muggleton, J. 1979. Non-random mating in wild populations of polymorphic *Adalia bipunctata. Heredity* 42:57–65.

Mukai, T. 1964. The Genetic Structure of Natural Populations of *Drosophila melanogaster*. I. Spontaneous mutation rate of polygenes controlling viability *Genetics* 50:1–19.

Mukai, T. and A. B. Burdick. 1959. Single gene heterosis associated with a second chromosome recessive lethal in *Drosophila melanogaster. Genetics* 44: 211–32.

Mukai, T. and I. Yoshikawa. 1964. Heterozygous effects of radiation-induced mutations on viability in homozygous and heterozygous genetic backgrounds in *Drosophila melanogaster. Japan. J. Genet.* 38:282–87.

Mukai, T., I. Yoshikawa, and K. Sano. 1966. The Genetic Structure of Natural Populations of *Drosophila melanogaster.* IV. Heterozygous effects of radiation-induced mutations on viability in various genetic backgrounds. *Genetics* 53:513-27.

Muller, H. J. 1918. Genetic variability, twin hybrids and constant hybrids, in a case of balanced lethal factors. *Genetics* 3:422-99.

_____ 1927. Artificial transmutation of the gene. *Science* 66:84-87.

_____ 1930. Radiation and genetics. *Amer. Natur.* 64:220-51.

_____ 1932a. Some genetic aspects of sex. *Amer. Natur.* 64:118-38.

_____ 1932b. Further studies on the nature and causes of gene mutations. *Proc. 6th Inter. Cong. Genet.* 1:213-55.

_____ 1940. Bearings of the Drosophila work on systematics. In J. Huxley, ed., *The New Systematics*, pp. 185-268. Oxford: Clarendon Press.

_____ 1950a. Radiation damage to the genetic material. I. Effects manifested mainly in descendants. II. Effects manifested mainly in exposed individuals. *Amer. Sci.* 38:33-59. 399-425.

_____ 1950b. Our load of mutations. *Amer. J. Hum. Genet.* 2:111-76.

_____ 1958. How much is evolution accelerated by sexual reproduction? *A.A.A.S. General Program,* p. 205.

Muller, H. J. and R. Falk. 1961. Are induced mutations in *Drosophila* over-dominant? I. Experimental design. *Genetics* 46:727-35.

Nabours, R. K. and L. L. Kingsley. 1934. The operations of a lethal factor in *Apotettix eurycephalus* (grouse locusts). *Genetics* 19:323-28.

Neel, J. V. and W. J. Schull. 1954. *Human Heredity.* Chicago: University of Chicago Press.

Neel, J. V., J. C. Wells, and H. A. Itano. 1951. Familial differences in the proportion of abnormal hemoglobin present in the sickle cell trait. *J. Clin. Invest.* 30:1120-24.

Nei, M. 1975. *Molecular Population Genetics and Evolution.* New York: American Elsevier.

Nei, M. and A. K. Roychoudhury. 1972. Gene differences between Caucasian, Negro, and Japanese populations. *Science* 177:434-36.

Nevo, E., T. Shimony, and M. Libni. 1977. Thermal selection of allozyme polymorphisms in barnacles. *Nature* 267:699-701.

Newcombe, H. B. 1979. Problems of assessing risks *versus* mutations. *Genetics* 92:s199-s201.

Nichols, B. P. and C. Yanofsky. 1979. Nucleotide sequences of *trpA* of *Salmonella typhimurium* and *Escherichia coli*: An evolutionary comparison. *Proc. Nat. Acad. Sci., U.S.* 76:5244-48.

Nicholson, A. J. 1957. The self-adjustment of populations to change. *Cold Spring Harbor Symp. Quant. Biol.* 22:153-73.

Nicholson, A. J. and V. A. Bailey. 1935. The balance of animal populations. Part 1. *Proc. Zool. Soc. Lond. Part 3*; pp. 551-98.

Norman, R. A. and S. Prakash. 1980. Variation in activities of amylase allozymes associated with chromosome inversions in *Drosophila pseudoobscura, D. persimilis,* and *D. miranda. Genetics* 95:187-209.

Novick, A. and L. Szilard. 1950. Experiments with the chemostat on spontaneous mutations of bacteria. *Proc. Nat. Acad. Sci., U.S.* 36:708-19.

—— 1951. Experiments on spontaneous and chemically induced mutations of bacteria growing in the chemostat. *Cold Spring Harbor Symp. Quant. Biol.* 16:337-43.

Nowak, E. 1971. [The range of expansion of animals and its causes] (in Polish). *Zesqyty Naukowe* 3:1-255 (published for the Smithsonian Institute and NSF by the Foreign Scientific Publications Dept of the Nat'l Center for Sci. Tech., and Econ. Info., U.S. Dept. of Comm.)

O'Brien, S. J. and R. J. MacIntyre. 1969. An analysis of gene-enzyme variability in natural populations of *Drosophila melanogaster* and *D. simulans*. *Amer. Natur.* 103:97-113.

O'Farrell, P. H. 1975. High resolution two-dimensional electrophoresis of proteins. *J. Biol. Chem.* 250:4007-21.

Ohta, T. and M. Kimura. 1971. Behavior of neutral mutants influenced by associated overdominant loci in finite populations. *Genetics* 69:247-60.

Orians, G. H. and M. F. Willson. 1964. Interspecific territories of birds. *Ecology* 45:736-45.

Oshima, C. and T. K. Watanabe. 1973. Fertility genes in natural populations of *Drosophila melanogaster*. I. Frequency, allelism and persistence of sterility genes. *Genetics* 74:351-61.

Owen, D. F. 1966. Polymorphism in pleistocene land snails. *Science* 152:71-72.

Paget, O. E. 1954. A cytological analysis of irradiated populations. *Amer. Natur.* 88:105-7.

Paik, Y. K. and K. C. Sung. 1969. Behavior of lethals in *Drosophila melanogaster* populations. *Japan. J. Genetics* 44:180-92.

Paine, R. T. 1965. Natural history, limiting factors and energetics of the opisthobranch, *Novanax inermis*. *Ecology* 46:603-19.

Palmiter, R. D., P. B. Moore, and E. R. Mulvihill. 1976. A significant lag in the induction of ovalbumin messenger RNA by steroid hormones: a receptor translocation hypothesis. *Cell* 8:557-72.

Parsons, P. A. 1965. Assortative mating for a metrical characteristic in *Drosophila*. *Heredity* 20:161-67.

—— 1980. Isofemale strains and evolutionary strategies in natural populations. *Evol. Biol.* 13:175-217.

Parsons, P. A. and W. F. Bodmer. 1961. The evolution of overdominance: Natural selection and heterozygote advantage. *Nature* 190:7-12.

Parsons, P. A., J. T. MacBean, and B. T. O. Lee. 1969. Polymorphism in natural populations for genes controlling radioresistance in *Drosophila*. *Genetics* 61:211-18.

Paterniani, E. 1969. Selection for reproductive isolation between two populations of maize, *Zea mays*. *Evolution* 23:534-47.

Pavan, C., Th. Dobzhansky, and H. Burla. 1950. Diurnal behavior of some neotropical species of *Drosophila*. *Ecology* 31:36-43.

Pearson, K. 1904. On a generalized theory of alternative inheritance, with special references to Mendel's laws. *Phil. Trans. Roy. Soc.,* A. 203:53-86.

Pearson, K. and A. Lee. 1903. On the laws of inheritance in man. 1. Inheritance of physical characters. *Biometrika* 2:357-462.

Pentzos-Daponte, A., E. Boesiger, and A. Kanellis, 1967. Frequences de genes mutants dans plusieurs populations naturelles de *Drosophila subobscura* de Grece. *Ann. de la Fac. Sci. Univ. Aristotelienne de Thessaloniki* 10:133–52.

Petit, C. 1958. Le déterminisme génétique et psycho-physiologique de la competiton sexuelle chez *Drosophila melanogaster. Bull. Biol. de Fr. et de Belgique* 92:248–329.

Pinsker, W. and D. Sperlich. 1979. Allozyme variation in natural populations of *Drosophila suboobscura* along a north-south gradient. *Genetica* 50:207–19.

Place, A. R. and D. A. Powers. 1979. Genetic variation and relative catalytic efficiencies: Lactate dehydrogenase B allozymes of *Fundulus heteroclitus. Proc. Nat. Acad. Sci., U.S.* 76:2354–58.

Poisson, S. D. 1837. Recherches sur la Probabilité des Jugements en Matiere Criminelle et en Matiere Civile, Précédées des Regles Generales du Calcul des Probabilités. Bachelier, Imprimeur-Libraire pour les Mathematiques, la Physique, etc. Paris.

Policansky, D. 1974. "Sex ratio," meiotic drive, and group selection in *Drosophila pseudoobscura. Amer. Natur.* 108:75–90.

_____ 1979. Fertility differences as a factor in the maintenance of the "sex-ratio" polymorphism in *Drosophila pseudoobscura. Amer. Natur.* 114:672–80.

Popescu, C. 1979. Natural selection in the industrial melanic psocid *Mesopsocus unipunctatus* (Mull.) (Insecta: psocoptera) in northern England. *Heredity* 42:133–42.

Powell, J. R. 1974. Temperature related genetic divergence in *Drosophila* body size. *J. Hered.* 65:257–58.

_____ 1975. Protein variation in natural populations of animals. *Evol. Biol.* 8:79–119.

_____ 1978. The founder-flush speciation theory: An experimental approach. *Evolution* 32:465–74.

Powell, J. R. and J. M. Lichtenfels. 1979. Population genetics of Drosophila amylase. 1. Genetic control of tissue-specific expression in *D. pseudoobscura. Genetics* 92:603–12.

Powell, J. R. and L. Morton. 1979. Inbreeding and mating patterns in *Drosophila pseudoobscura. Behav. Genet.* 9:425–29.

Powell, J. R. and C. E. Taylor. 1979. Genetic variation in ecologically diverse environments. *Amer. Sci.* 67:590–96.

Power, D. M. 1979. Evolution in peripheral isolated populations: *Carpodacus* finches on the California islands. *Evolution* 33:834–47.

Prevosti, A. 1955. Geographic variability in quantitative traits in populations of *Drosophila subobscura. Cold Spring Harbor Symp. Quant. Biol.* 20:294–99.

Rasmussen, I. E. 1960. New Mutants. *Dros. Inform. Serv.* 34:53.

Reed, S. C. and E. W. Reed. 1950. Natural selection in laboratory populations of *Drosophila.* 2. Competition between a white-eye gene and its wildtype allele. *Evolution* 4:34–42.

Reed, T. E. 1969. Caucasian genes in American Negroes. *Science* 165:762–68.

Rich, S. S., A. E. Bell, and S. P. Wilson. 1979. Genetic drift in small populations of *Tribolium*. *Evolution* 33:579–84.

Richardson, R. H. 1969. Migration and enzyme polymorphisms in natural populations of *Drosophila*. *Japan. J. Genet.* 44 (Suppl. 1):172–79.

——— 1970. Models and analyses of dispersal patterns. In K. Kojima, ed., *Mathematical Topics in Population Genetics*, pp. 79–103. New York: Springer-Verlag.

Richardson, R. H. and J. S. Johnston. 1975. Ecological specialization of Hawaiian *Drosophila*: Habitat selection in Kipuka Puaulu. *Oecologia* 21: 193–204.

Richmond, R. C. 1970. Non-Darwinian evolution: A critique. *Nature* 225: 1025–28.

Rindge, F. H. 1975. A revision of the new world Bistonini (Lepidoptera: Geometridae). *Bull. Amer. Mus. Nat. Hist.* 156:73–155.

Robertson, A. 1955. Selection in animals: Synthesis. *Cold Spring Harbor Symp. Quant. Biol.* 20:225–29.

Robertson, H. D., R. E. Webster, and N. D. Zinder. 1968. Purification and properties of ribonuclease III from *E. coli*. *J. Biol. Chem.* 243:82–91.

Roff, D. 1976. Stabilizing selection in *Drosophila melanogaster*: a comment. *J. Hered.* 67:245–46.

Rohlf, F. J. and G. D. Schnell. 1971. An investigation of the isolation-by-distance model. *Amer. Natur.* 105:295–324.

Roughgarden, J. 1979. *Theory of Population Genetics and Evolutionary Ecology: An Introduction.* New York: Macmillan.

Russell, P. J. and A. M. Srb. 1972. Dominance modifiers in *Neurospora crassa*: Phenocopy selection and influence on certain ascus mutants. *Genetics* 71: 233–45.

Sakai, K. 1955. Competition in plants and its relation to selection. *Cold Spring Harbor Symp. Quant. Biol.* 20:137–57.

Salceda, V. M. 1967. Recessive lethals in second chromosomes of *Drosophila melanogaster* with radiation histories. *Genetics* 57: 691–99.

Sammeta, K. P. V. and R. Levins. 1970. Genetics and ecology. *Ann. Rev. Genet.* 4:469–88.

Sankaranarayanan, K. 1964. Genetic loads in irradiated experimental populations of *Drosophila melanogaster*. *Genetics* 50:131–50.

——— 1965. Further data on the genetic loads in irradiated experimental populations of *Drosophila melanogaster*. *Genetics* 52:153–64.

——— 1966. Some components of the genetic loads in irradiated experimental populations of *Drosophila melanogaster*. *Genetics* 54:121–30.

Sarich, V. M. and A. C. Wilson. 1967. Immunological time scale for hominoid evolution. *Science* 158:1200–3.

Sarukhan, J. and J. L. Harper. 1973. Studies on plant demography: *Ranunculus repens* L., *R. bulbosus* L., and *R. acris* L. 1. Population flux and survivorship. *J. Ecol.* 61:675–716.

Sax, K. 1923. The association of size differences with seed-coat pattern and pigmentation in *Phaseolus vulgaris*. *Genetics* 8:552–60.

Schaal, B. A. and D. A. Levin. 1976. The demographic genetics of *Liatris cylindracea* Michx. (Compositae). *Amer. Natur.* 110:191–206.

Scharloo, W., M. den Boer, and M. S. Hoogmoed. 1967. Disruptive selection on sternopleural chaeta number. *Genet. Res.* 9:115–18.

Schmalhausen, I. I. 1949. *Factors of Evolution.* Philadelphia: Blakiston.

Scossiroli, R. E. 1959. Selezione artificiale per un carattere quantitativo in popolazioni di *Drosophila melanogaster* irradiate con raggi X. Commitoto Nazionale per Recher. Nucleari, Divisione Biologica (CNB-4). Rome.

Selander, R. K. 1976. Genic variation in natural populations. In F. J. Ayala, ed., *Molecular Evolution*, pp. 21–45. Sunderland, Mass.: Sinauer.

Shapiro, A. M. 1978. Letter to the Editor. *Amer. Sci.* 66:540–41.

Sheppard, P. M. 1952. A note on non-random mating in the moth *Panaxia dominula* (L.). *Heredity* 6:239–41.

———— 1969. Evolutionary genetics of animal populations: The study of natural populations. *Proc. 12th Inter. Cong. Genet.* 3:261–79.

Shinjo, A., Y. Mizuma, and S. Nishida. 1973. The effect of gamma-ray irradiation on the fitness in inbred line of Japanese quail. *Japan. Poul. Sci.* 10:226–31.

Sibley, C. G. 1954. Hybridization in the red-eyed towhees of Mexico. *Evolution* 8:252–90.

Sibley, C. G. and J. E. Ahlquist. 1981. The phylogeny and relationships of the ratite birds as indicated by DNA-DNA hybridization. *Proc. Second Inter. Cong. Syst. Evol. Biol.* (in press).

Sibley, C. G., K. W. Corbin, J. E. Ahlquist, and A. Ferguson. 1974. Birds. In C. A. Wright, ed., *Biochemical and Immunological Taxonomy of Animals*, pp. 89–176. New York: Academic Press.

Silander, J. A. 1979. Microevolution and clone structure in *Spartina patens.* *Science* 203:658–60.

Simmons, M. J. 1976. Heterozygous effects of irradiated chromosomes on viability in *Drosophila melanogaster. Genetics* 84:353–74.

———— 1978. *Genes in Populations* (review). *Bioscience* 28:788–89.

Singh, R. S., R. C. Lewontin, and A. A. Felton. 1976. Genetic heterogeneity within electrophoretic "alleles" of xanthine dehydrogenase in *Drosophila pseudoobscura. Genetics* 84:609–29.

Singh, S. M. and E. Zouros. 1978. Genetic variation associated with growth rate in the American oyster (*Crassostrea virginica*). *Evolution* 32:342–53.

Sinnott, E. W., L. C. Dunn, and Th. Dobzhansky. 1958. *Principles of Genetics.* New York: McGraw-Hill.

Slobodkin, L. B. 1968a. Systems analysis in ecology (a review). *Science* 159:416–17.

———— 1968b. Toward a predictive theory of evolution. In R. C. Lewontin, ed., *Population Biology and Evolution*, pp. 187–205. Syracuse, N. Y.: Syracuse University Press.

Smith, J. N. M. and R. Zach. 1979. Heritability of some morphological characters in a song sparrow population. *Evolution* 33:460–67.

Smithies, O. 1955. Zone electrophoresis in starch gels: group variations in the serum proteins of normal human adults. *Biochem. J.* 61:629–41.

Smouse, P. 1976. The implications of density-dependent population growth for frequency and density-dependent selection. *Amer. Natur.* 110:849–60.

Smouse, P. and K. Kasuda. 1977. The effects of genotypic frequency and population density on fitness differentials in *Escherichia coli*. *Genetics* 86: 399–411.

Snyder, T. P. and F. J. Ayala. 1979. Frequency-dependent selection at the *Pgm-1* locus of *Drosophila pseudoobscura*. *Genetics* 92:995–1003.

Sokoloff, A. 1965. Geographic variation of quantitative characters in populations of *Drosophila pseudoobscura*. *Evolution* 19:300–10.

_____ 1977. *The Biology of Tribolium*, vol. 3. Oxford: Clarendon Press.

Soulé, M. E. 1979. Heterozygosity and developmental stability: Another look. *Evolution* 33:396–401.

Souza, H. M. L. de, A. B. da Cunha, and E. P. dos Santos. 1970. Adaptive polymorphism of behavior evolved in laboratory populations of *Drosophila willistoni*. *Amer. Natur.* 104:175–89.

Sparrow, A. H. and A. F. Nauman. 1976. Evolution of genome size by DNA doublings. *Science* 192:524–29.

Spencer, W. P. 1947. Mutations in wild populations in *Drosophila*. *Advan. Genet.* 1:359–402.

Sperlich, D. and A. Karlik. 1968. Das Schicksal von markieten Letal chromosomen in mono- und polychromosomalen Experimentalpopulationen von *Drosophila melanogaster*. *Verh. Deutsch. Zool. Ges., Innsbruck* 12: 195–205.

Spiegelman, S. 1967. An *in vitro* analysis of a replicating molecule. *Amer. Sci.* 55:221–64.

Spiess, E. B. 1977. *Genes in Populations*. New York: Wiley.

Spoffard, J. B. 1969. Heterosis and the evolution of duplications. *Amer. Natur.* 103:407–32.

Sprague, G. F. 1941. The location of dominant favorable genes in maize by means of an inversion (abstract) *Genetics* 26:170.

_____ 1967. Plant Breeding. *Ann. Rev. Genet.* 1:269–94.

Srb. A. M. and M. Basl. 1972. Evidence for the differentiation of wild-type alleles in different species of *Neurospora*. *Genetics* 72:759–62.

Stalker, H. D. 1942. Sexual isolation studies in the species complex *Drosophila virilis*. *Genetics* 27:238–57.

_____ 1976. Chromosome studies in wild populations of *Drosophila melanogaster*. *Genetics* 82:323–47.

Stalker, H. D. and H. L. Carson. 1947. Morphological variation in natural populations of *Drosophila robusta* Sturtevant. *Evolution* 1:237–48.

_____ 1948. An altitudinal transect of *Drosophila robusta* Sturtevant. *Evolution* 2:295–305.

_____ 1949. Seasonal variation in the morphology of *Drosophila robusta* Sturtevant. *Evolution* 3:330–43.

Starlinger, P. and H. Saedler. 1972. Insertion mutations in microorganisms. Biochemie 54:177–85.

Stebbins, G. L. 1950. *Variation and Evolution in Plants*. New York: Columbia University Press.

Stern, C. 1968. *Genetic Mosaics and Other Essays.* Cambridge, Mass.: Harvard University Press.

Stern, C., G. Carson, M. Kinst, E. Novitski, and D. Uphoff. 1952. The viability of heterozygotes for lethals. *Genetics* 37:413-49.

Stone, W. S. 1955. Genetic and chromosomal variability in *Drosophila. Cold Spring Harbor Symp. Quant. Biol.* 20:256-70.

Stone, W. S., W. C. Guest, and F. D. Wilson. 1960. The evolutionary implications of the cytological polymorphism and phylogeny of the *virilis* group of *Drosophila. Proc. Nat. Acad. Sci., U.S.* 46:350-61.

"Student" 1907. On the error of counting with a haemacytometer. *Biometrika* 5:351-60.

Sturtevant, A. H. 1938. Essays on Evolution. III. On the origin of interspecific sterility. *Quant. Rev. Biol.* 13:333-35.

―――― 1965. *A History of Genetics.* New York: Harper and Row.

Sturtevant, A. H. and G. W. Beadle. 1939. *An Introduction to Genetics.* New York: Saunders. (Reissued 1962, Dover Publications, New York.)

Sturtevant, A. H. and Th. Dobzhansky. 1936. Inversions in the third chromosome of wild races of *Drosophila pseudoobscura*, and their use in the study of the history of the species. *Proc. Nat. Acad. Sci., U.S.* 22:448-50.

Sturtevant, A. H. and K. Mather. 1938. The inter-relations of inversions, heterosis, and recombination. *Amer. Natur.* 72:447-52.

Sumner, F. B. 1935. Evidence for the protective value of changeable coloration in fishes. *Amer. Natur.* 69:245-66.

Sumper, M. and R. Luce. 1975. Evidence for *de novo* production of self-replicating and environmentally adapted RNA structures by bacteriophage $Q\beta$ replicase. *Proc. Nat. Acad. Sci., U.S.* 72:162-66.

Sutton, W. S. 1903. The chromosomes in heredity. *Biol. Bull.* 4:231-51.

Suzuki, D. T. 1975. Temperature-sensitive mutations in *Drosophila melanogaster. Handbook of Genetics* 3:653-68.

Sved, J. A. 1979. The "hybrid dysgenesis" syndrome in *Drosophila melanogaster. Bioscience* 29:659-64.

Sved., J. A., T. E. Reed, and W. F. Bodmer. 1967. The number of balanced polymorphisms that can be maintained in a natural population. *Genetics* 55:469-81.

Symonds, N. and Ana Coelho. 1978. Role of the G segment in the growth of phage Mu. *Nature* 271:573-74.

Symposium: The effects of radiation on the hereditary fitness of mammalian populations. *Genetics* 50:1023-1217.

Tantawy, A. O. and G. S. Mallah. 1961. Studies on natural populations of *Drosophila.* 1. Heat resistance and geographic variation in *Drosophila melanogaster* and *D. simulans. Evolution* 15:1-14.

Tauber, C. A. and M. J. Tauber. 1977a. A genetic model for sympatric speciation through habitat diversification and seasonal isolation. *Nature* 268:702-5.

―――― 1977b. Sympatric speciation based on allelic changes at three loci: Evidence from natural populations in two habitats. *Science* 197:1298-99.

―――― 1978. Sympatric speciation: Evidence? (a reply). *Science* 200:346.

Tauber, C. A., M. J. Tauber, and J. R. Nichols. 1977. Two genes control seasonal isolation in sibling species. *Science* 197:592-93.

Taylor, C. E. and J. R. Powell. 1977. Microgeographic differentiation of chromosomal and enzyme polymorphisms in *Drosophila persimilis. Genetics* 85:681-95.

Temin, H. M. 1972. RNA-directed DNA synthesis. *Sci. Amer.* 226(1):24-33, 122.

Thoday, J. M. 1953. Components of fitness. *Symp. Soc. Exp. Biol.* 7:96-113.

—— 1955. Balance, heterozygosity and development stability. *Cold Spring Harbor Symp. Quant. Biol.* 20:318-26.

—— 1958. Effects of disruptive selection: The experimental production of a polymorphic population. *Nature* 181:1124-25.

—— 1964. Genetics and the integration of reproductive systems. In K. C. Highnam, ed., *Insect Reproduction*, pp. 108-19. *Symp. 2, Roy. Entol. Soc. London.*

—— 1972. Review Lecture: Disruptive selection. *Proc. Roy. Soc. Lond., Ser. B.* 182:109-43.

Thoday, J. M. and T. B. Boam. 1959. Effects of disruptive selection. 2. Polymorphism and divergence without isolation. *Heredity* 13:205-18.

Thoday, J. M. and J. B. Gibson. 1962. Isolation by disruptive selection. *Nature* 193:1164-66.

—— 1970. The probability of isolation by disruptive selection. *Amer. Natur.* 104:219-30.

Thomas, L. 1974. *The Lives of a Cell.* New York: Viking Press.

Tilman, D. 1977. Resource competition between planktonic algae: An experimental and theoretical approach. *Ecology* 58:338-48. (*see also* Titman, D.)

Timofeeff-Ressovsky, N. W. 1932. Mutations of the gene in different directions. *Proc. 6th Inter. Cong. Genet.* 1:308-30.

—— 1935. Auslösung von Vitalität mutationen durch Röntgenbestralung bei *Drosophila melanogaster. Nachs. Ges. Wiss. Gottingen, Biol. N.F.* 1:163-80.

—— 1939. Genetik and Evolution. *Z.i.A.V.* 76:158-219.

Tinkle, D. W. 1978. Reptile Ecology: a review. *Amer. Sci.* 66:623.

Titman, D. 1976. Ecological competition between algae: Experimental confirmation of resource-based competition theory. *Science* 192:463-65. (*see also* Tilman, D.)

Tobari, Y. N. and K. Kojima. 1972. A study of spontaneous mutation rates at ten loci detectable by starch gel electrophoresis in *Drosophila melanogaster. Genetics* 70:397-403.

Tsakas, S. C. 1977. Genetics of *Dacus olaea.* VIII. Selection for the amount of acetylcholinesterase after organophosphate treatment. *Evolution* 31:901-4.

Turner, J. R. G. 1977. Butterfly mimicry: The genetical evolution of an adaptation. *Evol. Biol.* 10:163-206.

Turner, J. R. G. and M. H. Williamson. 1968. Population size, natural selection, and genetic load. *Nature* 218:700.

Ullyett, G. D. 1950. Competition for food and allied phenomena in sheep-blowfly populations. *Phil. Trans. Roy. Soc. Lond. Ser. B.*, 234: 77-174.

U.N. Scientific Committee on the Effects of Atomic Radiation. 1962. New York: United Nations.

Utida, S. 1972. Density dependent polymorphism in the adult of *Callosobruchus maculatus* (Coleoptera, Bruchidae). *J. Stored Prod. Res.* 8:111-25.

Valenzuela, C. Y. and Z. Harb. 1977. Socioeconomic assortative mating in Santiago, Chile: A demonstration using stochastic matrices of mother-child relationships applied to ABO blood groups. *Soc. Biol.* 24:225-33.

Vann, E. G. 1966. The fate of X-ray induced chromosomal rearrangements introduced into laboratory populations of *Drosophila melanogaster. Amer. Natur.* 100:425-49.

Van Valen, L. 1973. A new evolutionary law. *Evol. Theor.* 1:1-30.

———— 1975. Life, death, and energy of a tree. *Biotropica* 7:259-69.

Vedbrat, S. S. and G. S. Whitt. 1975. Isozyme ontogeny of the mosquito, *Anopheles albimanus.* In C. L. Markert, ed., *Isozymes*, vol. 3, *Developmental Biology*, pp. 131-43. New York: Academic Press.

Vetukhiv, M. 1953. Viability of hybrids between local populations of *Drosophila pseudoobscura. Proc. Nat. Acad. Sci., U.S.* 39:30-34.

———— 1954. Integration of the genotype in local populations of three species of *Drosophila. Evolution* 8:241-51.

———— 1956. Fecundity of hybrids between geographic populations of *Drosophila pseudoobscura. Evolution* 10:139-46.

Voelker, R. A., H. E. Schaffer, and T. Mukai. 1980. Spontaneous allozyme mutations in *Drosophila melanogaster:* Rate of occurrence and nature of the mutants. *Genetics* 94:961-68.

Vrijenhoek, R. C., R. A. Angus, and R. J. Schultz, 1978. Variation and clonal structure in a unisexual fish. *Amer. Natur.* 112:41-55.

Waddington, C. H. 1953. Genetic assimilation of an acquired character. *Evolution* 7:118-26.

———— 1956. Genetic assimilation of the *bithorax* phenotype. *Evolution* 10:1-13.

———— 1975. *The Evolution of an Evolutionist.* Ithaca, N.Y.: Cornell University Press.

Wahlund, S. 1928. Zusammensetzung von Populationen und Korrelationserscheinungen von Standpunkt der Vererbungslehre aus betrachtet. *Hereditas* 11:65-106.

Wallace, B. 1948. Studies on "sex-ratio" in *Drosophila pseudoobscura.* 1. Selection and "sex-ratio." *Evolution* 2:189-217.

———— 1952. The estimation of adaptive values of experimental populations. *Evolution* 6:333-41.

———— 1954a. Genetic divergence of isolated populations of *Drosophila melanogaster. Proc. 9th Intern. Cong. Genet.* 2:761-64.

———— 1954b. Coadaptation and the gene arrangements of *Drosophila pseudoobscura. Int. Union Biol. Sci., Ser. B.* 15:67-94.

———— 1956. Studies on irradiated populations of *Drosophila melanogaster. J. Genet.* 54:280-93.

———— 1958. The average effect of radiation-induced mutations on viability in *Drosophila melanogaster. Evolution* 12:532-56.

―――― 1959a. The role of heterozygosity in *Drosophila* populations. *Proc. 10th Inter. Cong. Genet.* 1:408-19.

―――― 1959b. Studies of the relative fitnesses of experimental populations of *Drosophila melanogaster. Amer. Natur.* 93:295-314.

―――― 1959c. Influence of genetic systems on geographic distribution. *Cold Spring Harbor Symp. Quant. Biol.* 24:193-204.

―――― 1962. Temporal changes in the roles of lethal and semilethal chromosomes within populations of *Drosophila melanogaster. Amer. Natur.* 96: 247-56.

―――― 1963a. The annual invitation lecture. Genetic diversity, genetic uniformity, and heterosis. *Can J. Genet. and Cytol.* 5:239-53.

―――― 1963b. The elimination of an autosomal lethal from an experimental population of *Drosophila melanogaster. Amer. Natur.* 97:65-66.

―――― 1963c. Further data on the overdominance of induced mutations. *Genetics* 48:633-51.

―――― 1963d. Modes of reproduction and their genetic consequences. In W. D. Hanson and H. F. Robinson, eds., *Statistical Genetics and Plant Breeding,* pp. 3-20. Publ. 982. Washington, D.C.: NAS-NRC.

―――― 1964. *Population Genetics.* B. S. C. S. Pamphlet. Boston: Heath.

―――― 1966a. On the dispersal of *Drosophila. Amer. Natur.* 100:551-63.

―――― 1966b. Distance and the allelism of lethals in a tropical population of *Drosophila melanogaster. Amer. Natur.* 100:565-78.

―――― 1966c. The fate of *sepia* in small populations of *Drosophila melanogaster. Genetica* 37:29-36.

―――― 1966d. *Chromosomes, Giant Molecules, and Evolution.* New York: Norton.

―――― 1966e. Natural and radiation-induced chromosomal polymorphism in *Drosophila. Mutat. Res.* 3:194-200.

―――― 1968a. *Topics in Population Genetics.* New York: Norton.

―――― 1968b. Mutation rates for autosomal lethals in *Drosophila melanogaster. Genetics* 60:389-93.

―――― 1968c. Polymorphism, population size, and genetic load. In R. C. Lewontin, ed., *Population Biology and Evolution,* pp. 87-108. Syracuse, N.Y.: Syracuse University Press.

―――― 1970a. Observations on the microdispersion of *Drosophila melanogaster.* In M. K. Hecht and W. C. Steere, eds., *Essays in Evolution and Genetics.* New York: Appleton-Century-Crofts.

―――― 1970b. *Genetic Load: Its Biological and Conceptual Aspects.* Englewood Cliffs, N.J.: Prentice-Hall.

―――― 1974. Studies on intra- and inter-specific competition in *Drosophila. Ecology* 55:227-44.

―――― 1975a. At odds with a doomsday curve. *Sci. Dig.* 77:42-45.

―――― 1975b. Gene control mechanisms and their possible bearing on the neutralist-selectionist controversy. *Evolution* 29:193-202.

―――― 1975c. Hard and soft selection revisited. *Evolution* 29:465-73.

―――― 1976. The structure of gene control regions and its bearing on diverse aspects of population genetics. In S. Karlin and E. Nevo, eds., *Population Genetics and Ecology*, pp. 499–521. New York: Academic Press.

——— 1977a. Automatic culling and population fitness. *Evol. Biol.* 10:265–76.

——— 1977b. Recent thoughts on gene control and the fitness of populations. *Genetika* 9:323–34.

——— 1979a. The migration of a mutant gene into isolated populations of *Drosophila melanogaster. Genetica* 50:67–72.

——— 1979b. Population size, environment, and the maintenance of laboratory cultures of *Drosophila melanogaster. Genetika* 10:9–16.

Wallace, B. and T. L. Kass. 1974. On the structure of gene control regions. *Genetics* 77:541–58.

Wallace, B. and J. C. King. 1952. A genetic analysis of the adaptive values of populations. *Proc. Nat. Acad. Sci., U.S.* 38:706–15.

Wallace, B., J. C. King, C. V. Madden, B. Kaufmann, and E. C. McGunnigle. 1953. An analysis of variability arising through recombination. *Genetics* 38:272–307.

Wallace, B. and C. Madden. 1953. The frequencies of sub- and supervitals in experimental populations of *Drosophila melanogaster. Genetics* 38:456–70.

——— 1965. Studies on inbred strains of *Drosophila melanogaster. Amer. Natur.* 99:495–510.

Wallace, B. and A. M. Srb. 1964. *Adaptation.* 2d ed. Englewood Cliffs, N.J.: Prentice-Hall.

Wallace, B. and M. Vetukhiv. 1955. Adaptive organization of the gene pools of *Drosophila* populations. *Cold Spring Harbor Symp. Quant. Biol.* 20: 303–10.

Wallace, B., E. Zouros, and C. Krimbas. 1966. Frequencies of second and third chromosome lethals in a tropical population of *Drosophila melanogaster. Amer. Natur.* 100:245–51.

Ward, R. D. and P. D. N. Hebert. 1972. Variability of alcohol dehydrogenase activity in a natural population of *Drosophila melanogaster. Nature New Biol.* 236:243–44.

Wasserman, M. and H. R. Koepfer. 1977. Character displacement for sexual isolation between *Drosophila mojavensis* and *Drosophila arizonensis. Evolution* 31:812–23.

Watanabe, T. K. and M. Kawanishi. 1979. Mating preference and the direction of evolution in *Drosophila. Science* 205:906–7.

Watson, J. D. 1976. *Molecular Biology of the Gene*, 3d ed. Menlo Park, Calif.: Benjamin.

Watson, J. D. and F. H. C. Crick. 1953. A structure for deoxyribose nuclei acid. *Nature* 171:737–38.

Weinberg. W. 1908. Über den Nachweis der Vererbung beim Menschen. *Jahresh. Verein. L. vaterl. Naturk. in Württemberg* 64:368–82.

Wensink, P. C., D. J. Finnegan, J. E. Donelson, and D. S. Hogness. 1974. A system for mapping DNA sequences in the chromosomes of *Drosophila melanogaster. Cell* 3:315–25.

West, D. A. 1977. Melanism in *Biston* (Lepidoptera: Geometridae) in the rural central Appalachians. *Heredity* 39:75–81.

White, M. J. D. 1968. Models of speciation. *Science* 159:1065–70.

—— 1978. *Modes of Speciation.* San Francisco: Freeman.

White, M. J. D. and L. E. Andrew. 1962. Effects of chromosomal inversions on size and relative viability in the grasshopper *Moraba scurra.* In G. W. Leeper, ed., *The Evolution of Living Organisms,* pp. 94–101. Melbourne, Australia: Melbourne University Press.

Wickner, R. B. 1978. Twenty-six chromosomal genes needed to maintain the killer double-stranded RNA plasmid of *Saccharomyces cerevisiae. Genetics* 88:419–25.

Wiebe, G. A., F. C. Petr, and H. Stevens. 1963. Interplant competition between barley genotypes. In W. D. Hanson and H. F. Robinson, eds., *Statistical Genetics and Plant Breeding,* pp. 546–57. Publ. 982. Washington, D.C.: NAS-NRC.

Wiens, J. A. 1977. On competition and variable environments. *Amer. Sci.* 65: 590–97.

Willard, J. R. 1974. Horizontal and vertical dispersal of the California red scale, *Aonidiella aurantii* (Mask.) (Homoptera; Diaspididae), in the field. *Austr. J. Zool.* 22:531–48.

Williams, G. C. 1966. *Adaptation and Natural Selection.* Princeton, N.J.: Princeton University Press.

—— 1975. *Sex and Evolution.* Princeton, N.J.: Princeton University Press.

Wills, C. 1968. Three kinds of genetic variability in yeast populations. *Proc. Nat. Acad. Sci., U.S.* 61:937–44.

—— 1978. Rank-order selection is capable of maintaining all genetic polymorphisms. *Genetics* 89:403–17.

Wilson, A. C. 1975. Evolutionary importance of gene regulation. *Stadler Genet. Symp.* 7:117–34.

—— 1976. Gene regulation in evolution. In F. J. Ayala, ed., *Molecular Evolution,* pp. 225–34. Sunderland, Mass.: Sinauer.

Wilson, A. C., S. S. Carlson, and T. J. White. 1977. Biochemical evolution. *Ann. Rev. Biochem.* 46:573–639.

Wilson, A. C., V. M. Sarich, and L. R. Maxson. 1974. The importance of gene arrangement in evolution: Evidence from studies on rates of chromosomal, protein, and anatomical evolution. *Proc. Nat. Acad. Sci., U.S.* 71:3028–30.

Wilson, E. O. 1975. *Sociobiology: The new synthesis.* Cambridge, Mass.: Belknap Press.

—— 1978. *On Human Nature.* Cambridge, Mass.: Harvard University Press.

Wilson, E. O., T. Eisner, W. R. Briggs, R. E. Dickerson, R. L. Metzenberg, R. D. O'Brien, M. Susman, and W. E. Boggs. 1973. *Life on Earth.* Stamford, Conn.: Sinauer.

Wing, L. 1943. Spread of the starling and English sparrow. *Auk* 60:74–87.

Wit, C. T. de. 1960. On competition. *Versl. Landbouwkd Onderz.* 66:1–82.

Wright, S. 1931. Evolution in Mendelian populations. *Genetics* 16:97–159.

—— 1934. Physiological and evolutionary theories of dominance. *Amer. Natur.* 68:24–53.

—— 1964. The distribution of self-incompatibility alleles in populations. *Evolution* 18:609–19.

_____ 1968. Dispersion of *Drosophila pseudoobscura*. *Amer. Natur.* 102:81–84.

Wright, S. and Th. Dobzhansky. 1946. Genetics of natural populations. XII. Experimental reproduction of some of the changes caused by natural selection in certain populations of *Drosophila pseudoobscura. Genetics* 31: 125–56.

Wright, S., Th. Dobzhansky, and W. Hovanitz. 1942. Genetics of Natural Populations. VII. The allelism of lethals in the third chromosome of *Drosophila pseudoobscura. Genetics* 27:363–94.

Wright, T. R. F. 1963. The genetics of an esterase in *Drosophila melanogaster. Genetics* 48:787–801.

Wurtman, R. J. 1975. The effects of light on the human body. *Sci. Amer.* 233:69–77.

Yasuda, N. 1967. Distribution of matrimonial distance in Mishima district. *Ann. Report, Natl. Inst. Genet., Japan.* 18:68–70.

Yerington, A. P. and R. M. Warner. 1961. Flight distances of *Drosophila* determined with radioactive phosphorus. *J. Econ. Entom.* 54:425–28.

Yoda, K., T. Kira, H. Ogawa, and K. Hozumi. 1963. Self-thinning in overcrowded pure stands under cultivated and natural conditions (Intra specific competition among higher plants. 11) *J. Biol.* (Osaka City University) 14: 107–29.

Yokoyama, S. 1979. The rate of allelism of lethal genes in a geographically structured population. *Genetics* 93:245–62.

Yoon, J. S. and R. H. Richardson. 1976. Evolution of Hawaiian Drosophilidae. 2. Patterns and rates of chromosome evolution in an *Antopocerus* phylogeny. *Genetics* 83:827–43.

Young, J. F. and P. Palese. 1979. Evolution of human influenza A viruses in nature: Recombination contributes to genetic variation of H1N1 strains. *Proc. Nat. Acad. Sci., U.S.* 76:6547–51.

Young, J. P. W. 1979a. Enzyme polymorphism and cyclic parthenogenesis in *Daphnia magna.* 1. Selection and clonal diversity. *Genetics* 92:953–70.

_____ 1979b. Enzyme polymorphism and cyclic parthenogenesis in *Daphnia magna.* 2. Heterosis following sexual reproduction. *Genetics* 92:971–82.

Zieg, J., M. Silverman, M. Hilmen, M. Simon. 1977. Recombinational switch for gene expression. *Science* 196:170–72.

Zimmerman, E. C. 1958. Three hundred species of *Drosophila* in Hawaii? A challenge to geneticists and evolutionists. *Evolution* 12:557–58.

Zipkas, D. and M. Riley. 1975. Proposal concerning mechanism of evolution of the genome of *Escherichia coli. Proc. Nat. Acad. Sci., U.S.* 72:1354–58.

Zuckerkandl, E. 1963. Perspectives in molecular anthropology. In S. L. Washburn, ed., *Classification and Human Evolution*, pp. 243–72. Chicago: Aldine.

Index

680

Index